The Root Systems in Sustainable Agricultural Intensification

The Root Systems in Sustainable Agricultural Intensification

Edited by

Zed Rengel

UWA School of Agriculture and Environment, University of Western Australia
Perth, Australia

Ivica Djalovic

Institute of Field and Vegetable Crops
Novi Sad, Serbia

WILEY Blackwell

This edition first published 2021

Registered Office(s)
John Wiley & Sons, Inc., 111 River Street, Hoboken, NJ 07030, USA
John Wiley & Sons Ltd, The Atrium, Southern Gate, Chichester, West Sussex, PO19 8SQ, UK

Editorial Office
The Atrium, Southern Gate, Chichester, West Sussex, PO19 8SQ, UK

For details of our global editorial offices, customer services, and more information about Wiley products visit us at www.wiley.com.

Wiley also publishes its books in a variety of electronic formats and by print-on-demand. Some content that appears in standard print versions of this book may not be available in other formats.

Library of Congress Cataloging-in-Publication Data

Names: Rengel, Zdenko, editor. | Djalovic, Ivica, editor.
Title: The root systems in sustainable agricultural intensification / Zed Rengel, Ivica Djalovic.
Description: Hoboken, NJ : Wiley, [2021] | Includes index.
Identifiers: LCCN 2020056309 (print) | LCCN 2020056310 (ebook) | ISBN 9781119525400 (hardback) | ISBN 9781119525387 (adobe pdf) | ISBN 9781119525431 (epub)
Subjects: LCSH: Roots (Botany) | Plants–Water requirements. | Plant-soil relationships. | Sustainable agriculture.
Classification: LCC QK644 .R6566 2021 (print) | LCC QK644 (ebook) | DDC 581.4/98–dc23
LC record available at https://lccn.loc.gov/2020056309
LC ebook record available at https://lccn.loc.gov/2020056310

Cover Design: Wiley
Cover Image: © VR19/iStock/Getty Images

Set in 9.5/12.5pt STIXTwoText by SPi Global, Pondicherry, India

10 9 8 7 6 5 4 3 2 1

Contents

List of Contributors

Sebastian R.G.A. Blaser
Department of Soil System Science
Helmholtz-Centre for Environmental
Research GmbH – UFZ
Halle (Saale)
Germany

Gernot Bodner
Department of Crop Sciences
Division of Agronomy
University of Natural Resources and Life
Sciences BOKU Vienna
Tulln an der Donau
Austria

Cristian Dal Cortivo
Department of Agronomy, Food, Natural
Resources, Animals and the Environment
University of Padua
Legnaro - Padua
Italy

D.M.S.B. Dissanayaka
Graduate School of Biosphere Science
Hiroshima University
Higashihiroshima
Japan;
Department of Crop Science
Faculty of Agriculture
University of Peradeniya
Kandy
Sri Lanka

M. Djanaguiraman
Department of Crop Physiology
Tamil Nadu Agricultural University
Coimbatore
Tamil Nadu
India

Ian C. Dodd
Lancaster Environment Centre
Lancaster University
Lancaster
UK

Kun Du
Key Laboratory of Ecosystem Network Observation and Modeling, Institute of Geographic Sciences and Natural Resources Research
Chinese Academy of Sciences
Beijing
China;
Shandong Yucheng Agro-ecosystem National Observation and Research Station
Ministry of Science and Technology
Yucheng
China;
University of Chinese Academy of Sciences
Beijing
China

Muhammad Farooq
Department of Plant Sciences
College of Agricultural and Marine Sciences
Sultan Qaboos University
Al-Khoud
Oman;
Department of Agronomy
University of Agriculture
Faisalabad
Pakistan

Kemo Jin
Department of Plant Nutrition
National Academy of Agriculture Green Development
College of Resources and Environmental Sciences
Key Laboratory of Plant-Soil Interactions
Ministry of Education
China Agricultural University
Beijing
P.R. China

Katharina Keiblinger
Department of Forest- and Soil Sciences
Institute of Soil Research
University of Natural Resources and Life Sciences BOKU Vienna
Vienna
Austria

Nicolai Koebernick
Institute of Agricultural and Nutritional Sciences
Martin-Luther-University Halle-Wittenberg
Halle (Saale)
Germany

Guzel Kudoyarova
Department of Plant Physiology
Ufa Institute of Biology
Russian Academy of Sciences
Ufa
Russian Federation

Thomas W. Kuyper
Soil Biology Group
Wageningen University & Research
Wageningen
The Netherlands

Dong-Jin Lee
Department of Crop Sciences and Biotechnology
Dankook University
Cheonan-si
South Korea

Daniel Leitner
Simulationswerkstatt – Services in Computational Sciences
Linz
Austria

Peifang Leng
Key Laboratory of Ecosystem Network Observation and Modeling, Institute of Geographic Sciences and Natural Resources Research
Chinese Academy of Sciences
Beijing
China;
Shandong Yucheng Agro-ecosystem National Observation and Research Station
Ministry of Science and Technology
Yucheng
China;
University of Chinese Academy of Sciences
Beijing
China

Fadong Li
Key Laboratory of Ecosystem Network Observation and Modeling
Institute of Geographic Sciences and Natural Resources Research
Chinese Academy of Sciences
Beijing
China;
Shandong Yucheng Agro-ecosystem National Observation and Research Station
Ministry of Science and Technology
Yucheng
China;
University of Chinese Academy of Sciences
Beijing
China

Long Li
China Agricultural University
Beijing
China

Xiao-Fei Li
Institute of Cotton Research
Chinese Academy of Agricultural Sciences
Anyang, Henan
China

Yang Li
Division of Plant Sciences
University of Missouri
Columbia, MO
USA;
National Key Laboratory of Crop Genetics and Germplasm Enhancement
National Center for Soybean Improvement
Nanjing Agricultural University
Nanjing
China

Zhao Li
Key Laboratory of Ecosystem Network Observation and Modeling, Institute of Geographic Sciences and Natural Resources Research
Chinese Academy of Sciences
Beijing
China;
Shandong Yucheng Agro-ecosystem National Observation and Research Station
Ministry of Science and Technology
Yucheng
China;
University of Chinese Academy of Sciences
Beijing
China

Yan Li
National Key Laboratory of Crop Genetics and Germplasm Enhancement
National Center for Soybean Improvement
Nanjing Agricultural University
Nanjing
China

Shanbao Liu
Key Laboratory of Ecosystem Network Observation and Modeling
Institute of Geographic Sciences and Natural Resources Research
Chinese Academy of Sciences
Beijing
China;
Shandong Yucheng Agro-ecosystem National Observation and Research Station
Ministry of Science and Technology
Yucheng
China;
University of Chinese Academy of Sciences
Beijing
China

Yi-Xiang Liu
Yunnan Agricultural University
Kunming, Yunnan
China

Axel Mentler
Department of Forest- and Soil Sciences
Institute of Soil Research
University of Natural Resources and Life Sciences BOKU Vienna
Vienna
Austria

Mary N. Muchane
Botany Department
National Museums of Kenya
Nairobi
Kenya

Faisal Nadeem
Department of Agronomy
The University of Haripur
Haripur
Khyber Pakhtunkhwa, Pakistan

Alireza Nakhforoosh
Global Institute for Food Security
University of Saskatchewan
Saskatoon
Canada

Henry T. Nguyen
Division of Plant Sciences
University of Missouri
Columbia, MO
USA

Loïc Pagès
INRA UR PSH 1115 Centre PACA
Avignon cedex 9
France

Anna Panozzo
Department of Agronomy, Food, Natural Resources, Animals and the Environment
University of Padua
Legnaro - Padua
Italy

P.V.V. Prasad
Department of Agronomy; and Feed the Future Innovation Lab for Collaborative Research on Sustainable Intensification
Kansas State University
Manhattan, KS
USA

Yunfeng Qiao
Key Laboratory of Ecosystem Network Observation and Modeling, Institute of Geographic Sciences and Natural Resources Research
Chinese Academy of Sciences
Beijing
China;
Shandong Yucheng Agro-ecosystem National Observation and Research Station
Ministry of Science and Technology
Yucheng
China;
University of Chinese Academy of Sciences
Beijing
China

Abdul Rehman
Department of Crop Sciences and Biotechnology
Dankook University
Cheonan-si
South Korea

Zed Rengel
UWA School of Agriculture and Environment
University of Western Australia
Perth
Australia

Boris Rewald
Department of Forest and Soil Sciences
University of Natural Resources and Life Sciences
Vienna
Austria

Steffen Schlüter
Department of Soil System Science
Helmholtz-Centre for Environmental Research GmbH – UFZ
Halle (Saale)
Germany

Andrea Schnepf
Institute of Bio- and Geosciences IBG-3
Forschungszentrum
Jülich
Germany

Kadambot H.M. Siddique
The UWA Institute of Agriculture and School of Agriculture & Environment
The University of Western Australia
Perth, WA
Australia

Jianbo Shen
Department of Plant Nutrition
National Academy of Agriculture Green Development
College of Resources and Environmental Sciences
Key Laboratory of Plant-Soil Interactions
Ministry of Education
China Agricultural University
Beijing
P.R. China

Teofilo Vamerali
Department of Agronomy, Food, Natural Resources, Animals and the Environment
University of Padua
Legnaro - Padua
Italy

Doris Vetterlein
Department of Soil System Science
Helmholtz-Centre for Environmental Research GmbH – UFZ
Halle (Saale)
Germany;
Institute of Agricultural and Nutritional Sciences
Martin-Luther-University Halle-Wittenberg
Halle (Saale)
Germany

Giovanna Visioli
Department of Chemistry, Life Sciences and Environmental Sustainability
University of Parma
Parma
Italy

Liyang Wang
Department of Plant Nutrition
National Academy of Agriculture Green Development
College of Resources and Environmental Sciences
Key Laboratory of Plant-Soil Interactions
Ministry of Education
China Agricultural University
Beijing
P.R. China

Xin Wang
Department of Plant Nutrition
National Academy of Agriculture Green Development
College of Resources and Environmental Sciences
Key Laboratory of Plant-Soil Interactions
Ministry of Education
China Agricultural University
Beijing
P.R. China

Xinxin Wang
Mountain Area Research Institute
Agricultural University of Hebei
Baoding
P.R. China

Jun Wasaki
Graduate School of Integrated Sciences for Life
Hiroshima University
Higashihiroshima
Japan;
Graduate School of Biosphere Science
Hiroshima University
Higashihiroshima
Japan

Chuanyong Xiong
Department of Plant Nutrition
National Academy of Agriculture Green Development
College of Resources and Environmental Sciences
Key Laboratory of Plant-Soil Interactions
Ministry of Education
China Agricultural University
Beijing
P.R. China

Heng Ye
Division of Plant Sciences
University of Missouri
Columbia, MO
USA

Jiangsan Zhao
Division of biotechnology and Plant Health
Norwegian Institute of Bioeconomy Research (NIBIO)
Ås, Norway

Nong Zhu
Key Laboratory of Ecosystem Network Observation and Modeling, Institute of Geographic Sciences and Natural Resources Research
Chinese Academy of Sciences
Beijing
China;
Shandong Yucheng Agro-ecosystem National Observation and Research Station
Ministry of Science and Technology
Yucheng
China;
University of Chinese Academy of Sciences
Beijing
China

Preface

Major challenges lay ahead in providing food, feed, and fibre for increasing population on the planet using diminishing water and nutrient resources and being faced with pronounced climate change and variability. There are increasingly severe shortages of good-quality water to be used for irrigation as well as exacerbated frequency and severity of droughts in the areas reliant on rain-fed food production. Moreover, raw materials used in producing some fertilizers (e.g. P and K) are becoming scarce and expensive, and the high price of energy (production of N fertilizers has a high energy demand) combine to push fertilizer prices up. Food security is also threatened by declining average yield increases of staple crops in recent years, emphasizing the need for a shift in thinking about food production.

Sustainable agricultural intensification has been increasingly popular and important in expanding food production and enhancing efficiency of water and nutrient use in a range of agricultural systems, particularly in Africa and Asia. Increasing efficiency of water and nutrient use (i.e. increasing food production per unit of water and nutrient input) is crucial in (i) maintaining food security and food quality for increased global population and (ii) decreasing potentially negative environmental impacts of growing food, feed, and fibre.

Improved root systems capable of efficient acquisition of water and nutrients from soils underpin increased efficiency of utilization of soil resources that is essential for sustainable agricultural intensification of food production. The importance of roots has been recognized by scientists worldwide, with the number of published research papers dealing with roots increasing exponentially in the recent years, suggesting not just an increasing number of root-related studies, but also a flourishing number of scientists interested in, and dealing with, various aspects of root morphology and root system architecture and function. Indeed, improved root systems to underpin agricultural intensification need to be brought about as a collaborative effort of agronomists, crop physiologists, molecular biologists, soil and environmental scientists, modellers, breeders, and others.

This book summarizes the state-of-the-art knowledge in theoretical and practical aspects of agricultural management to enhance root system architecture and function that underpin efficient use of water and nutrient resources in various food, feed, and fibre agro-ecosystems, as well as minimize undesirable water and nutrient losses to the environment. A range of soil, plant, agronomy, farming systems, breeding, molecular, and other aspects are considered.

The book covers topics linked to global economic and social issues related to, or underpinning, sustainable agricultural intensification, discusses soil–root interactions from the molecular to the agroecosystem level, emphasizes the importance of plant–microbe relations, describes various strategies and mechanisms that support increased water- and nutrient-use efficiency, and brings together natural and agricultural systems to emphasize the continuity of traits and mechanisms. This book emphasizes agronomic and breeding (particularly phenotyping) advances aimed at enhancing root capacity to acquire resources, even from relatively inhospitable (e.g. dry) soil environments, to boost water- and nutrient-use efficiency. In addition, particular prominence was given to the modern technologies that breach a barrier of "invisibility" of soil-grown roots, as well as computer simulation modelling of the root system architecture and function to select for further testing only the potential "front-runner" (rather than all possible) solutions.

This book is intended to provide professionals and scholars (agronomists, horticulturalists, plant and soil scientists, breeders, soil microbiologists, but also those involved in designing the food and environmental policy and effecting land conservation) as well as advanced undergraduate and postgraduate students with an in-depth view of various aspects of root structure and function in intensive agricultural systems, underpinning efficient water and nutrient use in food, feed, and fibre production.

All chapters have been reviewed according to the standards of international scientific journals. We would like to thank the authors for patiently revising the chapters, often quite a few times, to meet the high standards. Given the fast pace of change in the knowledge and understanding of how root morphology and function as well as the root system architecture govern water- and nutrient-use efficiency, we would see the need to update the chapters in the near future.

Zed Rengel and Ivica Djalovic
September 2020

1

Sustainable Intensification

Meaning, Need, Components, and Role of Root Systems

P.V.V. Prasad[1], *M. Djanaguiraman*[2], *and Zed Rengel*[3]

[1] *Department of Agronomy; and Feed the Future Innovation Lab for Collaborative Research on Sustainable Intensification, Kansas State University, Manhattan, KS, USA*
[2] *Department of Crop Physiology, Tamil Nadu Agricultural University, Coimbatore, Tamil Nadu, India*
[3] *UWA School of Agriculture and Environment, University of Western Australia, Perth, Australia*

1.1 Introduction

The grand challenge for scholars, policymakers, and citizens is to eradicate poverty and hunger in all forms at all times for all people, and protect our planet from further degradation. There have been significant increases in the global population that has more than doubled in the last 50 years (from about 3.7 billion in 1970 to about 7.6 billion in 2018; FAO 2018). In the past five decades, cereal grain production increased from about 1200 million tons in 1970 to about 2650 million tons in 2018. However, even now about 821 million people – approximately one in every nine people in the world – lacks sufficient food to lead active and healthy life. About 30% of children under five years of age suffer from undernourishment in many parts of Africa and Asia.

All four dimensions of food security: availability (having sufficient quantities of food available); access (having adequate income or resources to access food); utilization/consumption (having adequate dietary intake and ability to absorb and use the nutrients from food); and stability (having availability and access to food at all times) need to be addressed. In general, high food availability leads to lower food insecurity and less undernourishment. However, adequate supply of food does not always lead to food security. Access to food is dependent on household income, food prices, physical infrastructure, and social structure including gender and power hierarchies. Food utilization focuses on nutritional aspects of food security. The ability of the human body to obtain energy and nutrients from food is dependent on diet diversity, nutritional quality, food safety, clean water and environment, and human health. Food stability ensures that people are not exposed to risks of losing access to food due to shocks from climate, financial and social factors, conflict, and/or governance. Limited availability, access, utilization, and/or stability

The Root Systems in Sustainable Agricultural Intensification, First Edition.
Edited by Zed Rengel and Ivica Djalovic.

impose tremendous economic and social costs on all countries, irrespective of their economic status.

Food production is related to the harvested area and productivity per unit of that land. The increases in total food production have come from three sources – expansion of agricultural land, rise in yields per unit land area, and increase in the number of crops per unit land area per unit time. In the last five decades, food production was driven primarily by increasing yields per unit land area. However, the annual rate of yield gains of major food grain crops has significantly slowed down in recent decades. There is an urgent need to increase the rate of yield gains to keep up with the expanded demand for food, feed, and fuel from rapidly increasing population. Food and nutritional security are key for ensuring a healthy population. At the contemporary rates of increase in global population and the current changes in consumer preferences for animal source protein/dairy, we will have to increase food production by about 60% to meet the needs of about 9.5 billion people by 2050.

Agriculture uses about 30% of total land area and about 70% of total freshwater withdrawals in the world through irrigation. Even though irrigated agriculture represents only 20% of cultivated land, it contributes up to 40% of global food production.

There have been significant increases in the global greenhouse gas emissions in the last 50 years. The carbon dioxide concentration in the air has increased from 325 $\mu mol\,mol^{-1}$ in 1970 to 410 $\mu mol\,mol^{-1}$ in 2018. In the same timeframe, the concentration of methane increased from 1.35 $\mu mol\,mol^{-1}$ to 1.85 $\mu mol\,mol^{-1}$, and concentration of nitrous oxide increased from 0.295 $\mu mol\,mol^{-1}$ to 0.330 $\mu mol\,mol^{-1}$. The increases in total greenhouse gases were evident in a range of economic sectors. The percentage contributions from the different sectors in the United States in 2017 were 28.9% from transportation, 27.5% from electricity generation, 22.2% from industry, 11.6% from commercial and residential use, and 9% from agriculture (EPA 2019). Hence, agriculture contributes relatively little compared to other sectors. Within the agriculture sector, key contributors are livestock, soils, and rice production. However, agriculture also offsets about 11% of greenhouse gases in the United States and thus is a net carbon sink. Agriculture is a part of the solution to minimize greenhouse emissions and issues related to loss of biodiversity and natural resource management. However, increasing demand for food, feed, fuel, and fibre from humans will put additional pressure on agriculture. Novel methods and broader outlook towards our agriculture and energy production is required to ensure that we are meeting the food demand of growing population as well as safeguarding our planet and natural resources for future generations.

The challenge of increasing food production by about 60% is not new; we were faced with a similar problem in the last century. Agriculture scholars and global community stood up to the challenge and transformed agriculture through Green Revolution that increased the productivity of major crops and livestock species through improved genetics and use of fertilizers, pesticides, irrigation water, and farm machinery. All of these practices led to intensification of agriculture and met the needs and demands of humans, particularly in terms of calorie availability (Pingali 2012). However, there were unintended consequences on natural resources (e.g. water and soil quality), soil degradation (e.g. sub-soil compaction, waterlogging, and micronutrient deficiencies), and chemical runoff leading to negative environmental impacts (Pingali and Rosegrant 1994). The environmental consequences were not caused directly by these technologies per se, but due to the policies that led to indiscriminate use and overuse of chemicals, and expansion of cultivation into marginal lands (Pingali 2012). The excessive reliance on chemical-based plant protection and expansion of monoculture of selected crop species also decreased crop, plant, and animal diversity (Singh 2000), including diversity of indigenous crop species (Nelson et al. 2019) and associated ecosystems services. Excessive and unsustainable use of chemical inputs made our agricultural systems less efficient (Murgai et al. 2001; Thrupp 2000).

Despite rapid increases in yield gains at the start of Green Revolution, the recent trends in annual yield gains of major food crops has stagnated in recent decades (Ray et al. 2012; Grassini et al. 2013). There is a need to reverse this trend and make our agricultural systems more diverse, efficient, and productive. There is a global desire for agriculture to produce more food without environmental harm and to make positive contributions to natural and social capital through sustainable agricultural practices (Pretty et al. 2018). This is evident in the development of the 2030 Sustainable Development Agenda with 17 Goals (SDGs) by the United Nations. The SDGs came into force in 2016 and were called into action by all countries to promote prosperity while protecting environment (United Nations 2016). This new agenda recognizes that ending poverty must go hand-in-hand with strategies that build economic growth and address a range of social needs, including education, health, social protection, and job opportunities, while tackling climate change and environmental protection. One such concept related to agriculture and food production gaining global acclaim is sustainable intensification (SI). It has now become central to the SDGs and the wider efforts to improve global food and nutritional security (SDG Platform 2017; Pretty et al. 2018).

1.2 Meaning of, and Need for, Sustainable Intensification

The focus of SI is to increase food production from existing farmland while minimizing the negative impact on environment and natural resources. The SI optimizes food production per unit area per unit time. The combination of the two terms (sustainable and intensification) suggests they are not mutually exclusive, and we can simultaneously increase food production and enhance ecosystems services. The SI is an agricultural process or system where valued outcomes of food production are increased while environmental outcomes are maintained or progressively improved (Pretty et al. 2018; Pretty 2018). There is a critical need for SI because there is limited availability of arable land, water, nutrient, and labour in agricultural sector. Furthermore, agriculture is facing existential threat from rapid increases in land degradation and rapidly changing climate and climate variability.

The concept of SI goes beyond just biophysical sciences and focuses on social sciences, capacity building of all players and institutions directly or indirectly involved in food production, and consumers (in terms of food choices and changing diets). It takes multidisciplinary approaches to address food and nutritional security and increasing resilience of global food production systems. The practices of SI should also encompass changes in consumer behaviour and diets (e.g. consumption of animal and plant products), reduction of food losses and food waste (pre- and post-harvest losses), malnutrition (over consumption of high calorie food; lower consumption of healthy, nutritious, and diverse whole foods), handling of demands, governance of land and markets, and managing growing population

(Smith 2013; Cook et al. 2015; Barnes 2016a, 2016b; Rockstrom et al. 2017; Scherer et al. 2018). Innovation in research, understanding barriers to adoption, knowledge sharing, education, and capacity building are the critical aspects of SI. However, SI is not a silver bullet to address global challenge of food production and consumption, but it takes a holistic approach, provides choices, evaluates trade-offs, builds on synergies, and creates an enabling environment for increasing food production using sustainable practices and for enhancing access to safe and nutritious food.

1.3 Components or Pillars of Sustainable Intensification

There are three broadly defined components or pillars of SI practices: genetic intensification, agro-ecological intensification, and socioeconomic intensification. Genetic intensification focuses on increasing genetic potential and breaking yield barriers, increasing resistance to abiotic (e.g. water, temperature, and nutrients) and biotic stresses (pests, diseases, and weed pressure), and enhancing nutritional and health benefits of plant- and animal-based foods. All forms of traditional and new approaches of selection, breeding, genetic manipulation, and modification, and gene editing should be considered and incorporated as necessary.

Agro-ecological intensification focuses on using ecological principles for enhancing access, utilization, and efficiency of aboveground (e.g. light, water, carbon dioxide, and nutrients) and belowground (e.g. stored water, nutrients, organic matter, and soil microbiome) resources across spatial and temporal scales, managing pests and diseases using integrated approaches (e.g. crop rotations, intercropping, mixed cropping, biological agents, and natural enemies), protecting and enhancing biodiversity (e.g. species of plants, microbes, animals, insects, birds, and specifically pollinators), increasing the system productivity per unit land per unit time (e.g. farming systems, crop rotations, integration of animals, poultry, livestock, fish, and aquaculture, agroforestry systems, integration of vegetables, fruits, and tree crops), minimizing greenhouse gas emissions, optimizing energy use, increasing nutrient recycling (with biomass, animals, birds, farm waste, household waste, rural to urban and vice versa, and human waste), incorporating conservation agriculture, organic farming, and permaculture, improving soil, water, and nutrient conservation, and enhancing carbon sequestration, wildlife management, and ecosystem services.

Socio-economic intensification focuses on using aspects related to economics (e.g. profit, income, savings, and return on investment), access to market (e.g. roads, infrastructure), commodity price (e.g. market price, trade, well-developed value chains), access to inputs (e.g. seed, breeds, water, inorganic fertilizers, pesticides, organic nutrients, machinery, labour, animal traction, fuel, and energy), social (e.g. farmers organizations, cooperatives, contractual services, and bargaining power), governance (e.g. local, regional, and national support systems and organizations, land tenure, and other provisions including subsidies), and building human and institutional capacity (e.g. social capital, knowledge, extension services, education, and health care). The overarching goal is to understand the needs, create enabling environment, and remove barriers to facilitate adoption and scaling up of SI practices to an extent that they will affect food and nutritional security.

1.4 Current Global Assessment of Sustainable Intensification

There are multiple stages or transitions for adoption or scaling up of SI practices: efficiency, substitution, and redesign. Efficiency focuses on better use of on-farm and imported resources within existing system configurations (e.g. all inputs and pre- and postharvest losses). Substitution focuses on the replacement of technologies and practices (e.g. new crops and animals; new varieties and breeds that are tolerant to drought, heat, and other abiotic stresses, resistant to pests and diseases, more nutritious, more responsive to inputs, and efficient in using resources). Redesign focuses on achieving (i) the impact at scale and (ii) composition and structure of agro-ecosystems for improving ecosystems services and biodiversity. Redesign requires a broader focus, depends on capacity building, social capital, institutional change, and knowledge sharing, fosters synergies, manages trade-offs, and mitigates negative and unintended consequences. Efficiency and substitution are steps towards SI, but the system redesign is essential to deliver optimum outcomes as ecological and economic conditions change (Pretty 2018; Pretty et al. 2018).

A recent global assessment of agricultural system redesign for SI has shown significant progress in terms of number of farms and number of hectares of land (Pretty et al. 2018). This study covered 400 SI projects, programmes, and initiatives worldwide (drawn from the literature and existing datasets, whereby selected practices were implemented at scales greater than 10 000 farms or hectares). They identified 47 SI initiatives in seven different system types (integrated pest management, conservation agriculture, integrated crop and biodiversity, pasture and forage, trees, irrigation management, and small or patch systems) that occupied 163 million farms (29% of all worldwide) on 453 million hectares (9% of the worldwide total). The authors concluded that SI may be approaching a tipping point where it can be transformative. However, it will need farmers and society to invest in SI, not just for the sake of sustainability, but also for livelihoods and profitability. Similarly, recent study (Pretty et al. 2020) concluded that about 8 million new social groups for sustainable agriculture and land management were formed in the last 20 years. This growth of rural social capital is leading to increased effectiveness of agricultural and land management services, with particular benefits for the previously marginalized people (Pretty et al. 2020).

1.5 Role of Root Systems in Selected Examples of Sustainable Intensification Practices

Keeping in view the focus of this book, we provide some specific examples of SI practices where plant roots play an important role in increasing food production through enhancing resource-use efficiency (both above-ground and below-ground resources) and increasing abiotic stress tolerance (water, nutrients, and salinity). Some of these practices include intercropping, agroforestry, crop rotations, perennial crops, conservation agriculture, increasing water productivity and drought tolerance, increasing access to nutrients, and enhancing nutrient-use efficiency and recycling of nutrients.

1.5.1 Intercropping Systems

Growing two or more crop species or genotypes together in the same field where they coexist for a period is referred to as intercropping (see also Chapters 11 and 12). Traditionally, intercropping practice is vital in the low-input or resource-limited and subsistence production systems in developing countries. However, it is gaining importance in both commercial and subsistence agriculture in developed and developing countries. The primary drive is to increase resource-use efficiency, and diversify and increase resilience of cropping systems. The combination of crop species in intercropping increases the overall productivity and captures the diversity of rooting patterns and root systems for efficient exploration of soil water and nutrients below-ground and enhances capture and use of light above-ground. Increased total yields in the intercropping system with improved resource-use efficiency with no or limited additional inputs makes it attractive SI practice.

Intercropping is a dominant farming system in some parts of the world. For example, in Latin America, marginal farmers grow >70% of beans (*Phaseolus vulgaris* L.) with maize (*Zea mays* L.), potatoes (*Solanum tuberosum* L.), or other crops. In sub-Saharan Africa, >80% of cowpea (*Vigna unguiculata* [L.] Walp.) is grown with cereal grain crops. Intercropping systems are increasingly important in many parts of Asia, including China.

Gao et al. (2010) observed that the roots in maize + soybean (*Glycine max* L. Merr.) intercropping exploited different zones. Maize roots grew deeper, while soybean roots were confined to the shallow soil layers. Yields from the intercropping system are closely associated with the root depth and spread, the traits that determine uptake and utilization of water and nutrients. Zhang et al. (2002) showed that during peak growth stages of wheat (*Triticum aestivum* L.) and faba bean (*Vicia faba* L.) the root biomass of each crop did not decrease, but there was decreased competition between the two crop species for water and nutrients, resulting in higher yields.

The intercrop (barley [*Hordeum vulgare* L.] + vetch [*Vicia sativa* L.]) and the sole cropping systems showed differential rooting patterns. In deep soil layers, intercropped barley had higher root intensity than sole-cropped barley, resulting in increased nitrogen uptake from deep layers (Ramirez-Garcia et al. 2014). Ramirez-Garcia et al. (2015) concluded that the root growth, biomass, and nitrogen uptake were greater in intercropping systems (barley + vetch) compared to sole cropping (barley). Compared with the sole-cropped maize, maize intercropped with wheat had higher root weight density, root length density, and root surface area density (Wang et al. 2018).

Intercropping is practiced in areas where water is limiting because these systems tend to have high water-use efficiency (Xu et al. 2008) due to complementary root distribution (Mao et al. 2012), hydraulic redistribution of water or reduced surface runoff (Van Duivenbooden et al. 2000), and mycorrhizal colonization (Prieto et al. 2012). Intercropping systems have varying degrees of root mixing, growth patterns, and yield benefits depending on the crop type, sowing ratio, growing conditions, and soil moisture availability (Adiku et al. 2001; Li et al. 2006).

1.5.2 Crop Rotations

Crop rotation is a practice whereby different crops are grown in series on the same land over time. In traditional continuous mono-cropping systems, the productivity declines over a period of time due to a range of reasons (e.g. exacerbated disease, pest and/or weed pressure). Mono-cropping systems are dependent on external inputs, particularly inorganic fertilizers and pesticides, which if used inappropriately may degrade soil quality and build pest resistance. Choosing diverse crops to rotate over time based on the hydro-agro-ecological conditions can enhance nutrient cycling and resource-use efficiency and decrease pest pressure. Most popular crop rotations include cereal-legume systems.

Cereal rotation with legumes minimizes the dependence of cereal crop on inorganic fertilizers due to biological nitrogen fixation in the legume-rhizobia symbiosis. Legume-based crop rotations increased soil nitrogen content and water-use efficiency (Gan et al. 2015), minimized pest and disease infestation (Kutcher et al. 2011; Kathiresan 2007), and increased the yield of subsequent cereals (Borrell et al. 2017). Variations in the root systems of crops in rotations promote diverse microflora and fauna in soils that can lead to improved soil aggregate stability and soil health. Some of the popular cereal-legume rotations around the world include maize–soybean, maize–common bean, maize–cowpea, wheat–clover (*Trifolium pratense* L.), rice (*Oryza sativa* L.)–mung bean (*Vigna radiata* L.), sorghum (*Sorghum bicolor* L. Moench)–cowpea, sorghum–groundnut (*Arachis hypogaea* L.), maize–soybean–winter wheat, and rice–mung bean–wheat. The selection of crops in rotations depends on environmental and soils conditions, and market demand.

Combinations of different legumes with different rooting patterns in rotation result in improved yields. Li et al. (2018) showed that in semiarid regions, the lentil (*Lens esculenta* L.)–chickpea–pea (*Pisum sativum* L.) rotation improved the root biomass of pea (44%) compared to continuous monoculture of pea. The diversified lentil–wheat–chickpea (*Cicer arietinum* L.) rotation improved the root biomass of chickpea (by 87%) compared with less diversified chickpea–wheat–chickpea and continuous chickpea systems. The lentil–wheat–chickpea rotation increased the chickpea nodule biomass by 1.6-fold compared to other systems (chickpea–wheat–chickpea or continuous chickpea). In contrast, there was no significant difference in the biomass of root and root + nodule of lentil in different rotations (continuous lentil, lentil–wheat–lentil or pea–wheat–lentil). These findings suggests that rotational effect on legume productivity and root growth varies with legume species and the frequency of a legume appearing in the rotation (Li et al. 2018). The higher root biomass in diversified crop rotation was related to lower root rot incidence and higher nitrogen uptake from the deeper soil layers, whereas the smaller root biomass in monoculture might be associated with auto-toxicity or allelopathy (Gealy et al. 2013).

A yield increase of $1.5\,t\,ha^{-1}$ was reported in wheat rotated with oilseed rape (*Brassica* sp.) for 14 years. However, such yield increase was absent in years with above-average precipitation during the crop-growing season and unfavourable preceding crop (wheat instead of oilseed rape) (Sieling and Christen 2015). Deep-rooted Brassicas used more nitrogen from the deep soil layers than cereals; moreover, Brassicas resulted in a higher level of residual nitrogen in the topsoil due to mineralization of fallen leaf material and other plant residues (Ryan et al. 2006; Dresboll et al. 2015).

Perennial crops/pastures such as lucerne (*Medicago sativa* L.) grown in sequence with annual crops could influence residual soil resources positively or negatively in dry environments (Sauer et al. 2002; Thorup-Kristensen 2006). Sometimes legumes could deplete more soil nitrogen than the nonlegume crop. However, the subsequent mineralization of nitrogen-rich residues usually adds to the pool of soil organic matter/nitrogen after harvest.

The amount of nitrogen deposition into soil via root exudation of legumes can be substantial. On average, 33% of the total plant nitrogen in legumes was found in roots (range 14–74%) (Wichern et al. 2008). A significant portion (64%) of root nitrogen was allocated to rhizodeposits (exudates and dead cells), and nitrogen present in rhizodeposits was readily available to plants; in contrast, thick roots take long time to decompose and therefore release nitrogen only slowly and over a long period of time (Urquiaga et al. 1998). The below-ground plant-derived nitrogen from soybean in West Africa was estimated at 16–23% of the total plant nitrogen, with 70% of the below-ground nitrogen found in the rhizodeposits (Laberge et al. 2009). Similarly, 52% of cowpea-derived nitrogen was found in the soil, and 85% of that nitrogen was in the rhizodeposits (Laberge et al. 2011). Thorup-Kristensen (2006) showed that deep-rooted crops recovered leached inorganic nitrogen and lifted the deep inorganic nitrogen up to shallow layers where annual crops had their roots.

1.5.3 Conservation Agriculture

Conservation agriculture generally includes (i) minimum tillage/soil disturbance, (ii) continuous and permanent soil cover with crop residues or live plants, and (iii) diversification of crop species grown in sequence (crop rotations) or associations (intercropping). Minimal mechanical soil disturbance is a generic term used to describe the agricultural system focused on conserving soil and water by reducing soil erosion and evaporation relative to conventional tillage. However, extent of conservation or minimum tillage is dependent on crop species, soil water content, type of soil, and climate. Conservation tillage practices benefit the systems via reducing soil erosion (by wind and water), improving soil structure, enhancing infiltration of water, conserving soil moisture, improving retention, storage and build-up of soil organic matter, improving soil quality and microbial populations, reducing costs associated with fuel, time, labour, and energy use, facilitating crop establishment, increasing crop yields, and in long term decreasing the amount of fertilizer use per unit area.

Use of continuous soil cover with crop residues or live mulches improves nutrient cycling, soil organic matter, soil biological activity, may contribute to carbon sequestration, and suppresses germination of weed seeds and growth of weeds. Under conservation agricultural practices, more than 30% of crop residue from the previous crop should be left in the field while planting the subsequent crop. As explained in the previous sections, intercropping and crop rotations – particularly with crops that have different root systems and rooting systems – improve water use through capturing water from different soil depths. Differential rooting patterns and root systems also facilitate nutrient uptake from different depths, increasing nutrient-use efficiency. Conservation agricultural practices often result in enhanced root growth due to improved soil structure and increased organic matter content, ultimately contributing to increased yields.

Soil physical properties are improved in the no-till compared with the tillage-based systems (Lal 1997). No-till practices significantly improved saturated and unsaturated hydraulic conductivity due to ensuring either continuity of pores or flow of water through few large pores (Allmaras et al. 1977; Benjamin 1993). More stable aggregates in the topsoil were recorded in the no-till systems, contributing to high total porosity (Lal et al. 2007). Long-term minimum tillage increased the concentrations of soil organic carbon and nitrogen within the aggregates in the topsoil (5–8 cm) after 37–40 years (Jacobs et al. 2009). Kargas et al. (2012) and McVay et al. (2006) observed that no-till plots retained more water in topsoil

(0–10 cm) than conventionally tilled plots via an increase in the storage pores (0.5–50 mm in diameter) and the amount of elongated transmission pores (50–500 mm in length).

The impacts of conservation agricultural practices on crop yield are often related to root growth and water- and nutrient-use efficiency (Boone and Veen 1994; Davis 1994; Lal 1993). Under no-till and reduced tillage practices, an increase in root length density was observed in the upper soil layers compared to conventional tillage. Malhi and Lemke (2007) reported an increase of 22% in root mass under no-tillage compared to conventional tillage. Increased cracks, worm channels, and a higher number of biopores facilitated root growth under no-tillage conditions (Francis and Knight 1993). In the no-till system, Lampurlanes et al. (2001) found increased soil moisture content in deeper soil layers and greater root growth than under conventional tillage.

Root growth under conservation agriculture is dependent on soil type and soil depth. Mosaddeghi et al. (2009) found that the root length density of maize was greater after traditional ploughing to 30 cm depth than in the no-till system. An increase in root length density under deep ploughing was due to the soil physical environment less restrictive for root growth compared to no-till (Cai et al. 2014). Deep tillage increased the root proliferation and rooting depth of maize, resulting in increased grain yields compared with no-tillage on silt loam soil (Varsa et al. 1997). Subsoil tillage promoted maize root growth and development in deeper soil layers, resulting in increased nutrient accumulation and yield compared to conventional tillage (Wang et al. 2012).

Yeboah et al. (2017) showed increased root length, root surface area, root diameter, and root volume up to soil depth of 50 cm in the no-till/straw-amended soils. These were attributed to the decreased soil bulk density and the increased soil moisture, which might have promoted root proliferation during the growing season. At all growth stages, most of the root characteristics in the top 50 cm of soil under no-tillage with straw retention were significantly better than under conventional tillage with no straw addition. The yield of maize under minimum tillage system was more sustainable compared with conventional tillage, and it was associated with better root development (Busari and Salako 2012, 2013). In the long-term and considering the energy and environmental perspectives, conservation tillage is more economical and sustainable in increasing and maintaining crop yields.

1.5.4 Agroforestry Systems

Land degradation, soil erosion, and decreasing soil fertility are major constraints in the arid and semi-arid regions of Africa. Diverse approaches to improving soil fertility in sub-Saharan Africa are crucial to food production and the resilient and sustainable livelihoods (Stewart et al. 2020) Topsoil erosion is a serious form of soil degradation due to long-term removal of surface cover, crop residues, and harsh environmental conditions (rainfall and temperature). Agroforestry is one of most common land-use systems across landscapes in sub-Saharan Africa. Agroforestry is a general term, where trees are grown in combination with food crops, forages, pastures or grassland systems to provide human food, livestock feed, and fuel. Simple agroforestry systems include alley cropping, intercropping, and hedgerow systems, whereas complex agroforestry systems comprise natural forest ecosystems along with crops. Inclusion of trees in cropping systems offers multiple benefits leading to increases in crop yield, biodiversity, and ecosystems services (Rice 2008; Oke and Odebiyi 2007).

One of the most common landscape in Sudano-Sahelian and sub-Saharan Africa are parklands, a form of agroforestry system with leguminous tree species (e.g. *Faiderbia albida* A. Chev., *Azadirachta indica* A. Juss., *Guiera senegalensis* J.F. Gmel, *Parkia biglobosa* (Jacq.) R. Br. Ex G. Don, and *Piliostigma reticulatum* (DC.) Hochst.). These tree species are unique due to their differential rooting patterns and phenology that benefit and support growth of annual crops or forages species under their crowns without competition. The tree species shed leaves during the crop-growing season and grow in off-season (reverse phenology). During the crop-growing season, trees do not compete for resources, allow ample radiation to the annual crops, and shed leaves that contribute to increased soil organic matter content and improved nutrient availability.

The tree canopy also keeps the temperature cooler through modifying the microclimates, decreases evapotranspiration, and retains greater soil moisture in the upper rooting zones. The increased soil water moisture in the upper rooting zones is associated with hydraulic lift and hydraulic redistribution, both of which are rely on the rooting systems. Hydraulic lift is a passive movement of water through the root systems of the trees from the deeper and wetter soil layers to the shallow and drier layers. Hydraulic lift happens vertically, but due to a large number of lateral roots, trees also redistribute water downwards or laterally from the moist regions to drier regions. Such bidirectional movement of soil water through tree root systems is termed hydraulic redistribution. These two traits (hydraulic lift and hydraulic redistribution) together often act as "bioirrigators" to the adjacent plants (Bayala and Prieto 2019), making agroforestry systems valuable and beneficial to crops, resulting in increased yields.

A recent structured and comprehensive review (1995–2014) assessing 350 journal articles from 23 countries on role of agroforestry systems in the semi-arid tropics regions showed that crop yields increased under tree-based systems in 68% of the studies, decreased in 18%, and 14% studies were unaffected (Kuyah et al. 2016). Tree canopies also provide shade for livestock to herd during summer and contribute soil nutrients. Some of the other benefits of the agroforestry systems were associated with nutrient cycling, regulation of pests and diseases, and enhanced soil fertility and biodiversity.

Beneficial impacts of successful agroforestry systems on the annual crops are associated with the differential root growth patterns. Trees take more nutrients and water from the soil for a longer time compared to annual crops. The roots of trees can extend more than 2 m deep; however, the rooting depth of annual plants rarely extends beyond 1 m. The deep rooting system of trees makes them more resilient to harsh environmental conditions. Higher above- and below-ground biomass of tress can reduce soil erosion, transfer organic inputs to soil microorganisms, and increase the amount of carbon stored in the soil.

Farmers across Africa grow leguminous trees along with maize, sorghum, or millet for enriching soil fertility (Garrity et al. 2010). Trees facilitate nutrient cycling and improve soil fertility and structure, resulting in enhanced water retention and decreased nutrient leaching (Lehmann et al. 1998; Bayala et al. 2008). Improved soil fertility is through increases in soil organic matter and biological nitrogen fixation by leguminous trees. Integration of agroforestry systems maintains year-round vegetative soil cover (Boffa 1999), increases nutrient availability through nitrogen fixation and nutrient cycling (Barnes and Fagg 2003), improves soil water infiltration (Chirwa et al. 2007), increases production of food, fodder, fuel, and fibre, rises farmers' income from tree products (Garrity 2004), and increases carbon storage both above- and below-ground (Makumba et al. 2007). The ideal root system of agroforestry trees includes low tree root density in the topsoil to lessen competition with associated crops, and large tree root mass in deeper layers resulting in soil carbon enrichment.

The perennial and annual species of the families Asteraceae, Poaceae, and Fabaceae were analyzed for root traits, and the results indicated that the perennials had lower specific root length, root nitrogen concentration, and mycorrhizal colonization, and higher root tissue density and coarser root systems, except in Fabaceae (Roumet et al. 2006). These findings indicated that the perennials had a conservative strategy compared with annuals, because high root length density, large root diameter, and low root nitrogen concentration would be the properties of the long-lived roots (Tjoelker et al. 2005). In contrast, the root system of annuals was categorized by (i) high specific root length, which may be related to quick rates of root elongation, high relative growth rate, and high nutrient and water uptake (Comas et al. 2002; Reich et al. 1998), as well as (ii) high root nitrogen concentration, which may reflect high nutrient uptake and metabolic activities (Tjoelker et al. 2005).

A meta-analysis of 94 studies in sub-Saharan Africa indicated that agroforestry increased maize yields from 0.7 to 2.5 $t\,ha^{-1}$ (Sileshi et al. 2008). Furthermore, this increase in maize yield was found to be more stable than in monoculture of maize at the recommended application of inorganic fertilizer (Sileshi et al. 2011). The indirect benefit of the leguminous trees was associated with the cycling of macro- and micronutrients via uptake from the deeper soil layers due to extensive and deep root systems (Mweta et al. 2007).

The organic nitrogen left in soil is stored in the soil organic matter pools (Schroth et al. 2001); this organic nitrogen is available to mycorrhizal fungi and earthworms (Sileshi and Mafongoya 2006), leading to enriched soil fertility and soil microbiome. Leguminous trees also increase water-use efficiency, as evidenced in maize intercropped with leguminous trees compared to maize monoculture (Chirwa et al. 2007).

1.5.5 Perennial Grain Crops and Pastures

The previous section on agroforestry dealt with perennial trees or shrubs species, their rooting systems and benefits to crop yields. This section focuses on the perennial grain crops and grass species. About 70% of the total cropland is planted to annual grain crops (cereals, legumes, and oilseeds) supplying majority of calories needed for humans. Three annual crops – rice, wheat, and maize – provide more than 60% of calorie needs of humans. Production of these annual crops is increasing in the marginal lands that are not suitable for them, leading to land degradation and a loss of ecosystems services, meaning these systems are not sustainable in longer term (Cassman et al. 2003; Glover et al. 2010, 2012). Such marginal lands are more suitable for perennial crops including those for grain, forage, or bioenergy. Advantage of these perennial grain crops and grass species is their deep and large root systems, with extensive rhizosphere that improves water and nutrient uptake, and enhances stress tolerance (Rasche et al. 2017). Other benefits of perennial grain crops and grasses summarized by Rasche et al. (2017) associated with roots include large rooting systems increase the functional effectiveness of rhizosphere microorganisms; consistent colonization by mycorrhizae; stimulation of beneficial soil microbiota by root turnover throughout the year; and inoculants becoming well established over several seasons due to high system stability. However, some of the risks include the pressure from the root-feeding and parasitic pests that may damage roots and provide opportunities for other pathogens to infect the root systems.

There has been significant progress in development of perennials grain crops, particularly in wheat, rye, sorghum, and rice through hybridization of high-performing domesticated annual species with closely related wild perennials (Cox et al. 2010; Ryan et al. 2018). Another approach has been through domestication of wild perennial plant species that can serve as new grain crops, such as *Helianthus maximiliani* and *Silphium integrifolium* that are relatives of common sunflower (*Helianthus annuus*; Van Tassel et al. 2013).

One of the most advanced, commercially marketed, domesticated perennial grain crops is intermediate wheat grass (*Thinopyrum intermedium*) that is commercially marketed as Kernza and is available in multiple commercial products (Lubofsky 2016; Baker 2017). This species has deep roots that extend down to 3 m, which is more than double the depth of annual wheat crop. It is the first commercial perennial grain crop introduced in the agricultural and food markets across the United States.

Highly productive perennial rice (e.g. PR23) was developed through a wide hybrid cross between annual rice (*Oryza sativa* L.) and perennial rice (*Oryza logistaminata* L.). A large number of on-farm tests over a period of 6 years in China showed that with 15 tons per hectare yields from two harvests per year, the perennial rice yields were similar or greater than those of the annual rice. Data from 45 site-years in China showed that perennial rice (PR23) were broadly adapted and accepted by farmers and were comparable to annual rice in terms of phenology, plant height, grain yield, and grain size, and had acceptable cooking quality (Huang et al. 2018). In addition, economic analyses indicated substantial labour saving in the perennial rice systems compared to annual rice.

Ryan et al. (2018) highlighted five strategies that can increase the multi-functionality of perennial grain crops: (i) perennial grain crops can be grown in extended rotations with annual crops, which would help in regeneration of soil health; (ii) perennial grains crops can be grown on sloped land or as a buffer along field edges, which would reduce soil erosion and protect water quality; (iii) vegetation from the perennials grain crops can be harvested or grazed by livestock, providing additional income; (iv) perennial cereals can be intercropped with legume forages, reducing the nitrogen requirements and increasing forage quality; and (v) perennials grain crops can be grown together as functionally diverse grain polycultures, which increases yield stability and provides multiple ecosystem services.

1.5.6 Nutrient-use Efficiency

Enhancing nutrient-use efficiency is one of the key aspects of SI to ensure that increasing yields are not accompanied by negative effects due to overuse of fertilizers. There has been a tremendous increase in yields of major food grain crops around the world in the last five decades due to improved genetics and use of chemical fertilizers. Fertilizer use has significantly increased globally, but particularly in Asia, Europe, and Americas, whereas there has been only marginal increases in use of inorganic fertilizers in sub-Saharan Africa, where average use is still less than 25 kg ha^{-1}. Low use in sub-Saharan Africa is primarily due to the cost and a lack of access to inorganic fertilizers as well as limited availability of organic resources to be used as soil amendments. There are large yield gaps (about 70%; difference between attainable and actual crop yield) for major food grain crops in sub-Saharan Africa.

In China, cereal grain yields have increased by 65% from 1980 to 2010; however, in the same period, use of fertilizer has increased by 512% (Zhang et al. 2011, 2012). Overuse and inefficient use of chemical fertilizers had led to low nutrient-use efficiency, degradation of soil health, and pollution of water resources. A recent large survey in sub-Saharan Africa highlighted the importance of using a systems approach and taking a broader view of soil fertility improvements utilizing all available options, including both inorganic and organic sources of nutrients and the farming systems approaches that cover biophysical and socio-economic aspects (Stewart et al. 2020).

Roots are critical for uptake of soil nutrients and maintaining a healthy rhizosphere. Root exudates play an important role in availability, mobilization, transformation, and efficient use of nutrients by plants. Plant roots have the capacity to modify rhizosphere chemical, physical, and biological properties through release of organic acid anions, phosphatases, protons, signalling molecules, and other compounds as well as via changing soil redox potential (Marschner 2012). In addition, plants can also regulate their root morphological traits to adapt to the changing soil environmental conditions. All of these processes directly or indirectly influence nutrient-use efficiency of plants.

There are multiple ways to enhance nutrient-use efficiency, including plant traits (e.g. genetics, root architecture, rhizosphere manipulation, and regulation), crop management (e.g. selection of crops, intercropping, and rotations), nutrient management (principles of 4Rs: right source, right rate, right time, and right place), and management of water, soil, and the root zone. We will limit our discussion to plant traits that are driven by root growth and root zone management, particularly with respect to nitrogen (N) and phosphorus (P).

Certain plant genotypes are more efficient in uptake, transport and/or utilization of N. The two major components of nitrogen-use efficiency are uptake efficiency and utilization efficiency. Morphology and architecture of root systems play an essential role in enhancing N-uptake efficiency. The uptake of N by roots depends on a degree to which the roots extend their absorptive area, that is, root morphology. Some of these morphological traits include increased root length density, increased root mass and depth, larger and deeper root systems, and enhanced lateral branching of roots. The deep roots absorb nitrate as it moves down the soil profile in the process of leaching. Lateral root growth can utilize the heterogeneously supplied soil nutrients, particularly under limited-N conditions. Utilization efficiency is partly related to storage and remobilization of N from senescing plant parts (leaves, stems, or roots) to grains. There are published reports on genetic variability in N-uptake efficiency and N-utilization efficiency in several grain crops (e.g. wheat, maize, barley, and rice) (see Malik and Rengel 2013). Both uptake and utilization efficiency contribute to higher N-use efficiency (NUE); however, the relative importance of the underlying traits in an individual genotype depends on the application of N (rate, time, place, and method), soil type, and environmental conditions. Similarly for P, which is relatively immobile in soil, and present mostly in the topsoil, important root traits are shallow roots with large number of laterals, root clusters, and adventitious roots (Lynch 2011).

Zhang et al. (2010) and Shen et al. (2013) in their reviews demonstrated how root zone management and maximizing root/rhizosphere efficiency provide a unique opportunity to increase crop productivity and nutrient-use efficiency and minimize an environmental impact. Some of the specific strategies include manipulating root systems (root development and size, root system architecture, and distribution), rhizosphere acidification, carboxylate exudation, microbial association with plants, rhizosphere interactions in terms of intercropping and crop rotations, localized application of nutrients, use of efficient crop genotypes, and synchronizing rhizosphere nutrient supply with the crop demands.

A comprehensive analysis of several maize (Ciampitti and Vyn 2012, 2013) and sorghum (Ciampitti and Prasad 2016) studies from different time intervals (old era: prior to 1990 for maize, and prior to 1980 for sorghum; transient era: from 1981 to 2000 for sorghum, and new era: after 1990 for maize, and after 2000 for sorghum) highlighted that historical change in NUE was primarily explained by changes in grain N percentage and in remobilized N that was primarily governed by the vegetative stage whole-plant uptake. Wu et al. (2019) evaluated different inbred lines of maize from different decades (1960s–2000s) and found that yield obtained per unit of N applied increased linearly over the last three decades (1980s–2000s) and it was accompanied by higher N accumulation in grains and greater N translocation efficiency from roots. The morphological properties of roots (dry weights, length, volume, surface area, weight density, and length density) tended to increase from 1960s to 1980s, but then decreased from 1980s to 2000s. They concluded that (i) the modern inbred lines produced fewer redundant roots in the topsoil (0–20 cm) with a larger proportion of roots in the deeper soil layers, and (ii) greater NUE of modern inbred lines was associated with greater physiological absorption capacity of roots rather than larger root size in normal or stress conditions.

1.5.7 Improved Water Productivity and Drought Tolerance

Fresh water resources used for irrigation are dwindling across the globe, causing significant yield losses throughout the food production systems. Agriculture is the largest user (about 70% of water withdrawal) of our freshwater resources. Frequent and prolonged occurrence of water stress (drought) during crop growing season due to irregular rainfall or lack of availability of irrigation is one of the most yield-limiting factor for crops. Global patterns of crop production losses associated with drought from 1983 to 2009 for key food crops – maize, rice, wheat, and soybean – were quantified (Kim et al. 2019). The authors found that three quarters of the global harvested area (about 454 million hectares) had drought-induced yield losses, with cumulative production losses corresponding to 166 billion U.S dollars. Per crop, harvested area that experienced yield losses was 161 million ha (45% of global harvested area) for wheat, 124 million ha (82%) for maize, 102 million ha (62%) for rice, and 67 million ha (91%) for soybean. In the United States, about 67% of crop losses in the last 50 years were associated with drought (Comas et al. 2013). It is projected that water shortages will become more severe in the future due to competition from agriculture, industry, and urban uses, compounded by growing population. Furthermore, global climate models estimate that increasing temperatures will further enhance the intensity and duration of drought stress in agricultural systems. Therefore, increasing water productivity (grain yield per unit of water used) and drought tolerance in major food crops is critical for SI and for addressing global food security.

Crop water productivity, defined as a measure of biophysical (e.g. yield or biomass) or economic gain from use of a unit of water consumed for crop production, is an important measure of resource-use efficiency. There are several factors that influence biophysical or economic gains from water applied, which ultimately influence water productivity. Some of these include crop species and genotype, amount of water applied, as well as agronomic, soil, and economic factors (Ali and Talukder 2008). However, we will limit our discussion to root traits that influence drought tolerance and plant capacity to absorb and utilize water efficiently.

In an agricultural context, drought tolerance is defined as the plant capacity to maintain productivity during conditions of water deficits or in water-limited environments (Tardieu 2012; Tardieu et al. 2018). Root architecture and function are critical for water uptake from the rhizosphere. Plant breeders and crop improvement programmes have focused on many root traits, such as root length, rooting depth, root growth angle, branching, number of lateral roots, distribution of roots down the soil profile, number and area of fine roots, and more (Lynch 2013; Bianco and Kepinski 2018). Plants have different root systems, and their growth and function vary at different plant growth stages. At the initial growth stages, the embryonic (primary and seminal) roots play an important role, whereas at later stages crown roots, hypocotyl roots, and lateral roots constitute a major portion of the root system. Lynch (2013) proposed root ideotype "steep, cheap, and deep" that integrates the root architecture, root anatomy, and root physiological traits to enhance and optimize water and N acquisition by maize. Steep refers to the root angle, cheap refers to the metabolic cost of root growth, and deep refers to the rooting depth. The premise of this ideotype is that, in general, more water and N would be available in deeper soil layers and that a deep rooting system would have laterals arising from seminal, crown, and brace roots that can acquire resources from shallow soil layers. He emphasized that this ideotype is applicable to low-input systems because of the importance of deep rooting for water acquisition and that these features are relevant to other cereal root systems beyond maize and root systems of dicots.

Large genetic variability exists for various root traits in different crops, including wheat (Narayanan et al. 2014; Narayanan and Prasad 2014; Djanaguiraman et al. 2019a, 2019b; Manschadi et al. 2006), maize (Hammer et al. 2009; Manavalan et al. 2011; Chimungu et al. 2014a, 2014b; Gao and Lynch 2016), rice (Henry et al. 2011; Uga et al. 2013), soybean (Prince et al. 2017; Manavalan et al. 2015), common bean (Ochao et al. 2006), and chickpea (Kashiwagi et al. 2015; Purushothaman et al. 2017). Palta and Turner (2018) in their recent overview emphasized that, despite the presence of large genetic variability, the role of roots in imparting drought tolerance and expression of the relevant traits under field conditions is highly complex. This is due to a strong interaction of roots with soil and other environmental factors (e.g. soil type, soil depth, soil physical and chemical properties, soil water holding capacity, and amount and distribution of rainfall). They argued that having an individual root trait for water uptake does not necessarily guarantee a root system would be efficient in capturing soil water (Palta and Turner 2018). However, their analysis shows that in addition to the direct impact of root traits on drought tolerance and yield, other plant traits such as early vigour, osmotic adjustment, and 'stay-green' that are associated with drought tolerance are imparted through increased root growth and water use. Palta and Turner (2018) conclude that breeding for root traits that increase drought tolerance is not easy and that there is no single marker or gene that would deliver drought tolerance in crop plants. However, specific roots traits in association with other drought-tolerance traits, and taking into consideration soil conditions and rainfall, can lead to higher and timely uptake of water and nutrients, which may be associated with higher grain yields under water-limited conditions.

1.6 Conclusions and Future Outlook

SI of farming systems is essential in addressing the global grand challenge of increasing production of and access to nutritious and safe food, while safeguarding the environment, minimizing food losses and waste, and creating/maintaining ecosystem services. There have been increases in the use of SI practices globally; however, tremendous opportunities remain to be utilized. Research on genetic, ecological, and social components of SI, along with creating the enabling environments through appropriate policy, is a prerequisite for success. Transdisciplinary approaches that integrate social and natural sciences and use participatory principles, with clear pathways towards improving health (soil, plant, animal, and human) can help with SI of global agriculture. For adoption and scaling of SI practices, it is critical to measure impacts across multiple domains (productivity, environment, economic, social, and human), so that synergies and tradeoffs in the system can be quantified. Developing and using a clear framework with standardized methods and metrics across domains with visualization tools that are easily accessible and available to all stakeholders is essential for scaling up and adoption of SI practices.

Plant roots play a critical role in multiple components of SI practices (e.g. intercropping, crop rotations, conservation agricultural practices, agroforestry systems, perennial grain crops and grasses, nutrient-use efficiency, water productivity, and water-use efficiency). These examples traverse both genetic and management aspects of agricultural production systems. However, our understanding of plant root systems is still limited and needs more attention. The challenges of understanding roots are in the complexities of phenotyping the underground/invisible systems and the dynamic nature of most root traits that are highly sensitive to soil and environmental conditions as well as rhizosphere properties. The recent advances in phenotyping facilities and genomic tools will help better understand and use the root systems for improving productivity, efficiency, and resilience of our cropping systems.

Acknowledgments

PVVP would like to thank the Feed the Future Innovation Lab for Collaborative Research on Sustainable Intensification (Grant no. AID-OAA-L-14-00006) funded by United States Agency for International Development. Contribution no. 20-100-B from the Kansas Agricultural Experiment Station.

References

Adiku, S.G.K., Ozier-Lafontaine, H., and Bajazet, T. (2001). Patterns of root growth and water uptake of a maize–cowpea mixture grown under greenhouse conditions. *Plant and Soil* 235: 85–94.

Ali, M.H. and Talukder, M.S.U. (2008). Increasing water productivity in crop production – a synthesis. *Agricultural Water Management* 95: 1201–1213.

Allmaras, R.R., Rickman, R.W., Ekin, L.G. et al. (1977). Chiselling influences on soil hydraulic properties. *Soil Science Society of America Journal* 41: 796–803.

Baker, B. (2017). Can modern agriculture be sustainable? Perennial polyculture holds promise. *BioScience* 67: 325–331.

Barnes, R.D. and Fagg, C.W. (2003). *Faidherbia albida*. Monograph and annotated bibliography. Tropical Forestry Papers No 41, Oxford Forestry Institute. Oxford, UK. p. 281

Barnes, A.P. (2016a). Quantifying ambivalence towards sustainable intensification: an exploration of UK public values. *Food Security* 8: 609–619.

Barnes, A.P. (2016b). Can't get there from here: attainable distance, sustainable intensification and full scale technical potential. *Regional Environmental Change* 16: 2269–2278.

Bayala, J., Heng, L.K., Noordwijk, M.V. et al. (2008). Hydraulic redistribution study in two native tree species of agroforestry parklands of West African dry savanna. *Acta Oecologia* 34: 370–378.

Bayala, J. and Prieto, I. (2019). Water acquisition, sharing and redistribution by roots: application to agroforestry systems. *Plant and Soil.* https://doi.org/10.1007/s11104-019-04173-z.

Benjamin, J.G. (1993). Tillage effects on near-surface soil hydraulic properties. *Soil and Tillage Research* 26: 277–288.

Bianco, M.D. and Kepinski, S. (2018). Building a future with root architecture. *Journal of Experimental Botany* 69: 5319–5323.

Boffa, J.M. (1999). Agroforestry Parklands in sub-Saharan Africa. FAO Conservation Guide 34, 254. Rome: Food and Agriculture Organization.

Boone, F.R. and Veen, B.W. (1994). Mechanisms of crop responses to soil compaction. In: Developments in Agricultural Engineering (eds. B.D. Soane and C. van Ouwerkerk). Netherlands: Elsevier.

Borrell, A.N., Shi, Y., Gan, Y. et al. (2017). Fungal diversity associated with pulses and its influence on the subsequent wheat crop in the Canadian prairies. *Plant and Soil* 414: 13–31.

Busari, M.A. and Salako, F.K. (2012). Effect of tillage and poultry manure application on soil infiltration rate and maize root growth in a sandy Alfisol. *Agro-Science Journal of Tropical Agriculture, Food, Environment and Extension* 11: 24–31.

Busari, M.A. and Salako, F.K. (2013). Effect of tillage, poultry manure and NPK fertilizer on soil chemical properties and maize yield on an Alfisol at Abeokuta, south-western Nigeria. *Nigerian Journal of Soil Science* 23: 206–218.

Cai, H.G., Ma, W., Zhang, X.Z. et al. (2014). Effect of subsoil tillage depth on nutrient accumulation root distribution and grain yield in spring maize. *Crop Journal* 2: 297–307.

Cassman, K.G., Doberman, A., Walters, D. et al. (2003). Meeting cereal demand while protecting natural resources and improving environmental quality. *Annual Review of Environment and Resources* 28: 315–358.

Chimungu, J.G., Brown, K.M., and Lynch, J.P. (2014a). Large root cortical cell size improves drought tolerance in maize (*Zea mays* L.). *Plant Physiology* 166: 2166–2178.

Chimungu, J.G., Brown, K.M., and Lynch, J.P. (2014b). Reduced root cortical cell file number improves drought tolerance in maize. *Plant Physiology* 166: 1943–1955.

Chirwa, P.W., Ong, C.K., Maghembe, J. et al. (2007). Soil water dynamics in intercropping systems containing *Gliricidia sepium*, pigeon pea and maize in southern Malawi. *Agroforestry Systems* 69: 29–43.

Ciampitti, I.A. and Vyn, T. (2012). Physiological perspective of changes over time in maize yield dependency on nitrogen uptake and associated nitrogen efficiencies: a review. *Field Crops Research* 133: 48–67.

Ciampitti, I.A. and Vyn, T. (2013). Grain nitrogen source changes over time in maize: a review. *Crop Science* 53: 366–377.

Ciampitti, I.A. and Prasad, P.V.V. (2016). Historical synthesis-analysis of changes in grain nitrogen dynamics in sorghum. *Frontiers in Plant Sciences* 7: 275.

Comas, L.H., Bouma, T.J., and Eissenstat, D.M. (2002). Linking root traits to potential growth rate in six temperate tree species. *Oecologia* 132: 34–43.

Comas, L.H., Becker, S.R., Cruiz, V.M.V. et al. (2013). Root traits contributing to plant productivity under drought. *Frontiers in Plant Science* 4: 442.

Cook, S., Silici, L., Adolph, B. et al. (2015). Sustainable Intensification Revisited. IIED Issue Paper. London: International Institute for the Environment and Development.

Cox, T.S., van Tassel, D.L., Cox, C.N. et al. (2010). Progress in breeding perennial grains. *Crop and Pasture Science* 61: 513–521.

Davis, J.G. (1994). Managing plant nutrients for optimum water use efficiency and water conservation. *Advances in Agronomy* 53: 85–120.

Djanaguiraman, M., Prasad, P.V.V., Kumari, J. et al. (2019a). Root length and root lipid composition contributes to drought tolerance of winter and spring wheat. *Plant and Soil* 439: 57–73.

Djanaguiraman, M., Prasad, P.V.V., Kumari, J. et al. (2019b). Alien chromosomal segment from *Aegilops speltoides* and *Dasypyrum villosum* increases drought tolerance in wheat via profuse and deep root system. *BMC Plant Biology* 19: 242.

Dresboll, D.B., Rasmussen, I.S., and Thorup-Kristensen, K. (2015). The significance of litter loss and root growth on N efficiency in normal and semi-dwarf winter oilseed rape genotypes. *Field Crops Research* 186: 166–178.

EPA (2019). Total U.S. greenhouse gas emissions by economic sector in 2017. All estimates from the Inventory of U.S. Greenhouse Gas Emissions and Sinks: 1990–2017. EPA430-R-19-001. www.epa.gov/ghgemissions/sources-greenhouse-gas-emissions

FAO (2018). Food and Agricultural Organization. Rome, Italy: Agricultural Statistics.

Francis, G.S. and Knight, T.L. (1993). Long-term effects of conventional and no-tillage on selected soil properties and crop yields in Canterbury, New Zealand. *Soil and Tillage Research* 26: 193–210.

Gan, Y., Hamel, C., O'Donovan, J.T. et al. (2015). Diversifying crop rotations with pulses enhances system productivity. *Scientific Reports* 5: 14625.

Gao, Y., Duan, A., and Qiu, X. (2010). Distribution of roots and root length density in a maize/soybean strip intercropping system. *Agriculture and Water Management* 98: 199–212.

Gao, Y. and Lynch, J. (2016). Reduced crown number improves water acquisition under water deficit stress in maize (*Zea mays* L.). *Journal of Experimental Botany* 67: 4545–4557.

Garrity, D.P. (2004). Agroforestry and the achievement of the millennium development goals. *Agroforestry Systems* 61: 5–17.

Garrity, D.P., Akinnifesi, F.K., Ajayi, O.C. et al. (2010). Evergreen agriculture: a robust approach to sustainable food security in Africa. *Food Security* 2: 197–214.

Gealy, D., Moldenhauer, K., and Duke, S. (2013). Root distribution and potential interactions between allelopathic rice, sprangletop (*Leptochloa* spp.) and barnyard grass (*Echinochloa crusgalli*) based on ^{13}C isotope discrimination analysis. *Journal of Chemical Ecology* 39: 186–203.

Glover, J.D., Reganold, J.P., Bell, L.W. et al. (2010). Increased food and ecosystems security via perennial grains. *Science* 5986: 1638–1639.

Glover, J.D., Reganold, J.P., and Cox, C.M. (2012). Plant perennials to save Africa's soil: integrating perennials with food crops could restore soil health and increase staple yields. *Nature* 486: 359–362.

Grassini, P., Eskridge, K.M., and Cassman, K.G. (2013). Distinguishing between yield advances and yield plateaus in historical crop production trends. *Nature Communication* 4: 2918.

Hammer, G.L., Dong, Z., McLean, G. et al. (2009). Can changes in canopy and/or root system architecture explain historical maize yield trends in the US corn belt? *Crop Science* 49: 299–213.

Henry, A., Gowda, V.P.R., Torres, R.O. et al. (2011). Variation in root system architecture and drought response in rice (*Oryza sativa*): phenotyping of the OryzaSNP panel in rainfed lowland fields. *Field Crops Research* 120: 205–214.

Huang, G., Qin, S., Zhang, S. et al. (2018). Performance, economics and potential impact of perennial rice PR23 relative to annual rice cultivars at multiple locations in Yunnan province of China. *Sustainability* 10: 1086.

Jacobs, A., Rauber, R., and Ludwig, B. (2009). Impact of reduced tillage on carbon and nitrogen storage of two Haplic Luvisols after 40 years. *Soil and Tillage Research* 102: 158–164.

Kargas, G., Kerkides, P., and Poulovassilis, A. (2012). Infiltration of rain water in semi-arid areas under three land surface treatments. *Soil and Tillage Research* 120: 15–24.

Kashiwagi, J., Krishnamurthy, L., Purushotaman, R. et al. (2015). Scope for improvement of yield under drought through the root traits in chickpea (*Cicer arietinum* L.). *Field Crops Research* 170: 47–54.

Kathiresan, R.M. (2007). Integration of elements of a farming system for sustainable weed and pest management in the tropics. *Crop Protection* 26: 424–429.

Kim, W., Iizumi, T., and Nishimori, M. (2019). Global patterns of crop production losses associated with drought from 1983 to 2009. *Journal of Applied Meteorology and Climatology* 58: 1233–1244.

Kutcher, H.R., Johnston, A.M., Bailey, K.L. et al. (2011). Managing crop losses from plant diseases with foliar fungicides, rotation and tillage on a black chernozem in Saskatchewan. Canada. *Field Crops Research* 124: 205–212.

Kuyah, S., Oborn, I., Jonsson, M. et al. (2016). Trees in agricultural landscapes enhance provision of ecosystem services in sub-Saharan Africa. *International Journal of Biodiversity Science, Ecosystem Services and Management* 12: 255–273.

Laberge, G., Franke, A.C., Ambus, P. et al. (2009). Nitrogen rhizodeposition from soybean (*Glycine max*) and its impact on nutrient budgets in two contrasting environments of the Guinean savannah zone of Nigeria. *Nutrient Cycling in Agroecosystems* 84: 49–58.

Laberge, G., Haussmann, B.I.G., Ambus, P. et al. (2011). Cowpea N rhizodeposition and its below-ground transfer to co-existing and to a subsequent millet crop on a sandy soil of the Sudano-Sahelian eco-zone. *Plant and Soil* 340: 369–382.

Lal, R. (1993). Tillage effects on soil degradation, soil resilience, soil quality and sustainability. *Soil and Tillage Research* 27: 1–8.

Lal, R. (1997). Long-term tillage and maize monoculture effects on a tropical Alfisol in western Nigeria. I. Crop yield and soil physical properties. *Soil and Tillage Research* 42: 145–160.

Lal, R., Reicosky, D.C., and Hanson, J.D. (2007). Evolution of the plow over 10,000 years and the rationale for no-till farming. *Soil and Tillage Research* 93: 1–12.

Lampurlanes, J., Angas, P., and Cantero-Martinez, C. (2001). Root growth: soil water content and yield of barley under different tillage systems on two soils in semiarid conditions. *Field Crops Research* 69: 27–40.

Lehmann, J., Peter, I., Steglich, C. et al. (1998). Below-ground interactions in dryland agroforestry. *Forest Ecological Management* 111: 157–169.

Li, L., Sun, J., Zhang, F. et al. (2006). Root distribution and interactions between intercropped species. *Oecologia* 147: 280–290.

Li, J., Liu, K., Zhang, J. et al. (2018). Soil-plant indices help explain legume response to crop rotation in a semiarid environment. *Frontiers in Plant Science* 9: 1488. https://doi.org/10.3389/fpls.2018.01488.

Lubofsky, E. (2016). The promise of perennials: working through the challenges of perennial grain crop development. *Crops, Soils, Agronomy, News* 61: 4–7.

Lynch, J.P. (2011). Root phenes for enhanced soil exploration and phosphorus acquisition: tools for future crops. *Plant Physiology* 156: 1041–1049.

Lynch, J.P. (2013). Steep, cheap and deep: an ideotype to optimize water and N acquisition by maize root systems. *Annals of Botany* 112: 347–357.

Makumba, W., Akinnifesi, F.K., Janssen, B. et al. (2007). Long-term impact of a *Gliricidia*-maize intercropping system on carbon sequestration in southern Malawi. *Agriculture Ecosystem Environment* 118: 237–243.

Malhi, S.S. and Lemke, R. (2007). Tillage, crop residue and N fertilizer effects on crop yield, nutrient uptake, soil quality and nitrous oxide gas emissions in a second 4-yr rotation cycle. *Soil and Tillage Research* 96: 269–283.

Malik, I. and Rengel, Z. (2013). Breeding for nitrogen efficient crops. In: Improving Water and Nutrient Use Efficiency in Food Production Systems (ed. Z. Rengel), 105–121. Ames, IA: Wiley-Blackwell.

Manavalan, L.P., Musket, T., and Nguyen, H.T. (2011). Natural genetic variation for root traits among diversity lines of maize (*Zea mays* L.). *Maydica* 56: 1717.

Manavalan, L.P., Prince, S.J., Musket, T.A. et al. (2015). Identification of novel QTL governing root architectural traits in a interspecific soybean population. *PLOS One* 10: e0120190.

Manschadi, A.M., Christopher, J., deVoil, P. et al. (2006). The role of root architecture traits in adaptation of wheat to water-limited environments. *Functional Plant Biology* 33: 823–837.

Mao, L.L., Zhang, L.Z., Li, W.Q. et al. (2012). Yield advantage and water saving in maize/pea intercrop. *Field Crops Research* 138: 11–20.

Marschner, P. (2012). Mineral Nutrition of Higher Plants, 3e. Academic Press: London.

McVay, A., Budde, J.A., Fabrizzi, K. et al. (2006). Management effects on soil physical properties in long-term tillage studies in Kansas. *Soil Science Society of America Journal* 70: 434–438.

Mosaddeghi, M.R., Mahboubi, A.A., and Safadoust, A. (2009). Short-term effects of tillage and manure on some soil physical properties and maize root growth in a sandy loam soil in western Iran. *Soil and Tillage Research* 104: 173–179.

Murgai, R., Ali, M., and Byerlee, D. (2001). Productivity growth and sustainability in post-green revolution agriculture: the case of the Indian and Pakistan Punjabs. *The World Bank Research Observer* 16: 199–218.

Mweta, D.E., Akinnifesi, F.K., Saka, J.D.K. et al. (2007). Green manure from prunings and mineral fertilizer affect phosphorus adsorption and uptake by maize crop in a *Gliricidia*-maize intercropping. *Science Research Essay* 2: 446–453.

Narayanan, S., Mohan, A., Gill, K.S. et al. (2014). Variability of root traits in spring wheat germplasm. *PLoS One* 9 (6): e100317.

Narayanan, S. and Prasad, P.V.V. (2014). Characterization of spring wheat association mapping panel for root traits. *Agronomy Journal* 106: 1593–1604.

Nelson, A.R.L.E., Ravichandran, K., and Antony, U. (2019). The impact of green revolution on indigenous crops in India. *Journal of Ethnic Foods* 6 (1): 8. https://doi.org/10.1186/s42779-019-0011-9.

Ochao, I., Blair, J., and Lynch, J.P. (2006). QTL analysis of adventitious root formation in common bean (*Phaseolus vulgaris* L.) under contrasting phosphorus availability. *Crop Science* 46: 1609–1621.

Oke, D.O. and Odebiyi, K.A. (2007). Traditional cocoa-based agroforestry and forest species conservation in Ondo State, Nigeria. *Agriculture Ecosystem Environment* 122: 305–311.

Palta, J.A. and Turner, N.C. (2018). Crop root systems traits cannot be seen as a silver bullet delivering drought resistance. *Plant and Soil* 439: 31–43. https://doi.org/10.1007/s11104-018-3864-6.

Pingali, P.L. and Rosegrant, M.W. (1994). Confronting the environmental consequences of the green revolution in Asia. EPTD Discussion Paper 2. Environmental and Production Technology Division, International Food Policy Research Institute, Washington, DC.

Pingali, P.L. (2012). Green revolution: impacts, limits, and the path ahead. *Proceedings of National Academy of Sciences U.S.A.* 109: 12302–12308.

Pretty, J. (2018). Intensification for redesigned and sustainable agricultural systems. *Science* 362: eaav0294.

Pretty, J., Benton, T.G., Bharucha, Z.P. et al. (2018). Global assessment of agricultural system redesign for sustainable intensification. *Nature Sustainability* 1: 441–446.

Pretty. J., Attwood, S., Bawden, R. et al. (2020). Assessment of the growth in social groups for sustainable agriculture and land management. *Global Sustainability* 3: e23.

Prieto, I., Armas, C., and Pugnaire, F.I. (2012). Water release through plant roots: new insights into its consequences at the plant and ecosystem level. *New Phytologist* 193: 830–841.

Prince, S.J., Murphy, M., Mutava, R.M. et al. (2017). Root xylem plasticity to improve water use and yield in water-stressed soybean. *Journal of Experimental Botany* 68: 2027–2031.

Purushothaman, R., Krishnamurthy, L., Upadhyaya, H.D. et al. (2017). Genotypic variation in soil water use and root distribution and their implications for drought tolerance in chickpea. *Functional Plant Biology* 44: 235–252.

Ramirez-Garcia, J., Martens, H.J., Quemada, M. et al. (2014). Intercropping effect on root growth and nitrogen uptake at different nitrogen levels. *Journal of Plant Ecology* 8: 380–389.

Ramirez-Garcia, J., Gabriel, J., Alonso-Ayuso, M. et al. (2015). Quantitative characterization of five cover crop species. *The Journal of Agricultural Science* 153: 1174–1185.

Rasche, F., Blagodatskaya, E., Emmerling, C. et al. (2017). A preview of perennial grain agriculture: knowledge gain from biotic interactions in natural and agricultural ecosystems. *Ecosphere* 8: e02048.

Ray, D.K., Ramankutty, N., Mueller, N.D. et al. (2012). Recent patterns of crop yield growth and stagnation. *Nature Communications* 3: 1293.

Reich, P.B., Walters, M.B., Tjoelker, M.G. et al. (1998). Photosynthesis and respiration rates depend on, leaf and root morphology in nine boreal tree species differing in relative growth rate. *Functional Ecology* 12: 395–405.

Rice, R.A. (2008). Agricultural intensification within agroforestry: the case of coffee and wood products. *Agriculture Ecosystems and Environment* 128: 212–218.

Rockstrom, J., Williams, J., Daily, G. et al. (2017). Sustainable intensification of agriculture for human prosperity and global sustainability. *Ambio* 46: 4–17.

Roumet, C., Urcelay, C., and Diaz, S. (2006). Suites of root traits differ between annual and perennial species growing in the field. *New Phytologist* 170: 357–368.

Ryan, M.H., Kirkegaard, J.A., and Angus, J.F. (2006). *Brassica* crops stimulate soil mineral N accumulation. *Australian Journal of Soil Research* 44: 367–377.

Ryan, M.R., Crews, T.E., Culman, S.W. et al. (2018). Managing for multifunctionality in perennial grain crops. *BioScience* 68: 294–304.

Sauer, S., Haussmann, W., and Harrach, T. (2002). Effective rooting depth, percolation water, and nitrate leaching in deeply developed loess soils of a water shortage area. *Journal of Plant Nutrition and Soil Science* 165: 269–273.

Scherer, L.A., Verburg, P.H., and Schulp, C.J.E. (2018). Opportunities for sustainable intensification in European agriculture. *Global Environmental Change* 48: 43–55.

Schroth, G., Salazar, E., and da Silva, J.P. (2001). Soil nitrogen mineralization under tree crops and a legume cover crop in multi-strata agroforestry in central Amazonia: spatial and temporal patterns. *Experimental Agriculture* 37: 253–267.

SDG Platform (2017). Sustainable development goals. www.sustianabledevelopment.un.org

Shen, J., Li, C., Mi, G. et al. (2013). Maximizing root/rhizosphere efficiency to improve productivity and nutrient use efficiency in intensive agriculture of China. *Journal of Experimental Botany* 64: 1181–1192.

Sieling, K. and Christen, O. (2015). Crop rotation effects on yield of oilseed rape, wheat and barley and residual effects on the subsequent wheat. *Archives of Agronomy Soil Science* 61: 1531–1549.

Sileshi, G. and Mafongoya, P.L. (2006). Long-term effect of legume–improved fallows on soil invertebrates and maize yield in eastern Zambia. *Agriculture Ecosystems and Environment* 115: 69–78.

Sileshi, G., Akinnifesi, F.K., Ajayi, O.C. et al. (2008). Meta-analysis of maize yield response to planted fallow and green manure legumes in sub-Saharan Africa. *Plant and Soil* 307: 1–19.

Sileshi, G.W., Akinnifesi, F.K., Ajayi, O.C. et al. (2011). Integration of legume trees in maize-based cropping systems improves rainfall use efficiency and crop yield stability. *Agriculture Water Management* 98: 1364–1372.

Singh, R. (2000). Environmental consequences of agricultural development: a case study from green revolution state of Haryana, India. *Agriculture, Ecosystem and Environment* 82: 97–103.

Smith, P. (2013). Delivering food security without increasing pressure on land. *Global Food Security* 2: 18–23.

Stewart, Z.P., Pierzynski, G.M., Middendorf, B.J. et al. (2020). Approaches to improve soil fertility in sub-Saharan Africa. *Journal of Experimental Botany* 71: 632–641.

Tardieu, F. (2012). Any trait or trait-related allele can confer drought tolerance: just design the right drought scenario. *Journal of Experimental Botany* 63: 25–31.

Tardieu, F., Simonneau, T., and Muller, B. (2018). The physiological basis of drought tolerance in crop plants: a scenario-dependent probabilistic approach. *Annual Review of Plant Biology* 69: 733–759.

Thorup-Kristensen, K. (2006). Effect of deep and shallow root systems on the dynamics of soil inorganic N during 3-year crop rotations. *Plant and Soil* 288: 233–248.

Thrupp, L.A. (2000). Linking agricultural biodiversity and food security: the valuable role of sustainable agriculture. *International Affairs* 76: 265–281.

Tjoelker, M.G., Craine, J.M., Wedin, D. et al. (2005). Linking leaf and root trait syndromes among 39 grassland and savannah species. *New Phytologist* 167: 493–508.

Uga, Y., Sugimoto, K., Ogawa, S. et al. (2013). Control of root system architecture by DEEPER ROOTING 1 increases rice yield under drought conditions. *Nature Genetics* 45: 1097–1102.

United Nations (2016). United Nations. www.un.org/sustainabledevelopment/development-agenda

Urquiaga, S., Cadisch, G., Alves, B.J.R. et al. (1998). Influence of decomposition of roots of tropical forage species on the availability of soil nitrogen. *Soil Biology and Biochemistry* 3: 2099–2106.

Van Duivenbooden, N., Pala, N., Studer, C. et al. (2000). Cropping systems and crop complementarity in dryland agriculture to increase soil water use efficiency: a review. NJAS-Wageningen. *Journal of Life Sciences* 48: 213–236.

Van Tassel, D.L., Asselin, S.R., Cox, S.A. et al. (2013). Evaluating perennial candidates for domestication: lessons from wild sunflower relatives. In: Perennial Crops for Food Security: Proceedings of the Food and Agricultural Organization of the United Nations – Expert Workshop (eds. C. Batellow, L.J. Wade, T.S. Cox, et al.), 112–140. United Nations.

Varsa, E.C., Chong, S.K., Abolaji, J.O. et al. (1997). Effect of deep tillage on soil physical characteristics and corn (*Zea mays* L.) root growth and production. *Soil and Tillage Research* 43: 219–228.

Wang, X.B., Wu, H.J., Dai, K. et al. (2012). Tillage and crop residue effects on rainfed wheat and maize production in northern China. *Field Crops Research* 132: 106–116.

Wang, Y., Qin, Y., Chai, Q. et al. (2018). Interspecies interactions in relation to root distribution across the rooting profile in wheat-maize intercropping under different plant densities. *Frontiers in Plant Science* 9: 483. https://doi.org/10.3389/fpls.2018.00483.

Wichern, F., Eberhardt, E., Mayer, J. et al. (2008). Nitrogen rhizodeposition in agricultural crops: methods, estimates and future prospects. *Soil Biology and Biochemistry* 40: 30–48.

Wu, Y., Wang, L., Bian, S. et al. (2019). Evolution of roots to improve water and nitrogen use efficiency in maize elite breeding lines released during different decades in China. *Agricultural Water Management* 216: 44–59.

Xu, B.C., Li, F.M., and Sham, L. (2008). Switchgrass and milkvetch intercropping under 2:1 row-replacement in semiarid region, northwest China: aboveground biomass and water use efficiency. *European Journal of Agronomy* 28: 485–492.

Yeboah, S., Lamptey, S., Zhang, S. et al. (2017). Conservation tillage practices optimizes root distribution and straw yield of spring wheat and field pea in dry areas. *Journal of Agricultural Sciences* 6: 37–48.

Zhang, E., Li, L., Huang, G. et al. (2002). Regulation of fertilizer application on yield and root growth of spring wheat-faba bean intercropping system. *China Journal of Applied Ecology* 13: 839–842.

Zhang, F.S., Shen, J.B., Zhang, J.L. et al. (2010). Rhizosphere processes and management for improving nutrient use efficiency and crop productivity: implications for China. *Advances in Agronomy* 107: 1–32.

Zhang, F.S., Cu, Z.L., Fan, M.S. et al. (2011). Integrated soil-crop system management: reducing environmental risk while increasing crop productivity and increasing nitrogen use efficiency in China. *Journal of Environmental Quality* 40: 1051–1057.

Zhang, F.S., Cui, Z.L., Chen, X.L. et al. (2012). Integrated nutrient management for food security and environmental quality in China. *Advances in Agronomy* 116: 1–40.

2

Genetic and Molecular Regulation of Root Growth and Development

Yang Li[1,2,], Heng Ye[1,*], Yan Li[2], and Henry T. Nguyen[1]*

[1] *Division of Plant Sciences, University of Missouri, Columbia, MO, USA*
[2] *National Key Laboratory of Crop Genetics and Germplasm Enhancement, National Center for Soybean Improvement, Nanjing Agricultural University, Nanjing, China*
* *These authors contributed equally to this work.*

2.1 Introduction

The rapid and continuous climate changes increase the frequency of extreme weather, such as higher temperature, altered precipitation patterns, and repeated occurrence of drought globally, which cause a significant reduction in crop yield and threaten global food security. It is evident that the losses in crop yield have been continuing to mount over the past decades due to these extreme weather patterns (Zheng 2010; Boyer et al. 2013; Lesk et al. 2016). It has been projected that these yield losses of major crops will continue to exacerbate, despite a progressive improvement in yield through breeding efforts and management practices since the 1960s (Iizumi et al. 2013;Lobell et al. 2014; Rosenzweig et al. 2014).

As a key organ of plants, roots play an essential role in nutrient and water uptake, supporting plant body, and interacting with root and rhizosphere microorganisms (Hopmans and Bristow 2002). Roots are also an important organ for nutrient and energy storage in some plant species. These roles of roots make them essential in adaption to the extreme weather patterns, such as drought and soil-flooding stresses (Nguyen et al. 1997; Ye et al. 2018b). The underground plant organ starts with the root initiation and develops into primary root and lateral roots (Osmont 2007; Prince et al. 2015, 2018). The traits such as root length, root diameter, and root growth direction shape the root system architecture (RSA), representing the spatial configuration of roots in the soil. RSA varies among species and also shows significant natural variation within species (Hochholdinger and Tuberosa 2009; Ulises et al. 2013). This genetic diversity of RSA can affect the stress resilience in plants (Valliyodan et al. 2017; Ye et al. 2018a).

RSA is controlled by many genes and biochemical processes involved in root initiation, growth, and development, enabling adaptation to the temporal and spatial changes in

The Root Systems in Sustainable Agricultural Intensification, First Edition.
Edited by Zed Rengel and Ivica Djalovic.

environmental conditions (Osmont 2007). After decades of study, significant progress in understanding the molecular basis of root initiation and development has been achieved, mainly through research in model plants *Arabidopsis* (*Arabidopsis thaliana*) and rice (*Oryza sativa*). In this chapter, we summarize the current knowledge about the genetic and molecular regulation of root initiation and development. The genetic diversity and natural variations in RSA in response to abiotic stresses are also discussed.

2.2 Molecular Regulation of Root Initiation and Development

2.2.1 Atlas of Genes Involved in Root Initiation and Development

Root initiation and development are regulated by complex networks and signalling pathways. The research advances in genome-wide study improved our understanding of the interactions between these networks. More than 22000 genes were localized in a global map of gene expression in *Arabidopsis* roots, and 15 different gene expression zones were mapped corresponding to different cell types and tissues (Birnbaum et al. 2003). A time series transcriptome analysis detected 7467 upregulated and 7503 downregulated genes during the adventitious root formation in green cuttings of highbush blueberry. About 35 of them were further confirmed to be involved in the auxin-induced pathway and adventitious root formation (An et al. 2020). The cell type-specific transcriptional signature is one of the way to predict cellular pathways in root cells. As a result, many cell type-specific transcriptome profiles have been revealed in roots, which are helpful in elucidating the complex gene regulatory networks at the cellular level (Brady et al. 2007; Dinneny et al. 2008; Gifford et al. 2008).

A protein expression map of *Arabidopsis* roots was also generated, comprising about 2000 identified proteins and their cell type-specific localization; these proteins are involved in the regulation of root initiation and development (Petricka et al. 2012a). The gene atlas draws a big picture of gene regulatory networks in different root cells and zones, which benefits the study of molecular regulation of root initiation and development.

2.2.2 Molecular Regulation of Primary Root Initiation

The primary root is the first visible organ that emerges out of the germinating seed (Grunewald et al. 2007). Root initiation starts at embryogenesis from a single extraembryonic suspensor cell in *Arabidopsis* (Tian et al. 2014). After a couple of rounds of expansions and divisions, the suspensor cells generate the quiescent centre (QC) and finally develop into the root apical meristem of the primary root (Dolan et al. 1993; De Smet et al. 2010a). The root apical meristem contains a set of stem cells that surround the QC. The QC is essential for specification of the stem cell niche and maintaining the undifferentiated state of stem cell initials (van den Berg et al. 1997). The fully functional root apical meristem is formed when the daughter cells are produced by the stem cells (Grunewald et al. 2007). On the upward and lateral sides of the QC, the stem cells generate the cortex, epidermis,

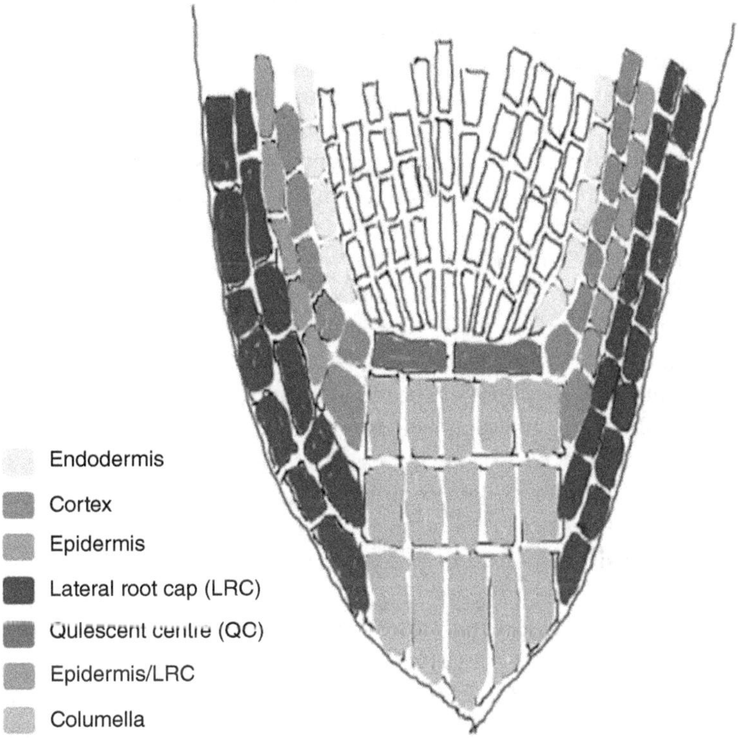

Figure 2.1 Simplified organization of the root apical meristem.

endodermis, and lateral root cap; on the downward side of the QC, stem cells generate the columella root cap (Figure 2.1). Stem cell daughters form distinct tissue layers along the longitudinal root axis (Petricka et al. 2012b).

There are two parallel pathways for the determination of the QC and root stem cell identity during the early embryogenesis. One is the SHORT ROOT/SCARECROW (SHR/SCR) pathway, and the other one is the PLETHORA (PLT) pathway (Sabatini et al. 2003; Aida et al. 2004; Galinha et al. 2007). *SHR* and *SCR* are the members of transcription factor family GIBBERELLIC-ACID INSENSITIVE (GAI), REPRESSOR of GAI (RGA), and SCR, whereas *PLT* encodes the APETALA2 (AP2)-domain transcription factors. *SHR* is expressed in the stele and transferred to the adjacent cells to activate *SCR* transcription. Both SHR and SCR interact with WUSCHEL-related homeobox 5 (WOX5) to regulate the determination of QC identity and the differentiation of root stem cells (Helariutta et al. 2000; Sabatini et al. 2003; Sarkar et al. 2007). WOX5, as a transcription factor, is involved in the balance between cell divisions and differentiation in QC during the embryogenesis (Haecker et al. 2004; Sarkar et al. 2007). Expression of *WOX5* is mainly regulated by the SHR/SCR pathway; hence, *WOX5* acts as a downstream effector of the SHR/SCR signalling pathway (Sarkar et al. 2007). Recently, a molecular pathway module of auxin – intermediate-clade WUSCHEL-RELATED HOMEOBOX (IC-WOX) – WUSCHEL-clade WOX (WC-WOX) was identified in root initiation and was supposed to be conserved in euphyllophytes (ferns and seed plants) (Yu et al. 2020). The *PLT* was highly expressed in the stem cell niche and acted as a dose-dependent master regulator of root development (Galinha et al. 2007).

The establishment of the hypophysis (suspensor cell) is crucial for the root apical meristem formation. Auxin plays a significant role during hypophysis specification. PIN-FORMED (PIN) proteins can transport auxin during the embryogenesis stage and act as positional signals for the primary root initiation (Friml et al. 2004; Weijers and Jürgens 2005). There are eight *PIN* genes in the *Arabidopsis* genome, and five of them have been studied (*PIN1, PIN2, PIN3, PIN4,* and *PIN7*) (Paponov et al. 2005). During the globular stage, the combination of PIN1, PIN4, and PIN7 can induce hypophysis specification via the accumulation of auxin in the uppermost suspensor cell (Friml et al. 2003). As an auxin-induced transcription factor, MONOPTEROS/AUXIN RESPONSE FACTOR5 (MP/ARF5) is required for hypophysis specification and primary root initiation. In *mp/arf5* mutant, the hypophysis was not established and primary roots were not formed (Hardtke and Berleth 1998). MP/ARF5 can be repressed by BODEN-LOS/INDOLE-3-ACETIC ACID INDUCIBLE12 (BDL/IAA12). In contrast, BDL protein can be degraded in response to auxin, leading to MP release. This balance between MP and BDL played an important role in the hypophysis specification process (Hamann et al. 2002; Weijers et al. 2006).

The *TARGET OF MP* (*TMO*) genes *TMO5* and *TMO7* are the crucial downstream targets of MP/ARF5 and can partially rescue the rootless phenotype of *mp/arf5* mutant. Under the regulation of MP/ARF5 transcription factor, TMO7 protein moved from its biosynthetic location to the nucleus of the hypophysis and started the *MP/ARF5*-dependent root initiation and formation (Schlereth et al. 2010).

In contrast to auxin-promoted-cell division, cytokinin plays many roles in root-cell differentiation. Generally, cytokinin regulates cell differentiation by inhibiting auxin transport and auxin responses between the meristem and root elongation zone (Chapman and Estelle 2009). Cytokinin-responsive transcription factor *ARABIDOPSIS RESPONSE REGULATOR1* (ARR1) and cytokinin receptor *ARABIDOPSIS HISTIDINE KINASE3* (AHK3) activated *SHORT HYPOCOTYL 2* (SHY2)/*IAA3* to regulate auxin transport by PIN proteins (Riefler et al. 2006; Ioio et al. 2008). Finally, cytokinin controlled the meristem by attenuating auxin-dependent stem-cell division in the *AHK3/ARR1, ARR12* pathway (Chapman and Estelle 2009). *OsNAC2* was reported to act as an upstream integrator of auxin and cytokinin signals to regulate cell division during root development by affecting CROWN ROOTLESS (CRL) and cyclin-dependent protein kinase (CDK) production in rice (Mao et al. 2020). Although many genes and networks involved in primary root initiation have been discovered, there are still many secrets in this biological process that need to be revealed.

2.2.3 Molecular Regulation of Lateral Root Initiation

With the growth and development of the primary root in *Arabidopsis*, the lateral roots start to initiate from the primary root. In *Arabidopsis* and other dicots, the lateral roots initiate from the pericycle cells adjacent to the xylem pole. A subset of specified pericycle cells, also called founder cells, form lateral root primordia (Blakely and Evans 1979). Seven typical stages and four key events including priming, initiation, primordium formation, and emergence were discovered during the lateral root formation (Malamy and Benfey 1997).

The auxin signalling is highly correlated with the priming and initiation specification of the lateral root primordium site in the basal meristem (De Smet et al. 2007). Before the lateral root initiation, auxin was already accumulated in founder cells; the maintenance of auxin level is crucial for lateral root organogenesis (Dubrovsky et al. 2008). Auxin can activate its downstream gene targets, such as *MP/ARF5*, to promote lateral root initiation through integrating cell divisions and the patterning process (Hamann et al. 2002; De Smet et al. 2010b). When the lateral root initiation has occurred, PIN-dependent auxin transport can regulate division of the founder cells to form lateral root primordium (Casimiro et al. 2001; Benková et al. 2003). Another auxin efflux carrier gene, *PIN3*, can regulate the auxin cycling between the endodermis and the pericycle to promote lateral root initiation (Marhavý et al. 2013). As a key auxin-responsive regulator, *SOLITARY-ROOT* (*SLR*)/*IAA14* gene was found to play a key role in lateral root initiation, as the *slr-1* mutation was identified to inhibit cell divisions of pericycle cells to suppress lateral root initiation (Fukaki et al. 2002). Further study showed that *SLR/IAA14* is involved in tissue-specific auxin responses in lateral root development by activating ARF7 and ARF19 proteins (Fukaki et al. 2005). ARF7/ARF19 can activate LATERAL ORGAN BOUNDARIES-DOMAIN/ASYMMETRIC LEAVES2-LIKE (LBD/ASLs) to promote lateral root formation (Okushima et al. 2007). In addition, another auxin response module controlled by BDL/IAA12-MP/ARF5 is sufficient to activate lateral root initiation (De Smet et al. 2010b). *IAA13, ARF19*, and *LBD1-8* were identified to be highly expressed in the cortex and lateral root primordia, and these genes function in the constitutive aerenchyma initiation and lateral root formation by AUX/IAA- and ARF-dependent auxin signalling (Yamauchi et al. 2019). For the initiation and primordium formation processes in the lateral roots, the receptor-like kinase ACR4 plays a key role in formative cell divisions and lateral root primordium specification (De Smet et al. 2008).

During the lateral root emergence process, the new lateral root primordium must break through the endodermis, cortex, and epidermis layers. The auxin influx carrier *LAX3*, expressed in cortical and epidermal cells, can promote cell separation to make way for the emergence of lateral root primordium (Swarup et al. 2008). *AtHB23* mediates the regulation of *LAX3* by ARF7/19 and plays distinct roles during the formation of lateral roots in *Arabidopsis* (Perotti et al. 2019).

Abscisic acid (ABA) also can modulate lateral root formation via auxin signalling. In the ABA signalling pathway, the *ABA INSENSITIVE3* (ABI3) gene is induced by auxin in the lateral root primordium, and *abi3* mutant plants show a reduced response to auxin and fewer lateral roots forming (Brady et al. 2003). The transcription factor MYB77 is involved in both auxin and ABA signalling pathways by interacting with ARF7 and the ABA receptor PYL8 to regulate lateral root formation (Shin et al. 2007; Zhao et al. 2014). A ZINC FINGER PROTEIN1 (AtZP1) works as a downstream gene of GLABRA2 (GL2) and negatively regulates root hair initiation and elongation by suppressing ROOT HAIR DEFECTIVE6 (RHD6), ROOT HAIR DEFECTIVE 6-LIKE 2 (*RSL4*), and *RSL2* transcription via the GL2/ZP1/RSL pathway (Han et al. 2020).

It can be concluded that the complex gene-networks regulated by a crosstalk among phytohormones are involved in lateral root initiation and development.

2.3 Regulation of RSA in Plants

2.3.1 Molecular Regulation of RSA in Plants

RSA is a composite concept and refers to the structure of the entire underground organ of a plant. It mainly refers to root structure and shape (Berntson 1994; Orman-Ligeza et al. 2014). As the structure of root system, RSA mainly defines the assembly and properties of the different root segments, linking to the root developmental processes such as the elongation, direction, and production of new roots (Waisel 2002; Hochholdinger et al. 2004). As the shape of root system, RSA mainly defines the spatial distribution of roots and major functions of the root system such as nutrient capture, water uptake, and plant anchorage (Gregory et al. 2003; Orman-Ligeza et al. 2014).

Different plants have different types of RSA. Generally, RSA of dicots, such as *Arabidopsis* and soybean (*Glycine max* and *Glycine soja*), is mainly composed of the primary root and lateral roots arising from the primary root; RSA of monocots, such as rice and maize (*Zea mays* L.), is mainly composed of the primary root, seminal roots, crown roots, and lateral roots (Figure 2.2) (Atkinson et al. 2014; Wachsman et al. 2015).

RSA is a complex trait, controlled by many genes and involved in different biological pathways. A novel regulatory factor of *BREVIS RADIX* (BRX), mapped in *Arabidopsis*, that can regulate cell proliferation and elongation in the growth zone of the root tip explained about 80% of the phenotypic variation in primary root elongation (Mouchel et al. 2004). The first root-specific expansion gene, *GmEXP1*, had an important role in the primary root elongation in soybean (Lee et al. 2003). In rice, a *DEEPER ROOTING 1* (DRO1) gene was underlying the major quantitative trait locus (QTL) for deep rooting (Uga et al. 2011). Further research found that *DRO1* increased the root growth angle to direct root growth downward and was involved in auxin pathways in root tips (Uga et al. 2013). In addition,

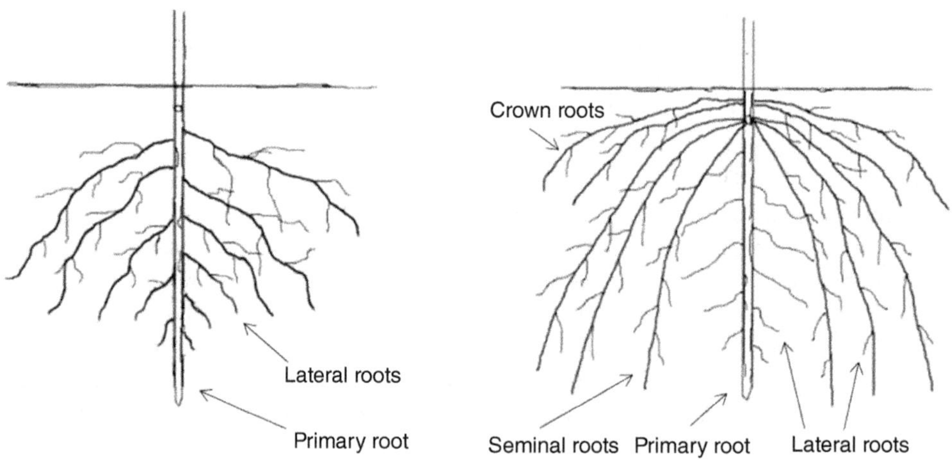

Figure 2.2 Simplified models of the root system architecture of dicots and monocots. (a) The root system architecture of dicots containing a primary root and lateral roots; (b) the root system architecture of monocots containing a primary root, lateral roots, seminal roots, and crown roots.

loss-of-function of a guanine nucleotide exchange factor for ADP-ribosylation factor (*gnom1*) led to a reduction in lateral root numbers and partial loss of root gravitropism in rice (Liu et al. 2009a).

Crown root is the indispensable root class in the rice and maize RSA. Many genes have been identified to control crown roots in these two crops. In rice, *Crown rootless1* (Crl1) was characterized as a positive regulator of the crown and lateral root formation through the regulation of the auxin-induced *ARF* gene (Inukai et al. 2005). *Adventitious rootless1* (ARL1) promotes auxin-mediated cell dedifferentiation and division to form crown roots in rice (Liu et al. 2005). In maize, two auxin-responsive genes, *rootless concerning crown and seminal roots* (*RTCS*) and *RTCS-LIKE* (*RTCL*), were identified to initiate and maintain the seminal and crown root primordia formation (Taramino et al. 2007). Independent of *RTCS*, the *deficient in lateral root formation* (*LRT1*) locus also functions in the regulation of lateral and crown roots at the early postembryonic root initiation stage in maize (Hochholdinger and Feix 1998).

Although there has been a lot of progress understanding the regulation of root traits such as the primary and lateral root initiation, root growth and elongation, our current knowledge on molecular regulation of RSA is still very limited; more efforts should be directed to deciphering the root structures to help understand the underground half of plants.

2.3.2 Responses of RSA to Abiotic Stresses

Crop yield is significantly limited by abiotic stresses, such as drought, salinity, and waterlogging. Plants have evolved abiotic stress adaptation mechanisms under natural selection, and abundant phenotypic diversity was acquired during plant evolution (Mickelbart et al. 2015). As an important stress-resilience organ, roots also show large phenotypic variation that leads to differential resistance to abiotic stresses.

During the last decades, a large number of QTL controlling the RSA parameters have been detected for drought resistance in rice (Zheng et al. 2003; Kato et al. 2008; Courtois et al. 2009; Ding et al. 2011). The most important gene *DRO1* was mapped in rice, and the natural variation of 1-bp deletion within exon 4 was identified in the parent line IR64. *Dro1* regulates rooting angles to develop deeper roots to avoid drought in rice (Uga et al. 2011, 2013). *OsESG1* was involved in early crown root development by controlling auxin response and distribution, and drought resistance by regulating the antioxidant activities in rice (Pan et al. 2020).

The genetic control of 21 root traits was analysed via genome-wide association study (GWAS) of 529 rice accessions; in total, 110 significant association loci were detected, and many genes related to root traits, such as *DRO1* and *WOX11*, were found in these loci (Li et al. 2017). Variations were found in root penetration capacity, osmotic adjustment, and dehydration tolerance among *indica* lines adapted to lowlands (Chandra Babu et al. 2001). Root penetration capacity was related to drought tolerance, and several QTL were detected for root penetration capacity, root thickness, penetrating root number, and total root number; these QTL explained 8–19% of the phenotypic variation (Ray et al. 1996; Zheng et al. 2000).

In maize, *ZmVPP1* was detected by GWAS in 367 different inbred lines under drought stress. The *ZmVPP1* tolerant allele was associated with better drought tolerance than the sensitive one as confirmed by comparing morphological phenotypes of the near-isogenic lines (NILs) under drought stress. The transgenic maize with the *ZmVPP1* tolerant allele showed an increased number of lateral roots, larger root mass, and improved drought tolerance than the wild type (Wang et al. 2016).

The *EXOCYST70A3* gene was mapped by GWAS in 215 diverse *Arabidopsis* accessions. Allelic variation of *EXOCYST70A3* led to the alteration of root gravitropic responses and root growth angle, resulting in a drought-resistant phenotype with different RSA depth profiles (Ogura et al. 2019).

For waterlogging stress, the major QTL *qWT_Gm03* was mapped in a population derived from an elite/exotic soybean cross. The tolerant allele of *qWT_Gm03* improved total root length, number of root tips, and number of adventitious roots, thus enhancing soybean tolerance to waterlogging stress (Ye et al. 2018b).

Phosphate is a plant macronutrient. Primary root and lateral roots play crucial roles in response to the phosphate-starvation stress. In *Arabidopsis*, 73 natural accessions exhibited large phenotypic variations in primary root length and the lateral root number in response to phosphate starvation (Chevalier et al. 2003). *ATNRT1.5* promoted primary root elongation and lateral root growth under low phosphate supply in *Arabidopsis* (Cui et al. 2019). In rice, *OsMYB2P-1* and *OsARF16* were associated with improved primary root length and tiller number under phosphate starvation (Dai et al. 2012; Shen et al. 2013). Overexpression of *GmEXPB2* in soybean and *Arabidopsis* also improved root length under phosphate starvation (Guo et al. 2011; Zhou et al. 2014). The cytokinin response factor (*BnaCRF8*) had negative regulation in root architecture and plant growth by modifying phosphate homeostasis-related genes (Wang et al. 2020).

Several QTL have been mapped for tolerance to low phosphorus supply in maize inbred lines; these QTL were associated with the lateral root branching and length, root hair length, and seminal root length and number (Zhu et al. 2005a,b, 2006). In rice, *PSTOL1* was the first cloned gene contributing to the natural variation in tolerance to phosphate starvation through increased total root length and surface area under phosphorus deficiency (Gamuyao et al. 2012). In soybean, *GmACP1* was found (by combining linkage disequilibrium analysis and GWAS) to be associated with tolerance to phosphorus starvation. The candidate-gene association analysis indicated that six natural *GmACP1* polymorphisms explained 33% of the phenotypic variation in tolerance to phosphate starvation, and transgenic lines with overexpressed *GmACP1* had higher root dry weight and tolerance to phosphorus deficiency than the control lines (Zhang et al. 2014).

In addition to phosphate starvation stress, RSA was also found to be influenced by nitrogen supply. Two major QTL were mapped for nitrogen-deficiency tolerance in rice, and both of them were associated with root length (Obara et al. 2010; Anis et al. 2019). In *Arabidopsis*, the MADS Box gene *ANR1* is a key determinant for suppressing the rate of lateral root elongation in high NO_3^- supply (Zhang and Forde 1998). The miR393/AFB3 is a unique nitrogen-responsive combination controlling RSA in *Arabidopsis*. AFB3 plays a specific role in increasing primary root length and decreasing lateral root density in response to nitrate supply (Vidal et al. 2010).

Roots are also involved in responses to micronutrient deficiency stresses, such as boron and iron deficiency. In *Arabidopsis*, low boron supply inhibited primary root growth but not that of the lateral roots (Martín-Rejano et al. 2011). Two genes (*OsPIP2;4* and *OsPIP2;7*) were associated with increased primary root length in transgenic *Arabidopsis* lines under low boron supply (Kumar et al. 2014). Wheat (*Triticum aestivum* L.) genotypes vary greatly in their adaptation to low boron supply (Rerkasem and Jamjod 2004). Two major loci were found to control wheat response to boron deficiency, and they are utilized widely in the wheat breeding program (Jamjod et al. 2004).

Iron deficiency chlorosis is a big yield-limiting problem in soybean grown in the upper Midwest of USA (Naeve 2006). Two loci were found to be associated with iron deficiency chlorosis in two independent soybean populations (Wang et al. 2008). A total of 15 candidate genes were mapped in the vicinity (<500 kb) of significant markers for tolerance to iron deficiency in soybean (Mamidi et al. 2011). An RNA-Seq study showed that the genes involved in phytohormone signalling and iron uptake play key roles in the early soybean response to iron deficiency (Lauter et al. 2014).

Metal toxicity stress, particularly caused by excessive aluminium concentration in acid soils, is a significant growth-limiting factor threatening crop production. High concentrations of aluminium in the soil solution significantly inhibited primary root elongation in plants (Hoekenga et al. 2006; Sawaki et al. 2009; Sun et al. 2009; Liu et al. 2016). The Al^{3+}-activated malate transporter (*ALMT*) and multidrug and toxic compound extrusion (*MATE*) genes can improve root growth under aluminium toxicity by coding for anion channels conducting malate and citrate exudation into the rhizosphere soil to chelate aluminium ions (Hoekenga et al. 2006; Magalhaes et al. 2007; Liu et al. 2009b; Liang et al. 2013). Three QTL were identified through bi-parental QTL mapping and GWAS to be associated with aluminium tolerance in rice (Famoso et al. 2011). A gene regulating aluminium ion transport in roots, *Nrat1*, was found to sequester aluminium into the vacuole in rice (Famoso et al. 2011; Li et al. 2014).

Salinity caused by high concentration of NaCl in soil is another major ion stress. Salt stress can reduce the primary root length and modulate root growth direction by reducing the gravitropic response (Sun et al. 2008). The *Salt Overly Sensitive 1* (SOS1) and *high affinity* K^+ *transporter* (*HKT1*) genes can improve root growth under salt stress in both *Arabidopsis* and soybean (Shi et al. 2000, 2003; Mäser et al. 2002; Chen et al. 2014). The *NAC* (NAM, no apical meristem; ATAF, *Arabidopsis* transcription activation factor; and CUC, cup-shaped cotyledon) genes also contribute to salt tolerance by improving root growth in rice and soybean (Hao et al. 2011; Liu et al. 2014). The first cloned gene for natural variation in crop salt tolerance was *SKC1* (QTL of Shoot K^+ Concentration) in rice, which encoded an HKT-type transporter involved in regulating K^+/Na^+ homeostasis under salt stress (Lin et al. 2004; Ren et al. 2005). Using the whole-genome sequencing, a novel salt tolerance gene *GmCHX1* was identified in wild soybean (Qi et al. 2014). *GmCHX1* encoded a sodium ion transporter that moved Na^+ from the xylem to root tissues to exclude Na^+ from the above-ground tissues (Guan et al. 2014; Qi et al. 2014).

Copper is an essential transition metal, but excessive uptake can cause oxidative stress. In *Arabidopsis* seedlings, copper toxicity can inhibit root elongation by affecting cell elongation in a dose-dependent manner (Pasternak et al. 2005). One major QTL associated with copper tolerance was identified in *Arabidopsis*, explaining 52% of the phenotypic variations (Kobayashi et al. 2008). The tolerant allele of this QTL significantly improved root elongation under high copper supply (Kobayashi et al. 2008). The *heavy metal ATPase 5* (HMA5) was identified to regulate copper compartmentalization and detoxification in roots (Andrés-Colás et al. 2006; Kobayashi et al. 2008). In wheat, three loci were identified for copper tolerance by genetic and physical mapping, and *Vrn-A1* in the vernalization pathway was suggested as the candidate gene underlying the copper-tolerance QTL on chromosome 5A (Bálint et al. 2009).

2.4 Future Perspectives

In the second part of the twentieth century, the Green Revolution allowed the food production to keep pace with the worldwide population growth (Khush 2001). The Green Revolution boosted crop yields mainly by introducing dwarf genotypes of wheat and then rice adapted to high fertilization rates without lodging (Lynch 2007). Currently, we need the second Green Revolution to meet the goal of increased food production to address food security for the expanding world population. The first Green Revolution was based on crop responses to high soil fertility, but the second Green Revolution is proposed to be based on crop tolerance to various abiotic stress, such as drought, flooding, ion toxicity, and low soil fertility (Pathan et al. 2007; Ye et al. 2018a,b).

Roots, as the underground half of plant, play a crucial role in the growth, development, and yield of crops by directly facing the majority of abiotic stresses (Den Herder et al. 2010). Considerable insights into root initiation, development and structure have been acquired from the model plants, such as *Arabidopsis* and rice. However, this knowledge is far from sufficient to enable optimization of the root systems to various stress environments to ensure high crop productivity in less-than-ideal soil-climate conditions.

High-throughput phenotyping platforms that assess RSA are essential to advancing our knowledge of root tolerance to various stresses. High-throughput phenotyping of RSA has focused on field-based approaches in crops such as maize (Trachsel et al. 2011; York and Lynch 2015), rice (Wissuwa et al. 2016), and wheat (Richard et al. 2015), with a recent shift to legume species, including soybean, common bean, and cowpea (Manavalan et al. 2015; Prince et al. 2015; Burridge et al. 2016). The adoption of the high-throughput image analysis of RSA using digital imaging of root traits improved the efficiency and accuracy of the phenotyping of RSA (Bucksch et al. 2014;Das et al. 2015). New root phenotyping platforms, such as magnetic resonance imaging, positron emission tomography, and X-ray computer tomography allow 3D root phenotyping in soil over time, facilitating more sophisticated real-time monitoring and measuring of various root architectural parameters (Atkinson et al. 2019).

Benefiting from the next-generation sequencing technologies, the construction of high-resolution genomic/genetic maps and development of DNA markers for phenotypic traits is becoming cost-effective. Allelic mining for the root traits in the accumulating genomic information can be useful. The targeted genome editing by the CRISPR/Cas9 system (Jacobs et al. 2015; Sun et al. 2015; Wang et al. 2017; Bao et al. 2019; Char et al. 2019) can enable targeted modification of the alleles associated with RSA enhancements. Integration of these genomic resources and technologies into the next-generation breeding strategies will accelerate the genetic improvement of crop yield under stress conditions via enhanced root systems optimized for specific growth environments.

References

Aida, M., Beis, D., Heidstra, R. et al. (2004). The PLETHORA genes mediate patterning of the *Arabidopsis* root stem cell niche. *Cell* 119: 109–120.

An, H.S., Zhang, J.Y., Xu, F.J. et al. (2020). Transcriptomic profiling and discovery of key genes involved in adventitious root formation from green cuttings of highbush blueberry (*Vaccinium corymbosum* L.). *BMC Plant Biology* 20: 182.

Andrés-Colás, N., Sancenón, V., Rodríguez-Navarro, S. et al. (2006). The *Arabidopsis* heavy metal P-type ATPase HMA5 interacts with metallochaperones and functions in copper detoxification of roots. *The Plant Journal* 45: 225–236.

Anis, G.B., Zhang, Y., Islam, A. et al. (2019). RDWN6 ^{XB}, a major quantitative trait locus positively enhances root system architecture under nitrogen deficiency in rice. *BMC Plant Biology* 19: 12.

Atkinson, J.A., Rasmussen, A., Traini, R. et al. (2014). Branching out in roots: uncovering form, function, and regulation. *Plant Physiology* 166: 538–550.

Atkinson, J.A., Pound, M.P., Bennett, M.J. et al. (2019). Uncovering the hidden half of plants using new advances in root phenotyping. *Current Opinion in Biotechnology* 55: 1–8.

Bálint, A.F., Szira, F., Röder, M.S. et al. (2009). Mapping of loci affecting copper tolerance in wheat—the possible impact of the vernalization gene *Vrn-A1*. *Environmental and Experimental Botany* 65: 369–375.

Bao, A., Burritt, D.J., Chen, H. et al. (2019). The CRISPR/Cas9 system and its applications in crop genome editing. *Critical Reviews in Biotechnology* 39: 321–336.

Benková, E., Michniewicz, M., Sauer, M. et al. (2003). Local, efflux-dependent auxin gradients as a common module for plant organ formation. *Cell* 115: 591–602.

van den Berg, C., Willemsen, V., Hendriks, G. et al. (1997). Short-range control of cell differentiation in the *Arabidopsis* root meristem. *Nature* 390: 287–289.

Berntson, G. (1994). Modelling root architecture: are there tradeoffs between efficiency and potential of resource acquisition? *New Phytologist* 127: 483–493.

Birnbaum, K., Shasha, D.E., Wang, J.Y. et al. (2003). A gene expression map of the *Arabidopsis* root. *Science* 302: 1956–1960.

Blakely, L. and Evans, T. (1979). Cell dynamics studies on the pericycle of radish seedling roots. *Plant Science Letters* 14: 79–83.

Boyer, J., Byrne, P., Cassman, K. et al. (2013). The US drought of 2012 in perspective: a call to action. *Global Food Security* 2: 139–143.

Brady, S.M., Sarkar, S.F., Bonetta, D. et al. (2003). The ABSCISIC ACID INSENSITIVE 3 (ABI3) gene is modulated by farnesylation and is involved in auxin signaling and lateral root development in *Arabidopsis*. *The Plant Journal* 34: 67–75.

Brady, S.M., Orlando, D.A., Lee, J.-Y. et al. (2007). A high-resolution root spatiotemporal map reveals dominant expression patterns. *Science* 318: 801–806.

Bucksch, A., Burridge, J., York, L.M. et al. (2014). Image-based high-throughput field phenotyping of crop roots. *Plant Physiology* 166: 470–486.

Burridge, J., Jochua, C.N., Bucksch, A. et al. (2016). Legume shovelomics: high—throughput phenotyping of common bean (*Phaseolus vulgaris* L.) and cowpea (*Vigna unguiculata subsp, unguiculata*) root architecture in the field. *Field Crops Research* 192: 21–32.

Casimiro, I., Marchant, A., Bhalerao, R.P. et al. (2001). Auxin transport promotes *Arabidopsis* lateral root initiation. *The Plant Cell* 13: 843–852.

Chandra Babu, R., Shashidhar, H., Lilley, J. et al. (2001). Variation in root penetration ability, osmotic adjustment and dehydration tolerance among accessions of rice adapted to rainfed lowland and upland ecosystems. *Plant Breeding* 120: 233–238.

Chapman, E.J. and Estelle, M. (2009). Cytokinin and auxin intersection in root meristems. *Genome Biology* 10: 210.

Char, S.N., Wei, J., Mu, Q. et al. (2019). An Agrobacterium-delivered CRISPR/Cas9 system for targeted mutagenesis in sorghum. *Plant Biotechnology Journal* 18: 319–321.

Chen, H., Chen, X., Gu, H. et al. (2014). *GmHKT1;4*, a novel soybean gene regulating Na^+/K^+ ratio in roots enhances salt tolerance in transgenic plants. *Plant Growth Regulation* 73: 299–308.

Chevalier, F., Pata, M., Nacry, P. et al. (2003). Effects of phosphate availability on the root system architecture: large-scale analysis of the natural variation between *Arabidopsis* accessions. *Plant, Cell and Environment* 26: 1839–1850.

Courtois, B., Ahmadi, N., Khowaja, F. et al. (2009). Rice root genetic architecture: meta-analysis from a drought QTL database. *Rice* 2: 115–128.

Cui, Y.-N., Li, X.-T., Yuan, J.-Z. et al. (2019). Nitrate transporter NPF7.3/NRT1.5 plays an essential role in regulating phosphate deficiency responses in *Arabidopsis. Biochemical and Biophysical Research Communications* 508: 314–319.

Dai, X., Wang, Y., Yang, A. et al. (2012). *OsMYB2P-1*, an R2R3 MYB transcription factor, is involved in the regulation of phosphate-starvation responses and root architecture in rice. *Plant Physiology* 159: 169–183.

Das, A., Schneider, H., Burridge, J. et al. (2015). Digital imaging of root traits (DIRT): a high-throughput computing and collaboration platform for field-based root phenomics. *Plant Methods* 11: 51.

De Smet, I., Tetsumura, T., De Rybel, B. et al. (2007). Auxin-dependent regulation of lateral root positioning in the basal meristem of *Arabidopsis. Development* 134: 681–690.

De Smet, I., Vassileva, V., De Rybel, B. et al. (2008). Receptor-like kinase ACR4 restricts formative cell divisions in the *Arabidopsis* root. *Science* 322: 594–597.

De Smet, I., Lau, S., Mayer, U. et al. (2010a). Embryogenesis–the humble beginnings of plant life. *The Plant Journal* 61: 959–970.

De Smet, I., Lau, S., Voß, U. et al. (2010b). Bimodular auxin response controls organogenesis in *Arabidopsis. Proceedings of the National Academy of Sciences of the United States of America* 107: 2705–2710.

Den Herder, G., Van Isterdael, G., Beeckman, T. et al. (2010). The roots of a new green revolution. *Trends in Plant Science* 15: 600–607.

Ding, X., Li, X., and Xiong, L. (2011). Evaluation of near-isogenic lines for drought resistance QTL and fine mapping of a locus affecting flag leaf width, spikelet number, and root volume in rice. *Theoretical and Applied Genetics* 123: 815–826.

Dinneny, J.R., Long, T.A., Wang, J.Y. et al. (2008). Cell identity mediates the response of *Arabidopsis* roots to abiotic stress. *Science* 320: 942–945.

Dolan, L., Janmaat, K., Willemsen, V. et al. (1993). Cellular organisation of the *Arabidopsis thaliana* root. *Development* 119: 71–84.

Dubrovsky, J.G., Sauer, M., Napsucialy-Mendivil, S. et al. (2008). Auxin acts as a local morphogenetic trigger to specify lateral root founder cells. *Proceedings of the National Academy of Sciences of the United States of America* 105: 8790–8794.

Famoso, A.N., Zhao, K., Clark, R.T. et al. (2011). Genetic architecture of aluminum tolerance in rice (*Oryza sativa*) determined through genome-wide association analysis and QTL mapping. *Plos Genetics* 7: e1002221.

Friml, J., Vieten, A., Sauer, M. et al. (2003). Efflux-dependent auxin gradients establish the apical–basal axis of *Arabidopsis*. *Nature* 426: 147–153.

Friml, J., Yang, X., Michniewicz, M. et al. (2004). A PINOID-dependent binary switch in apical-basal PIN polar targeting directs auxin efflux. *Science* 306: 862–865.

Fukaki, H., Tameda, S., Masuda, H. et al. (2002). Lateral root formation is blocked by a gain-of-function mutation in the SOLITARY-ROOT/IAA14 gene of *Arabidopsis*. *The Plant Journal* 29: 153–168.

Fukaki, H., Nakao, Y., Okushima, Y. et al. (2005). Tissue-specific expression of stabilized SOLITARY-ROOT/IAA14 alters lateral root development in *Arabidopsis*. *The Plant Journal* 44: 382–395.

Galinha, C., Hofhuis, H., Luijten, M. et al. (2007). PLETHORA proteins as dose-dependent master regulators of *Arabidopsis* root development. *Nature* 449: 1053–1057.

Gamuyao, R., Chin, J.H., Pariasca-Tanaka, J. et al. (2012). The protein kinase Pstol1 from traditional rice confers tolerance of phosphorus deficiency. *Nature* 488: 535–539.

Gifford, M.L., Dean, A., Gutierrez, R.A. et al. (2008). Cell-specific nitrogen responses mediate developmental plasticity. *Proceedings of the National Academy of Sciences of the United States of America* 105: 803–808.

Gregory, P.J., Hutchison, D., Read, D.B. et al. (2003). Non-invasive imaging of roots with high resolution X-ray micro-tomography. In: Roots: The Dynamic Interface between Plants and the Earth (ed. J. Abe), 351–359. Dordrecht: Springer.

Grunewald, W., Parizot, B., Inzé, D. et al. (2007). Developmental biology of roots: one common pathway for all angiosperms. *The International Journal of Developmental Biology* 1: 212–225.

Guan, R., Qu, Y., Guo, Y. et al. (2014). Salinity tolerance in soybean is modulated by natural variation in *GmSALT3*. *The Plant Journal* 80: 937–950.

Guo, W., Zhao, J., Li, X. et al. (2011). A soybean β-expansin gene *GmEXPB2* intrinsically involved in root system architecture responses to abiotic stresses. *The Plant Journal* 66: 541–552.

Haecker, A., Groß-Hardt, R., Geiges, B. et al. (2004). Expression dynamics of WOX genes mark cell fate decisions during early embryonic patterning in *Arabidopsis thaliana*. *Development* 131: 657–668.

Hamann, T., Benkova, E., Bäurle, I. et al. (2002). The *Arabidopsis* BODENLOS gene encodes an auxin response protein inhibiting MONOPTEROS-mediated embryo patterning. *Genes and Development* 16: 1610–1615.

Han, G.L., Wei, X.C., Dong, X.X. et al. (2020). *Arabidopsis* ZINC FINGER PROTEIN1 acts downstream of GL2 to repress root hair initiation and elongation by directly suppressing bHLH genes. *The Plant Cell* 32: 206–225.

Hao, Y.J., Wei, W., Song, Q.X. et al. (2011). Soybean NAC transcription factors promote abiotic stress tolerance and lateral root formation in transgenic plants. *The Plant Journal* 68: 302–313.

Hardtke, C.S. and Berleth, T. (1998). The *Arabidopsis* gene MONOPTEROS encodes a transcription factor mediating embryo axis formation and vascular development. *The EMBO Journal* 17: 1405–1411.

Helariutta, Y., Fukaki, H., Wysocka-Diller, J. et al. (2000). The SHORT-ROOT gene controls radial patterning of the *Arabidopsis* root through radial signaling. *Cell* 101: 555–567.

Hochholdinger, F. and Feix, G. (1998). Early post-embryonic root formation is specifically affected in the maize mutant *lrt1*. *The Plant Journal* 16: 247–255.

Hochholdinger, F. and Tuberosa, R. (2009). Genetic and genomic dissection of maize root development and architecture. *Current Opinion in Plant Biology* 12: 172–177.

Hochholdinger, F., Park, W.J., Sauer, M. et al. (2004). From weeds to crops: genetic analysis of root development in cereals. *Trends in Plant Science* 9: 42–48.

Hoekenga, O.A., Maron, L.G., Piñeros, M.A. et al. (2006). *AtALMT1*, which encodes a malate transporter, is identified as one of several genes critical for aluminum tolerance in *Arabidopsis*. *Proceedings of the National Academy of Sciences of the United States of America* 103: 9738–9743.

Hopmans, J.W. and Bristow, K.L. (2002). Current capabilities and future needs of root water and nutrient uptake modeling. *Advances in Agronomy* 77: 103–183.

Iizumi, T., Sakuma, H., Yokozawa, M. et al. (2013). Prediction of seasonal climate-induced variations in global food production. *Nature Climate Change* 3: 904–908.

Inukai, Y., Sakamoto, T., Ueguchi-Tanaka, M. et al. (2005). Crown rootless1, which is essential for crown root formation in rice, is a target of an AUXIN RESPONSE FACTOR in auxin signaling. *The Plant Cell* 17: 1387–1396.

Ioio, R.D., Nakamura, K., Moubayidin, L. et al. (2008). A genetic framework for the control of cell division and differentiation in the root meristem. *Science* 322: 1380–1384.

Jacobs, T.B., LaFayette, P.R., Schmitz, R.J. et al. (2015). Targeted genome modifications in soybean with CRISPR/Cas9. *BMC Biotechnology* 15: 16.

Jamjod, S., Niruntrayagul, S., and Rerkasem, B. (2004). Genetic control of boron efficiency in wheat (*Triticum aestivum* L.). *Euphytica* 135: 21–27.

Kato, Y., Hirotsu, S., Nemoto, K. et al. (2008). Identification of QTLs controlling rice drought tolerance at seedling stage in hydroponic culture. *Euphytica* 160: 423–430.

Khush, G.S. (2001). Green revolution: the way forward. *Nature Reviews Genetics* 2: 815–822.

Kobayashi, Y., Kuroda, K., Kimura, K. et al. (2008). Amino acid polymorphisms in strictly conserved domains of a P-type ATPase HMA5 are involved in the mechanism of copper tolerance variation in *Arabidopsis*. *Plant Physiology* 148: 969–980.

Kumar, K., Mosa, K.A., Chhikara, S. et al. (2014). Two rice plasma membrane intrinsic proteins, OsPIP2;4 and OsPIP2;7, are involved in transport and providing tolerance to boron toxicity. *Planta* 239: 187–198.

Lauter, A.N.M., Peiffer, G.A., Yin, T. et al. (2014). Identification of candidate genes involved in early iron deficiency chlorosis signaling in soybean (*Glycine max*) roots and leaves. *BMC Genomics* 15: 702.

Lee, D.-K., Ahn, J.H., Song, S.-K. et al. (2003). Expression of an expansin gene is correlated with root elongation in soybean. *Plant Physiology* 131: 985–997.

Lesk, C., Rowhani, P., and Ramankutty, N. (2016). Influence of extreme weather disasters on global crop production. *Nature* 529: 84–87.

Li, J.-Y., Liu, J., Dong, D. et al. (2014). Natural variation underlies alterations in Nramp aluminum transporter (NRAT1) expression and function that play a key role in rice aluminum tolerance. *Proceedings of the National Academy of Sciences of the United States of America* 111: 6503–6508.

Li, X., Guo, Z., Lv, Y. et al. (2017). Genetic control of the root system in rice under normal and drought stress conditions by genome-wide association study. *Plos Genetics* 13: e1006889.

Liang, C., Piñeros, M.A., Tian, J. et al. (2013). Low pH, aluminum, and phosphorus coordinately regulate malate exudation through *GmALMT1* to improve soybean adaptation to acid soils. *Plant Physiology* 161: 1347–1361.

Lin, H., Zhu, M., Yano, M. et al. (2004). QTLs for Na^+ and K^+ uptake of the shoots and roots controlling rice salt tolerance. *Theoretical and Applied Genetics* 108: 253–260.

Liu, H., Wang, S., Yu, X. et al. (2005). ARL1, a LOB-domain protein required for adventitious root formation in rice. *The Plant Journal* 43: 47–56.

Liu, J., Magalhaes, J.V., Shaff, J. et al. (2009a). Aluminum-activated citrate and malate transporters from the MATE and ALMT families function independently to confer *Arabidopsis* aluminum tolerance. *The Plant Journal* 57: 389–399.

Liu, S., Wang, J., Wang, L. et al. (2009b). Adventitious root formation in rice requires *OsGNOM1* and is mediated by the *OsPINs* family. *Cell Research* 19: 1110–1119.

Liu, G., Li, X., Jin, S. et al. (2014). Overexpression of rice NAC gene SNAC1 improves drought and salt tolerance by enhancing root development and reducing transpiration rate in transgenic cotton. *Plos One* 9: e86895.

Liu, J., Li, Y., Wang, W. et al. (2016). Genome-wide analysis of MATE transporters and expression patterns of a subgroup of MATE genes in response to aluminum toxicity in soybean. *BMC Genomics* 17: 223.

Lobell, D.B., Roberts, M.J., Schlenker, W. et al. (2014). Greater sensitivity to drought accompanies maize yield increase in the US Midwest. *Science* 344: 516–519.

Lynch, J.P. (2007). Roots of the second green revolution. *Australian Journal of Botany* 55: 493–512.

Magalhaes, J.V., Liu, J., Guimaraes, C.T. et al. (2007). A gene in the multidrug and toxic compound extrusion (MATE) family confers aluminum tolerance in sorghum. *Nature Genetics* 39: 1156–1161.

Malamy, J.E. and Benfey, P.N. (1997). Organization and cell differentiation in lateral roots of *Arabidopsis thaliana*. *Development* 124: 33–44.

Mamidi, S., Chikara, S., Goos, R.J. et al. (2011). Genome-wide association analysis identifies candidate genes associated with iron deficiency chlorosis in soybean. *The Plant Genome* 4: 154–164.

Manavalan, L.P., Prince, S.J., Musket, T.A. et al. (2015). Identification of novel QTL governing root architectural traits in an interspecific soybean population. *Plos One* 10: e0120490.

Mao, C.J., He, J.M., Liu, L.N. et al. (2020). *OsNAC2* integrates auxin and cytokinin pathways to modulate rice root development. *Plant Biotechnology Journal* 18: 429–442.

Marhavý, P., Vanstraelen, M., De Rybel, B. et al. (2013). Auxin reflux between the endodermis and pericycle promotes lateral root initiation. *The EMBO Journal* 32: 149–158.

Martín-Rejano, E.M., Camacho-Cristóbal, J.J., Herrera-Rodríguez, M.B. et al. (2011). Auxin and ethylene are involved in the responses of root system architecture to low boron supply in *Arabidopsis* seedlings. *Physiologia Plantarum* 142: 170–178.

Mäser, P., Eckelman, B., Vaidyanathan, R. et al. (2002). Altered shoot/root Na^+ distribution and bifurcating salt sensitivity in *Arabidopsis* by genetic disruption of the Na^+ transporter *AtHKT1*. *FEBS Letters* 531: 157–161.

Mickelbart, M.V., Hasegawa, P.M., and Bailey-Serres, J. (2015). Genetic mechanisms of abiotic stress tolerance that translate to crop yield stability. *Nature Reviews Genetics* 16: 237–251.

Mouchel, C.F., Briggs, G.C., and Hardtke, C.S. (2004). Natural genetic variation in *Arabidopsis* identifies BREVIS RADIX, a novel regulator of cell proliferation and elongation in the root. *Genes and Development* 18: 700–714.

Naeve, S.L. (2006). Iron deficiency chlorosis in soybean. *Agronomy Journal* 98: 1575–1581.

Nguyen, H.T., Babu, R.C., and Blum, A. (1997). Breeding for drought resistance in rice: physiology and molecular genetics considerations. *Crop Science* 37: 1426–1434.

Obara, M., Tamura, W., Ebitani, T. et al. (2010). Fine-mapping of *qRL6.1*, a major QTL for root length of rice seedlings grown under a wide range of NH_4^+ concentrations in hydroponic conditions. *Theoretical and Applied Genetics* 121: 535–547.

Ogura, T., Goeschl, C., Filiault, D. et al. (2019). Root system depth in *Arabidopsis* is shaped by EXOCYST70A3 via the dynamic modulation of auxin transport. *Cell* 178: 400–412.

Okushima, Y., Fukaki, H., Onoda, M. et al. (2007). ARF7 and ARF19 regulate lateral root formation via direct activation of LBD/ASL genes in *Arabidopsis*. *The Plant Cell* 19: 118–130.

Orman-Ligeza, B., Civava, R., de Dorlodot, S. et al. (2014). Root system architecture. In: Root Engineering (eds. A. Morte and A. Varma), 39–56. Berlin: Springer.

Osmont, K.S. (2007). Hidden branches: developments in root system architecture. *Annual Review of Plant Biology* 58: 93–113.

Pan, J., Li, Z., Wang, Q. et al. (2020). An S-domain receptor-like kinase, *OsESG1*, regulates early crown root development and drought resistance in rice. *Plant Science* 290: 110318.

Paponov, I.A., Teale, W.D., Trebar, M. et al. (2005). The PIN auxin efflux facilitators: evolutionary and functional perspectives. *Trends in Plant Science* 10: 170–177.

Pasternak, T., Rudas, V., Potters, G. et al. (2005). Morphogenic effects of abiotic stress: reorientation of growth in *Arabidopsis thaliana* seedlings. *Environmental and Experimental Botany* 53: 299–314.

Pathan, M.S., Lee, J.-D., Shannon, J.G. et al. (2007). Recent advances in breeding for drought and salt stress tolerance in soybean. In: Advances in Molecular Breeding Toward Drought and Salt Tolerant Crops (eds. M.A. Jenks, P.M. Hasegawa and S.M. Jain), 739–773. Dordrecht: Springer.

Perotti, M.F., Ribone, P.A., Cabello, J.V. et al. (2019). *AtHB23* participates in the gene regulatory network controlling root branching, and reveals differences between secondary and tertiary roots. *The Plant Journal* 100: 1224–1236.

Petricka, J.J., Schauer, M.A., Megraw, M. et al. (2012a). The protein expression landscape of the *Arabidopsis* root. *Proceedings of the National Academy of Sciences of the United States of America* 109: 6811–6818.

Petricka, J.J., Winter, C.M., and Benfey, P.N. (2012b). Control of *Arabidopsis* root development. *Annual Review of Plant Biology* 63: 563–590.

Prince, S.J., Song, L., Qiu, D. et al. (2015). Genetic variants in root architecture-related genes in a *Glycine soja* accession, a potential resource to improve cultivated soybean. *BMC Genomics* 16: 132.

Prince, S.J., Valliyodan, B., Ye, H. et al. (2018). Understanding genetic control of root system architecture in soybean: insights into the genetic basis of lateral root number. *Plant, Cell and Environment* 42: 212–229.

Qi, X., Li, M.-W., Xie, M. et al. (2014). Identification of a novel salt tolerance gene in wild soybean by whole-genome sequencing. *Nature Communications* 5: 4340.

Ray, J., Yu, L., McCouch, S. et al. (1996). Mapping quantitative trait loci associated with root penetration ability in rice (*Oryza sativa* L.). *Theoretical and Applied Genetics* 92: 627–636.

Ren, Z.-H., Gao, J.-P., Li, L.-G. et al. (2005). A rice quantitative trait locus for salt tolerance encodes a sodium transporter. *Nature Genetics* 37: 1141–1146.

Rerkasem, B. and Jamjod, S. (2004). Boron deficiency in wheat: a review. *Field Crops Research* 89: 173–186.

Richard, C.A., Hickey, L.T., Fletcher, S. et al. (2015). High-throughput phenotyping of seminal root traits in wheat. *Plant Methods* 11: 13.

Riefler, M., Novak, O., Strnad, M. et al. (2006). *Arabidopsis* cytokinin receptor mutants reveal functions in shoot growth, leaf senescence, seed size, germination, root development, and cytokinin metabolism. *The Plant Cell* 18: 40–54.

Rosenzweig, C., Elliott, J., Deryng, D. et al. (2014). Assessing agricultural risks of climate change in the 21st century in a global gridded crop model intercomparison. *Proceedings of the National Academy of Sciences of the United States of America* 111: 3268–3273.

Sabatini, S., Heidstra, R., Wildwater, M. et al. (2003). SCARECROW is involved in positioning the stem cell niche in the *Arabidopsis* root meristem. *Genes and Development* 17: 354–358.

Sarkar, A.K., Luijten, M., Miyashima, S. et al. (2007). Conserved factors regulate signalling in *Arabidopsis thaliana* shoot and root stem cell organizers. *Nature* 446: 811–814.

Sawaki, Y., Iuchi, S., Kobayashi, Y. et al. (2009). STOP1 regulates multiple genes that protect *Arabidopsis* from proton and aluminum toxicities. *Plant Physiology* 150: 281–294.

Schlereth, A., Möller, B., Liu, W. et al. (2010). MONOPTEROS controls embryonic root initiation by regulating a mobile transcription factor. *Nature* 464: 913–916.

Shen, C., Wang, S., Zhang, S. et al. (2013). *OsARF16*, a transcription factor, is required for auxin and phosphate starvation response in rice (*Oryza sativa* L.). *Plant, Cell and Environment* 36: 607–620.

Shi, H., Ishitani, M., Kim, C. et al. (2000). The *Arabidopsis thaliana* salt tolerance gene SOS1 encodes a putative Na^+/H^+ antiporter. *Proceedings of the National Academy of Sciences of the United States of America* 97: 6896–6901.

Shi, H., Lee, B.-h., Wu, S.-J. et al. (2003). Overexpression of a plasma membrane Na^+/H^+ antiporter gene improves salt tolerance in *Arabidopsis thaliana*. *Nature Biotechnology* 21: 81–85.

Shin, R., Burch, A.Y., Huppert, K.A. et al. (2007). The *Arabidopsis* transcription factor MYB77 modulates auxin signal transduction. *The Plant Cell* 19: 2440–2453.

Sun, F., Zhang, W., Hu, H. et al. (2008). Salt modulates gravity signaling pathway to regulate growth direction of primary roots in *Arabidopsis*. *Plant Physiology* 146: 178–188.

Sun, P., Tian, Q.-Y., Chen, J. et al. (2009). Aluminium-induced inhibition of root elongation in *Arabidopsis* is mediated by ethylene and auxin. *Journal of Experimental Botany* 61: 347–356.

Sun, X., Hu, Z., Chen, R. et al. (2015). Targeted mutagenesis in soybean using the CRISPR-Cas9 system. *Scientific Reports* 5: 10342.

Swarup, K., Benková, E., Swarup, R. et al. (2008). The auxin influx carrier LAX3 promotes lateral root emergence. *Nature Cell Biology* 10: 946–954.

Taramino, G., Sauer, M., Stauffer, J.L. Jr. et al. (2007). The maize (*Zea mays L.*) RTCS gene encodes a LOB domain protein that is a key regulator of embryonic seminal and post-embryonic shoot-borne root initiation. *The Plant Journal* 50: 649–659.

Tian, H., Jia, Y., Niu, T. et al. (2014). The key players of the primary root growth and development also function in lateral roots in *Arabidopsis*. *Plant Cell Reports* 33: 745–753.

Trachsel, S., Kaeppler, S.M., Brown, K.M. et al. (2011). Shovelomics: high throughput phenotyping of maize (*Zea mays L.*) root architecture in the field. *Plant and Soil* 341: 75–87.

Uga, Y., Okuno, K., and Yano, M. (2011). Dro1, a major QTL involved in deep rooting of rice under upland field conditions. *Journal of Experimental Botany* 62: 2485–2494.

Uga, Y., Sugimoto, K., Ogawa, S. et al. (2013). Control of root system architecture by DEEPER ROOTING 1 increases rice yield under drought conditions. *Nature Genetics* 45: 1097–1102.

Ulises, R., Angelica, C.J., Daniela, R. et al. (2013). Integration of responses within and across *Arabidopsis* natural accessions uncovers loci controlling root systems architecture. *Proceedings of the National Academy of Sciences of the United States of America* 110: 15133–15138.

Valliyodan, B., Ye, H., Song, L. et al. (2017). Genetic diversity and genomic strategies for improving drought and waterlogging tolerance in soybeans. *Journal of Experimental Botany* 68: 1835–1849.

Vidal, E.A., Araus, V., Lu, C. et al. (2010). Nitrate-responsive miR393/AFB3 regulatory module controls root system architecture in *Arabidopsis thaliana*. *Proceedings of the National Academy of Sciences of the United States of America* 107: 4477–4482.

Wachsman, G., Sparks, E.E., and Benfey, P.N. (2015). Genes and networks regulating root anatomy and architecture. *New Phytologist* 208: 26–38.

Waisel, Y. (2002). Aeroponics: a tool for root research under minimal environmental restrictions. In: Plant Roots: The Hidden Half (eds. Y. Waisel, A. Eshel and U. Kafkafi), 323–331. New York, NY: Marcel Dekker.

Wang, J., McClean, P.E., Lee, R. et al. (2008). Association mapping of iron deficiency chlorosis loci in soybean (*Glycine max* L. Merr.) advanced breeding lines. *Theoretical and Applied Genetics* 116: 777–787.

Wang, X., Wang, H., Liu, S. et al. (2016). Genetic variation in *ZmVPP1* contributes to drought tolerance in maize seedlings. *Nature Genetics* 48: 1233–1241.

Wang, Y., Meng, Z., Liang, C. et al. (2017). Increased lateral root formation by CRISPR/Cas9-mediated editing of arginase genes in cotton. *Science China. Life Sciences* 60: 524–527.

Wang, S., Zhang, H., Shi, L. et al. (2020). Genome-wide dissection of the *CRF* gene family in *Brassica napus* indicates that *BnaCRF8s* specifically regulate root architecture and phosphate homeostasis against phosphate fluctuation in plants. *International Journal of Molecular Sciences* 21: 3660.

Weijers, D. and Jürgens, G. (2005). Auxin and embryo axis formation: the ends in sight? *Current Opinion in Plant Biology* 8: 32–37.

Weijers, D., Schlereth, A., Ehrismann, J.S. et al. (2006). Auxin triggers transient local signaling for cell specification in *Arabidopsis* embryogenesis. *Developmental Cell* 10: 265–270.

Wissuwa, M., Kretzschmar, T., and Rose, T.J. (2016). From promise to application: root traits for enhanced nutrient capture in rice breeding. *Journal of Experimental Botany* 67: 3605–3615.

Yamauchi, T., Tanaka, A., Inahashi, H. et al. (2019). Fine control of aerenchyma and lateral root development through AUX/IAA- and ARF-dependent auxin signaling. *Proceedings of the National Academy of Sciences of the United States of America* 116: 20770–20775.

Ye, H., Roorkiwal, M., Valliyodan, B. et al. (2018a). Genetic diversity of root system architecture in response to drought stress in grain legumes. *Journal of Experimental Botany* 69: 3267–3277.

Ye, H., Song, L., Chen, H. et al. (2018b). A major natural genetic variation associated with root system architecture and plasticity improves waterlogging tolerance and yield in soybean. *Plant, Cell and Environment* 41: 2169–2182.

York, L.M. and Lynch, J.P. (2015). Intensive field phenotyping of maize (*Zea mays* L.) root crowns identifies phenes and phene integration associated with plant growth and nitrogen acquisition. *Journal of Experimental Botany* 66: 5493–5505.

Yu, J., Zhang, Y.Y., Liu, W. et al. (2020). Molecular evolution of auxin-mediated root initiation in plants. *Molecular Biology and Evolution* 37: 1387–1393.

Zhang, H. and Forde, B.G. (1998). An *Arabidopsis* MADS box gene that controls nutrient-induced changes in root architecture. *Science* 279: 407–409.

Zhang, D., Song, H., Cheng, H. et al. (2014). The acid phosphatase-encoding gene *GmACP1* contributes to soybean tolerance to low-phosphorus stress. *Plos Genetics* 10: e1004061.

Zhao, Y., Xing, L., Wang, X. et al. (2014). The ABA receptor PYL8 promotes lateral root growth by enhancing MYB77-dependent transcription of auxin-responsive genes. *Science Signaling.* 7: ra53.

Zheng, S.J. (2010). Crop production on acidic soils: overcoming aluminium toxicity and phosphorus deficiency. *Annals of Botany* 106: 183–184.

Zheng, H.-G., Babu, M., Chandra, R. et al. (2000). Quantitative trait loci for root-penetration ability and root thickness in rice: comparison of genetic backgrounds. *Genome* 43: 53–61.

Zheng, B., Yang, L., Zhang, W. et al. (2003). Mapping QTLs and candidate genes for rice root traits under different water-supply conditions and comparative analysis across three populations. *Theoretical and Applied Genetics* 107: 1505–1515.

Zhou, J., Xie, J., Liao, H. et al. (2014). Overexpression of β-expansin gene *GmEXPB2* improves phosphorus efficiency in soybean. *Physiologia Plantarum* 150: 194–204.

Zhu, J., Kaeppler, S.M., and Lynch, J.P. (2005a). Mapping of QTL controlling root hair length in maize (*Zea mays* L.) under phosphorus deficiency. *Plant and Soil* 270: 299–310.

Zhu, J., Kaeppler, S.M., and Lynch, J.P. (2005b). Mapping of QTLs for lateral root branching and length in maize (*Zea mays* L.) under differential phosphorus supply. *Theoretical and Applied Genetics* 111: 688–695.

Zhu, J., Mickelson, S.M., Kaeppler, S.M. et al. (2006). Detection of quantitative trait loci for seminal root traits in maize (*Zea mays* L.) seedlings grown under differential phosphorus levels. *Theoretical and Applied Genetics* 113: 1–10.

3

Plant Roots for Sustainable Soil Structure Management in Cropping Systems

Gernot Bodner[1], Axel Mentler[2], and Katharina Keiblinger[2]

[1] *Department of Crop Sciences, Division of Agronomy, University of Natural Resources and Life Sciences BOKU Vienna, Tulln an der Donau, Austria*
[2] *Department of Forest- and Soil Sciences, Institute of Soil Research, University of Natural Resources and Life Sciences BOKU Vienna, Vienna, Austria*

3.1 Introduction

Sustainable intensification (SI) of crop production envisions a new paradigm of post Green Revolution agronomy where efficient use of natural resources confers high yields while reducing dependence on external inputs and minimizing environmental losses (Collette et al. 2011). In line with conservation agriculture, a fundamental aim of SI is the minimum mechanical disturbance of soil to conserve organic matter, structure, and overall soil health (Hobbs et al. 2007). Soil structure and plant rooting are closely related indicators of soil health/quality that essentially determine the soil functionality regarding crop production and reflect the sustainability of soil management practices (Doran and Zeiss 2000; Arshad and Martin 2002).

Globally, soil erosion is the primary source of human-induced soil degradation, with 56% of erosion caused by water (Oldeman 1992). Soil erosion is closely linked to a loss of soil organic carbon (SOC) (Lal 2003) that is commonly considered a key indicator of soil quality (Obalum et al. 2017), responding sensitively to both land use and climate change (Scharlemann et al. 2014). Degradation of soil structure is considered a primary cause in a downward spiral of declining soil quality, leading to a loss of organic carbon, accelerated erosion, and exacerbated nutrient leaching (Lal 2015). A recent assessment of the status of world soil resources also revealed that structure degradation is actively involved in five (compaction, sealing, erosion, organic matter decline, and salinization) out of eight major global soil threats (FAO/ITPS 2015). However, in their review on soil quality indicators, Bünemann et al. (2018) reported aggregation to be used in less than 20% and structural stability in less than 30%, of soil quality assessment approaches. Thus, although global estimates of soil degradation such as erosion indirectly indicate a severe decline in soil

The Root Systems in Sustainable Agricultural Intensification, First Edition.
Edited by Zed Rengel and Ivica Djalovic.

structure, there are neither methods nor models that currently allow mapping of the state of soil structure.

In the context of SI, the focus is on the soil productivity function (Blum 2005). Soil structure, in combination with texture and organic matter, determines the pore system that controls the soil water, gas, temperature, and nutrient regime and thereby the capacity of soil to sustain biomass growth. In managed ecosystems such as agricultural fields, the importance of soil structure furthermore stems from soil vulnerability to disturbance. This is best indicated by soil compaction, whereby a single heavy machinery passage on wet soil can severely and persistently degrade structure (e.g. decrease air-filled porosity, increase penetration resistance) particularly in the subsoil (Bullock et al. 1985; Yavuzcan et al. 2005).

Structural features, as a result of pedogenic processes, are also used to distinguish certain soil diagnostic horizons that characterize particular soil types and, in some cases, explain their low productivity, e.g. anthraquic, petrocalcic, and petrogypsic or petroplinthic horizons with massive structures that impede root penetration (IUSS Working Group WRB 2014).

Aggregates, the secondary building blocks constituting the soil solid phase, are a product of particle size distribution (and clay mineralogy) plus organic-binding agents (Jastrow 1996). Here, we will focus on the role of plant roots in soil structural dynamics, including some aspects of root-mediated microbial effects. Roots and soil structure constitute a feedback loop (Angers and Caron 1998), whereby root-induced structure formation interacts with root growth and architecture (Figure 3.1). This chapter aims to provide an overview of the role of roots in soil structure formation and stabilizations. The effects of structural soil properties on root growth, architecture, and metabolism are reviewed elsewhere (Bengough et al. 2005).

Oldeman (1992) estimated 68 Mha of soil to be physically degraded by compaction, sealing, and crusting (4% of degraded soils worldwide), with 33 Mha in Europe, where heavy

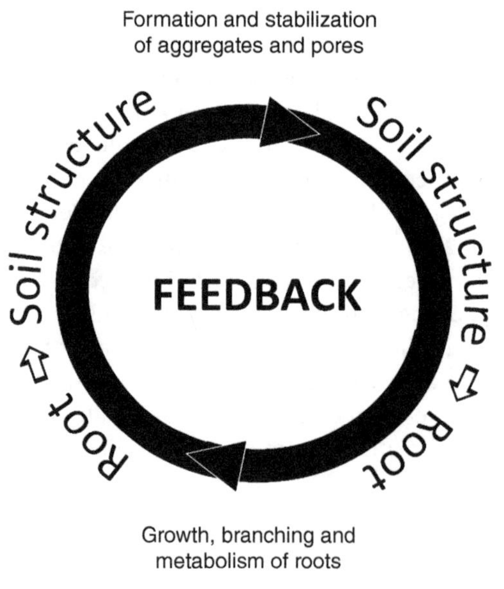

Figure 3.1 Feedback between plant roots and soil structure. Roots influence the formation and stabilization of soil aggregates and pores, which in turn shape the growth and architecture of the root system.

machinery was identified as the primary cause. The influence of soil physical conditions on root growth has received increasing attention in plant research and sustainable crop production with the awareness of soil compaction as stress (Hamza and Anderson 2005). On the other hand, awareness of the role of roots in structure formation/stabilization and as a visual indicator of soil structural quality is still confined mainly to soil science. Only a comparatively small proportion of root research has been dedicated to structural effects (1.5% of Scopus-listed root studies from 2008 to 2019) compared to 8.0% on mycorrhizae and 12.2% of published research on soil organic matter. With the essential role of soil structure in crop growth, the increasing challenge of mitigating physical degradation, and the novel methodological opportunities for studying roots, we consider it crucial to increase the attention of crop science and root research towards root-mediated management options to influence soil structure.

This chapter first gives a short overview of roots as sensitive indicators of soil structural quality in different field assessment schemes and provides some examples of root systems in adverse soil structures, including the indicator plants specific to compacted soils. Beyond field assessment, we also provide a short overview of laboratory methods for quantifying relevant aspects of soil structure. This leads to the role of plant roots in soil structure formation and stabilization following the aggregate hierarchy model of Tisdall and Oades (1982) and its application to soil pores (Elliott and Coleman 1988). Finally, we discuss relevant fields and opportunities for root-based management measures in sustainable soil use in cropping systems.

3.2 Methodological Approaches to the Root–Soil Structure Relations

3.2.1 Roots and Soil Structure Assessment in Field Surveys

The objective of a field soil survey for agricultural purposes is an overall rating of soil quality and productivity. On the one hand, soil quality and productivity are the results of pedogenic factors (climate, parent material, topography, and age; Jenny 1994) that determine inherent soil properties (e.g. texture, profile depth) beyond the influence of management. On the other hand, the state of soil also reflects management influences. Some assessment schemes combine both types of indicators, e.g. the Muencheberger Soil Quality Rating (Mueller et al. 2012), but others focus on management-sensitive parameters relevant for decision support to optimize agricultural operations.

In this context, soil structure and plant rooting are key indicators. For example, the New Zealand Visual Soil Assessment (Shepherd 2000) attributes the highest weighting factor of 3 to soil structure and porosity among all the soil criteria included, whereas the size and development of the root system are used as plant indicator with a weighting factor of 2. Rooting depth and barriers to root penetration are also used by the USDA Soil Survey (Soil Science Division Staff 2017) as well as the German Muencheberger Soil Quality Rating, the latter even including assessment of biopores (induced by roots and/or earthworms). Ball et al. (2007) developed a field test specifically for structural quality assessment based on the break-up of an extracted soil block into aggregates: beyond aggregate shape and size,

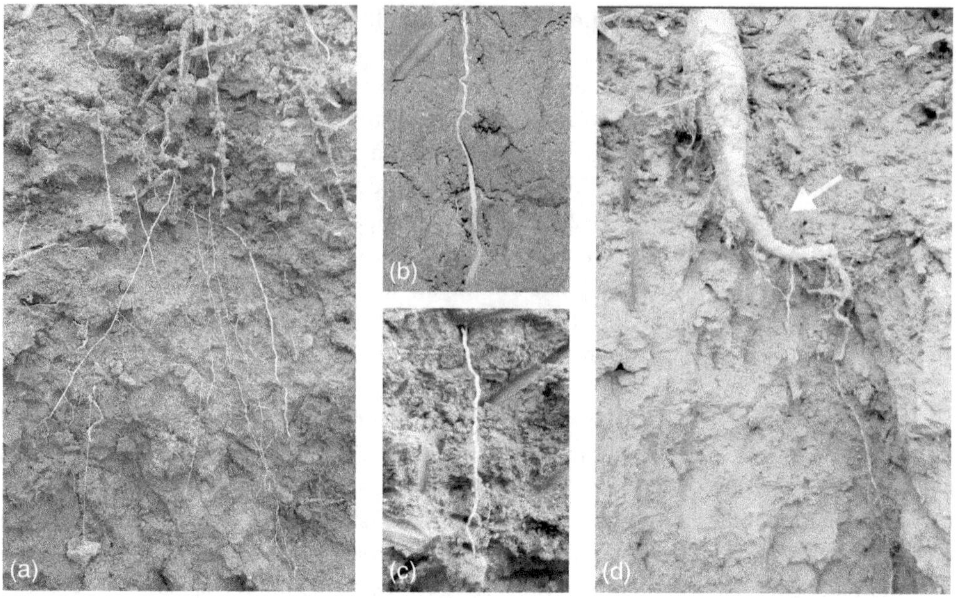

Figure 3.2 Roots as visual indicators of adverse soil structure. (a) sparse rooting (winter wheat, growth stage BBCH 29) in dense topsoil (Endocalcic Chernozem, 15% sand, 38% silt, and 47% clay, with angular blocky structure); (b) and (c) root deformation – flattening (B: winter wheat in Endocalcic Chernozem; C: soya bean in Chromic Luvisol; 18% sand, 37% silt, and 45% clay, with massive structure); (d) taproot bending and thinning in response to a compacted layer (*Raphanus sativus* var. *oleiformis* in Calcic Chernozem, 29% sand, 52% silt, and 19% clay, crumby structure with a plough pan at 25 cm soil depth indicated by white arrow). *Source:* Gernot Bodner, Institute of Agronomy, University of Natural Resources and Life Sciences (BOKU), Vienna, Austria.

scoring also includes root criteria such as bundling, thickening, and deflections. Figure 3.2 gives some examples of rooting patterns applicable in the visual soil assessment to identify adverse structural properties.

Guimarães et al. (2013) demonstrated that visual scoring of soil structure could predict several key physical properties such as bulk density, air-filled porosity, penetration resistance, and least-limiting water range. In addition, more subtle differences in pore network characteristics, quantified via X-ray computer tomography (CT) imaging of soil cores from different tillage treatments (plough vs. direct seeding in sandy loam soil), could be inferred from visual scoring (Garbout et al. 2013a).

The interplay between roots and soil structure is also revealed by indicator plants (Ellenberg 1974) that support soil quality assessment in the field. Figure 3.3 gives an example of the relation between the percent coverage of indicator plants for soil compaction in grassland swards and measured values of penetration resistance (results for bulk density are similar). The species used as indicators provided a significant prediction of compaction risk, which was highest for swards under intensive cattle grazing.

Adverse soil structure (induced either pedogenically or by management) can impede drainage, with waterlogging inducing temporary anoxia in the root zone (Drewry et al. 2008). Indicator species for soil compaction are adapted to these soil conditions. This is (partially) explained by specific root characteristics such as root aerenchyma, providing

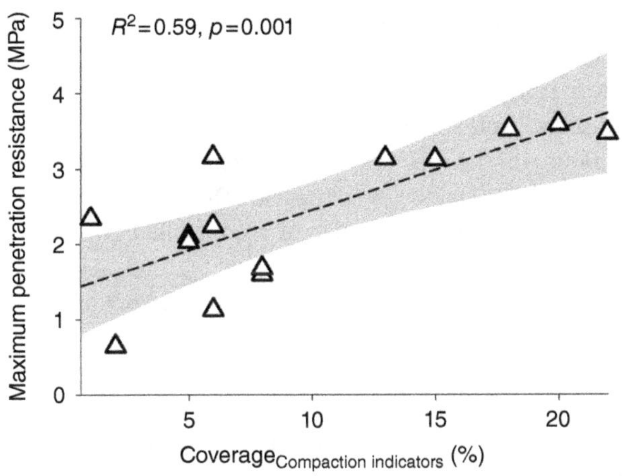

Figure 3.3 Relationship between percent coverage of indicator plants for soil compaction in grasslands (cumulative coverage of *Ranunculus repens, Plantago major* ssp. *major, Polygonum aviculare agg., Scorzoneroides autumnalis, Poa annua,* and *Poa supina*) and maximum penetration resistance in the topsoil (0–30 cm). Measurements were performed at three grassland sites in Austria (Obertrum am See, 47°56′N, 13°04′E; Anthering, 47°52′N, 13°00′E; and Seekirchen am Wallersee, 47°53′N, 13°08′E) with soil types being Cambisol and Gleysol. Triangles represent sites with different management intensity and corresponding compaction from cattle trampling (pasture) and machinery (meadow). Grey shaded area indicates the 95% confidence limits.

aeration to root tissues under anoxic conditions (e.g. *Ranunculus repens, Plantago major* ssp. *major*), or shallow root systems that spread in the topsoil above horizons influenced by stagnant water (e.g. *Poa annua*, also *Bellis perennis* not included in Figure 3.3; Bohner et al. 2018). Bartelheimer and Poschlod (2016) provided a general review of the relevance of root properties in relation to Ellenberg indicator values of plants (Ellenberg et al. 1992). They found root weight per total dry matter as a responsive trait in indicator species for soil acidity (Ellenberg reaction value R), rooting depth and root/shoot ratio for drought, adventitious root formation for waterlogging (Ellenberg moisture value M), and specific root length for soil fertility (Ellenberg nutrient value N).

Soil survey (visual soil scores) and plant ecology (indicator plants) provide qualitative/semi-quantitative evidence for the combined role of roots and soil structure in shaping soil quality. To further unravel their linkage and feedback relations via root-mediated processes of aggregation and pore formation, direct analytical methods are required.

3.2.2 Tailored Approaches to Measuring Root-induced Soil Structure

An extensive overview of measurement methods for soil structure and structural porosity is given in Dane and Topp (2002) and Amezketa (1999). Here, we focus on ultrasonic aggregate dispersion and CT imaging only because these methods were developed/applied with a particular focus on quantifying the linkage between structural properties and organic agents of aggregation such as roots. As postulated by the hierarchy model (Section 3.3) of Tisdall and Oades (1982), differently sized aggregates are the result of distinct binding forces with variable aggregating strength. The traditional methods of slaking of aggregates by water (Kemper and Koch 1966) were intended to simulate their stability against a raindrop impact and thereby to provide relevant information for soil conservation against erosion. The size methods use the mechanical force of rotating sieves (with or without water) to break-up a soil sample into differently sized aggregates (Nimmo and Perkins 2002) and provide information on the mechanical disaggregation as a result of soil tillage implements (Boydaş and Turgut 2007). However, these methods do not control the energy amount applied for aggregate disruption. The use of a controlled amount of energy from ultrasonic waves for slaking aggregates was first suggested by Imeson and Vis (1984) and later refined for low-energy sonication by Mayer et al. (2002) and Schomakers et al. (2011). The amount of ultrasonic energy applied to aggregates is intended to approximate the energy of binding agents for different size fractions. The resulting distribution of aggregate size classes thereby reflects differences in the structure formation/stabilization processes. Figure 3.4 shows an advanced measurement setup for ultrasonic aggregate dispersion that can be combined with (i) a sedigraph to quantify microaggregate fractions, or (ii) a peristaltic pump that is linked to a cuvette for spectrophotometric determination of dissolved organic carbon, thereby simultaneously quantifying simple soluble aggregates bound to low-molecular-weight fractions.

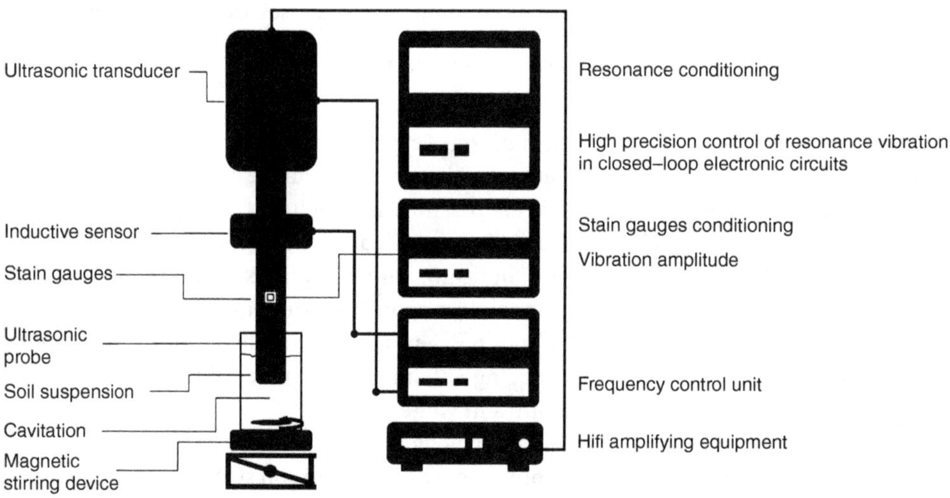

Figure 3.4 Setup of an ultrasonic fatigue testing equipment, developed by the Institute of Physics and Material Sciences, University of Natural Resources and Life Sciences (BOKU), Vienna, with the capacity to generate very low vibration amplitudes with high accuracy and vibrations below the cavitation limit. The equipment consists of a 48.3 cm rackmount installation for the high-precision control of resonance vibration in closed-loop electronic circuits and a possibility to use different (non-standard) mechanical setups to guarantee the accuracy of the vibration amplitude. Ultrasonic probes with large diameters (30–40 mm) can be used to allow the generation of approximately planar ultrasonic waves with defined acoustic pressure. The vibration amplitude of a cylindrical ultrasonic probe is measured with an induction coil and serves to control the magnitude of loading. The vibration amplitude is controlled and kept constant at a pre-selected value with very high accuracy. In addition to the vibration amplitude, the duration of sonification can be selected at the beginning of the dispersion experiment. *Source:* Axel Mentler, Institute of Soil Research, University of Natural Resources and Life Sciences (BOKU), Vienna, Austria. © 2020, University of Natural Resources and Life Sciences (BOKU).

Low-energy sonication has been successfully applied to study tillage impact (Kasper et al. 2009), the effects of different land use/vegetation cover (An et al. 2010), influences from decomposing crop residues (Zhu et al. 2017), and turnover times of soil carbon fractions (Poeplau et al. 2018). These results demonstrated a close relationship between differently sized aggregates and variable concentration/quality of organic carbon fractions. Therefore, low-energy sonication is particularly relevant to identifying the role of root-related binding agents at distinct aggregate size fractions.

Beyond such combined physical and chemical analytical methods, imaging – for qualitative visualization, hypotheses development, and quantification of (small scale) structures – has long been used to highlight the interplay between roots and soil structure (for a recent review see Chenu et al. 2015). The aggregation of mineral particles on roots and mycorrhizal hyphae was identified visually via scanning and transmission electron microscopy (SEM, TEM), leading Tisdall and Oades (1979) to postulate the model of aggregate hierarchy. Similarly, Amellal et al. (1998) used ultrathin sections and TEM to unravel the influence of bacterial exopolysaccharides (EPSs) in the wheat root rhizosphere on aggregation and their effects on structural porosity. Environmental scanning electron microscopy (ESEM) was applied recently to quantify differences in the shape of aggregates derived from the root, microbial, and abiotic processes of soil structure formation (Blankinship et al. 2016).

In recent years, soil and plant sciences increasingly have been using 3D imaging technologies, particularly X-ray CT, to improve spatial resolution and allow in situ characterization of aggregate and pore shapes, including observation of temporal dynamics (Helliwell et al. 2013). Current μCT scans provide images with a voxel size of <10 μm that allows quantifying fine geometric features, including micropores in the diameter range of 3–50 μm (Zhao et al. 2017). Using synchrotron μCT imaging revealed that roots and soil microbes actively create their surrounding microhabitats by rearranging pore geometry and influencing aggregate shape (Feeney et al. 2006). In particular, the modification of structural porosity and pore connectivity by barley root hairs could be visualized and quantified by synchrotron μCT (voxel size 3.2 μm) to estimate changes in rhizosphere hydraulic properties (Koebernick et al. 2017). Rabbi et al. (2018a) showed the role of structural changes in soil in the rhizosheath of chickpea with enhanced drought resistance using a 16-μm voxel size μCT.

In situ measurements of pore geometry as well as root architecture are a remarkable advance by the current high-resolution CT imaging compared to the previous methods. The importance of time-resolved measurements of aggregates for better understanding the relative contribution of binding agents as highlighted by Degens (1997) would suggest CT imaging can also focus on in situ monitoring of the 'solid phase' of soil structure.

3.3 Roots in the Aggregate and Pore Hierarchies

Since the landmark paper (Jarvis et al. 2012) of Tisdall and Oades (1982), soil structure (except structures dominated by Fe/Al-[hydr]oxides) has been understood as a hierarchical arrangement of differently sized building blocks (aggregates) bound by size-specific binding agents that determine their distinctive stability/turnover. Comprehensive reviews on the history of soil aggregation have demonstrated that, in spite of further details and slight modifications, the model of Tisdall and Oades still constitutes the fundamental framework for understanding soil structure (Lal and Shukla 2004; Six et al. 2004).

The original aggregate hierarchy model distinguishes three major size classes: macroaggregates >250 μm, microaggregates between 20 and 250 μm, and silt/clay-sized particles <20 μm. Further, subdivision within the microaggregate fraction has been suggested at 53 μm to separate smaller from larger microaggregates (Denef et al. 2007). Other authors, based on recent evidence of small-sized composite particle, extend the microaggregate fraction from 20 to 2 μm, while defining mineral clusters of <2 μm as nanoaggregates (Tisdall 1996; Krause et al. 2018; Rowley et al. 2018; Totsche et al. 2018).

A complementary model for the soil pore system with four hierarchical pore categories: (i) macropores; (ii) pores between macroaggregates, (iii) pores between microaggregates within macroaggregates, and (iv) pores within microaggregates, has been suggested by Elliott and Coleman (1988).

In the following part, we briefly cover the key processes of aggregate formation and stabilization, focusing on the particular role of roots in the aggregate (Section 3.3.1) and pore hierarchy (Section 3.3.2).

3.3.1 Aggregate Hierarchy

The hierarchical constitution of soil structure is expressed by the following stepwise aggregation processes (modified after Lal and Shukla 2004; Figure 3.5):

$$\begin{aligned}&\text{Clay particle}\left(<2\ \mu\text{m}\right)\rightleftarrows \text{floccule/quasi-crystal/domain}\left(2-20\ \mu\text{m}\right)\\&\rightleftarrows \text{microaggregate}\left(20-250\ \mu\text{m}\right)\\&\rightleftarrows \text{macroaggregate}\left(250-2000\ \mu\text{m}\right)\rightleftarrows \text{clod}\left(>2000\ \mu\text{m}\right)\end{aligned} \tag{3.1}$$

In most soils, organic matter is the essential driver for all steps of aggregation beyond the flocculation of primary particles, except in soils dominated by iron and aluminium oxyhydrates (Tisdall 1996); however, see Vrdoljak and Sposito (2002) for the aggregating role of organic matter in an Oxisol A-horizon.

3.3.1.1 Clay Domains <20 μm

Totsche et al. (2018) and Rowley et al. (2018) identified associated particle domains even in the fraction extending from the clay (<2 μm) to the silt (<20 μm) sizes. Thus, even in the very small size fractions, the primary particles as well as the composite building units were found, leading Yudina et al. (2018) to propose the term elementary soil particle (ESP) for such compounds as distinguished from similarly sized single mineral primary particles.

Formation of tightly bound domains of phyllosilicates requires the approximation of their charged surfaces. The attraction is facilitated by flocculating agents/conditions that reduce the extension of the diffusive double layer (salt concentration, cation charge, hydration, and pH; Chibowski 2011), leading to their association in specific configurations (face-to-face, face-to-edge, and edge-to-edge; Sergeyev et al. 1980). In addition, precipitates of Fe and Al oxides and hydroxides were attributed to the size class (<20 μm) of building units (Totsche et al. 2018).

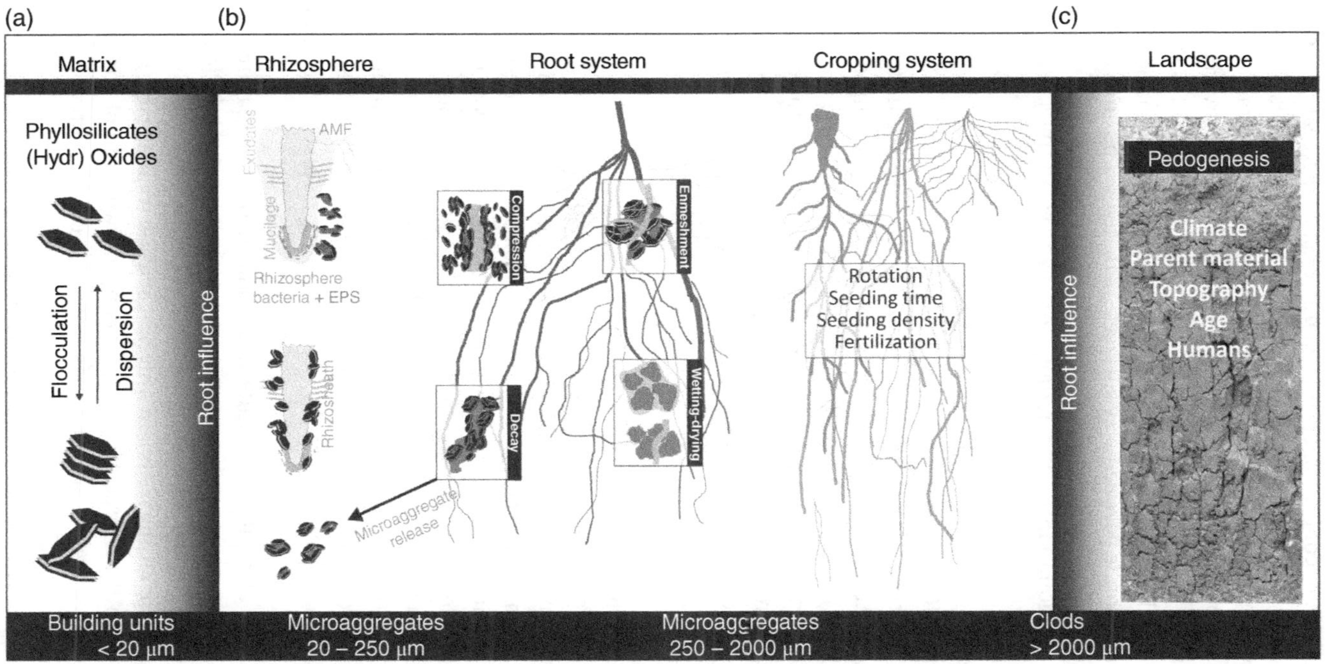

Figure 3.5 Scales of soil aggregate hierarchy, size range with major influence arising from plant roots and relevant pathways, and factors for root-mediated aggregate formation and stabilization. (a) Flocculation/dispersion as the primary process forming building blocks of mineral particles at the smallest size scale. (b) Domain of major root influences on soil aggregation at the rhizosphere, root system, and cropping system scales. Rhizosphere processes influencing microaggregation with particles bound by root and microbial substances (EPS, extracellular polymeric substances; AMF, arbuscular mycorrhiza fungi) and attached to the root surface forming rhizosheaths; root system effects on macroaggregation via biotic and abiotic processes; and cropping system effects via management impacts on below-ground processes. (c) Large scale influences on structural properties driven by pedogenic factors.

3.3.1.2 Small Microaggregates (20–50 μm)

When approaching the silt-sized fraction (20 μm), an interplay of organic substances with polyvalent cations has been conceived as the characteristic aggregating agent between the primary/elementary soil particles (Tisdall 1996). Such composites of C–P–OM, with C being a clay-sized primary/elementary particle, P polyvalent metal (Fe, Al, Ca) and OM organic matter (or organo–metal complex), were described by Edwards and Bremner (1967) as low-hierarchy units combining to form microaggregates $([C\text{–}P\text{–}OM]_x)_y$ of 50 μm size. The negatively charged organic substances such as polysaccharides or aromatic humic material act as bonds between clay surfaces mediated via positive cation charges (e.g. Ca^{2+}; Rowley et al. 2018). The organic carbon bonds at this hierarchical scale are considered to be predominantly of microbial origin (Tisdall 1996; Stewart et al. 2007).

3.3.1.3 Large Microaggregates (50–250 μm)

For larger microaggregates, Six et al. (2000) and Gale et al. (2000) have postulated the mechanism of microaggregate within macroaggregate formation, driven by the turnover of particulate organic carbon (POM): coarse POM within macroaggregates (coarse iPOM) is transformed by microbial decomposition towards finer iPOM that constitutes the core of new microaggregates. These microaggregates are released upon decay of the successively destabilized macroaggregate shells upon decomposition of their transient organic-binding agents.

3.3.1.4 Macroaggregates (>250 μm)

This aggregate size fraction is formed upon decomposition of fresh (shoot- and root-derived) plant material, with young POM constituting the core of aggregation, while labile organic substances, released upon decomposition of fresh inputs, stabilize the adhesion of particles, and/or smaller composite building units (Degens 1997; Puget et al. 2000; Poirier et al. 2014). The transient-binding agents at the macroaggregate scale are easily decomposable organic glues, including microbial and root exudates (Degens 1997). Stabilization of larger macroaggregate structures and clods/crumbs (>2000 μm) against disruption along the inter-aggregate failure plains is increasingly mediated by transient enmeshment via fine roots and fungal hyphae ('sticky-string bag'; Angers and Caron 1998; Miller and Jastrow 2000).

At the pedon scale, structural attributes constitute important distinctive features for soil classification. Vegetation types are recognized to interact with the main soil-forming factors (climate, parent material, topography, and age; Crocker 1952), e.g. via quantity and quality of plant litter input (including roots). In order to focus on (manageable) field-scale structural processes within the aggregate hierarchy concept, we will not consider further root effects at the scale of pedogenic processes (for a review see Verboom and Pate 2006).

At the different hierarchical levels of soil structure, three modes of root contribution to aggregate formation and stabilization are distinguished: (i) direct influences via living roots, root residues, and rhizodeposition, (ii) indirect influences via the root/rhizosphere microbiome, and (iii) abiotic root zone influences.

3.3.1.5 Direct Root Effects on Aggregation

Plant roots are considered the predominant source of SOC (soil organic carbon), with an average contribution of root carbon to SOC formation 2.4 times that of shoots (Rasse et al. 2005). According to these authors, only about one-fourth of this difference is due to higher chemical recalcitrance, whereas physicochemical stabilization (organo–mineral interactions and aggregate occlusion) is the main reason for the higher proportion of SOC from root origin. Root input is estimated to constitute between 20% (cereals) and 50% (pasture) of assimilated carbon (Kuzyakov and Domanski 2000). Rhizodeposition from living roots comprises a wide range of organic compounds originating from sloughed-off border cells at the root tip as well as root exudation (Kögel-Knabner 2002; Dennis et al. 2010; Kögel-Knabner 2017). Hütsch et al. (2002) and Nguyen (2009) found that only a minor proportion (2–5%) of rhizodeposition was conserved in the soil, while most input was quickly cycled through microbial catabolism and anabolism (see below). Maize root exudates were allocated mostly (79%) to the fraction of water-soluble carbon comprising 64% carbohydrates, 22% amino acids/amides, and 14% organic acid anions. The predominant form of rhizodeposits (sugars vs. organic acid anions) influences their gelling vs. dispersing effects on soil particles. For example, the higher proportion of negatively charged organic acid anions in barley rhizodeposits tended to disperse soil, whereas the sugar-dominated exudates of chia and maize had a gelling effect (Naveed et al. 2017).

Despite quick microbial turnover, the gelatinous root mucilage – in combination with root hairs and local drying – can still directly act as an aggregating agent between mineral particles (mainly via hydrogen bonds between carboxyl groups and mineral surfaces; Figure 3.6).

The gelling effect on aggregate formation is especially evident in species forming rhizosheaths, with soil particles strongly adhering to root axes, mainly in the apical/root hair zone, mediated by mucigels (Watt et al. 1993; Brown et al. 2017; Pang et al. 2017). An increase in stable organo-–mineral associations in the size fraction 2–20 μm at the root–soil interface of maize (bound by humified root tissues; e.g. fine-root cell walls) has been observed using TEM (Watteau et al. 2006).

Rillig et al. (2015) differentiated between the traits relevant to the aggregate formation (e.g. root exudate C:N) and stabilization. Some compounds that increase aggregate stability are rendering aggregate surfaces hydrophobic (e.g. polygalacturonic acid, PGA, as a mucilage analog; Czarnes et al. 2000). PGA also strongly glues clay particles, making them highly stable against fracturing (Zhang et al. 2008). The stabilizing effect of organic carbon in the rhizosphere decreases from more complex to simpler compounds (modelled exudate > mucilage > PGA > glucose; Traoré et al. 2000). Abiven et al. (2009), in a literature review on aggregate stabilization by different organic inputs, categorized mucilage as a strongly stabilizing agent with fast aggregative capacity.

Partially decomposed plant residues (particulate organic matter, POM) act as organic cores of aggregate formation (Six et al. 2000). Macroaggregate POM cores originate predominantly from structural [cell wall] biopolymers (e.g. cellulose, hemicellulose, pectin, and glycoproteins) of root detritus. Root residues supplied between 56% and 80% of POM (physically stabilized POM) (Cambardella and Elliott 1992), whereas shoots made a minor contribution to new POM in field-grown maize and soya bean (Mazzilli et al. 2015). Kölbl

Figure 3.6 Environmental scanning electron micrographs (ESEM) showing the effect of applying different ultrasound energies to soil aggregates. (a) and (b): 2.2 J ml^{-1}; (c) and (d): 6.7 J ml^{-1}). (a) and (c): Cambisols. (b) and (d): Podzols. Different energy inputs are needed to disperse the organic matter from the mineral compounds of the soil by sonification. With higher dispersion energy inputs (c), the organic fraction is peeled from the aggregates (compared to [a]). In contrast, the aggregates in the Podzol show hardly any effects, even with higher dispersion energy ([d] vs. [b]). *Source:* ESEM images provided by Herwig Mayer, Institute of Physics and Material Sciences, University of Natural Resources and Life Sciences (BOKU), Vienna, Austria.

and Kögel-Knabner (2004), using the aryl C/O-alkyl C ratio as an indicator of POM decomposition, showed that higher clay content increased the process of POM occlusion and stabilization within aggregates.

Organic inputs and their microbial turnover drive the formation of aggregates, whereas entanglement by living roots and hyphae mainly acts to stabilize existing aggregates (Lehmann and Rillig 2015). Application of path analysis (i) demonstrated the vital contribution of roots and mycorrhizal hyphae to stabilization of macroaggregates, and (ii) confirmed entanglement as a major binding agent for the largest size fraction in agreement with the aggregate hierarchy model (Jastrow et al. 1998). Hyphae build an extensive network extending beyond the rhizosphere and penetrate soil pores accessible only to fine roots/root hairs. Therefore, the fungal contribution to enmeshment is considered more important than direct root contribution (Ritz and Young 2004). However, some authors found root hairs to be better related to aggregation than hyphae in a sandy soil (Moreno-Espíndola et al. 2007).

3.3.1.6 Rhizosphere Microbiome Effects on Aggregation

Formation of microaggregates and organo–mineral/metal complexes is mainly mediated by soil micro-organisms decomposing the primary plant metabolites (organic acids, sugars, and amino acids) in microbial hotspots such as the rhizosphere and the detritusphere (Six et al. 2004; Castellano et al. 2015; Kuzyakov and Blagodatskaya 2015; Kumar et al. 2017; Kögel-Knabner 2017). However, this does not exclude that microaggregation also occurs around occluded root carbon cores (e.g. suberin and lignin; probably upon macroaggregate decay), with organo–mineral/metal complexes arising directly from interaction of root mucilage (e.g. uronic acid) and mineral surfaces/metal ions (see review by Poirier et al. 2018). The rhizosphere and mycorrhizosphere (Andrade et al. 1998) still constitute the major microbial hotspots causally related to enhanced aggregation and structural differentiation from bulk soil (Whalley et al. 2005). Rhizodeposition and microbial activity are tightly interlinked, e.g. microbes regulate exudation at the root tip by maintaining diffusive gradients between root cells and rhizosphere soil (Canarini et al. 2019). Concurrently, secondary metabolites in root exudates shape the microbial community structure in the rhizosphere (Zhalnina et al. 2018). In addition, changes in root biomass have been shown to alter the structure of soil food webs, dynamics of belowground C cycling, and the mean residence time of different SOC pools (Reid et al. 2012). This suggests the feedback relation between roots and soil micro-organisms, acting upon the processes of soil aggregation.

There appear to be two key reasons for microbial dominance in the small-scale aggregation processes (microaggregates and organo–mineral complexes): (i) labile organic root compounds released at carbon input hotspots (e.g. rhizosphere) are degraded/transformed by microbes before interacting with mineral particles, (ii) microbes access small-sized pore spaces between soil particles where easily degradable by-products of microbial metabolism are physically stabilized. Estimates of the limiting pore size still allowing access to bacteria are 1–2 μm (Watt et al. 2006), i.e. the size of intra-microaggregate pores. Stabilization vs. degradation of labile carbon sources depends on their intrinsic chemical nature, soil/solute parameters, and spatial separation from decomposers (Marschner and Kalbitz 2003). Thus, it can be hypothesized that the primary input of labile compounds at microbial hotspots such as rhizosphere exudates, predominately composed of highly biodegradable compounds and in close contact with an abundant microbial community, is physico-chemically stabilized by aggregates/mineral surfaces mainly upon microbial processing (see the Microbial Efficiency-Matrix Stabilization, MEMS, concept; Cotrufo et al. 2013). Exceptions are less abundant compounds with lower degradability (e.g. aromatic and hydrophobic compounds) and soil/solute properties inhibiting microbial degradation (low pH and high concentration of Al^{3+} and heavy metals). With ongoing microbial turnover of organic inputs and the related aggregating processes, it can be assumed that spatial heterogeneity (distinct microsites) is successively increasing (also depending on pore-water content and diffusion paths), thereby further enhancing physico-chemical stabilization of microbial products at the less accessible and/or degradation-limiting (e.g. low oxygen) microsites (Ebrahimi and Or 2015). Indeed, soil organic matter research provides increasing evidence for the key role of spatial separation between carbon sources and decomposers to explain turnover dynamics (e.g. 'regulatory gate hypothesis'; Kemmitt et al. 2008; Dungait et al. 2012; Schimel and Schaeffer 2012). In contrast, the surface-exposed recalcitrant chemical forms (including lignin) lack protection from decomposition, even those bound to mineral particles, and hence do not accumulate in soil due to microbial accessibility (Dungait et al. 2012; Hernandez-Soriano et al. 2018).

Mummey et al. (2006) and Upton et al. (2019) showed that differently sized aggregate fractions host distinct microbial communities, with greatest microbial diversity in microaggregates, particularly at specific times (hot moments; Kuzyakov and Blagodatskaya 2015) of vigorous plant growth, i.e. with living roots priming microbial decomposers by abundant sources of easily degradable carbon. Self-engineering by rhizosphere organisms of such localized microhabitats with distinctive aggregate structures and physical properties was shown by the synchrotron μCT techniques (Feeney et al. 2006).

Among microbial compounds, extracellular biopolymers have been described frequently as the key drivers of microaggregation (Chenu 1995). Amellal et al. (1998) characterized an EPS-exuding bacterial strain in the wheat rhizosphere (in soil adhering to roots) that strongly facilitated aggregation; this strain was not established in the bulk soil. A similar process was described for sunflower (Alami et al. 2000). The extracellular microbial biopolymers act at various scales of aggregation (including primary particles bonding into organo–mineral complexes) and also serve as glues for bacterial biofilms on mineral and litter surfaces (Costa et al. 2018).

An important role of symbiotic and saprophytic fungi in soil aggregation in the root zone is probably also related to the aspect of spatial distribution; fungal hyphae can spread into soil beyond single hotspots dominated by bacteria (Kuzyakov and Blagodatskaya 2015; Lehmann and Rillig 2015). Mycorrhizal hyphae are a source of carbon compounds with strong aggregating properties (e.g. glomalin; Rillig 2004), but they also mechanically stabilize aggregates via enmeshment (Chantigny et al. 1997). A shift in soil microbial community composition from bacterial to fungal dominance enhances carbon sequestration due to enhanced stabilization of (macro)aggregates with occluded POM (Chiu et al. 2006; Six et al. 2006). Simpson et al. (2004) suggested that, in no-tillage soils, higher topsoil carbon contents are related to fungal hyphae that stabilize macroaggregates and thereby protect microbial carbon located in microaggregates inside macroaggregates.

3.3.1.7 Root-Induced Abiotic Aggregation

Abiotic root effects on aggregate formation and stabilization are related to (i) mechanical compression, (ii) wetting-drying, and (iii) chemical changes in the rhizosphere with relevance to flocculation/dispersion (pH and solute composition/concentration).

Aggregation requires the spatial vicinity of reactive surfaces to establish the respective bonds. Dense root systems not only distribute organic-binding agents more homogeneously in the soil (Degens 1997), but they also enhance local compression around roots, which was found to modify the shape of mineral particle packing concomitantly with the release of organic-binding agents (Dorioz et al. 1993; Aravena et al. 2010). For example, dense clay packing around maize roots was found to increase bulk density up to 1.8 $g\,cm^{-3}$ based on backscattered electron scanning images (Bruand et al. 1996). In addition, fragmentation of macroaggregates by roots growing into inter-aggregate pores can modify aggregate size distribution, increasing smaller-sized fractions (Angers and Caron 1998). However, the recent X-ray computed tomography studies indicated that the physical consequences of root penetration are complex, and increased density is not necessarily a general pattern (Koebernick et al. 2019).

Other drivers of local compression/reorientation are wetting–drying processes induced by roots, which influence aggregate formation and stabilization. Soil drying by root water uptake has often been found to cause the formation of microcracks, resulting in smaller-sized aggregates with increased stability/tensile strength (e.g. Semmel et al. 1990). There are different interacting effects between organic-binding agents and root-induced soil drying in the rhizosphere. Czarnes et al. (2000) demonstrated enhanced aggregation of particles at root surfaces (rhizosheath) by mucilage upon soil drying. Albalasmeh and Ghezzehei (2014) related these observations to higher exudation under dry conditions, transport of exudates along moisture gradients towards binding sites, and a higher concentration of organic glues upon drying, contributing to stronger inter-particles bonds. Fast wetting can destabilize aggregates due to pressure from entrapped air. The root and microbial organic compounds with hydrophobic properties slow rewetting, thereby enhancing aggregate stability (Chenu et al. 2000; Czarnes et al. 2000).

Rhizosphere soil differs from bulk soil in pH, osmotic potential, and solute concentrations, depending on the balance between transport (diffusion and mass flow) vs. uptake/efflux by roots of ions (including fertilizer input), protons, and organic compounds (Vetterlein and Jahn 2004; Hinsinger et al. 2006). Rhizosphere pH affects electrical charges on clay surfaces and functional groups of organic-binding agents, whereas solute concentrations (Ca^{2+} and Mg^{2+} vs. K^+ and Na^+) influence flocculation/dispersion. For example, at lower rhizosphere pH, bacterial polysaccharides have an enhanced capacity to flocculate colloidal clays (Labille et al. 2005). Higher concentrations of divalent cations were found to result in stronger sorption of the anionic bacterial polysaccharide xanthan on clay surfaces (Dontsova and Bigham 2005).

3.3.2 Roots and Structural Porosity

Different approaches are used to classify the soil pore system (see Lal and Shukla 2004, for an overview), delineating distinct classes according to size (e.g. macro-, meso-, micro-, ultramicro-, and cryptopores; Soil Science Glossary Terms Committee 2008), function (e.g. residual, storage, drainage, and aeration), and/or origin (e.g. textural, structural, and biological). Pore-size distribution often reveals the existence of a structural flow domain that is separated from the texture-related matrix domain by a distinct minimum (bimodal soils; Kutílek 2004). The hydraulic properties of soils, an essential attribute of soil health (Stewart et al. 2018), are thus the typical result of primary, texture-related pore spaces (particle size distribution; Arya and Paris 1981), and secondary, structure-related pore spaces (aggregate size distribution; Pachepsky and Rawls 2003).

Elliott and Coleman (1988), following the hierarchy model of soil structure, postulated that aggregate size classes are mirrored in corresponding pore-size classes and defined four categories of pores: large macropores (root channels, earthworm holes, and shrinkage cracks), inter-macroaggregate, inter-microaggregate, and intra-microaggregate pores. Based on this understanding of structural porosity bound to the processes of aggregate formation and turnover, plant roots are an essential factor shaping pore-size distribution, pore geometry, and temporal pore dynamics.

3.3.2.1 Root Effects on Pore-Size Distribution

Figure 3.7 gives examples of the influence of distinct root systems of cover crops (fine vs. coarse rooting types) on the pore-size distribution, described by a log-normal distribution function (Kosugi 1996). The pore volume of specific size classes depended on the rooting type: coarse root systems (large average diameter of root axes) had a higher impact on macropores (with functional relevance for drainage/infiltration), whereas root systems dominated by fine axes increased the volume of storage pores with an equivalent diameter <30 μm.

Using measurements from a column experiment, Scholl et al. (2014) described pore evolution of an initially unplanted soil under the influence of two cover crops (tap-rooted mustard and fibrous-rooted rye) using the pore-evolution model presented by Leij et al. (2002). Growing plant roots changed the initially narrow pore-size distribution of the homogenized silt-loam soil (2-mm-sieved soil) towards a more heterogeneous pore system (i.e. with an increased standard deviation of the log-normal distribution). This form of the pore-size distribution function was related to the development of a structure-related secondary pore system (Horn et al. 1994). Another expression of root influence on the shape of soil pore-size distribution is a multimodal pattern arising from the multiple S-shaped retention curves (Holtham et al. 2007; Kodešová et al. 2006).

Even these complex bi/multimodal pore-size distribution functions might obscure some of the subtle root effects on the pore system. Holtham et al. (2007) therefore applied a pore network model (Pore-Cor) to describe their observations of root-induced changes in the soil hydraulic properties. They revealed that localized structural changes in pore geometry near roots (pore throat size and local autocorrelation) could have a substantial impact on water transport (saturated conductivity). This indicates that, in addition to changes in the volumetric frequency of pore-size classes, the use of geometric descriptors is relevant to capture fully root influences on the pore system.

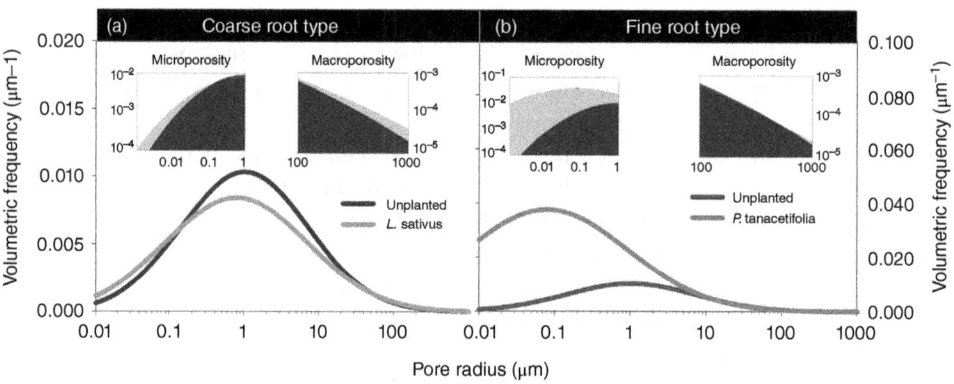

Figure 3.7 Example of the influence of root systems on the pore-size distribution in a silt loam soil. (a) Coarse rooting type (*L. sativa*) vs. unplanted control with large macropore volume; (b) Fine rooting type (*P. tanacetifolia*) vs. unplanted control with large micropore volume. Inserts show the macropore and micropore ranges on a log–log scale. Notice that due to the differences in the prevailing pore radius spans between a and b, different scale of volumetric frequency on the *y*-axis is chosen. *Source:* Modified from Bodner et al. (2014).

3.3.2.2 Root Effects on Pore Geometry

Geometric and topological parameters (e.g. tortuosity and connectivity; Vervoort and Cattle 2003) describe single-pore scale properties that can substantially influence Darcian scale water flow (e.g. via anisotropy of pore-water flow; Stewart et al. 2006). There have been substantial advances in measuring and quantifying such microscopic and localized root effects and their hydraulic consequences with the improvement in X-ray CT resolution and the use of pore-network models (Helliwell et al. 2013).

Comparison of pore characteristics of aggregates (approximately 4 mm in diameter) in the rhizosphere soil vs. bulk soil demonstrated that rhizosphere aggregates had a higher number and connectivity of large pores and also a larger wetting angle compared to the bulk soil, which contributed to different water retention characteristics between the two domains (drainage at higher matrix potential values; Whalley et al. 2005).

In conventionally tilled vs. grassland soils, distinctive pore geometry was found at the aggregate scale. Grassland soil with higher rooting density and less mechanical disturbance had significantly larger pore-throat radius and surface area, higher frequency of elongated pores, and lower tortuosity, thus facilitating gas and water transport (Peth et al. 2008).

The extent of differentiation between the rhizosphere and bulk soils, however, depends on crop species (wheat < maize or barley; Whalley et al. 2005). In addition, soil texture influences the effect of roots on pore geometry in the rhizosphere: stronger changes occur in soils with higher clay content due to stronger aggregating processes (Daly et al. 2015).

The rhizosphere hydraulic properties can contribute to differences in drought tolerance, which was demonstrated for two contrasting chickpea cultivars: the drought-tolerant genotype had a wider extension of its rhizosphere with higher porosity and more anisotropic hydraulic permeability (radial > axial direction; Rabbi et al. 2018b). The main factor suggested to explain the observed changes in soil hydraulic properties in the rhizosphere of the drought-tolerant cultivar was a more intense rhizodeposition, in line with the findings of Carminati and Vetterlein (2012) who measured the distinct hydraulic behaviour of mucilage.

Several studies investigated the geometry of root-induced biopores, comparing tap-rooted (chicory and lucerne) and fibrous-rooted species (tall fescue) (Pagenkemper et al. 2013; Uteau et al. 2013). Notably, chicory had the highest continuity of pore channels, with strong interconnectivity between pores via lateral channels. In deeper soil layers (>75 cm), root-related effects on soil porosity were strongest for lucerne among the three investigated species.

Changes in land use also shape characteristic pore geometries, which was evidenced by comparing arable land (row crops: maize–soya bean rotation) and adjacent buffer strips (grass and agroforestry; Udawatta et al. 2008). The buffer strips had higher pore-body volume and pore-throat radii, enhanced pore connectivity (coordination number), and higher length/lower tortuosity of paths among pores. These differences were interpreted to be a result of higher root growth and organic matter accumulation in the buffer strips compared with the arable soil.

In summary, the use of geometric features for pore characterization reveals that roots tend to enhance pore connectivity and elongation and reduce tortuosity of water flow pathways. Nevertheless, the extent of root effects is influenced by basic soil properties (texture) and differs among plant species/genotypes. Therefore, the expected impact on macroscopic hydraulic properties (water retention and conductivity) can be of variable importance.

3.3.2.3 Pore Surface and Rhizodeposition Effects on Soil Hydraulic Properties

In addition to pore-size distribution and geometry, water transport also depends on the surface properties of pore walls (wetting angle and surface tension). Water repellency of root-induced or root-influenced pore systems has been described by several authors (Hallett et al. 2003; Haas et al. 2018). The mucilage may change the rhizosphere soil properties between hydrophilic and hydrophobic upon wetting and drying (Carminati and Vetterlein 2012; Moradi et al. 2012). Hydrophobicity of pore walls may cause (i) preferential flow, (ii) delayed rewetting (increasing aggregate stability; cf. Section 3.3.1), and (iii) reduced water loss from root tissues into dry soil. Mao et al. (2016) characterized the chemical nature and origin (leaves/needles or roots) of hydrophobic compounds in the water-repellent sandy soils and showed they had mainly root origin (ω-hydroxy fatty acids and α,ω-dicarboxylic acids).

Rhizodeposits/mucilage can act as a hydrophilic gel, causing differences in the retention properties between rhizosphere and bulk soils (Bengough 2012; Carminati et al. 2016; Brax et al. 2017). Based on neutron radiography images, Carminati et al. (2010) demonstrated the hysteretic behaviour of the rhizosphere: upon drying rhizosphere soil stayed wetter, but the situation was reversed upon rewetting. The high concentration of rhizodeposits around the roots of transpiring plants reduced the drop in soil water content with increasing matric potential, facilitating sustained water flow towards the root surface and uptake by plants (Ghezzehei and Albalasmeh 2015).

3.3.2.4 Root-induced Temporal Pore Dynamics

The structure-related soil properties vary with time due to the environment-dependent (moisture and temperature) processes and turnover of organic-binding agents (Garbout et al. 2013b); for example, De Gryze et al. (2006) calculated turnover times to be 30 days for macroaggregates and 88 days for microaggregates in a silt loam soil. Therefore, it is evident that structural porosity changes over time, which contrasts the frequent assumption in water-flow modelling of hydraulic characteristics (i.e. pore-size distribution) being stable material properties. Figure 3.8 provides examples of the temporal evolution of pore-size distribution after mechanical soil disturbance with and without cover crops, and for the root-dependent stabilization of the pore system after stubble tillage.

Although several studies have demonstrated temporal changes in soil hydraulic properties/pore-size distribution, there is still a lack of quantification of the specific role of roots. The enhanced aggregation of rooted soil in an outflow experiment in planted vs. unplanted columns (Scholl et al. 2014) showed an increasing heterogeneity of the pore space in rooted columns, which could be modelled by the pore evolution model of Leij et al. (2002). However, the specific modes of root influence were not modelled explicitly. Recently, an attempt was made to model the effect of pore clogging by roots on the soil hydraulic properties (Ng et al. 2016). In accordance with empirical observations, a lower void ratio of rooted silty sand soil in the model (reduced proportionally to the root volume ratio) shifted the air-entry value to significantly higher matric potentials (from 1 to 3 kPa).

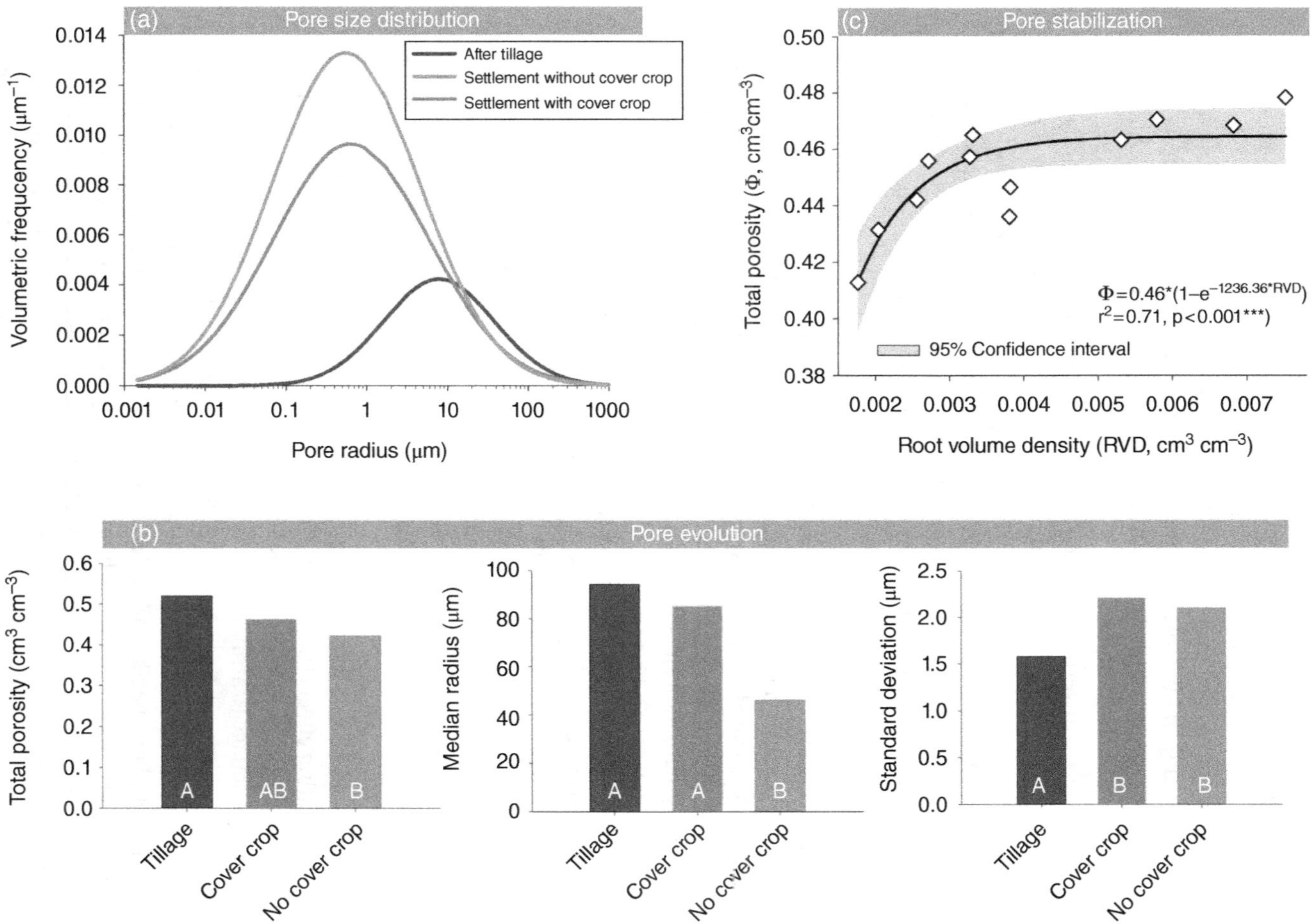

Figure 3.8 Pore evolution in relation to cover crops (mean of two cover crop treatments not differing significantly between each other: *Sinapis alba* L. and *Secale cereale* L.) and root volume density. (a) Change in pore-size distribution 18 months after tillage with and without the integration of cover crops; (b) Parameters capturing the soil pore dynamics shown in a (significant treatment differences are indicated by different uppercase letters). *Source:* Bodner et al. (2013). © 2013, Elseveir. (c) Stabilization of total porosity as a function of root volume density of cover crops. Source: Bodner et al. (2014). Licensed under CC BY.

Generally, the advance in imaging technologies with increasingly detailed root-structure data, including temporal dynamics of soil aggregates (Garbout et al. 2013b), still has not produced a comprehensive quantitative framework for the assessment of root–pore interactions. The multiple potential ways of how roots can modify the pore system and their varying extent in differently textured soils make a mechanistic model development challenging.

The relevant modes of root influence on the soil pore system can be grouped into direct effects (e.g. pore clogging and biopore formation), indirect abiotic effects (e.g. mechanical compression and drying), and biochemical effects. Direct root effects differ between living/growing and decaying roots (Gish and Jury 1983). Roots growing into pre-existing pores can clog pore channels. However, the extent of pore closure by roots has rarely been documented directly. In a column experiment, Hua et al. (2014) found that water flow was restricted in the early stages of root growth, but roots opened new pore spaces during later growth stages. Similar behaviour was predicted by a modelling study of Scanlan and Hinz (2010), with the hydraulic effects of root in-growth into existing pores depending on the ratio between root and pore radii and being most relevant near saturation. Jotisankasa and Sirirattanachat (2017) suggested a root biomass density threshold ($6.5\,kg\,m^{-3}$ in their study with a grass species on a silt loam soil) above which pore clogging would reduce soil permeability, whereas lower root densities could increase permeability due to enhanced formation of drying cracks. In contrast to the effects of in-growth of living roots, a large number of studies investigated the hydraulic impact of biopores remaining after root decay. Commonly, they found an increase in hydraulic conductivity, infiltration, and preferential flow through the stable root-induced macropores (Mitchell et al. 1995; Devitt and Smith 2002; Ghestem et al. 2011).

Abiotic effects can originate from axial and radial pressures exerted by roots during soil penetration. While potentially enlarging existing pores (e.g. in case of secondary thickening), this causes densification and reduction in porosity in the adjacent rhizosphere soil (Dexter 1987; Whalley et al. 2004; Aravena et al. 2010). Another indirect abiotic effect of roots on pore formation is due to crack formation and microfissuring as a consequence of wetting-drying (Young 1998; Whalley et al., 2005). Such drying cracks may also originate from a diameter reduction in dehydrating roots leading to air gaps at the root–soil interface (Carminati et al. 2009).

As already discussed above, roots influence hydraulic properties biochemically via (i) their general aggregating effects (Lado et al. 2004), (ii) water repellency of the root-derived compounds in pore channels, and (iii) distinct hydraulic properties of root-derived organic input, particularly mucilage (Carminati and Vetterlein 2012).

In summary, the most consistent effect of densely rooted soil on the Darcian-scale hydraulic properties (particularly near-saturated conductivity and air-filled porosity), reported for different plant species and various soil types, appears to be related to a higher volume of elongated (bio)macropores with low tortuosity that are left by decaying roots. A range of localized root effects on the rhizosphere hydraulic properties (e.g. densification, anisotropy, and mucilage) are increasingly recognized as the relevant factors for plant water supply in dry soils. Still, their impact on other Darcian scale water flow processes (e.g. infiltration, retention characteristics, and drainage) is more soil type/species–dependent and thus more complex to manage in a cropping situation.

3.4 Plant Roots for Soil-Structure Management

We first introduce a general approach to crop management as a rationale for the importance we attribute to roots. Then, we provide examples of how root systems can contribute to optimum soil (structural) conditions and how to integrate them into the standard agronomic measures.

3.4.1 Crop Management Analogue to Physiological Breeding

Crop production takes place under conditions of environmental uncertainty in the available growth factors. This uncertainty/variability is exacerbated by climate change (Mearns et al. 1997). The higher yield variability (Iizumi and Ramankutty 2016) also implies an increasing sustainability challenge. For example, fertilizer recommendations are based on the expected yield (Lory and Scharf 2003). Increasing rainfall uncertainty thus leads to a higher risk of inappropriate fertilizer dose accompanied potentially by higher losses (e.g. leaching and gaseous emissions). Sustainable crop management largely depends on optimizing amounts and timing of inputs/operations in the context of environmental uncertainty. Weather is an intrinsically unforeseeable boundary condition for farmers' decision-making (in spite of improved forecast models). A decrease in uncertainty, therefore, depends mainly on managing soil conditions that can either buffer or amplify climatic signals (Folberth et al. 2016).

For advances in plant breeding, Reynolds et al. (2009) pointed to the importance of crop physiology (physiological breeding) to identify yield-relevant traits for specific target environments and thereby increase rates of genetic gains through a trait-based crossing. We argue that SI of crop production requires a similar (eco)physiological basis: agronomic ecophysiology identifies the environmental conditions that drive the expression of yield-relevant traits and unravels how these conditions can be modified in a targeted way. Figure 3.9 gives an overview of (i) the environmental drivers controlling shoot and root traits that determine crop growth/yield potential and (ii) standard management measures that are applied (mostly based on farmer experience only) to control the relevant trait-environment interactions.

At the shoot level, crop management aims to optimize the insertion of the crop into a given soil-climate situation. The practices (e.g. seeding time and density, crop rotation, cultivar selection, and crop protection) are typically adapted to the site-specific average atmospheric settings (climate) with relatively few options to buffer uncertainty (see e.g. Bodner et al. 2015 for a detailed discussion on water-limited cropping systems). In contrast, at the root level, crop management modifies the environment (root zone) to optimize the plant growth factors/conditions. Ecophysiological understanding of processes at the soil–root interface can minimize uncertainties in the translation of the soil management decisions (e.g. fertilization and tillage) into desired plant responses. Based on Körner (2015), who provided evidence that the source (photosynthetic) activity depends on the sink strength (with turgor pressure-driven cell elongation as a primary signal), the root zone management turns into a fundamental agronomic lever to control shoot growth and yield potential.

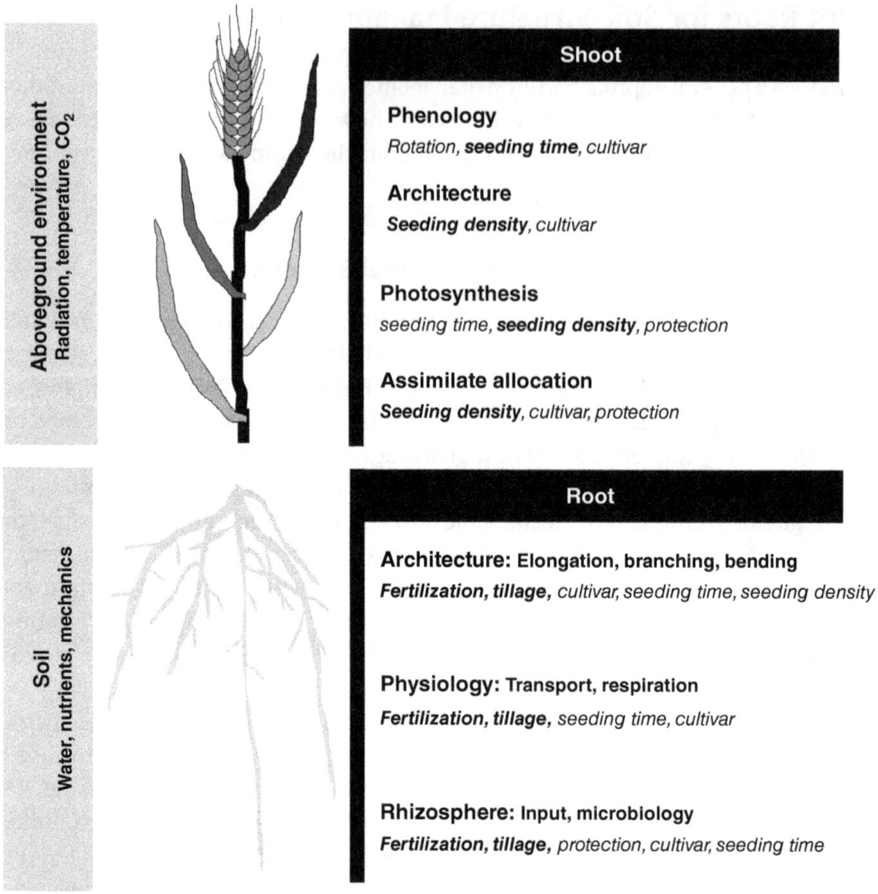

Figure 3.9 Above-ground and soil-related factors driving shoot and root growth via different functional trait groups and the standard agronomic measures to optimize crop performance in a given environment (bold: agronomic measures with the strongest influence on a given group of traits).

3.4.2 Roots for Soil Priming ('Bio-tillage')

The exploitation of root systems to create optimum soil structural conditions for crop growth is an obvious target considering the fundamental role of root systems in soil aggregation and porosity. This strategy is known as 'soil priming', with plant species defined as 'primer plants' (Yunusa and Newton 2003), or biological drilling (Cresswell and Kirkegaard 1995).

The disruptive/loosening mechanical forces from tillage implements are much higher than the forces exerted by plant roots (e.g. ≈20 N for ploughing vs. ≈0.1 N for a 1-mm-diameter root; Kolb et al., 2017; Ucgul et al. 2017). However, there is no tillage implement to create elongated and highly connective pores similar to biological agents (e.g. roots and earthworms; Pagliai et al. 2004). There are cases where 'biological tillage' via plant roots changes soil structure in a way that cannot be attained by the mere utilization of mechanical implements: (i) subsoil resource access, (ii) soil compaction, (iii) no-tillage agriculture, and (iv) soil erosion control.

3.4.2.1 Access to Subsoil Resources

Biological macropores constitute only a small proportion (<1%) of total soil porosity (Gaiser et al. 2013). However, 85–100% of wheat roots in a dense soil were clustered in cracks or biopores when growing into the subsoil (deeper than 0.6 m).

Roots growing into subsoil via large biopores could have insufficient soil contact (Stirzaker et al. 1996). However, for wheat roots, it was found that, when growing in macropores of biological origin, they interacted with surrounding soil by root hairs; in contrast, when grown in cracks, they did not form root hairs (White and Kirkegaard 2010). This was likely related to the high microbial activity at the biopore walls providing roots with readily available nutrient sources (Figure 3.10, see also Banfield et al. 2017).

Crops differ in the way their roots established contact with subsoil when growing in biopores: barley roots were observed to attach onto biopore walls by the strong formation

Figure 3.10 Wheat roots growing in a biopore filled with nutrient-rich earthworm cast. The bunched root axes exploit the nutrient hotspot by producing an intensive net of root hairs. *Source:* Gernot Bodner, Institute of Agronomy, University of Natural Resources and Life Sciences (BOKU), Vienna, Austria.

of root hairs, whereas oilseed rape used lateral roots to penetrate into the subsoil (Athmann et al. 2013). Interestingly, plant roots appear to be capable of sensing and preferentially growing towards biopores in compacted soil in which they face high mechanical impedance (Colombi et al. 2017a).

Primer plant species create different types of biopore systems, which provides some opportunity for targeted management. Nuttall et al. (2008) suggested that primer plants with highly branched root systems, e.g. *Lotus corniculatus* L. or *Hedysarum coronarium* L., leave a network of fine biopores that allows better subsoil water extraction by a subsequent wheat crop compared with the less branched and coarser biopore network after lucerne. Comparison of biopore density among three fodder crops (lucerne, chicory, and tall fescue) showed that the tap-rooted species were more effective than the fibrous-rooted grass (Han et al. 2015a). Root length densities of subsequent crops (winter barley and winter oilseed rape) in the subsoil were higher after tap-rooted chicory than the fibrous-rooted pre-crops (tall fescue or oat; Perkons et al. 2014). Particularly, dense soil layers (Luvisol with Bt horizon in the studies of Perkons et al. 2014 and Han et al. 2015b) require tap-rooted species to facilitate penetration into the subsoil.

3.4.2.2 Roots to Alleviate Soil Compaction

Soil compaction is an increasing sustainability problem on arable land, caused by excessive use of heavy machinery on structurally unstable soils (e.g. moist, with low organic matter) (Hamza and Anderson 2005). Compaction is commonly defined as the soil physical thresholds that impede root growth/activity (penetration resistance >2 MPa and air-filled porosity $<0.10\,cm\,cm^{-3}$); for example, see van Lier and Gubiani (2015) for discussion on these thresholds in relation to the least limiting water range concept of da Silva and Kay (1997).

There has been extensive research on the capacity of roots to grow into dense soil layers (see reviews by Bengough and Mullins 1990; Unger and Kaspar 1994; Bengough et al. 2005). The measured values of maximum axial roots pressure range from ≈0.24 to ≈0.58 MPa (Misra et al. 1986; Clark and Barraclough 1999). Root axes with larger diameter can better penetrate dense soil layers due to higher bending stiffness under mechanical stress (Clark et al. 2008).

Cover crops, such as radish (*Raphanus sativus* L.), have a high capacity to penetrate compacted soil, creating biopores for root growth of subsequent crops, e.g. soya bean (Williams and Weil 2004). Comparing different cover crops, Chen and Weil (2010) found radish to be most effective in growing through the dense soil layers. Comparison of cover crops for tropical conditions (*Crotalaria juncea, Crotalaria spectabilis, Helianthus annuus, Pennisetum americanum*, and *Sorghum bicolor* grown on a Ferralsol in Brazil) showed higher sensitivity of the C4 grass species to soil compaction, but they maintained higher root length density below the compacted layer, compared with the dicots (Rosolem et al. 2002). For primary crops, differences in their capacity to access subsoil through the compacted soil layers were found not only among species but also among genotypes (e.g. for soya bean and maize, Bushamuka and Zobel 1998; for rice, Clark et al. 2008).

Opportunities for biological alleviation of soil compaction by means of primer plant roots strongly depend on a degree of compaction (Figure 3.11). Short-growth cover crops are an important building block of sustainable agriculture to conserve good soil structure

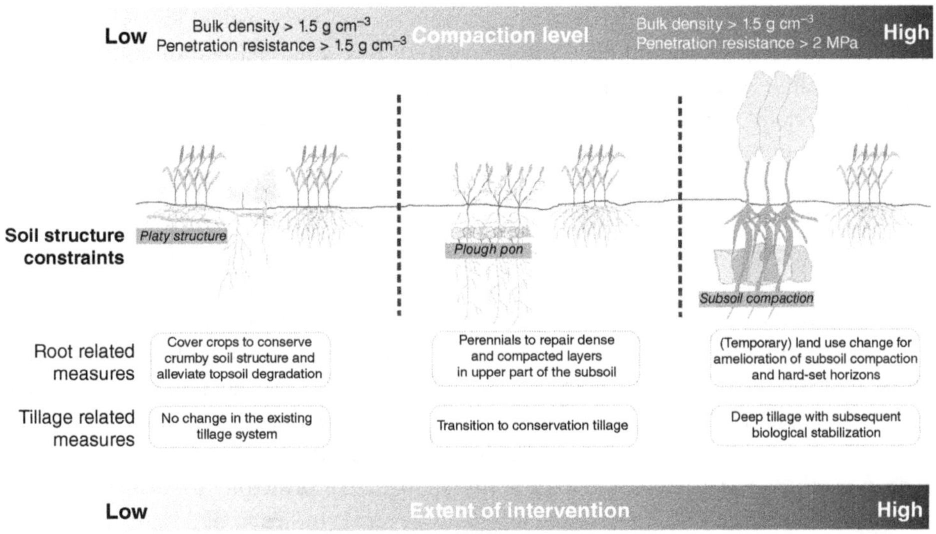

Figure 3.11 Root- and tillage-related measures to mitigate various degrees of soil compaction. With increasing compaction, the extent of intervention required becomes more extreme. Structure conservation and alleviation of slight topsoil degradation are achieved by cover crops integrated into a rotation and tillage system. An intermediate degree of structure degradation (e.g. plough pan) can be treated with deep-rooted perennials (e.g. lucerne, white clover, and ryegrass mixtures) and subsequent transition to conservation tillage. High compaction (subsoil) requires soil amelioration via (cautiously applied) deep tillage with biological stabilization and/or land use changes to woody species for sustained soil improvement (e.g. timber plantations for biomass production).

and eventually improve relatively minor degradation (platy structure) in the topsoil, with tap-rooted species being more effective than fibrous root systems. Perennial (tap-rooted) forage crops, cultivated for at least two seasons, can improve structural properties of dense soil layers (e.g. plough pan and clay lessivation/accumulation) and ameliorate intermediate compaction in the upper parts of subsoil by biopore formation (Lesturgez et al. 2004; McCallum et al. 2004; Han et al. 2015b). Priming of strongly compacted subsoil or hard-set horizons at higher depth requires intensive intervention using woody species (Yunusa et al. 2002; Bartens et al. 2008) to improve penetrability of these layers for subsequent crops, or may require a combination of mechanical and biological measures.

3.4.2.3 Roots and Tillage Systems

In the cropping systems with regular tillage for soil loosening, plant roots are relevant in stabilizing post-tillage structures against re-settling (cf. Section 3.3.2; Figure 3.8). Here, we discuss further the two special cases: (i) mechanical subsoil loosening, and (ii) no-tillage systems.

Mechanical loosening (e.g. subsoiling and deep tillage) of subsoil to manage induced compaction (Hamza and Anderson 2005) or naturally hard-set horizons (Daniells 2012) is an intensive tillage intervention to facilitate root growth. The two major conclusions from a recent meta-analysis concerning the deep tillage effects on yield were (i) deep tillage on soils with root-restricting layers improved yield, particularly in dry environments, and

(ii) on silt-dominated (loess) soils, deep tillage could lead to serious yield losses (Schneider et al. 2017a). This confirms that deep tillage is a high-risk soil amelioration measure with potentially positive effects that may turn into detrimental damages. Even when properly practiced (especially timing with regard to soil moisture being lower than the plastic limit; Keller and Dexter 2012), the loosening effects of deep tillage in soils with low clay content are transient, and re-compaction to the initial state may occur in 2–3 years (Van Ouwerkerk and Raats 1986; Carter et al. 1996).

Evidence from reclaimed (brown coal mining) sandy soil revealed that the establishment of deep-rooting lucerne was more successful than subsoiling to restore stably the structure-related hydraulic properties (saturated conductivity; Krümmelbein and Raab 2012). In a soil rehabilitation trial on degraded loamy soils in Bolivia, Barber and Navarro (1994) combined subsoiling and different cover crops. Subsoiling had no effect on cash crop yield and did not contribute to restoring soil fertility, whereas two *Panicum maximum* cover crop varieties were most effective in improving yield. Although the combination of subsoiling and deep-rooting perennials is a commonly recognized principle in extension (Van Ouwerkerk and Raats 1986), there are hardly any (recent) studies to highlight which crop species/root traits are best suited to stabilize the mechanical loosening effect. Recognizing potentially positive impacts on some soils, as evidenced in the meta-analysis by Schneider et al. (2017a), additional specific studies on the (root) biological requirements for sustainable subsoiling practices are needed.

No-tillage systems are the other extreme to intensive deep tillage, with crop production practiced without mechanical soil disturbance for stubble removal and soil loosening. One of the main reasons for farmers to adopt no tillage is the optimization of workload between harvest and seeding of subsequent crops, which is particularly critical for large-sized farms (Lahmar 2010). The effect of no tillage on yield is controversial because meta-analyses indicate a slight decrease in yield (5.1–8.5%), particularly in humid regions (Van den Putte et al. 2010; Pittelkow et al. 2015). In contrast, in water-limited environments, enhanced deepwater storage via no tillage has been found to stabilize and improve yields (Lampurlanés et al. 2016). Among other reasons (e.g. weeds management and crop establishment), the two major soil-related causes appear to be related to compacted topsoil: reduced volume of functionally relevant macropores for soil aeration and delayed warming in spring (Rasmussen 1999; Tebrügge and Düring 1999). Again, this is particularly evident in soils with low clay content and low soil organic matter (Figure 3.12).

In no-tillage systems, macropore volume depends largely on biopore formation. However, in some cases, even in long-term no tillage, no build-up of large biopores was found in no till compared to tilled soil (Wuest 2001, using a photographic method). Most soils under no tillage tend to develop compact structure due to wetting–drying-driven coalescence that, in spite of the existence of biopores, can constrain root growth (Cockroft and Olsson 2000). A high (>40 $g\,kg^{-1}$) C_{org} is required to prevent such a natural (abiotic) tendency for hard setting. The transition from mechanical-loosening-based cropping systems towards no tillage of arable soil thus requires a particular emphasis on biological properties contributing to structure formation (Jin et al. 2017). Beyond priming effects via root/earthworm-induced biopores for subsoil access, the comprehensive management efforts to increase organic carbon content via crop rotation (including cover crops) and (organic) fertilization are

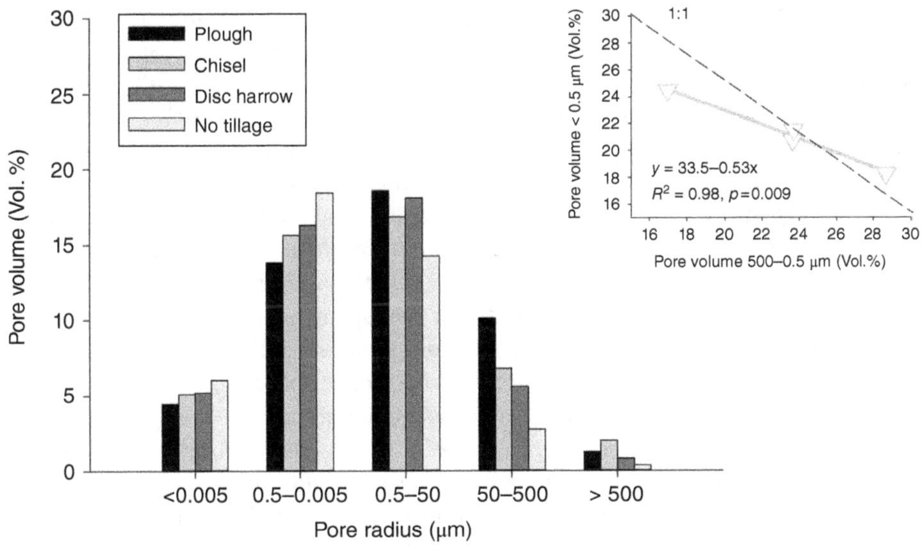

Figure 3.12 Shift in pore sizes towards smaller-sized classes (pore classification according to Greenland (1977) with reduced-tillage intensity in a silt loam soil [sand 23.8%, silt 54.6%, and clay 21.6%; soil organic matter 25 g kg^{-1}]). The insert shows that the no-till treatment resulted in a significant shift in pore sizes: from larger (500–0.5 μm) towards smaller pores (<0.5 μm). Measurements (evaporation method) and curve fitting of the soil water retention curve followed Schindler et al. (2010). *Source:* From Weninger et al. (2019). © 2019, CSIRO Publishing.

required to ensure a crumby/loose topsoil structure necessary for uniform crop establishment and vigorous (early) crop growth (Kadžienė et al. 2011; Abdollahi and Munkholm 2014; Abdollahi et al. 2014).

3.4.2.4 Roots for Soil Erosion Control

Water erosion is the most wide spread (56%) cause of human-induced soil degradation, estimated to affect 1100 Mha of arable land worldwide (Oldeman 1992). Management measures to mitigate soil erosion focus on three terms in the RUSLE (revised universal soil loss equation) framework (Renard et al. 1991): (i) Soil surface protection by mulch/living canopy coverage against the impact of erosive rainfalls (cover-management C factor), (ii) protective measures changing the slope structure (P factor), and (iii) increased rainfall infiltration into the soil (lower soil erodability K factor).

Optimizing mulch management (i.e. erosion mitigation via C-factor) is the most common approach to soil conservation and is highly effective, with an increase in percentage cover exponentially reducing runoff/soil detachment (Zuazo and Pleguezuelo 2008). Although other management factors, such as contour farming, terracing, strip cropping, and hedgerows (P-factor effects), can be highly effective, they imply strong intervention into field structures, which limits their adoption by farmers (Panagos et al. 2015). Plant root effects are of relevance concerning the K-factor. Firstly, root-induced improvement of aggregate stability improves the infiltration of (otherwise erosive) rainfall. Water erosion decreases with increasing root density and follows a similar exponential reduction as observed for surface cover, with roots being particularly effective against rill and gully erosion, whereas surface cover was valuable in controlling interrill and splash erosion (Gyssels et al. 2005). An erosion-reducing potential is higher for species with fine fibrous root architectures compared to tap-rooted species (Baets et al. 2007). Densely rooted topsoil has higher shear strength to better resist soil detachment compared to soils without root reinforcement (Mamo and Bubenzer 2001) or significantly improved aggregate stability (Zhou and Shangguan 2007). The high efficiency of fibrous- and fine-rooted species points to their role (in combination with mycorrhizae) as a structure-stabilizing agent at the scale of macroaggregates (otherwise having low stability) via enmeshment (Mardhiah et al. 2016). It can be hypothesized that fertilizer management (placement of fertilizers) provides a means to manipulate topsoil root proliferation and thereby enhance soil stabilization against erosion via improved root enmeshment (Ola et al. 2015).

Roots act via biopore channels. This effect is higher for coarse-rooted species. For example, a modelling study on cover crop root effects demonstrated that runoff/erosion mitigation was most effective with legumes (*Vicia sativa* and *Lathyrus sativus*) forming enlarged pore channels that enhanced rainfall infiltration (Yu et al. 2016). This root effect is most effective in case of intensive rainfalls exceeding soil infiltrability and leading to ponding/saturated conditions where these large, otherwise air-filled, pore channels contribute to large water inflow (by the fourth power of their radius; Nimmo 2004). Therefore, targeting the hydraulic function of plant roots for runoff/erosion control requires tap-rooted species that create large and elongated pores that are more effective in runoff mitigation than those formed by species with a fibrous root system (Mitchell et al. 1995).

Effective erosion control by roots is a combination of mechanical and hydraulic effects. Despite uncertainties regarding accurate quantitative prediction, the mechanical improvement of soil shear strength and aggregate stability by dense fibrous root systems (most commonly found in grasses) appear to play a dominant role in root-mediated erosion control, with the highest impact in non-sandy soils and soils in which detachment is dominated by rill erosion (Vannoppen et al. 2015).

3.4.3 Crop Rotation and Cultivar Selection based on Root Properties

Farmers make decisions on which species to integrate into their crop rotations predominantly based on market considerations and phytosanitary aspects. Nevertheless, a preceding crop has an essential impact on the yield of a succeeding crop (e.g. for Brassicas, Smith et al. 2004; for legumes, Preissel et al. 2015). Influences from the preceding on the subsequent crops are manifold (e.g. water, nutrients, weeds, pests, and diseases), also including the soil structural state left after harvest – a result of the biological (including roots) and management (tillage and harvesting techniques) impacts. It is suggested that root-related aggregate/pore processes should be included in decisions on crop rotation and cultivar selection to facilitate targeted management of soil structure (Ball et al. 2005). Here, particular attention is given to cover crops as a flexible element in crop rotations and because they are integrated easily into a range of cropping systems (Blanco-Canqui et al. 2015).

The structural impact of roots from any preceding crop depends primarily on input quantity (e.g. biomass and density) and differs strongly among common crop species (for a comprehensive overview see Kutschera et al. 2009) and management conditions (e.g. fertilization rates). Strong environmental interactions, however, make any general statements shaky: cereals and pulses tend to have a more even root distribution throughout the soil profile, whereas oilseed crop root systems are more surface-concentrated (Fan et al. 2016) and produce lower root biomass input (at least in Nordic cropping systems; Pietola and Alakukku 2005).

The root system type matters in the soil structure management. Several studies point to the positive preceding crop effect of tap-rooted species via macropore-related soil properties. The relevant mechanisms have been reported above, e.g. enhanced subsoil access by primary crops after soil priming by cover crops (e.g. forage radish and chicory; Chen and Weil 2011; Han et al. 2015a) and higher saturated hydraulic conductivity and rainfall infiltration under tap-rooted primary crops (e.g. lupin, lucerne, and faba bean; Chan and Heenan 1996; Blair et al. 2006; Masri and Ryan 2006). There is, however, a lack of comparative studies with regard to the species (root system) effects on soil structure for better-targeted decision-making. Figure 3.13 provides only a qualitative estimate of the aggregating and hydraulic effects of different species grown as primary and/or cover crops. This estimate of species usability in the soil structure management is based on the interspecies comparisons available (reported above; also the study of Bodner et al. (2014) on different cover crop species). Furthermore, the grouping is based on the assumption that (i) tap-root systems increase hydraulic effects mainly due to enhanced macroporosity, (ii) dense fibrous root systems improve aggregation due to better soil enmeshment, and (iii) perennials exert stronger effects than annual species.

Beyond the single species effects, it has been reported that structural properties also improve with a higher diversity of crop rotations as well as with species mixtures (Munkholm et al. 2013; Ehrman and Ritz 2014; Nunes et al. 2018). This points to a complementarity advantage of various species growing in a field (either in sequence or

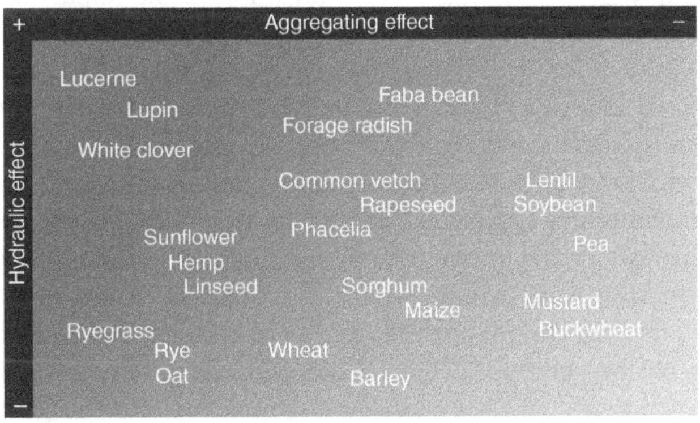

Figure 3.13 A qualitative estimate of species effects on the two main aspects of soil structure: aggregate formation/stabilization and hydraulics (modification of pore volume, size, and geometry). The scheme provides an estimate of species usability in targeted soil structure management.

together), which could be mediated by higher microbial activity/diversity in the multi-species cropping systems (Tiemann et al. 2015; D'Acunto et al. 2018). Diverse species, such as grass–clover mixtures (Van Eekeren et al. 2009), might also better combine different mechanisms governing aggregation, e.g. exudate-driven aggregate formation with enmeshment-mediated stabilization. Recent studies demonstrated improved soil physical properties (aggregate stability, hydraulic conductivity, and soil strength) in highly diverse grassland ecosystems and the opportunity to combine distinct root traits to achieve the specific soil physical management targets (Gould et al. 2016). In crop production, species mixtures in cover cropping are the promising candidates to increase diversity and enhance soil biological functions (Chavarría et al. 2016). In addition, intercropping (see also Chapters 11 and 12) takes advantage of complementary root-zone effects (mainly regarding nutrients; e.g. maize-legume intercropping) and has been found to improve soil structure (Garland et al. 2017). However, below-ground interactions are still poorly understood in root ecology because of complex feedbacks among species, genotypes, and environment. Thus, research has to advance substantially to allow management recommendations beyond mere speculative assumptions and hypotheses.

With species-specific structural effects still insufficiently documented/understood, it is not surprising that there has been hardly any systematic research on genotypic differences in root–soil structure interactions. Generally, agronomic measures to improve soil structure might be more effective than breeding (e.g. for subsoil resource acquisition: preceding crop/fallow management vs. genotype differences in rooting depth, Thorup-Kristensen and Kirkegaard 2016). Breeding for genotype-mediated modification of soil structure is complicated: a given (root) trait to be selected for would not be a sole breeding target per se, but just one factor in the interplay among soil physical and microbial properties that together shape the structural state of the soil. However, with an improved understanding of key root traits involved in the aggregation/structural porosity and the advances in root phenotyping, the systematic investigation of existing variability among genotypes might reveal avenues for improvement over existing germplasm (see e.g. Kell 2011 regarding root breeding for carbon sequestration).

There are some emerging examples of relevant genotypic variation in the impact on soil structure. Colombi et al. (2017b) discovered that variability in root tip geometry among 14 wheat genotypes explained differences in penetrating compacted soil. George et al. (2014) found quantitative trait locus (QTL) for rhizosheath weight in barley that were associated with enhanced phosphorus uptake under drought. In contrast, Lehmann and Rillig (2013) did not find significant effects of barley breeding lines (differing in fine root length) on water-stable aggregates and aggregate-size distribution. Similarly, Fang et al. (2018), assessing the soil physical impact of two contrasting rice genotypes, did not find significant genotype effects. However, using synchrotron imaging, Koebernick et al. (2018) revealed differences in rhizosphere porosity and pore connectivity in soil with two extremely contrasting barley genotypes (hairless mutant vs. genotype with root hairs).

These results point to a need for further investigations with larger sets of genotypes to obtain a better estimate of the significance of genotypic effects on soil physical properties and their potential for further exploitation in managing soil structure in SI of cropping systems.

3.5 Conclusions

Following Cockroft and Olsson (2000), we consider soil structure as a fundamental management frontier in SI of cropping systems. Managing soil structure requires a complex system understanding and is beyond simple input-driven solutions available in other areas of agronomic intensification (e.g. plant nutrition). This is evident from the variable results obtained by different tillage systems aimed at manipulating soil structure to optimize the physical conditions for crop growth. A mere variation of mechanical soil loosening intensity is insufficient to manage a biologically driven soil property. With plant roots being a key component in the soil structure formation (aggregates and pores) and stabilization, managing soil structure essentially means managing the biophysical and biochemical aspects of root–soil interactions.

Soil organic matter research has substantially advanced our knowledge on root processes involved in aggregation at different hierarchical levels. In the context of novel imaging technologies applied to plant–soil systems, new evidence has been collected regarding root-induced pore formation and rhizosphere hydraulic modifications. Despite these fundamental insights, however, their use in underpinning agronomic approaches for root-mediated soil structure management is slow. Some progress has been made in the identification of tap-rooted species with high efficiency in root–biopore creation that improve access to subsoil resources. In addition, quantification of soil erosion mitigation via species with dense fibrous root systems is now available. For the complex feedback between plant roots and soil structure, advances in (pore-scale) modelling would be decidedly necessary. This would improve our capacity to predict explicitly the expected changes in aggregation and pore structure from an agricultural intervention, such as crop rotation change or modification of the tillage regime.

Crop production traditionally has aimed to achieve a crumby and loose soil structure to facilitate root growth and microbial nutrient cycling. Roots are the key levers in a dynamic below-ground interplay among climate, soil texture, organic matter, soil (micro)organisms, and plants to achieve this optimum state of the soil. Integration of fundamental soil and root sciences into an agronomic framework of management measures can advance the design of crop rotations, cultivars, tillage systems, fertilization, and plant protection methods to manage soil structure as the biophysical side of soil fertility in SI.

References

Abdollahi, L. and Munkholm, L.J. (2014). Tillage system and cover crop effects on soil quality: I. Chemical, mechanical, and biological properties. *Soil Science Society of America Journal* 78: 262–270.

Abdollahi, L., Munkholm, L.J., and Garbout, A. (2014). Tillage system and cover crop effects on soil quality: II. Pore characteristics. *Soil Science Society of America Journal* 78: 271–279.

Abiven, S., Menasseri, S., and Chenu, C. (2009). The effects of organic inputs over time on soil aggregate stability–a literature analysis. *Soil Biology and Biochemistry* 41: 1–12.

Alami, Y., Achouak, W., Marol, C. et al. (2000). Rhizosphere soil aggregation and plant growth promotion of sunflowers by an exopolysaccharide-producing *Rhizobium* sp. strain isolated from sunflower roots. *Applied and Environmental Microbiology* 66: 3393–3398.

Albalasmeh, A.A. and Ghezzehei, T.A. (2014). Interplay between soil drying and root exudation in rhizosheath development. *Plant and Soil* 374: 739–751.

Amellal, N., Burtin, G., Bartoli, F. et al. (1998). Colonization of wheat roots by an exopolysaccharide-producing pantoea agglomerans strain and its effect on rhizosphere soil aggregation. *Applied and Environmental Microbiology* 64: 3740–3747.

Amezketa, E. (1999). Soil aggregate stability: a review. *Journal of Sustainable Agriculture* 14: 83–151.

An, S., Mentler, A., Mayer, H. et al. (2010). Soil aggregation, aggregate stability, organic carbon and nitrogen in different soil aggregate fractions under forest and shrub vegetation on the Loess Plateau, China. *Catena* 81: 226–233.

Andrade, G., Mihara, K.L., Linderman, R.G. et al. (1998). Soil aggregation status and rhizobacteria in the mycorrhizosphere. *Plant and Soil* 202: 89–96.

Angers, D.A. and Caron, J. (1998). Plant-induced changes in soil structure: processes and feedbacks. *Biogeochemistry* 42: 55–72.

Aravena, J.E., Berli, M., Ghezzehei, T.A. et al. (2010). Effects of root-induced compaction on rhizosphere hydraulic properties-X-ray microtomography imaging and numerical simulations. *Environmental Science and Technology* 45: 425–431.

Arshad, M.A. and Martin, S. (2002). Identifying critical limits for soil quality indicators in agro-ecosystems. *Agriculture, Ecosystems and Environment* 88: 153–160.

Arya, L.M. and Paris, J.F. (1981). A physicoempirical model to predict the soil moisture characteristic from particle-size distribution and bulk density data. *Soil Science Society of America Journal* 45: 1023–1030.

Athmann, M., Kautz, T., Pude, R. et al. (2013). Root growth in biopores – evaluation with in situ endoscopy. *Plant and Soil* 371: 179–190.

Baets, S.D., Poesen, J., Knapen, A. et al. (2007). Impact of root architecture on the erosion-reducing potential of roots during concentrated flow. *Earth Surface Processes and Landforms: The Journal of the British Geomorphological Research Group* 32: 1323–1345.

Ball, B.C., Bingham, I., Rees, R.M. et al. (2005). The role of crop rotations in determining soil structure and crop growth conditions. *Canadian Journal of Soil Science* 85: 557–577.

Ball, B.C., Batey, T., and Munkholm, L.J. (2007). Field assessment of soil structural quality–a development of the Peerlkamp test. *Soil Use and Management* 23: 329–337.

Banfield, C.C., Dippold, M.A., Pausch, J. et al. (2017). Biopore history determines the microbial community composition in subsoil hotspots. *Biology and Fertility of Soils* 53: 573–588.

Barber, R.G. and Navarro, F. (1994). The rehabilitation of degraded soils in Eastern Bolivia by subsoiling and the incorporation of cover crops. *Land Degradation and Development* 5: 247–259.

Bartelheimer, M. and Poschlod, P. (2016). Functional characterizations of E llenberg indicator values–a review on ecophysiological determinants. *Functional Ecology* 30: 506–516.

Bartens, J., Day, S.D., Harris, J.R. et al. (2008). Can urban tree roots improve infiltration through compacted subsoils for stormwater management? *Journal of Environmental Quality* 37: 2048–2057.

Bengough, A.G. (2012). Water dynamics of the root zone: rhizosphere biophysics and its control on soil hydrology. *Vadose Zone Journal* 11 (2): vzj2011.0111.

Bengough, A.G. and Mullins, C.E. (1990). Mechanical impedance to root growth: a review of experimental techniques and root growth responses. *Journal of Soil Science* 41: 341–358.

Bengough, A.G., Bransby, M.F., Hans, J. et al. (2005). Root responses to soil physical conditions; growth dynamics from field to cell. *Journal of Experimental Botany* 57: 437–447.

Blair, N., Faulkner, R.D., Till, A.R. et al. (2006). Long–term management impacts on soil C, N and physical fertility: Part III: tamworth crop rotation experiment. *Soil and Tillage Research* 91: 48–56.

Blanco-Canqui, H., Shaver, T.M., Lindquist, J.L. et al. (2015). Cover crops and ecosystem services: insights from studies in temperate soils. *Agronomy Journal* 107: 2449–2474.

Blankinship, J.C., Fonte, S.J., Six, J. et al. (2016). Plant versus microbial controls on soil aggregate stability in a seasonally dry ecosystem. *Geoderma* 272: 39–50.

Blum, W.E.H. (2005). Functions of soil for society and the environment. *Reviews in Environmental Science and Bio/Technology* 4: 75–79.

Bodner, G., Scholl, P., Loiskandl, W. et al. (2013). Environmental and management influences on temporal variability of near saturated soil hydraulic properties. *Geoderma* 204: 120–129.

Bodner, G., Leitner, D., and Kaul, H.P. (2014). Coarse and fine root plants affect pore size distributions differently. *Plant and Soil* 380: 133–151.

Bodner, G., Nakhforoosh, A., and Kaul, H.P. (2015). Management of crop water under drought: a review. *Agronomy for Sustainable Development* 35: 401–442.

Bohner, A., Gehmacher, P., Bodner, G. et al. (2018). Bodenverdichtung im dauergrünland und ihre auswirkung auf die grünland vegetation. *Die Bodenkultur: Journal of Land Management, Food and Environment* 68: 113–129.

Boydaş, M.G. and Turgut, N. (2007). Effect of tillage implements and operating speeds on soil physical properties and wheat emergence. *Turkish Journal of Agriculture and Forestry* 31: 399–412.

Brax, M., Buchmann, C., and Schaumann, G.E. (2017). Biohydrogel induced soil–water interactions: how to untangle the gel effect? A review. *Journal of Plant Nutrition and Soil Science* 180: 121141.

Brown, L.K., George, T.S., Neugebauer, K. et al. (2017). The rhizosheath – a potential trait for future agricultural sustainability occurs in orders throughout the angiosperms. *Plant and Soil* 418: 115–128.

Bruand, A., Cousin, I., Nicoullaud, B. et al. (1996). Backscattered electron scanning images of soil porosity for analyzing soil compaction around roots. *Soil Science Society of America Journal* 60: 895–901.

Bullock, P., Newman, A.C.D., and Thomasson, A.J. (1985). Porosity aspects of the regeneration of soil structure after compaction. *Soil and Tillage Research* 5: 325–341.

Bünemann, E.K., Bongiorno, G., Bai, Z. et al. (2018). Soil quality–a critical review. *Soil Biology and Biochemistry* 120: 105–125.

Bushamuka, V.N. and Zobel, R.W. (1998). Differential genotypic and root type penetration of compacted soil layers. *Crop Science* 38: 776–781.

Cambardella, C.A. and Elliott, E.T. (1992). Particulate soil organic-matter changes across a grassland cultivation sequence. *Soil Science Society of America Journal* 56: 777–783.

Canarini, A., Wanek, W., Merchant, A. et al. (2019). Root exudation of primary metabolites: mechanisms and their roles in plant responses to environmental stimuli. *Frontiers in Plant Science* 10: 157.

Carminati, A. and Vetterlein, D. (2012). Plasticity of rhizosphere hydraulic properties as a key for efficient utilization of scarce resources. *Annals of Botany* 112: 277–290.

Carminati, A., Vetterlein, D., Weller, U. et al. (2009). When roots lose contact. *Vadose Zone Journal* 8: 805–809.

Carminati, A., Moradi, A.B., Vetterlein, D. et al. (2010). Dynamics of soil water content in the rhizosphere. *Plant and Soil* 332: 163–176.

Carminati, A., Kroener, E., Ahmed, M.A. et al. (2016). Water for carbon, carbon for water. *Vadose Zone Journal* 15 (2): vzj2015.04.0060.

Carter, M.R., Holmstrom, D.A., Cochrane, L.M. et al. (1996). Persistence of deep loosening of naturally compacted subsoils in Nova Scotia. *Canadian Journal of Soil Science* 76: 541–547.

Castellano, M.J., Mueller, K.E., Olk, D.C. et al. (2015). Integrating plant litter quality, soil organic matter stabilization, and the carbon saturation concept. *Global Change Biology* 21: 3200–3209.

Chan, K.Y. and Heenan, D.P. (1996). The influence of crop rotation on soil structure and soil physical properties under conventional tillage. *Soil and Tillage Research* 37: 113–125.

Chantigny, M.H., Angers, D.A., Prévost, D. et al. (1997). Soil aggregation and fungal and bacterial biomass under annual and perennial cropping systems. *Soil Science Society of America Journal* 61: 262–267.

Chavarría, D.N., Verdenelli, R.A., Muñoz, E.J. et al. (2016). Soil microbial functionality in response to the inclusion of cover crop mixtures in agricultural systems. *Spanish Journal of Agricultural Research* 14 (2): e0304.

Chen, G. and Weil, R.R. (2010). Penetration of cover crop roots through compacted soils. *Plant and Soil* 331: 31–43.

Chen, G. and Weil, R.R. (2011). Root growth and yield of maize as affected by soil compaction and cover crops. *Soil and Tillage Research* 117: 17–27.

Chenu, C. (1995). Extracellular polysaccharides: an interface between microorganisms and soil constituents. *Environmental Impact of Soil Component Interactions* 1: 217–233.

Chenu, C., Le Bissonnais, Y., and Arrouays, D. (2000). Organic matter influence on clay wettability and soil aggregate stability. *Soil Science Society of America Journal* 64: 1479–1486.

Chenu, C., Rumpel, C., and Lehmann, J. (2015). Methods for studying soil organic matter: nature, dynamics, spatial accessibility, and interactions with minerals. In: Soil Microbiology, Ecology and Biochemistry, 4e (ed. E.A. Paul), 383–419. London, UK: Academic Press.

Chibowski, E. (2011). Flocculation and dispersion phenomena in soils. In: Encyclopedia of Agrophysics. Encyclopedia of Earth Sciences Series (eds. J. Gliński, J. Horabik and J. Lipiec), 301–304. Dordrecht: Springer.

Chiu, C.Y., Chen, T.H., Imberger, K. et al. (2006). Particle size fractionation of fungal and bacterial biomass in subalpine grassland and forest soils. *Geoderma* 130: 265–271.

Clark, L.J. and Barraclough, P.B. (1999). Do dicotyledons generate greater maximum axial root growth pressures than monocotyledons? *Journal of Experimental Botany* 50: 1263–1266.

Clark, L.J., Price, A.H., Steele, K.A. et al. (2008). Evidence from near-isogenic lines that root penetration increases with root diameter and bending stiffness in rice. *Functional Plant Biology* 35: 1163–1171.

Cockroft, B. and Olsson, K.A. (2000). Degradation of soil structure due to coalescence of aggregates in no-till, no-traffic beds in irrigated crops. *Soil Research* 38: 61–70.

Collette, L., Hodgkin, T., Kassam, A.H. et al. (2011). Save and Grow: A Policymaker's Guide to Sustainable Intensification of Smallholder Crop Production. FAO.

Colombi, T., Braun, S., Keller, T. et al. (2017a). Artificial macropores attract crop roots and enhance plant productivity on compacted soils. *Science of The Total Environment* 574: 1283–1293.

Colombi, T., Kirchgessner, N., Walter, A. et al. (2017b). Root tip shape governs root elongation rate under increased soil strength. *Plant Physiology* 174: 2289–2301.

Costa, O.Y.A., Raaijmakers, J.M., and Kuramae, E.E. (2018). Microbial extracellular polymeric substances: ecological function and impact on soil aggregation. *Frontiers in Microbiology* 9: 1636. https://doi.org/10.3389/fmicb.2018.01636.

Cotrufo, M.F., Wallenstein, M.D., Boot, C.M. et al. (2013). The microbial efficiency – matrix stabilization (MEMS) framework integrates plant litter decomposition with soil organic matter stabilization: do labile plant inputs form stable soil organic matter? *Global Change Biology* 19: 988–995.

Cresswell, H.P. and Kirkegaard, J.A. (1995). Subsoil amelioration by plant-roots-the process and the evidence. *Soil Research* 33: 221–239.

Crocker, R.L. (1952). Soil genesis and the pedogenic factors. *The Quarterly Review of Biology* 27: 139–168.

Czarnes, S., Hallett, P.D., Bengough, A.G. et al. (2000). Root-and microbial-derived mucilages affect soil structure and water transport. *European Journal of Soil Science* 51: 435–443.

D'Acunto, L., Andrade, J.F., Poggio, S.L. et al. (2018). Diversifying crop rotation increased metabolic soil diversity and activity of the microbial community. *Agriculture, Ecosystems and Environment* 257: 159–164.

Daly, K.R., Mooney, S.J., Bennett, M.J. et al. (2015). Assessing the influence of the rhizosphere on soil hydraulic properties using X-ray computed tomography and numerical modelling. *Journal of Experimental Botany* 66: 2305–2314.

Dane, J.H. and Topp, C.G. (2002). Methods of Soil Analysis: Part 4 Physical Methods. SSSA Book Ser. 5.4. Madison, WI: SSSA.

Daniells, I.G. (2012). Hardsetting soils: a review. *Soil Research* 50: 349–359.

De Gryze, S., Six, J., and Merckx, R. (2006). Quantifying water-stable soil aggregate turnover and its implication for soil organic matter dynamics in a model study. *European Journal of Soil Science* 57: 693–707.

Degens, B.P. (1997). Macro-aggregation of soils by biological bonding and binding mechanisms and the factors affecting these: a review. *Soil Research* 35: 431–460.

Denef, K., Zotarelli, L., Boddey, R.M. et al. (2007). Microaggregate-associated carbon as a diagnostic fraction for management-induced changes in soil organic carbon in two oxisols. *Soil Biology and Biochemistry* 39: 1165–1172.

Dennis, P.G., Miller, A.J., and Hirsch, P.R. (2010). Are root exudates more important than other sources of rhizodeposits in structuring rhizosphere bacterial communities? *FEMS Microbiology Ecology* 72: 313–327.

Devitt, D.A. and Smith, S.D. (2002). Root channel macropores enhance downward movement of water in a Mojave Desert ecosystem. *Journal of Arid Environments* 50: 99–108.

Dexter, A.R. (1987). Compression of soil around roots. *Plant and Soil* 97: 401–406.

Dontsova, K.M. and Bigham, J.M. (2005). Anionic polysaccharide sorption by clay minerals. *Soil Science Society of America Journal* 69: 1026–1035.

Doran, J.W. and Zeiss, M.R. (2000). Soil health and sustainability: managing the biotic component of soil quality. *Applied Soil Ecology* 15: 3–11.

Dorioz, J.M., Robert, M., and Chenu, C. (1993). The role of roots, fungi and bacteria on clay particle organization. An experimental approach. In: Soil Structure/Soil Biota Interrelationships (eds. L. Brussaard and M.J. Kooistra), 179–194. Amsterdam: Elsevier.

Drewry, J.J., Cameron, K.C., and Buchan, G.D. (2008). Pasture yield and soil physical property responses to soil compaction from treading and grazing – a review. *Soil Research* 46: 237–256.

Dungait, J.A.J., Hopkins, D.W., Gregory, A.S. et al. (2012). Soil organic matter turnover is governed by accessibility not recalcitrance. *Global Change Biology* 18: 1781–1796.

Ebrahimi, A. and Or, D. (2015). Hydration and diffusion processes shape microbial community organization and function in model soil aggregates. *Water Resources Research* 51: 9804–9827.

Edwards, A.P. and Bremner, J.M. (1967). Microaggregates in soils. *Journal of Soil Science* 18: 64–73.

Ehrman, J. and Ritz, K. (2014). Plant: soil interactions in temperate multi-cropping production systems. *Plant and Soil* 376: 1–29.

Ellenberg, H. (1974). Indicator values of vascular plants in central Europe. *Scripta Geobotanica* 9: 97.

Ellenberg, H., Weber, H.E., Düll, R. et al. (1992). Zeigerwerte von Pflanzen in Mitteleuropa, 2nd ed. *Scripta Geobotanica* 18: 1–258.

Elliott, E.T. and Coleman, D.C. (1988). Let the soil work for us. *Ecological Bulletins* 39: 23–32.

Fan, J., McConkey, B., Wang, H. et al. (2016). Root distribution by depth for temperate agricultural crops. *Field Crops Research* 189: 68–74.

Fang, H., Zhou, H., Norton, G.J. et al. (2018). Interaction between contrasting rice genotypes and soil physical conditions induced by hydraulic stresses typical of alternate wetting and drying irrigation of soil. *Plant and Soil* 430: 233–243.

FAO/ITPS (2015). Status of the World's Soil Resources (SWSR) – Technical Summary. Rome, Italy: Food and Agriculture Organization of the United Nations and Intergovernmental Technical Panel on Soils.

Feeney, D.S., Crawford, J.W., Daniell, T. et al. (2006). Three-dimensional microorganization of the soil–root–microbe system. *Microbial Ecology* 52: 151–158.

Folberth, C., Skalský, R., Moltchanova, E. et al. (2016). Uncertainty in soil data can outweigh climate impact signals in global crop yield simulations. *Nature Communications* 7: 11872. https://doi.org/10.1038/ncomms11872.

Gaiser, T., Perkons, U., Küpper, P.M. et al. (2013). Modeling biopore effects on root growth and biomass production on soils with pronounced sub-soil clay accumulation. *Ecological Modelling* 256: 6–15.

Gale, W.J., Cambardella, C.A., Bailey, T.B. et al. (2000). Root-derived carbon and the formation and stabilization of aggregates. *Soil Science Society of America Journal* 64: 201–207.

Garbout, A., Munkholm, L.J., Hansen, S.B. et al. (2013a). Tillage effects on topsoil structural quality assessed using X-ray CT, soil cores and visual soil evaluation. *Soil and Tillage Research* 128: 104–109.

Garbout, A., Munkholm, L.J., Hansen, S.B. et al. (2013b). Temporal dynamics for soil aggregates determined using X-ray CT scanning. *Geoderma* 204: 15–22.

Garland, G., Bünemann, E.K., Oberson, A. et al. (2017). Plant-mediated rhizospheric interactions in maize-pigeon pea intercropping enhance soil aggregation and organic phosphorus storage. *Plant and Soil* 415: 37–55.

George, T.S., Brown, L.K., Ramsay, L. et al. (2014). Understanding the genetic control and physiological traits associated with rhizosheath production by barley (*Hordeum vulgare*). *New Phytologist* 203: 195–205.

Ghestem, M., Sidle, R.C., and Stokes, A. (2011). The influence of plant root systems on subsurface flow: implications for slope stability. *Bioscience* 61: 869–879.

Ghezzehei, T.A. and Albalasmeh, A.A. (2015). Spatial distribution of rhizodeposits provides built-in water potential gradient in the rhizosphere. *Ecological Modelling* 298: 53–63.

Gish, T.J. and Jury, W.A. (1983). Effect of plant roots and root channels on solute transport. *Transactions of the ASAE* 26: 440–444.

Gould, I.J., Quinton, J.N., Weigelt, A. et al. (2016). Plant diversity and root traits benefit physical properties key to soil function in grasslands. *Ecology Letters* 19: 1140–1149.

Greenland, D.J. (1977). Soil damage by intensive arable cultivation: temporary or permanent. *Philosophical Transactions of the Royal Society of London. B, Biological Sciences* 281: 193–208.

Guimarães, R.M.L., Ball, B.C., Tormena, C.A. et al. (2013). Relating visual evaluation of soil structure to other physical properties in soils of contrasting texture and management. *Soil and Tillage Research* 127: 92–99.

Gyssels, G., Poesen, J., Bochet, E. et al. (2005). Impact of plant roots on the resistance of soils to erosion by water: a review. *Progress in Physical Geography* 29: 189–217.

Haas, C., Gerke, H.H., Ellerbrock, R.H. et al. (2018). Relating soil organic matter composition to soil water repellency for soil biopore surfaces different in history from two Bt horizons of a Haplic Luvisol. *Ecohydrology* 11: e1949.

Hallett, P.D., Gordon, D.C., and Bengough, A.G. (2003). Plant influence on rhizosphere hydraulic properties: direct measurements using a miniaturized infiltrometer. *New Phytologist* 157: 597–603.

Hamza, M.A. and Anderson, W.K. (2005). Soil compaction in cropping systems: a review of the nature, causes and possible solutions. *Soil and Tillage Research* 82: 121–145.

Han, E., Kautz, T., Perkons, U. et al. (2015a). Quantification of soil biopore density after perennial fodder cropping. *Plant and Soil* 394: 73–85.

Han, E., Kautz, T., Perkons, U. et al. (2015b). Root growth dynamics inside and outside of soil biopores as affected by crop sequence determined with the profile wall method. *Biology and Fertility of Soils* 51: 847–856.

Helliwell, J.R., Sturrock, C.J., Grayling, K.M. et al. (2013). Applications of X-ray computed tomography for examining biophysical interactions and structural development in soil systems: a review. *European Journal of Soil Science* 64: 279–297. https://doi.org/10.1111/ejss.12028.

Hernandez-Soriano, M.C., Dalal, R.C., Warren, F.J. et al. (2018). Soil organic carbon stabilization: mapping carbon speciation from intact microaggregates. *Environmental Science and Technology* 52: 12275–12284.

Hinsinger, P., Plassard, C., and Jaillard, B. (2006). Rhizosphere: a new frontier for soil biogeochemistry. *Journal of Geochemical Exploration* 88: 210–213.

Hobbs, P.R., Sayre, K., and Gupta, R. (2007). The role of conservation agriculture in sustainable agriculture. *Philosophical Transactions of the Royal Society B: Biological Sciences* 363: 543–555.

Holtham, D.A.L., Matthews, G.P., and Scholefield, D.S. (2007). Measurement and simulation of void structure and hydraulic changes caused by root-induced soil structuring under white clover compared to ryegrass. *Geoderma* 142: 142–151.

Horn, R., Taubner, H., Wuttke, M. et al. (1994). Soil physical properties related to soil structure. *Soil and Tillage Research* 30: 187–216.

Hua, G.F., Zhao, Z.W., Kong, J. et al. (2014). Effects of plant roots on the hydraulic performance during the clogging process in mesocosm vertical flow constructed wetlands. *Environmental Science and Pollution Research* 21: 13017–13026.

Hütsch, B.W., Augustin, J., and Merbach, W. (2002). Plant rhizodeposition – an important source for carbon turnover in soils. *Journal of Plant Nutrition and Soil Science* 165: 397–407.

Iizumi, T. and Ramankutty, N. (2016). Changes in yield variability of major crops for 1981–2010 explained by climate change. *Environmental Research Letters* 11: 034003.

Imeson, A.C. and Vis, M. (1984). Assessing soil aggregate stability by water-drop impact and ultrasonic dispersion. *Geoderma* 34: 185–200.

IUSS Working Group WRB (2014). World Reference Base for Soil Resources 2014 International Soil Classification System for Naming Soils and Creating Legends for Soil Maps. Rome: FAO.

Jarvis, S., Tisdall, J., Oades, M. et al. (2012). Landmark papers. *European Journal of Soil Science* 63: 1–21.

Jastrow, J.D. (1996). Soil aggregate formation and the accrual of particulate and mineral-associated organic matter. *Soil Biology and Biochemistry* 28: 665–676.

Jastrow, J.D., Miller, R.M., and Lussenhop, J. (1998). Contributions of interacting biological mechanisms to soil aggregate stabilization in restored prairie. *Soil Biology and Biochemistry* 30: 905–916.

Jenny, H. (1994). Factors of Soil Formation: A System of Quantitative Pedology. New York, NY: Dover Publications Inc.

Jin, K., White, P.J., Whalley, W.R. et al. (2017). Shaping an optimal soil by root–soil interaction. *Trends in Plant Science* 22: 823–829.

Jotisankasa, A. and Sirirattanachat, T. (2017). Effects of grass roots on soil-water retention curve and permeability function. *Canadian Geotechnical Journal* 54: 1612–1622.

Kadžienė, G., Munkholm, L.J., and Mutegi, J.K. (2011). Root growth conditions in the topsoil as affected by tillage intensity. *Geoderma* 166: 66–73.

Kasper, M., Buchan, G.D., Mentler, A. et al. (2009). Influence of soil tillage systems on aggregate stability and the distribution of C and N in different aggregate fractions. *Soil and Tillage Research* 105: 192–199.

Kell, D.B. (2011). Breeding crop plants with deep roots: their role in sustainable carbon, nutrient and water sequestration. *Annals of Botany* 108: 407–418.

Keller, T. and Dexter, A.R. (2012). Plastic limits of agricultural soils as functions of soil texture and organic matter content. *Soil Research* 50: 7–17.

Kemmitt, S.J., Lanyon, C.V., Waite, I.S. et al. (2008). Mineralization of native soil organic matter is not regulated by the size, activity or composition of the soil microbial biomass – a new perspective. *Soil Biology and Biochemistry* 40: 61–73.

Kemper, W.D. and Koch, E.J. (1966). Aggregate stability of soils from western United States and Canada. Measurement procedure, correlation with soil constituents. USDA Technical Bulletin 1355, Agricultural Research Service, USDA.

Kodešová, R., Kodeš, V., Žigová, A. et al. (2006). Impact of plant roots and soil organisms on soil micromorphology and hydraulic properties. *Biologia* 61: S339–S343.

Koebernick, N., Daly, K.R., Keyes, S.D. et al. (2017). High-resolution synchrotron imaging shows that root hairs influence rhizosphere soil structure formation. *New Phytologist* 216: 124–135.

Koebernick, N., Naveed, M., Daly, K. et al. (2018). Root induced compaction alleviation by root hairs--visualization with synchrotron imaging. *EGU General Assembly Conference Abstracts* 20: 8410.

Koebernick, N., Daly, K.R., Keyes, S.D. et al. (2019). Imaging microstructure of the barley rhizosphere: particle packing and root hair influences. *New Phytologist* 221: 1878–1889.

Kögel-Knabner, I. (2002). The macromolecular organic composition of plant and microbial residues as inputs to soil organic matter. *Soil Biology and Biochemistry* 34: 139–162.

Kögel-Knabner, I. (2017). The macromolecular organic composition of plant and microbial residues as inputs to soil organic matter: fourteen years on. *Soil Biology and Biochemistry* 105: A3–A8.

Kolb, E., Legué, V., and Bogeat-Triboulot, M.B. (2017). Physical root–soil interactions. *Physical Biology* 14: 065004.

Kölbl, A. and Kögel-Knabner, I. (2004). Content and composition of free and occluded particulate organic matter in a differently textured arable Cambisol as revealed by solid-state 13C NMR spectroscopy. *Journal of Plant Nutrition and Soil Science* 167: 45–53.

Körner, C. (2015). Paradigm shift in plant growth control. *Current Opinion in Plant Biology* 25: 107–114.

Kosugi, K. (1996). Lognormal distribution model for unsaturated soil hydraulic properties. *Water Resources Research* 32: 2697–2703.

Krause, L., Rodionov, A., Schweizer, S.A. et al. (2018). Microaggregate stability and storage of organic carbon is affected by clay content in arable Luvisols. *Soil and Tillage Research* 182: 123–129.

Krümmelbein, J. and Raab, T. (2012). Development of soil physical parameters in agricultural reclamation after brown coal mining within the first four years. *Soil and Tillage Research* 125: 109–115.

Kumar, A., Dorodnikov, M., Splettstößer, T. et al. (2017). Effects of maize roots on aggregate stability and enzyme activities in soil. *Geoderma* 306: 50–57.

Kutílek, M. (2004). Soil hydraulic properties as related to soil structure. *Soil and Tillage Research* 75: 175–184.

Kutschera, L., Lichtenegger, E., and Sobotik, M. (2009). Wurzelatlas der Kulturpflanzen gemäßigter Gebiete mit Arten des Feldgemüsebaues. Frankfurt am Main: DLG-Verlag.

Kuzyakov, Y. and Blagodatskaya, E. (2015). Microbial hotspots and hot moments in soil: concept and review. *Soil Biology and Biochemistry* 83: 184–199.

Kuzyakov, Y. and Domanski, G. (2000). Carbon input by plants into the soil. *Review. Journal of Plant Nutrition and Soil Science* 163: 421–431.

Labille, J., Thomas, F., Milas, M. et al. (2005). Flocculation of colloidal clay by bacterial polysaccharides: effect of macromolecule charge and structure. *Journal of Colloid and Interface Science* 284: 149–156.

Lado, M., Paz, A., and Ben-Hur, M. (2004). Organic matter and aggregate size interactions in infiltration, seal formation, and soil loss. *Soil Science Society of America Journal* 68: 935–942.

Lahmar, R. (2010). Adoption of conservation agriculture in Europe: lessons of the KASSA project. *Land Use Policy* 27: 4–10.

Lal, R. (2003). Soil erosion and the global carbon budget. *Environment International* 29: 437–450.

Lal, R. (2015). Restoring soil quality to mitigate soil degradation. *Sustainability* 7: 5875–5895.

Lal, R. and Shukla, M.K. (2004). Principles of Soil Physics. New York, NY: CRC Press, Marcel Dekker Inc.

Lampurlanés, J., Plaza-Bonilla, D., Álvaro-Fuentes, J. et al. (2016). Long-term analysis of soil water conservation and crop yield under different tillage systems in Mediterranean rainfed conditions. *Field Crops Research* 189: 59–67.

Lehmann, A. and Rillig, M.C. (2013). Are there temporal trends in root architecture and soil aggregation for *Hordeum vulgare* breeding lines? *Applied Soil Ecology* 65: 31–34.

Lehmann, A. and Rillig, M.C. (2015). Understanding mechanisms of soil biota involvement in soil aggregation: a way forward with saprobic fungi? *Soil Biology and Biochemistry* 88: 298–302.

Leij, F.J., Ghezzehei, T.A., and Or, D. (2002). Modeling the dynamics of the soil pore-size distribution. *Soil and Tillage Research* 64: 61–78.

Lesturgez, G., Poss, R., Hartmann, C. et al. (2004). Roots of Stylosanthes hamata create macropores in the compact layer of a sandy soil. *Plant and Soil* 260: 101–109.

van Lier, Q.D.J. and Gubiani, P.I. (2015). Beyond the "least limiting water range": rethinking soil physics research in Brazil. *Revista Brasileira de Ciência do Solo* 39: 925–939.

Lory, J.A. and Scharf, P.C. (2003). Yield goal versus delta yield for predicting fertilizer nitrogen need in corn. *Agronomy Journal* 95: 994–999.

Mamo, M. and Bubenzer, G.D. (2001). Detachment rate, soil erodibility, and soil strength as influenced by living plant roots part II: Field study. *Transactions of the ASAE* 44: 1175.

Mao, J., Nierop, K.G.J., Rietkerk, M. et al. (2016). The influence of vegetation on soil water repellency-markers and soil hydrophobicity. *Science of the Total Environment* 566: 608–620.

Mardhiah, U., Caruso, T., Gurnell, A. et al. (2016). Arbuscular mycorrhizal fungal hyphae reduce soil erosion by surface water flow in a greenhouse experiment. *Applied Soil Ecology* 99: 137–140.

Marschner, B. and Kalbitz, K. (2003). Controls of bioavailability and biodegradability of dissolved organic matter in soils. *Geoderma* 113: 211–235.

Masri, Z. and Ryan, J. (2006). Soil organic matter and related physical properties in a Mediterranean wheat-based rotation trial. *Soil and Tillage Research* 87: 146–154.

Mayer, H., Mentler, A., Papakyriacou, M. et al. (2002). Influence of vibration amplitude on the ultrasonic dispersion of soils. *International Agrophysics* 16: 53–60.

Mazzilli, S.R., Kemanian, A.R., Ernst, O.R. et al. (2015). Greater humification of belowground than aboveground biomass carbon into particulate soil organic matter in no-till corn and soybean crops. *Soil Biology and Biochemistry* 85: 22–30.

McCallum, M.H., Kirkegaard, J.A., Green, T.W. et al. (2004). Improved subsoil macroporosity following perennial pastures. *Australian Journal of Experimental Agriculture* 44: 299–307.

Mearns, L.O., Rosenzweig, C., and Goldberg, R. (1997). Mean and variance change in climate scenarios: methods, agricultural applications, and measures of uncertainty. *Climatic Change* 35: 367–396.

Miller, R.M. and Jastrow, J.D. (2000). Mycorrhizal fungi influence soil structure. In: Arbuscular Mycorrhizas: Physiology and Function (eds. Y. Kapulnik and D.D. Douds Jr.), 3–18. Dordrecht: Springer.

Misra, R.K., Dexter, A.R., and Alston, A.M. (1986). Maximum axial and radial growth pressures of plant roots. *Plant and Soil* 95: 315–326.

Mitchell, A.R., Ellsworth, T.R., and Meek, B.D. (1995). Effect of root systems on preferential flow in swelling soil. *Communications in Soil Science and Plant Analysis* 26: 2655–2666.

Moradi, A.B., Carminati, A., Lamparter, A. et al. (2012). Is the rhizosphere temporarily water repellent? *Vadose Zone Journal* 11: vzj2011.0120. https://doi.org/10.2136/vzj2011.0120.

Moreno-Espíndola, I.P., Rivera-Becerril, F., de Jesús Ferrara-Guerrero, M., and De León-González, F. (2007). Role of root-hairs and hyphae in adhesion of sand particles. *Soil Biology and Biochemistry* 39: 2520–2526.

Mueller, L., Schindler, U., Shepherd, T.G. et al. (2012). A framework for assessing agricultural soil quality on a global scale. *Archives of Agronomy and Soil Science* 58: S76–S82.

Mummey, D., Holben, W., and Six, J. (2006). Spatial stratification of soil bacterial populations in aggregates of diverse soils. *Microbial Ecology* 51: 404–411.

Munkholm, L.J., Heck, R.J., and Deen, B. (2013). Long-term rotation and tillage effects on soil structure and crop yield. *Soil and Tillage Research* 127: 85–91.

Naveed, M., Brown, L.K., and Raffan, A.C. (2017). Plant exudates may stabilize or weaken soil depending on species, origin and time. *European Journal of Soil Science* 68: 806–816.

Ng, C.W.W., Ni, J.J., and Leung, A.K. (2016). A new and simple water retention model for root-permeated soils. *Géotechnique Letters* 6: 106–111.

Nguyen, C. (2009). Rhizodeposition of organic C by plant: mechanisms and controls. In: Sustainable Agriculture (eds. E. Lichtfouse, M. Navarrete, P. Debaeke, et al.), 97–123. Dordrecht: Springer.

Nimmo, J.R. (2004). Porosity and pore size distribution. *Encyclopedia of Soils in the Environment* 3: 295–303.

Nimmo, J.R. and Perkins, K.S. (2002). Aggregate stability and size distribution. In: Methods of Soil Analysis: Part 4 Physical Methods. SSSA Book Ser. 5.4 (eds. J.H. Dane and C.G. Topp), 317–328. Madison, WI: SSSA.

Nunes, M.R., van Es, H.M., Schindelbeck, R. et al. (2018). No-till and cropping system diversification improve soil health and crop yield. *Geoderma* 328: 30–43.

Nuttall, J.G., Davies, S.L., Armstrong, R.A. et al. (2008). Testing the primer-plant concept: wheat yields can be increased on alkaline sodic soils when an effective primer phase is used. *Australian Journal of Agricultural Research* 59: 331–338.

Obalum, S.E., Chibuike, G.U., Peth, S. et al. (2017). Soil organic matter as sole indicator of soil degradation. *Environmental Monitoring and Assessment* 189: 176.

Ola, A., Dodd, I.C., and Quinton, J.N. (2015). Can we manipulate root system architecture to control soil erosion? *Soil* 1: 603–612.

Oldeman, L.R. (1992). Global extent of soil degradation. In: Bi-Annual Report 1991–1992/ ISRIC (ed. ISCRI), 19–36. Wageningen: ISRIC.

Pachepsky, Y.A. and Rawls, W.J. (2003). Soil structure and pedotransfer functions. *European Journal of Soil Science* 54: 443–452.

Pagenkemper, S.K., Peth, S., Puschmann, D.U. et al. (2013). Effects of root-induced biopores on pore space architecture investigated with industrial X-Ray computed tomography. In:

Soil–Water–Root Processes: Advances in Tomography and Imaging, SSSA Special Publication 56 (eds. S.H. Anderson and J.W. Hopmans), 69–96. Madison, WI: SSSA.

Pagliai, M., Vignozzi, N., and Pellegrini, S. (2004). Soil structure and the effect of management practices. *Soil and Tillage Research* 79: 131–143.

Panagos, P., Borrelli, P., Meusburger, K. et al. (2015). Modelling the effect of support practices (P-factor) on the reduction of soil erosion by water at European scale. *Environmental Science and Policy* 51: 23–34.

Pang, J., Ryan, M.H., Siddique, K.H.M. et al. (2017). Unwrapping the rhizosheath. *Plant and Soil* 418: 129–139.

Perkons, U., Kautz, T., Uteau, D. et al. (2014). Root-length densities of various annual crops following crops with contrasting root systems. *Soil and Tillage Research* 137: 50–57.

Peth, S., Horn, R., Beckmann, F. et al. (2008). Three-dimensional quantification of intra-aggregate pore-space features using synchrotron-radiation-based microtomography. *Soil Science Society of America Journal* 72: 897–907.

Pietola, L. and Alakukku, L. (2005). Root growth dynamics and biomass input by Nordic annual field crops. *Agriculture, Ecosystems and Environment* 108: 135–144.

Pittelkow, C.M., Linquist, B.A., Lundy, M.E. et al. (2015). When does no-till yield more? A global meta-analysis. *Field Crops Research* 183: 156–168.

Poeplau, C., Don, A., Six, J. et al. (2018). Isolating organic carbon fractions with varying turnover rates in temperate agricultural soils–a comprehensive method comparison. *Soil Biology and Biochemistry* 125: 10–26.

Poirier, V., Angers, D.A., and Whalen, J.K. (2014). Formation of millimetric-scale aggregates and associated retention of 13C–15N-labelled residues are greater in subsoil than topsoil. *Soil Biology and Biochemistry* 75: 45–53.

Poirier, V., Roumet, C., and Munson, A.D. (2018). The root of the matter: linking root traits and soil organic matter stabilization processes. *Soil Biology and Biochemistry* 120: 246–259.

Preissel, S., Reckling, M., Schläfke, N. et al. (2015). Magnitude and farm-economic value of grain legume pre-crop benefits in Europe: a review. *Field Crops Research* 175: 64–79.

Puget, P., Chenu, C., and Balesdent, J. (2000). Dynamics of soil organic matter associated with particle-size fractions of water-stable aggregates. *European Journal of Soil Science* 51: 595–605.

Rabbi, S.M.F., Tighe, M.K., Knox, O. et al. (2018a). The impact of carbon addition on the organisation of rhizosheath of chickpea. *Scientific Reports* 8: 18028. https://doi.org/10.1038/s41598-018-36958-0.

Rabbi, S.M.F., Tighe, M.K., Flavel, R.J. et al. (2018b). Plant roots redesign the rhizosphere to alter the three-dimensional physical architecture and water dynamics. *New Phytologist* 219: 542–550.

Rasmussen, K.J. (1999). Impact of ploughless soil tillage on yield and soil quality: a Scandinavian review. *Soil and Tillage Research* 53: 3–14.

Rasse, D.P., Rumpel, C., and Dignac, M.F. (2005). Is soil carbon mostly root carbon? Mechanisms for a specific stabilisation. *Plant and Soil* 269: 341–356.

Reid, J.P., Adair, E.C., Hobbie, S.E. et al. (2012). Biodiversity, nitrogen deposition, and CO_2 affect grassland soil carbon cycling but not storage. *Ecosystems* 15: 580–590.

Renard, K.G., Foster, G.R., Weesies, G.A. et al. (1991). RUSLE: revised universal soil loss equation. *Journal of Soil and Water Conservation* 46: 30–33.

Reynolds, M., Manes, Y., Izanloo, A. et al. (2009). Phenotyping approaches for physiological breeding and gene discovery in wheat. *Annals of Applied Biology* 155: 309–320.

Rillig, M.C. (2004). Arbuscular mycorrhizae, glomalin, and soil aggregation. *Canadian Journal of Soil Science* 84: 355–363.

Rillig, M.C., Aguilar-Trigueros, C.A., Bergmann, J. et al. (2015). Plant root and mycorrhizal fungal traits for understanding soil aggregation. *New Phytologist* 205: 1385–1388.

Ritz, K. and Young, I.M. (2004). Interactions between soil structure and fungi. *Mycologist* 18: 52–59.

Rosolem, C.A., Foloni, J.S.S., and Tiritan, C.S. (2002). Root growth and nutrient accumulation in cover crops as affected by soil compaction. *Soil and Tillage Research* 65: 109–115.

Rowley, M.C., Grand, S., and Verrecchia, É.P. (2018). Calcium-mediated stabilisation of soil organic carbon. *Biogeochemistry* 137: 27–49.

Scanlan, C.A. and Hinz, C. (2010). Insights into the processes and effects of root-induced changes to soil hydraulic properties. In: World Congress of Soil Science, Soil Solutions for a Changing World, vol. 2 (eds. R. Gilkes and N. Prakongkep), 41–44. Brisbane: International Union of Soil Sciences.

Scharlemann, J.P.W., Tanner, E.V.J., Hiederer, R. et al. (2014). Global soil carbon: understanding and managing the largest terrestrial carbon pool. *Carbon Management* 5: 81–91.

Schimel, J. and Schaeffer, S.M. (2012). Microbial control over carbon cycling in soil. *Frontiers in Microbiology* 3: 348.

Schindler, U., Durner, W., von Unold, G. et al. (2010). The evaporation method: extending the measurement range of soil hydraulic properties using the air-entry pressure of the ceramic cup. *Journal of Plant Nutrition and Soil Science* 173: 563–572.

Schneider, F., Don, A., Hennings, I. et al. (2017a). The effect of deep tillage on crop yield–what do we really know? *Soil and Tillage Research* 174: 193–204.

Scholl, P., Leitner, D., Kammerer, G. et al. (2014). Root induced changes of effective 1D hydraulic properties in a soil column. *Plant and Soil* 381: 193–213.

Schomakers, J., Mentler, A., Steurer, T. et al. (2011). Characterization of soil aggregate stability using low intensity ultrasonic vibrations. *International Agrophysics* 25: 165–172.

Semmel, H., Horn, R., Hell, U. et al. (1990). The dynamics of soil aggregate formation and the effect on soil physical properties. *Soil Technology* 3: 113–129.

Sergeyev, Y.M., Grabowska-Olszewska, B., Osipov, V.I. et al. (1980). The classification of microstructures of clay soils. *Journal of Microscopy* 120: 237–260.

Shepherd, G. (2000). Visual Soil Assessment. Volume 1. Field guide for cropping and pastoral grazing on flat to rolling country.horizons.mw & Landcare Research, Palmerston North. 84.

da Silva, A.P. and Kay, B.D. (1997). Estimating the least limiting water range of soils from properties and management. *Soil Science Society of America Journal* 61: 877–883.

Simpson, R.T., Frey, S.D., Six, J. et al. (2004). Preferential accumulation of microbial carbon in aggregate structures of no-tillage soils. *Soil Science Society of America Journal* 68: 1249–1255.

Six, J., Elliott, E.T., and Paustian, K. (2000). Soil macroaggregate turnover and microaggregate formation: a mechanism for C sequestration under no-tillage agriculture. *Soil Biology and Biochemistry* 32: 2099–2103.

Six, J., Bossuyt, H., Degryze, S. et al. (2004). A history of research on the link between (micro) aggregates, soil biota, and soil organic matter dynamics. *Soil and Tillage Research* 79: 7–31.

Six, J., Frey, S.D., Thiet, R.K. et al. (2006). Bacterial and fungal contributions to carbon sequestration in agroecosystems. *Soil Science Society of America Journal* 70: 555–569.

Smith, B.J., Kirkegaard, J.A., and Howe, G.N. (2004). Impacts of Brassica break-crops on soil biology and yield of following wheat crops. *Australian Journal of Agricultural Research* 55: 1–11.

Soil Science Division Staff (2017). Soil survey manual. In: USDA Handbook 18 (eds. C. Ditzler, K. Scheffe and H.C. Monger). Washington, DC: Government Printing Office.

Soil Science Glossary Terms Committee (2008). Glossary of Soil Science Terms. Madison, WI: Soil Science Society of America.

Stewart, M.L., Ward, A.L., and Rector, D.R. (2006). A study of pore geometry effects on anisotropy in hydraulic permeability using the lattice-Boltzmann method. *Advances in Water Resources* 29: 1328–1340.

Stewart, C.E., Paustian, K., Conant, R.T. et al. (2007). Soil carbon saturation: concept, evidence and evaluation. *Biogeochemistry* 86: 19–31.

Stewart, R.D., Jian, J., Gyawali, A.J. et al. (2018). What we talk about when we talk about soil health. *Agricultural and Environmental Letters* 3: 180033.

Stirzaker, R.J., Passioura, J.B., and Wilms, Y. (1996). Soil structure and plant growth: impact of bulk density and biopores. *Plant and Soil* 185: 151–162.

Tebrügge, F. and Düring, R.A. (1999). Reducing tillage intensity – a review of results from a long-term study in Germany. *Soil and Tillage Research* 53: 15–28.

Thorup-Kristensen, K. and Kirkegaard, J. (2016). Root system-based limits to agricultural productivity and efficiency: the farming systems context. *Annals of Botany* 118: 573–592.

Tiemann, L.K., Grandy, A.S., Atkinson, E.E. et al. (2015). Crop rotational diversity enhances belowground communities and functions in an agroecosystem. *Ecology Letters* 18: 761–771.

Tisdall, J.M. (1996). Formation of soil aggregates and accumulation of soil organic matter. In: Structure and Organic Matter Storage in Agricultural Soils (eds. M.R. Carter and B.A. Stewart), 57–96. Boca Raton, FL: Lewis.

Tisdall, J.M. and Oades, J.M. (1979). Stabilization of soil aggregates by the root systems of ryegrass. *Soil Research* 17: 429–441.

Tisdall, J.M. and Oades, J.M. (1982). Organic matter and water-stable aggregates in soils. *Journal of Soil Science* 33: 141–163.

Totsche, K.U., Amelung, W., Gerzabek, M.H. et al. (2018). Microaggregates in soils. *Journal of Plant Nutrition and Soil Science* 181: 104–136.

Traoré, O., Groleau-Renaud, V., and Plantureux, S. (2000). Effect of root mucilage and modelled root exudates on soil structure. *European Journal of Soil Science* 51: 575–581.

Ucgul, M., Saunders, C., and Fielke, J.M. (2017). Discrete element modelling of tillage forces and soil movement of a one-third scale mouldboard plough. *Biosystems Engineering* 155: 44–54.

Udawatta, R.P., Gantzer, C.J., Anderson, S.H. et al. (2008). Agroforestry and grass buffer effects on pore characteristics measured by high-resolution X-ray computed tomography. *Soil Science Society of America Journal* 72: 295–304.

Unger, P.W. and Kaspar, T.C. (1994). Soil compaction and root growth: a review. *Agronomy Journal* 86: 759–766.

Upton, R.N., Bach, E.M., and Hofmockel, K.S. (2019). Spatio-temporal microbial community dynamics within soil aggregates. *Soil Biology and Biochemistry* 132: 58–68.

Uteau, D., Pagenkemper, S.K., Peth, S. et al. (2013). Root and time dependent soil structure formation and its influence on gas transport in the subsoil. *Soil and Tillage Research* 132: 69–76.

Van den Putte, A., Govers, G., Diels, J. et al. (2010). Assessing the effect of soil tillage on crop growth: a meta-regression analysis on European crop yields under conservation agriculture. *European Journal of Agronomy* 33: 231–241.

Van Eekeren, N., van Liere, D., de Vries, F. et al. (2009). A mixture of grass and clover combines the positive effects of both plant species on selected soil biota. *Applied Soil Ecology* 42: 254–263.

Van Ouwerkerk, C. and Raats, P.A.C. (1986). Experiences with deep tillage in the Netherlands. *Soil and Tillage Research* 7: 273–283.

Vannoppen, W., Vanmaercke, M., De Baets, S. et al. (2015). A review of the mechanical effects of plant roots on concentrated flow erosion rates. *Earth-Science Reviews* 150: 666–678.

Verboom, W.H. and Pate, J.S. (2006). Bioengineering of soil profiles in semiarid ecosystems: the "phytotarium" concept. A review. *Plant and Soil* 289: 71–102.

Vervoort, R.W. and Cattle, S.R. (2003). Linking hydraulic conductivity and tortuosity parameters to pore space geometry and pore-size distribution. *Journal of Hydrology* 272: 36–49.

Vetterlein, D. and Jahn, R. (2004). Gradients in soil solution composition between bulk soil and rhizosphere–in situ measurement with changing soil water content. *Plant and Soil* 258: 307–327.

Vrdoljak, G. and Sposito, G. (2002). Soil aggregate hierarchy in a Brazilian Oxisol. In: Developments in Soil Science, vol. 28 (eds. A. Violante, P.M. Huang, J.-M. Bollag and L. Gianfreda), 197–217. Elsevier.

Watt, M., McCully, M.E., and Jeffree, C.E. (1993). Plant and bacterial mucilages of the maize rhizosphere: comparison of their soil binding properties and histochemistry in a model system. *Plant and Soil* 151: 151–165.

Watt, M., Silk, W.K., and Passioura, J.B. (2006). Rates of root and organism growth, soil conditions, and temporal and spatial development of the rhizosphere. *Annals of Botany* 97: 839–855.

Watteau, F., Villemin, G., Burtin, G. et al. (2006). Root impact on the stability and types of micro-aggregates in silty soil under maize. *European Journal of Soil Science* 57: 247–257.

Weninger, T., Kreiselmeier, J., Chandrasekhar, P. et al. (2019). Effects of tillage intensity on pore system and physical quality of silt-textured soils detected by multiple methods. *Soil Research* 57: 703–711.

Whalley, W.R., Leeds-Harrison, P.B., Leech, P.K. et al. (2004). The hydraulic properties of soil at root-soil interface. *Soil Science* 169: 90–99.

Whalley, W.R., Riseley, B., Leeds-Harrison, P.B. et al. (2005). Structural differences between bulk and rhizosphere soil. *European Journal of Soil Science* 56: 353–360.

White, R.G. and Kirkegaard, J.A. (2010). The distribution and abundance of wheat roots in a dense, structured subsoil–implications for water uptake. *Plant, Cell and Environment* 33: 133–148.

Williams, S.M. and Weil, R.R. (2004). Crop cover root channels may alleviate soil compaction effects on soybean crop. *Soil Science Society of America Journal* 68: 1403–1409.

Wuest, S.B. (2001). Soil biopore estimation: effects of tillage, nitrogen, and photographic resolution. *Soil and Tillage Research* 62: 111–116.

Yavuzcan, H.G., Matthies, D., and Auernhammer, H. (2005). Vulnerability of Bavarian silty loam soil to compaction under heavy wheel traffic: impacts of tillage method and soil water content. *Soil and Tillage Research* 84: 200–215.

Young, I.M. (1998). Biophysical interactions at the root–soil interface: a review. *The Journal of Agricultural Science* 130: 1–7.

Yu, Y., Loiskandl, W., Kaul, H.P. et al. (2016). Estimation of runoff mitigation by morphologically different cover crop root systems. *Journal of Hydrology* 538: 667–676.

Yudina, A.V., Fomin, D.S., and Kotelnikova, A.D. (2018). From the notion of elementary soil particle to the particle-size and microaggregate-size distribution analyses: a review. *Eurasian Soil Science* 51: 1326–1347.

Yunusa, I.A.M. and Newton, P.J. (2003). Plants for amelioration of subsoil constraints and hydrological control: the primer-plant concept. *Plant and Soil* 257: 261–281.

Yunusa, I.A.M., Mele, P.M., Rab, M.A. et al. (2002). Priming of soil structural and hydrological properties by native woody species, annual crops, and a permanent pasture. *Soil Research* 40: 207–219.

Zhalnina, K., Louie, K.B., Hao, Z. et al. (2018). Dynamic root exudate chemistry and microbial substrate preferences drive patterns in rhizosphere microbial community assembly. *Nature Microbiology* 3: 470.

Zhang, B., Hallett, P.D., and Zhang, G. (2008). Increase in the fracture toughness and bond energy of clay by a root exudate. *European Journal of Soil Science* 59: 855–862.

Zhao, D., Xu, M., Liu, G. et al. (2017). Quantification of soil aggregate microstructure on abandoned cropland during vegetative succession using synchrotron radiation-based micro-computed tomography. *Soil and Tillage Research* 165: 239–246.

Zhou, Z.C. and Shangguan, Z.P. (2007). The effects of ryegrass roots and shoots on loess erosion under simulated rainfall. *Catena* 70: 350–355.

Zhu, Z., Angers, D.A., Field, J. et al. (2017). Using ultrasonic energy to elucidate the effects of decomposing plant residues on soil aggregation. *Soil and Tillage Research* 167: 1–8.

Zuazo, V.H. and Pleguezuelo, C.R. (2008). Soil-erosion and runoff prevention by plant covers. A review. *Agronomy and Sustainable Development* 28: 65–86.

4

Dynamics of Root Systems in Crop and Pasture Genotypes over the Last 100 Years

Lessons Learned

Abdul Rehman[1], *Muhammad Farooq*[2,3], *Dong-Jin Lee*[1], *and Kadambot H.M. Siddique*[4]

[1] *Department of Crop Sciences and Biotechnology, Dankook University, Cheonan-si, South Korea*
[2] *Department of Plant Sciences, College of Agricultural and Marine Sciences, Sultan Qaboos University, Al-Khoud, Oman*
[3] *Department of Agronomy, University of Agriculture, Faisalabad, Pakistan*
[4] *The UWA Institute of Agriculture and School of Agriculture & Environment, The University of Western Australia, Perth, WA, Australia*

4.1 Introduction

Land plants form the base of the food chain, supporting terrestrial life. They also provide medicine, shelter, fuel, and industrial material to humans (White et al. 2013). More than 90% of the world's food supply comes from about 100 land plant species (Meyer et al. 2012). Plants need water and nutrients for normal growth, yield, and nutritional quality, which they acquire through roots (White and Brown 2010). Roots are an integral plant component for anchorage, water, and nutrient acquisition, and the synthesis of plant hormones, amino acids, and organic acids (Yang et al. 2004).

Research on root biology is not new; early reports have been published almost 150 years ago (e.g. Hanstein 1870; Vines 1888; Pfeffer 1894). Despite enduring observations and rigorous research over the years, root dynamics and biology have received less attention from leading plant scientists than shoots (Waisel et al. 2002; Meister et al. 2014). The first symposium on "root structure and function," organized about half a century ago (1971) in Slovakia, turned the attention of plant scientists to this neglected plant part. The proceedings of this symposium were published three years later (Kolek 1974).

By 2011, around 7000 published papers mentioned "root" in their title (http://webofknowledge.com/WOS), and the number is increasing. Many specialized societies have been formed by researchers interested in root-related research. Professor Eleonore Kutschera and her research group established the International Society of Root Research (ISRR) in 1982. The society has organized ten symposia to date (in 1982, 1988, 1991, 1994, 1996, 2001, 2009, 2012, 2015, and 2018) on different aspects of root research. The 10th Symposium of the ISRR, Exposing the Hidden Half – Root Research at the Forefront of Science, was held from 8–12 July 2018 in Israel (https://www.rootresearch.org/index.php?index=57). The Japanese

The Root Systems in Sustainable Agricultural Intensification, First Edition.
Edited by Zed Rengel and Ivica Djalovic.

Society for Root Research was established 36 years ago, and their online journal "Plant Root" (http://www.plantroot.org/) started in 2007.

Despite being below-ground and the difficulties associated with characterizing them, roots are a crucial component of our biosphere for nutrient and water extraction from the soil for plant use (Meister et al. 2014). There have been quite a few studies exploring the genetics of root system architecture (RSA) for developing crop cultivars with enhanced water and nutrient uptake (e.g. Zhu et al. 2010). In the last three decades, quantitative trait locus (QTL) responsible for RSA regulation have been identified (Kamoshita et al. 2002; Li et al. 2005). Efficient and nondestructive root phenotyping platforms that precisely capture and reflect RSA need to be developed. Moreover, these platforms should continuously monitor the effects of different edaphic conditions on root growth and development and have an efficient throughput system to evaluate large numbers of crop and pasture genotypes (Paez-Garcia et al. 2015).

This chapter discusses research on root development, RSA dynamics, root ideotypes, and the genetics of RSA in crop and pasture plants in the last 100 years. Moreover, phenotyping platforms developed for RSA analysis are described.

4.2 Root Morphology and Physiology

Root physiology and morphology are linked to shoot growth and development. Roots are the main organ in vascular plants, having a highly differentiated structure comprising the endodermis, pericycle, cortex, epidermis, root cap, and lateral roots that provide anchorage and allow water absorption and mineral uptake (Seago and Fernando 2013; Rogers and Benfey 2015). The radicle (primary root) is the first structure that protrudes from the germinating seed of most plants (Clowes 1961).

Cannon (1949) classified roots into three groups based on their ontogenesis (Figure 4.1; primary, nodal, and lateral roots). Later, Fitter (1987) and Wahid (2000) classified roots based on topology (branch structure) and root activity, respectively. However, the Cannon's root classification based on development (Figure 4.1; Cannon 1949) is still helpful for understanding the growth and development of RSA and for differentiating dicot and monocot species. The root system in dicot species originates from primary roots (that grow throughout the plant life cycle) and lateral branching (primary root system) (Sinnot 1960; Cannon 1949). The extension of lateral roots, relative to the primary root axis in dicots, varies between taproot, fasciculate, and dispersed root system morphologies (Figure 4.1; Kutschera 1960).

In monocots, the fibrous root system comprises primary and seminal roots (embryonic) and lateral- and shoot-borne roots (post-embryonic) (Hochholdinger and Tuberosa 2009; Coudert et al. 2010). Primary root growth is determined, ceasing after the emergence of several lateral and adventitious roots that form a multiple-branched fibrous root system (Hodge et al. 2009). The primary root is formed from the embryonic axis, whereas seminal roots are formed from the scutellar node. Embryonic roots are crucial during early seedling development, whereas post-embryonic roots play a key role in the later growth stages (Hochholdinger and Tuberosa 2009; Coudert et al. 2010).

Primary root formation occurs during embryogenesis through hypophysis specification (single extra-embryonic suspensor cell), giving rise to the quiescent centre after asymmetric

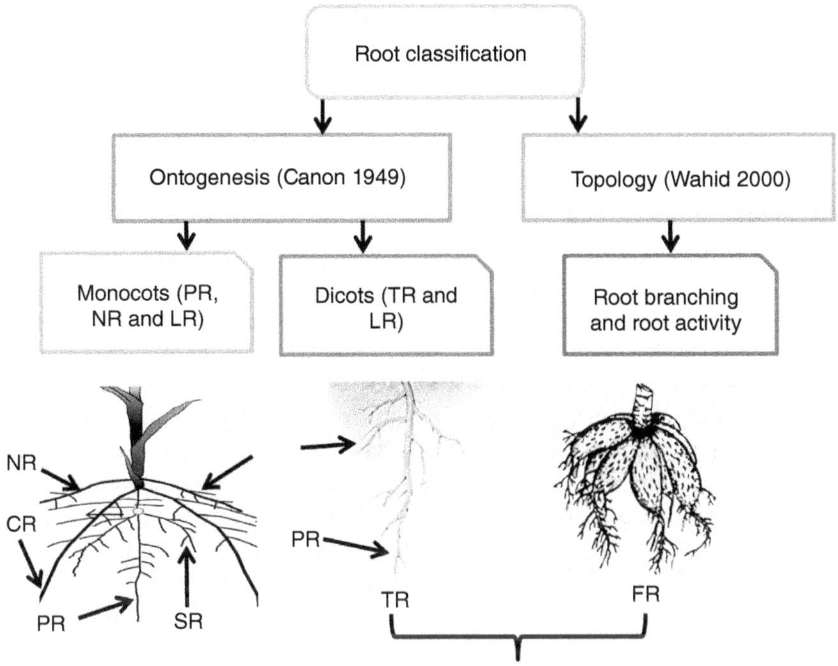

Figure 4.1 Root classifications. Most widely used root classification is based on root ontogenesis (Cannon 1949). NR, nodal roots; LR, lateral roots; CR, crown roots; PR, primary roots; SR, seminal roots; TR, taproot; FR, fasciculate root. *Source:* Based on Cannon (1949).

cell divisions, which ultimately generate the meristem of the primary root (De Smet et al. 2010a). The stem cells surrounding the quiescent centre divide asymmetrically to generate various cell and tissue types in the primary root (De Smet and Beeckman 2011). Von Nägeli (1858) introduced the term meristem for a group of cells that form a complete organ, e.g. shoot apical meristem (dome-shaped or flat structure located on the shoot axis tip, lying between the primordia of developing organs; Esau 1965) and root apical meristem (subapical structure with a layer of cells protecting its apex that gives rise to the root cap).

Lateral roots form post-embryonically and largely determine RSA. Lateral root formation is initiated from the pericycle (heterogenous tissue comprising two types of cells, those associated with the phloem that are quiescent and those linked to the xylem that are involved in cell divisions associated with lateral root primordia formation) (Parizot et al. 2008) and is followed by highly coordinated asymmetric divisions that give rise to lateral root primordia that transform into post-embryonic lateral roots emerging from the primary root (Péret et al. 2009).

Primary and seminal roots develop a root framework to explore the soil for nutrients and water and are crucial for early, vigorous plant growth (Sinnot 1960). Plant RSA is controlled by internal processes (organogenesis) and external factors (biotic or abiotic) (Malamy 2005). Moreover, land plants produce different types of roots based on functional requirements for survival, including aerial, hair, cluster, contractile, parasitic, and storage roots in the specific environments (Gregory and Kirkegaard 2017).

Nodal roots originate from above-ground shoot nodes, whereas crown roots originate from below-ground shoot nodes; both root types are involved in water and nutrient uptake and lodging resistance. Around 120 years ago, Engler described unusual root morphology in plants of proteaceae family which were latter termed proteoid roots by Pernal in 1960s (Watt and Evans 1999). Proteoid roots form one or more clusters along the length of roots. Lamont (1974) identified new root morphology (dauciform roots) in *Cyathochaete avenacea*; these roots differ greatly in shape and size from other families and are short-lived and may become endomycorrhizal (Shane et al. 2005).

Plant root structure can vary with changes in the direction/rate, growth, and proliferation of cells. However, the signals associated with these changes are poorly known, despite some studies on lateral root initiation and development (Blakely et al. 1982; Casero et al. 1993). Blakely et al. (1982) were the first to suggest that lateral root primordia could develop in radish (*Raphanus sativus* L.) from a small group of pericycle cells, which continue dividing after emerging from the root apical meristem. Later, other studies confirmed that the first anti-clinal division of pericycle cells starts the lateral root primordia development in sunflower (*Helianthus annuus* L.), maize (*Zea mays* L.) (Casero et al. 1993), and carrot (*Daucus carota* L.) (Lloret et al. 1989). Moreover, lateral root primordia initiation occurs in the meristem of some species, whereas others have lateral root primordia at distance from the parent root apex, i.e. outside the elongation zone. Demchenko and Demchenko (1996) observed that wheat (*Triticum aestivum* L.) root proliferation continues during lateral root primordia initiation, with the resumption of DNA synthesis in the endodermis, pericycle, and stellar parenchyma nuclei, which continues closer to the root tip compared to where the first dividing cells appear in these tissues, to establish a new growth site after completion of several cell divisions.

Phytohormones play an important role in the regulation of root growth in crop and pasture species. The presence and role of auxins in roots have been under discussion since the 1930s when first identified and applied to roots (Thimann 1936). Later, Bridges et al. (1973) and Elliott and Greenwood (1974) confirmed the presence of auxins in maize in the root apex, stele, and cortical cells, with the highest concentration (fresh weight basis) in stellar tissues (Bridges et al. 1973). About 10 years ago, Vanneste and Friml (2009) demonstrated that auxins initiate vascular tissue formation through the induction of a chain of plant developmental responses. Moreover, Mähönen et al. (2006) reported that cytokinins are involved in vascular patterning and act together with auxins in proliferation and differentiation of cell lineages that form root vascular tissues. Two decades ago, Riou-Khamlichi (1999) suggested that these phytohormones (auxins and cytokinins) not only regulate cell patterning, but also promote cell division. De Rybel (2013) discovered that *TMO5-LHW* (transcription factor complex), acting downstream of auxin, synchronizes cell division (periclinal) in the vasculature, which is important for radial patterning in roots.

Water access influences root dynamics in soil. In a recent study, Ahmadi et al. (2018) demonstrated that water availability significantly influenced RSA in wheat because drought stress decreased root branching and root length and increased specific root length, root:shoot ratio, and root tissue density. Nutrient availability in the soil also influences RSA dynamics; Weaver and Clements (1938) found that when nutrients were unevenly distributed in soil, roots preferred to develop in nutrient-rich patches. Wiersum (1958) reported that the inorganic, nutrient-rich, and ion (nitrate and phosphate)-specific exchange resins stimulated lateral root formation in pea (*Pisum sativum* L.). Similar responses of lateral root initiation and extension to nitrate were reported in barley (*Hordeum vulgare* L.; Drew et al. 1973) and wheat (Hackett 1972). Moreover, Drew (1975) found that high phosphate supply enhanced first- and second-order lateral root initiation and extension, relative to the low-phosphate patches. Craine et al. (2003) studied the relationship between RSA and resources (water and nitrogen) in 11 grassland plants (grasses, legumes, and forbs) and found that fine root biomass strongly influenced water availability and water and nitrogen uptake.

Phosphorus deficiency modulated RSA by increasing the number and growth of lateral roots, but high P increased the primary root length (Liao et al. 2001). Low P supply alters plant sensitivity to auxin via the TIR1 (transport inhibitor response 1) receptor and the *ARF19* transcription factor, which control auxin-responsive genes related to pericycle cell division and lateral root development (Pérez-Torres et al. 2008).

Nitrate supply also influenced the RSA as a uniform supply of high NO_3^- (50 mM) inhibited lateral roots in Arabidopsis (*Arabidopsis thaliana* L.), but low NO_3^- supply stimulated lateral roots (Zhang et al. 1999). In contrast, the high supply of NO_3^- stimulated lateral root formation in Arabidopsis (Linkohr et al. 2002).

In summary, the ontogenesis-based classification is mostly used for roots of field crops and pastures, whereas the topology-based classification is more suitable for tree species. The morphology and physiology of RSA are controlled by internal (phytohormones and genes) and external factors (biotic and abiotic), as plants develop special types of root morphology to adapt to specific soil and environmental conditions. Furthermore, soil physico-chemical properties (e.g. mechanical impedance, nutrient content, pH, etc.) influence the morphology, anatomy, and physiology of RSA. Research on root structure and physiology has excelled in the last four decades, but there are many mechanisms influencing root morphology and physiology that remain unknown.

4.3 Key Elements of Root System Architecture Relevant to Crop and Pasture Genotypes

4.3.1 Rooting Depth

Rooting depth is a crucial element of RSA and is determined by axial root growth in terms of the direction (defining the shape) and elongation (defining the dynamics of soil colonization) of roots. Rooting depth and root distribution in soil horizons are influenced by the edaphic, environmental, and genetic factors (Rogers and Benfey 2015).

Soil compaction restricts root growth and crop yield. A less compact soil produced 36% longer root length, 27% higher root length density, 24% greater root dry weight, and 19% higher grain yield in wheat than a compact (nonripped) soil (Chen et al. 2014). Phytohormones also influence root elongation under mechanical resistance because ethylene production increases with mechanical impedance, which affects root elongation and diameter (Sarquis et al. 1991).

Batra et al. (1975) suggested that auxin was transported acropetally in roots and regulated cell elongation in the proximal zone of root apical meristem. Acropetal transport of auxin takes place principally in the vasculature of the central cylinder (Shaw and Wilkins 1974). Van Overbeek (1966) opined that gibberellins are involved in root elongation, and Whaley and Kephart (1957) observed an increase in maize root growth with the application of gibberellic acid (GA). More recently, Achard et al. (2009) reported that GA was not only involved in cell elongation in the root elongation zone, but that it regulated the size of the meristem. However, Khan (1969) found that abscisic acid (ABA) inhibited barley root growth, and El-Antably and Larsen (1974) reported ABA-induced root growth inhibition in garden cress (*Lepidium sativum* L.) and faba bean (*Vicia faba* L.).

Mouchel et al. (2004) identified the locus *BREVIS RADIX* (*BRX*) as responsible for 80% of the root length variation in tested Arabidopsis plants and further demonstrated that auxin enhanced *BRX* expression (Mouchel et al. 2006). Scacchi et al. (2009) affirmed the role of *BRX* in auxin translocation to the nucleus. In the last 10 years, significant progress has been made in understanding the molecular mechanisms involved in auxin action (Chapman and Estelle 2009; Perrot-Rechenmann 2010) (see also Chapter 2).

Rooting depth is influenced by soil (Table 4.1) (water and nutrients) and environmental conditions (temperature). McMichael and Burke (2002) reported that the optimal temperature for root growth varied among plant species (17–35 °C) depending on their origin. Recently, Rogers and Benfey (2015) reviewed rooting depth in some pasture and crop species; root length density in the topsoil layer (0–0.1 m) was 10 cm cm^{-3} L_V (root length/unit soil volume) for cereals, 20 cm cm^{-3} for grasses, and 1–2 cm cm^{-3} for other crops. For instance, grasslands have greater rooting depth than crops on deep soils; the rooting depth for tropical and temperate grasslands were ~15 m and ~2.6 m, respectively, but only 2.1 m for crops (Rogers and Benfey 2015).

Different tropisms (phototropism, gravitropism, and hydrotropism) influence the root trajectory, and the root cap is the principal sensor of these tropisms (Kaneyasu et al. 2007). In soil, the root tip moves downward during elongation, while the root part adjacent to the elongation zone is anchored to the soil. During this time, the root must overcome the mechanical impedance – as rooting depth increases, the roots need more force to move through the soil, which increases root diameter and reduces root elongation/penetration (Hodge et al. 2009). Rooting depth is low on shallow soils or soils with chemical and physical restrictions that limit root growth. The benefits of deep roots in conferring drought avoidance are not evident on saline/saline-sodic soils or acidic soils in tropical areas, where physical and chemical properties restrict deep root growth (Rogers and Benfey 2015).

Clark et al. (2003) proposed that the capacity of roots to penetrate high-strength soils was directly associated with their diameter. Materechera et al. (1991) observed that dicots with larger root diameters had a greater rooting depth than grasses or monocots with smaller root diameters.

Recently, Yang et al. (2017) studied the root morphology of temperate pasture legumes in relation to P uptake. They reported a three to sixfold range in specific root length (79–281 m g^{-1} root) and a 6.1-fold range for root hair length (0.12–0.75 mm) in the tested pasture species. The most common pastures – subterranean clover (*Trifolium subterraneum* L.) and lucerne (*Medicago sativa* L.) – had lower specific root length and shorter root hairs than grasses, whereas the specific root length and root hair length of biserrula (*Biserrula pelecinus* L.), yellow serradella (*Ornithopus compressus* L.), and French serradella

Table 4.1 Influence of nutrient availability on root length.

Species	Nutrient/water supply	Root length	Reference
Arabidopsis Col-0	Low P (1 μM)	↓ Primary RL	López-Bucio et al. (2003)
	High P (1 mM)	↑ Primary RL	López-Bucio et al. (2003)
	Low P (1 mmol m^{-3})	↓ Primary RL	Bates and Lynch (1996)
	High P (1000 mmol m^{-3})	↑ Primary RL	Bates and Lynch (1996)
	Low SO_4^{2-}	↑ RL	Kutz et al. (2002)
	High SO_4^{2-}	Shallower RL than SO_4^{2-}-starved plants	Kutz et al. (2002)
Roundhead bush clover (*Lespedeza capitata* Michx.)	↓ Inorganic N uptake No difference in water acquisition	↑ Fine root depth, biomass	Craine et al. (2003)
Dalea villosa (*Petalostemum villosum*)	↑ N mineralization in soil solution		Craine et al. (2003)
Quackgrass (*Agropyron repens* L. Beauv)	Acquisition of water and N from deep layers and low N patches	↓ N fine roots	Craine et al. (2003)
Big bluestem grass (*Andropogon gerardii* Vitman)		↑ Root biomass density	Craine et al. (2003)
Bluestem (*Schizachyrium scoparium* (Michx.) Nash)			Craine et al. (2003)
Blue grama (*Bouteloua gracilis* Willd. ex Kunth)	↓ Water and inorganic N uptake	Shallow fine roots	Craine et al. (2003)
Prairie junegrass (*Koeleria cristata* (L.) Pers.)			Craine et al. (2003)
Common meadow grass (*Poa pratensis* L.)			Craine et al. (2003)
Jarrow (*Achillea millefolium* L.)	↓ Water and inorganic N uptake	↓ SRL	Craine et al. (2003)
Indian balm (*Anemone cylindrica* Gray.)		Shallow fine root biomass	Craine et al. (2003)
Stiff goldenrod (*Solidago rigida* L.)			Craine et al. (2003)
Alpine pennygrass (*Thlaspi caerulescens* L.) (Zn hyper-accumulator)	250–1000 mg Zn kg^{-1} soil	↑ RL in Zn-enriched patch	Whiting et al. (2000)

(*Continued*)

Table 4.1 (Continued)

Species	Nutrient/water supply	Root length	Reference
Common bean	P-efficient genotype	↓ RL	Liao et al. (2001)
	P-inefficient genotype	↑ RL	Liao et al. (2001)
Sunnhemp	Low P (0 mg kg^{-1} soil)	RL 38.8% ↑ at high P (200 mg kg^{-1} soil)	Fageria et al. (2015)
Pigeonpea		RL 7.3% ↓ at high P (200 mg kg^{-1} soil)	Fageria et al. (2015)
Lablab		RL 32.2% ↓ at high P (200 mg kg^{-1} soil)	Fageria et al. (2015)
Grey mucuna bean		RL 38.2% ↓ at high P (200 mg kg^{-1} soil)	Fageria et al. (2015)
White jack bean		RL 31.2% ↓ at high P (200 mg kg^{-1} soil)	Fageria et al. (2015)
Rice	N-efficient genotypes	↑ RL and RLD, root biomass, ↑ ROA, ↑ N uptake, ↑ GY	Ju et al. (2015)
	Drought-resistant rice genotype	↑ RL, ↑RLD, ↑ RDW, ↑ ROA, ↑ WP, ↑GY	Chu et al. (2014)
Wheat	Deep P placement	↑ RL, ↑ RDW, ↑ WUE, ↑ GY	Kang et al. (2014)
	Deep band placement of P	↑ RL, ↑ RLD, ↑ RDW, ↑ WUE, ↑ GY	Kang et al. (2014)
White lupin (*Lupinus albus* L.)	Low P	Proteoid root formation	Johnson et al. (1996)

↑, increase; ↓, decrease; RL, root length; SRL, specific root length; RLD, root length density; RDW, root dry weight; WP, water productivity; WUE, water use efficiency; ROA, root oxidation activity; GY, grain yield.

(*Ornithopus sativus* Brot.) were similar to grasses. Furthermore, specific root length was strongly correlated with root diameter (Yang et al. 2017).

In soybean, high root length was positively associated with high P and N uptake under normal and drought stress conditions (He et al. 2017). Liao et al. (2001) observed that a P-inefficient genotype of common bean (*Phaseolus vulgaris* L.) produced deeper roots under P starvation, whereas a P-efficient genotype produced more lateral roots with shallow primary root length. They further reported that the P response regulated the root geotropic response and that genotype P efficiency was related to root foraging in shallow depths during P stress (Table 4.1). Recently, Fageria et al. (2015) reported that a high P supply (200 mg P kg^{-1} soil) increased specific root length in some tropical legumes (pigeonpea [*Cajanus cajan* L. Millspaugh], lablab [*Dolichos lablab* L.], grey mucuna bean [*Mucuna cinereum* L.], and white jack bean [*Canavalia ensiformis* (L.) DC.]), but a low P supply increased specific root length in sunn hemp (*Crotalaria juncea* L.).

Rooting depth influences crop yield and N uptake. Ju et al. (2015) observed that N-efficient rice (*Oryza sativa* L.) genotypes had deeper roots and higher root length density and biomass than N-inefficient genotypes, resulting in increased N uptake and grain yield. Rice cultivars with high root length, root length density, root dry weight, and root antioxidant activity had increased water productivity and grain yields than those with lower values (Table 4.1; Chu et al. 2014).

In summary, there is a large variation in rooting depth among plant species and crop genotypes. Root length is influenced by plant hormones (auxins, gibberellins, and ABA). Moreover, soil temperature and nutrient and water supply influence rooting depth. Plants with deep roots tolerate drought stress better than those with shallow roots and can acquire water and nutrients from deep soil layers.

4.3.2 Root Hairs

Root hairs, as defined by Gilroy and Jones (2000), are structures projecting from the root surface that help in water and nutrient acquisition and soil anchorage. They also increase distribution of root exudates (Ryan et al. 2001), enhance root surface area for resource acquisition (Table 4.2; Wasson et al. 2012), explore soil patches in macro-structured soils (Misra et al. 1988), indirectly improve soil health, and serve as the site for microbial interactions with plants (Wasson et al. 2012). Root hairs are unicellular, and alterations in length and number enhance root functions depending on the soil properties (Meister et al. 2014). Root hairs have distinctive features and functions and can be a new model for understanding plant biology. Root hairs contribute to cell determination, expansion, and differentiation, nutrient and water uptake, colonization by N_2-fixing microbes, and nodule formation (Libault et al. 2010). Root water channels (Brechenmacher et al. 2009) as well as nitrogen (Kojima et al. 2007), phosphate (Brechenmacher et al. 2009), and sulphate (Maruyama-Nakashita et al. 2004) transporters are present on root hairs, which also take up auxins (AUXIN-RESISTANT1) for root elongation (Cho et al. 2007). Moreover, Pb and Cd exudation from root hairs can contribute to resistance to heavy metal toxicity (Kim et al. 2007).

Research on root hairs started more than 125 years ago (Jamieson 1893). Snow (1905) reviewed the studies on root hair biology up to the early 1900s. Later, Farr (1927) studied root elongation in rice, oats (*Avena sativa* L.), and brassicas (*Brassica oleracea* L.) and concluded that the average length of hairless cells was greater than that of hair-forming cells. He opined that differences in cortical and epidermal cell elongation played a key role in hair production. Miller (1938) discussed the physiology of root hairs. In the last two decades, many studies have examined root hair structure (Maruyama-Nakashita et al. 2004; Brechenmacher et al. 2009; Meister et al. 2014), functioning (Misra et al. 1988; Ryan et al. 2001; Cho et al. 2007; Wasson et al. 2012), and genetics (Xu et al. 2005; Costa and Shaw 2006; Tominaga-Wada et al. 2009; see also Chapter 2) in relation to their role in nutrient acquisition from low-nutrient soils, water and nutrient transporters, soil anchorage, organic compound and phytohormone exudation, and genes and transcription factors controlling root hair development.

Nye (1966) highlighted the role of root hairs in P acquisition. Phosphorus is a soil-immobile nutrient; roots with long root hairs (in the absence of mycorrhizal associations) have enhanced P uptake (Gahoonia et al. 1997). Furthermore, Baylis (1975) and St John (1980) suggested that nonmycorrhizal plants had longer root hairs than those with mycorrhizal associations. Indeed, nonmycorrhizal plants have evolved long root hairs to

Table 4.2 Influence of nutrient availability on root hair growth.

Species	Nutrient/water supply	Root hair growth	Reference
Arabidopsis	Low P (1 mmol m^{-3})	↑ RHL, ↑ RHD	Bates and Lynch (1996)
	High P (1000 mmol m^{-3})	↓ RHL, ↓ RHD	Bates and Lynch (1996)
	Low P	↑ RHL, ↑ RHD	López-Bucio et al. (2003)
	Low Fe	↑ RHL, ↑ RHD	López-Bucio et al. (2003)
Tufted grass (*Holcus lanatus* L.)	Low NO_3^- (0.02 mol m^{-3})	↑ RHL, ↓ RHD, ↓ SRHL	Robinson and Rorison (1987)
	Low NH_4^+ (0.02 mol m^{-3})	↓ RHL, ↑ RHD, ↑ SRHL	Robinson and Rorison (1987)
Wavy hair-grass (*Deschampsia flexuosa* (L.) Trin.)	Low-medium NO_3^- (0.02–0.2 mol m^{-3})	↑ RHL, ↑ RHD, ↑ SRHL	Robinson and Rorison (1987)
	Low NH_4^+ (0.02 mol m^{-3})	↑ RHL, ↑ RHD, ↑ SRHL > NO_3^- (0.02 mol m^{-3})	Robinson and Rorison (1987)
Meadow grass (*Poa annuua* L.)	Low NO_3^- (0.02 mol m^{-3})	↑ RHL, ↑ RHD, ↑ SRHL	Robinson and Rorison (1987)
	Low NH_4^+ (0.02 mol m^{-3})	↑ RHL, ↑ RHD, ↑ SRHL > NO_3^- (0.02 mol m^{-3})	Robinson and Rorison (1987)
Perennial ryegrass (*Lolium perenne* L.) cv. S.23	Low NO_3^- (0.02 mol m^{-3})	↓ RHL, ↑ RHD, ↑ SRHL > NH_4^+ (0.2 mol m^{-3})	Robinson and Rorison (1987)
	Low NH_4^+ (0.2 mol m^{-3})	↑ RHL than NO_3^- ↑ RHD, ↑ SRHL < NO_3^- (0.02 mol m^{-3})	Robinson and Rorison (1987)
Perennial ryegrass cv. Norlea	Low NO_3^- (0.02 mol m^{-3})	↑ RHL, ↑ RHD, ↑ SRHL	Robinson and Rorison (1987)
	Low NH_4^+ (0.02 mol m^{-3})	↑ RHL, ↑ RHD, ↑ SRHL	Robinson and Rorison (1987)
Rape (*Brassica napus* L.)	High P (>100 μM)	↓ RHL and low frequency of root hairs	Foehse and Jungk (1983)
Spinach (*Spinacia oleracea* L.)			Foehse and Jungk (1983)
Tomato (*Solanum lycopersicum* L.)			Foehse and Jungk (1983)
Rape	Low Pi (<10 μM)	↑ RHL, ↑ RHD	Foehse and Jungk (1983)
Spinach			Foehse and Jungk (1983)
Tomato			Foehse and Jungk (1983)
Spinach	0.01 mM NO_3^-	Sevenfold ↑ RHL than at 1 mM NO_3^-	Foehse and Jungk (1983)

↑, increase; ↓, decrease; RHL, root hair length; RHD, root hair density; SHRL, specific root hair length.

take up water and nutrients (reviewed by Parniske 2008). For example, Purnell (1960) observed that some Proteaceae species have long root hairs, lack mycorrhizae, and have cluster-type roots (proteoid roots). These long root hairs help take up nutrients from the soil. Similar cluster-type roots were also observed in a number of native legumes (family Fabaceae) (Adams et al. 2002).

Root hair growth is vital for increasing the surface area of root hairs to enhance nutrient and water absorption and soil interactions. Root tip growth is a steady process of exocytosis of cell wall compounds and plasma membrane vesicles at the extending cell apex (Libault et al. 2010). Moreover, the rearrangement of cytoskeleton components, i.e. myosins (Peremyslov et al. 2008) and actin (ACT2 and ACT7) (Bao et al. 2001; Tominaga-Wada et al. 2009), is crucial for root hair elongation (Libault et al. 2010). Higher Ca^{2+} ionic activity in the cytosol at the root hair apex, fluctuations in extracellular and intracellular pH, reactive oxygen species (ROS) generation, and phospholipid signalling have been suggested as the regulators of root growth and polarity (reviewed by Libault et al. 2010). The increased ionic activity of Ca^{2+} in the cytosol, due to its release from mitochondria after actin filament disruption, indicated that normal polymerization and depolymerization of actin filaments regulated Ca^{2+} storage in the cytoplasm of root hair cells (Wang et al. 2010). Moreover, an increase in the length and number of root hairs can improve water and nutrient acquisition in crop and pasture species. For instance, Brown et al. (2013) showed that increasing root hair length and longevity (rather than root density) enhanced P uptake. In legumes, root hairs help with N_2 fixation by responding to the Nod factors via altering the depolarization of the plasma membrane, ion fluxes, Ca spiking, and early expression of the nodulin gene (Amor et al. 2003). The Nod factors induce the expression of the nodulin genes and promote cell division in the cortical cells to form nodule primordia (Amor et al. 2003).

The determination and differentiation of root hairs are dependent on cell position and is controlled by a signal exchange between cortical and epidermal cells (N and H cells) of neighbouring roots, and between N cells that differentiate into root hair and H cells that would form root hair or remain hairless (Libault et al. 2010), leading to the expression of *GLABRA2* (*GL2*) in N cells through the modification of chromatin that alters acetylation of histone and ultimately results in differentiation of hairless cells (Costa and Shaw 2006; Xu et al. 2005). Galway et al. (1994), Di Cristina et al. (1996), and Masucci et al. (1996) suggested that the *TRANSPARENT TESTA GLABRA* (TTG) and *GLABRA2* (*GL2*) genes suppress hair formation in differentiating hairless cells and act as a negative transcriptional regulator of root hair development. Wada et al. (1997) found that the *MYB* homologue determines epidermal cell differentiation in root hair formation. Moreover, *GL2* may be involved in the regulation of cell differentiation by changing cell wall composition as it controls the *xyloglucan endotransglucosylase 17* (*XTH17*) and *cellulose synthase5* (*CESA5*) expression (Tominaga-Wada et al. 2009).

Root hairs increase the root surface area and take up nutrients and water by their numerous water channels and ion transporters present on the root hair surface. Moreover, in legume crops, root hairs are important for N_2 fixation. In the last two decades, many genes have been identified that control the formation and elongation of root hairs. These genes can be exploited to improve RSA for efficient nutrient and water acquisition.

4.3.3 Root Branching

Root architecture is determined by the root branching pattern and the rate of growth and elongation of individual roots, both of which are not solely under genetic control, but are highly plastic and influenced by external and internal factors (Forde and Lorenzo 2001). However, 25 years ago, only a few studies had reported the early stages of lateral root initiation (Casero et al. 1993; Demchenko and Demchenko 1996). Torrey (1965) reviewed the physiology of lateral root development. Goebel (1928) and Voronin (1964) reported that all ancestral root branching forms initiated lateral roots, and development started with some endogenous compounds, most likely phytohormones in the vascular cylinder. Street (1969) discussed the role of auxins in lateral root initiation and their interaction with cytokinins and gibberellins. He concluded that ABA inhibited the lateral root initiation in excised tomato root cultures, but potential inhibition of primary root extension remained unclear. McCully (1975) discussed the anatomical aspects of root branching and reported that the position of lateral roots in relation to primary roots was significant and helped in understanding the evolution of organ development (Barlow 1984; Barlow and Adam 1988; Barlow et al. 2004). Esau (1977) suggested that the parent roots in most flowering plants served as the main axis from which root branching started and that lateral root initiation took place in the pericycle (outer layer of vasculature). Lateral root formation from the parental apical meristem has been identified in several plant families, including Cucurbitaceae (Gulyaev 1964), Polygonaceae (O'Dell and Foard 1969), and Convolvulaceae (Seago 1973) – or from proximal roots to the root elongation zone some distance from the root apex in specific crops, e.g. Brassicaceae family (Ilina et al. 2018). Celenza et al. (1995) identified some genes involved in the lateral root initiation and development in Arabidopsis.

Plant nutrient availability in soil (e.g. NO_3^-, NH_4^+, P, Zn, etc.) influences root branching (Table 4.3), with research advancing in the last two decades (Zhang et al. 1999; Forde and Lorenzo 2001; Haines 2002). Genes and transcription factors involved in lateral root initiation and regulation have been identified (Benková et al. 2003; De Smet 2012) (see also Chapter 2). Lateral roots respond to *ACR4* (receptor-like kinase) to regulate the number of progenitor cells in the root apical meristem; while the root apical meristem is a permanent organ, the lateral root generation takes place when needed (De Smet et al. 2008). The origin of primary and lateral roots differs, but is similar at a mechanistic level. For instance, auxin is needed for the initiation of primary and lateral roots (Péret et al. 2009; De Smet et al. 2010b; Lavenus et al. 2013). Auxin regulates primary root initiation by positively regulating the *PIN-FORMED 1* (*PIN1*) expression, which acts as an auxin efflux carrier and promotes auxin transport from the embryo to the hypophysis precursor, thus providing a positional signal for hypophysis specification (Benková et al. 2003). Moreover, *PIN-1* dependent auxin is critical for lateral root initiation (Benková et al. 2003). Several auxin responses related to molecular components play a vital role in lateral root initiation (De Smet 2012). Cytokinins also play an important role in lateral root initiation by antagonizing the function of auxins. Disturbances in cytokinin synthesis, perception, and signal transduction influence organogenesis in lateral root development (Bielach et al. 2012). Furthermore, ABA inhibits lateral root development (reversibly) after emergence from the parent root and before lateral root meristem activation (De Smet et al. 2003). Ethylene is also involved in lateral root formation. For instance, ethylene production induced under

Table 4.3 Influence of nutrient availability on root branching.

Species	Nutrient/water supply	Root branching	Reference
Arabidopsis Col-0	Low P (1 μM)	Three to fivefold ↑ LRN and LRD	López-Bucio et al. (2003)
	High P (1 mM)	Three to fivefold ↓ LRN and LRD	López-Bucio et al. (2003)
	Low P (1 mmol m^{-3})	↑ LRN (twofold)	Bates and Lynch (1996)
	High P (1000 mmol m^{-3})	↓ LRN	Bates and Lynch (1996)
	NO_3^- (1 mM KNO_3)	↑ LRL	Zhang et al. (1999)
	NO_3^- (50 mM KNO_3)	↓ LRL	Zhang et al. (1999)
Alpine pennygrass (*Thlaspi caerulescens* L.) (Zn hyper-accumulator)	250–1000 mg Zn kg^{-1} soil	↑ LRL and LRN in Zn-enriched patch	Whiting et al. (2000)
	No Zn	↓ LRL ↓ RDW	Haines (2002)
	Heterogenous Zn supply	↑ LRL ↑ RDW	Haines (2002)
	Homogenous Zn	↑ LRL < heterogenous Zn, ↓ RDW	Haines (2002)
Barley (*Hordeum vulgare* L.)	High NO_3^-	↑ LR initiation and extension	Drew et al. (1973); Drew (1975)
	High NH_4^+	In the absence of NO_3^-, ↑ LR stimulation	Drew (1975)
	High PO_4	↑ LR formation	Drew (1975)
	High K	Intermediate response to LR formation	Drew (1975)
Wheat	High NO_3^-	↑ Lateral root initiation and extension	Hackett (1972)
Common bean	No P		
	High P (1 mM)		

↑, increase; ↓, decrease; LR, lateral roots; LRN, lateral root number; LRD, lateral root density; RDW, root dry weight.

flooding or other treatments increased lateral root development (Mergemann and Sauter 2000) because ethylene enhanced auxin synthesis and transport to the elongation zone, which activated a local auxin response and inhibited cell elongation.

Nutrient availability in the soil also affects lateral root formation and growth; NO_3^- and phosphate are the most influential inorganic ions involved in lateral root stimulation in pea (Wiersum 1958). Similar responses have been observed in wheat (Hackett 1972) and barley (Drew et al. 1973) in response to NO_3^- availability. Liao et al. (2001) demonstrated that P availability influenced RSA in bean roots; the basal root angle in the P-efficient genotype was shallower under low compared with high P availability. Moreover, under low P supply, root length increased and fewer lateral roots were formed, whereas the inverse was true under high P supply (Miller 1938; Liao et al. 2001). In a recent study, He et al. (2017) demonstrated that under drought stress increased adventitious and lateral root branching enhanced P and N uptake, which was positively associated with grain yield.

Root branching is a key component of RSA because it increases root surface area and volume. Auxins are the principal phytohormones involved in lateral root initiation. The *PIN-1* gene has been identified as a regulator of lateral root initiation. Other plant hormones (cytokinins and ABA) influence lateral root initiation. Lateral root number and length are also influenced by soil nutrient availability. Increased lateral root growth and density can improve the acquisition of soil-immobile nutrients.

4.3.4 Root Architectural Ideotypes

Improving crop yields has become increasingly important in this era of climate change, especially on marginal lands. Improved yields of many plant species have occurred by targeting above-ground plant components while largely ignoring the roots. The concept of ideotype breeding was coined by Donald (1968) to combine plant traits that contribute to increasing crop yields. He found a flaw in the yield selection approach that did not explain the reasons for yield improvement. He suggested studying root traits and how they contribute to yield and then combining those traits through breeding. Ideotype breeding successfully contributed to higher grain yields in maize (Mock and Pearce 1975), wheat (Reynolds et al. 1994), and common bean (McClean et al. 2011) and has received considerable attention in the last 20 years (Dunbabin et al. 2003; Lynch 2011, 2013; White et al. 2013). Vigorous root architecture is crucial for plant fitness, but advances in altering RSA to improve crop productivity have been slow (Comas et al. 2013). Spatial and temporal distribution of RSA influences the source/sink relationship and thus directly governs crop yield in the standard and stressed environments (Uga et al. 2013; Rogers and Benfey 2015). Future breeding programmes based on root-centric ideotypes may be a relatively quick strategy to enhance plant resource-use efficiencies.

Plants need nutrients for growth and development (Havlin et al. 2004); however, soil physico-chemical properties and RSA traits influence water and nutrient acquisition from the soil and thus plant growth and productivity (Chen et al. 2014). Recently, He et al. (2017) have suggested that RSA traits play a key role in N, P, and water uptake under conditions of low P and moisture availability. Root architectural traits imparting efficiency in resource acquisition from the soil are critical for breeding of crop plants for efficient water and nutrient uptake (Chen et al. 2014). Therefore, the development of genotypes with suitable RSA traits will improve crop yields on marginal or poorly fertile soils (Lynch 2011).

Different studies have proposed a steep (steep growth angle), cheap (low metabolic investment), and deep (long primary root with few but deep lateral roots) root ideotype for N acquisition (Dunbabin et al. 2003; Lynch 2013; White et al. 2013) with early root vigour, high root biomass and density, aerenchyma cells in the cortex, large surface area (root branching and root hairs), high water uptake, a strong association with N_2-fixing microbes, and enhanced exudation of nitrification inhibitors (White et al. 2013). Subbarao et al. (2012) proposed a high root length density ideotype that can proliferate in NH_4^+-rich patches and release a nitrification inhibitor to reduce its conversion to NO_3^- and maintain NH_4^+ availability in the root zone.

For enhanced P uptake on low P soils, a topsoil foraging (more lateral roots at shallow depth) ideotype has been effective in soybean (*Glycine max* L.) (Wang et al. 2010) and common bean cultivars (Lynch 2011). Brown et al. (2013) indicated that P uptake could be enhanced by increasing root hair length and longevity rather than enlarging root density. Root hairs are responsible for ~90% of P acquisition under low-P conditions and, in some crop and pasture species, they can contribute two-thirds of the root surface area (Raghothama 2005). Similarly, K-efficient root ideotypes have deeper roots, greater root density, biomass, and root hairs, increased organic acid anion and H^+ exudation, strong capacity to explore high-K patches, and increased capacity of root cells to take up K and water (White et al. 2013). White and Greenwood (2013) suggested that a topsoil foraging ideotype was effective in uptake of poorly mobile Cu, Mn, and Ni, which are relatively enriched in the topsoil. However, this ideotype is not ideal for Zn and Fe because they are more evenly distributed and more phytoavailable in the soil profile. A root ideotype with root biomass that is evenly spread throughout the soil profile and capable of proliferatic lateral roots in nutrient-rich patches is suitable for Mn, Fe, and Zn uptake (White and Greenwood 2013).

In the changing climate, crop genotypes with high water-use efficiency to boost yields under reduced water supply are needed in the irrigated systems (Thorup-Kristensen and Kirkegaard 2016). Irrigated root ideotypes should proliferate into the nutrient- and water-rich zones near the water source, have varied axial root angles at shallow depth and restricted lateral foraging in the wet zone, salt/waterlogging tolerance, and transplant vigour (Schmidt and Gaudin 2017).

In conclusion, characterizing root traits specific to irrigated and rain-fed cropping systems and understanding soil-root interactions, root ecology, and soil mineralogy will contribute to the development of root ideotypes that can be used in breeding programmes to develop crop cultivars suited to specific environments.

4.4 Genetics of Root System Architecture

The development of RSA shows evidence of developmental instability (Forde 2009), resulting in little success in the characterization of RSA genetics and its potential use in plant breeding. For instance, in field crops such as wheat, maize, soybean, and rice, some success has been achieved in the characterization of genes controlling RSA (Table 4.1). Most of the research on root genes can be grouped on the basis of discovery techniques (QTL, mutant) or response variables (phytohormones, nutrients, and water). However, these studies have not been combined systematically to understand the genetics of RSA (Jung and McCouch 2013).

Research on the genetics of RSA has been underway for 25 years (Kawahara et al. 1995; Meijer et al. 1997) and has advanced in wheat, maize, rice, and soybean, but is still an area requiring numerous discoveries (Meister et al. 2014). Several genes involved in root formation have been identified. For instance, Schizorhiza (SCZ), belonging to the family of heat shock transcription factors, is needed to establish the progenitor cells, which produce support tissue in the embryonic root meristem (Pernas et al. 2010). The *SCZ* gene plays a role in deciding the fate of epidermal cells in the CEI (cortex endodermis initial) daughter cells and supernumerary periclinal cell division in roots (Mylona et al. 2002). Two genes, *SCM* (Kwak and Schiefelbein 2008) and zinc finger *JKD* (Hassan et al. 2010), play a critical role in the initial epidermal cell patterning in the root meristem.

The genes involved in root cap elongation (Willemsen et al. 2008) and maturation (Bennett et al. 2010) have been identified. Wiśniewska et al. (2006) suggested that the *PIN* family proteins controlled auxin distribution and gravitropism. Liscum and Briggs (1995) identified the *NONPHOTOTROPIC HYPOCOTYL 1/PHOTOTROPIN 1* genes involved in the negative phototropism, and Kiss et al. (2003) detected "phytochrome A and B" genes associated with the negative phototropism.

Most of the genes identified to date are homeobox-containing genes (Table 4.4; Kawahara et al. 1995; Meijer et al. 1997) and phytohormone-responsive genes, especially auxin-induced genes (Walker and Key 1982; Kitomi et al. 2011). Imin et al. (2006) discovered that auxin addition to *Medicago truncatula* induced the AP2-domain comprising *PLETHORA 1* and *2*, homeobox domain *WOX5*, and *BABY BOOM1* transcription factor, whereas cytokinin addition reduced the expression of these transcription factors, indicating the role of these genes in the progenitor cell formation in roots. Furthermore, homeodomain *WOX5* is expressed in the quiescent centre of root apical meristem in Arabidopsis and plays a crucial role in root apical meristem formation during root development (embryonic and post-embryonic root formation) (Xu et al. 2006).

Using breeding or transgenic approaches to target suppression or overexpression of identified genes can modify root architecture. For instance, in rice, many genes control RSA, including auxin-regulated crown rootless (*crl5*) and adventitious rootless (*arl1*) that encode the lateral organ boundaries (LOB) domain transcription factor conserved across dicots and monocots (Kitomi et al. 2011), the auxin biosynthesis gene *OsYUCC1* (Yamamoto et al. 2007), and ubiquitin ligase homologous to *AtCAND* (Wang et al. 2011). Furthermore, the loci that regulate root expansion include glucosamine-6-P acetyl-transferase encoded by *GNA1* (Rebouillat et al. 2009; Coudert et al. 2010), *Osglu3-1* (a putative membrane-bound endo-1,4-beta-glucanase, Zhang et al. 2012), *OsRPK1* (a Ca^{2+}-independent Ser/Thr kinase, Zou et al. 2014), and other candidates such as WUSCHEL-related homeobox 3A (*OsWOX3A*) and *OsARF16* (an integrator of phosphate starvation and auxin response) (Yoo et al. 2013; Shen et al. 2013), which control lateral root development (see also Chapter 2). Chen et al. (1998) discovered that *AGRAVITROPIC 1 (AGR1)* influenced gravitropism because it was involved in polar auxin transport.

In maize, the genetics of RSA can be manipulated by suppression or overexpression of certain genes. For instance, the LOB domain protein is encoded by the genes rootless concerning crown and seminal roots (RTCS) and its downstream target auxin-response factor *ARF34* that regulate nodal root formation in maize (Majer et al. 2012). Root hair mutants *root hairless 1* and 3 (*RTH1* [code sec3 homology of yeast] and *RTH3* [GPI anchor COBRA protein]) affect root hair elongation (Hochholdinger and Tuberosa 2009), whereas *RTH5* is responsible for coding NADPH oxidase specific for monocots (Nestler et al. 2014). Furthermore, lateral root 1 *LTR1*, short lateral root *SLR1*, and *SLR2* loci influence lateral root development in maize (Hochholdinger et al. 2005). The *RUM1* locus in maize encodes an auxin response regulator that affects lateral and seminal root initiation (Von Behrens et al. 2011).

Table 4.4 Genes involved in regulation of root growth and nutrient and water acquisition.

Gene/QTL	Development process	Function	Reference
AGR1	Root gravitropism	Auxin signalling	Chen et al. (1998)
SHY2	Lateral root growth and gravitropism	Encode IAA3	Tian and Reed (1999)
PIN1	Primary root and lateral root initiation	Auxin signalling	Benková et al. (2003)
BRX	Root elongation	Auxin translocation towards nucleus	Mouchel et al. (2004, 2006); Scacchi et al. (2009)
RTCS	Shoot-borne and seminal root initiation	Auxin signalling	Taramino et al. (2007)
Rtcl	Shoot-borne root elongation	Auxin signalling	Xu et al. (2015)
rum1	Seminal and lateral root initiation	Auxin signalling	von Behrens et al. (2011)
bige1	Number of seminal and shoot-borne roots	Multidrug and toxin extrusion transport	Suzuki et al. (2015)
rth1	Root hair elongation	Exocytotic vesicle fusion	Wen et al. (2005)
rth3		Cellulose organization	Hochholdinger et al. (2008)
rth5		Cell wall loosening	Nestler et al. (2014)
rth6		Cellulose synthesis	Li et al. (2016)
GL2	Root hair development	Cell differentiation Cellulose synthesis	Galway (1994); Di Cristina et al. (1996); Masucci et al. (1996)
WOX5	Root development	RAM formation	Xu et al. (2006)
QTL	Drought resistance	Nodal angle	Mace et al. (2012)
QTL	Drought resistance	Nodal angle	Christopher et al. (2013)
QTL	Growth in low nutrient conditions	Lateral and axial root growth under low P and N	Ingram et al. (2012)
LTN1	P acquisition	Crucial Pi starvation signalling component	Hu et al. (2011)
PIP2,7	Water uptake	Increase water uptake	Li et al. (2008)
DRO1	Drought tolerance	Steeper root angles, higher rooting depth, and more robust seedling gravitropic responses	Uga et al. (2013)
SULTR1, SULTR2	Influx of sulphate	Upregulation during sulphate starvation	Maruyama-Nakashita et al. (2004)
PDR8	Heavy metal tolerance	Efflux of lead and cadmium	Kim et al. (2007)
NIP5; 1	Drought tolerance	Enhanced water uptake	Takano et al. (2006)
PT1	P acquisition	Enhanced expression under P starvation	Chiou et al. (2001)

(*Continued*)

Table 4.4 (Continued)

Gene/QTL	Development process	Function	Reference
NRT2;1	N acquisition	Up-regulation during NO_3^- starvation	Ono et al. (2000)
CM1	NO_3^- uptake	Encodes NO_3^--inducible NO_3^- transporter	Tsay et al. (1993)
AtNRT1	NO_3^- uptake	Limitation of lateral root growth	Zhang and Forde (1998); Krouk et al. (2010)
AtOCT1	NO_3^- uptake	Lateral foraging in high inorganic N patches	Lelandais-Briere et al. (2007)

Many genes controlling nutrient uptake have been identified in the last 20 years (e.g. Zhang and Forde 1998; Maruyama-Nakashita et al. 2004; Krouk et al. 2010). For instance, *LTN1* and *PT1* are involved in P uptake and homeostasis; their expression is enhanced under P starvation (Chiou et al. 2001; Hu et al. 2011). Likewise, enhanced expression of *AtNRT1* increased NO_3^- uptake from deep soil layers by limiting lateral root growth and increasing rooting depth (Zhang and Forde 1998; Krouk et al. 2010). The *AtOCT1* is linked to enhanced lateral foraging in inorganic N-rich patches (Lelandais-Brière et al. 2007). Maruyama-Nakashita et al. (2004) observed *SULTR1* and *SULTR2* upregulation under SO_4^{2-} starvation.

Many QTL have been identified in maize that influences RSA and yield sustainability across different genetic backgrounds and water availability scenarios (Landi et al. 2010). Other QTL have been reported that control root length, root density, and root surface area (Cai et al. 2012). Most of these genetic loci have not been targeted for transgenic manipulation; it should be borne in mind that the effects on RSA may be accompanied by pleiotropic effects (Bian et al. 2012).

In conclusion, research on the genetics of RSA has advanced in the last three decades, but with only a few plant species being explored comprehensively. In recent years, many genes and transcription factors controlling plant hormones, cell division, cell elongation, root apical meristem, and nutrient and water uptake have been identified. These findings suggest the need for precision in transgene expression and a better understanding of genetic and functional interactions among loci to inform gene selections, which can help in the development of plant genotypes with improved RSA.

4.5 Conclusions and Perspectives

Roots are an integral plant component, being involved in nutrient and water uptake. The physiology and morphology of roots are complex and flexible and are influenced by edaphic and environmental conditions. Studies on root biology started almost 150 years ago. In the second quarter of the twentieth century, research focused on the role of plant hormones in root growth and lateral root initiation. However, rapid advances in root research have occurred in the last 50 years, following the confirmation of the role of phytohormones (auxins, cytokinins, and ethylene) in root development.

Understanding the genetics of RSA has advanced in the last 30 years. Application of molecular and biotechnology approaches has helped elucidate the genetics and mechanisms of root development, phytohormonal regulation, and resource acquisition. However, the genetics of RSA has been identified fully in a few field crops only. The application of new phenotyping technologies in root research has helped identify potential root ideotypes with desirable traits.

In future, high-throughput field proxy screening methods need to be developed with the more efficient image and data analyzing software to understand RSA under natural conditions. Furthermore, a root phenotying platform needs to be developed that can measure root cell/tissue anatomy quickly with easy sample preparation. Moreover, research on genetics and the mechanisms of root growth regulation and water and nutrient acquisition in pastures and crops is needed.

References

Achard, P., Gusti, A., Cheminant, S. et al. (2009). Gibberellin signaling controls cell proliferation rate in *Arabidopsis*. *Current Biology* 19: 1188–1193.

Adams, M.A., Bell, T.L., and Pate, J.S. (2002). Phosphorus sources and availability modify growth and distribution of root clusters and nodules of native Australian legumes. *Plant, Cell & Environment* 25: 837–850.

Ahmadi, J., Pou-Aboughadareh, A., Fabriki-Ourang, S. et al. (2018). Screening wheat germplasm for seedling root architectural traits under contrasting water regimes: potential sources of variability for drought adaptation. *Archives of Agronomy and Soil Science* 64: 1351–1365.

Amor, B.B., Sidney, L., Shaw, S.L. et al. (2003). The NFP locus of *Medicago truncatula* controls an early step of Nod factor signal transduction upstream of a rapid calcium flux and root hair deformation. *The Plant Journal* 34: 495–506.

Bao, Y., Kost, B., and Chua, N.H. (2001). Reduced expression of α-tubulin genes in *Arabidopsis thaliana* specifically affects root growth and morphology, root hair development and root gravitropism. *The Plant Journal* 28: 145–157.

Barlow, P.W. (1984). Positional controls in root development. In: Positional Controls in Plant Development (eds. P.W. Barlow and D.J. Carr), 281–318. Cambridge University Press.

Barlow, P.W. and Adam, J.S. (1988). The position and growth of lateral roots on cultured root axes of tomato, *Lycopersicon esculentum* (Solanaceae). *Plant Systematics and Evolution* 158: 141–154.

Barlow, P.W., Volkmann, D., and Baluška, F. (2004). Polarity in roots. In: Polarity in Plants (ed. K. Lindsey), 192–241. Oxford: Blackwell Publishing.

Bates, T.R. and Lynch, J.P. (1996). Stimulation of root hair elongation in *Arabidopsis thaliana* by low phosphorus availability. *Plant, Cell & Environment* 19: 529–538.

Batra, M.W., Edwards, K.L., and Scott, T.K. (1975). Auxin transport in roots: its characteristics and relationship to growth. In: The Development and Functions of Roots (eds. J.G. Torrey and J.G. Clarkson), 299–325. New York: Academic Press.

Baylis, G.T.S. (1975). The magnoloid mycorrhiza and mycotrophy in root systems derived from it. In: Endomycorrhizae (eds. F.E. Sanders, B. Mosse and P.B. Tinker), 373–389. New York: Academic Press.

Benková, E., Michniewicz, M., Sauer, M. et al. (2003). Local, efflux-dependent auxin gradients as a common module for plant organ formation. *Cell* 115: 591–602.

Bennett, T., van den Toorn, A., Sanchez-Perez, G.F. et al. (2010). SOMBRERO, BEARSKIN1, and BEARSKIN2 regulate root cap maturation in *Arabidopsis*. *Plant Cell* 22: 640–654.

Bian, H., Xie, Y., Guo, F. et al. (2012). Distinctive expression patterns and roles of the miRNA393/TIR1 homolog module in regulating flag leaf inclination and primary and crown root growth in rice (*Oryza sativa*). *New Phytologist* 196: 149–161.

Bielach, A., Podlešáková, K., Marhavý, P. et al. (2012). Spatiotemporal regulation of lateral root organogenesis in *Arabidopsis* by cytokinin. *Plant Cell* 1: 100–112.

Blakely, L.M., Durham, M., Evans, T.A. et al. (1982). Experimental studies on lateral root formation in radish seedling roots. 1. General methods, developmental stages and spontaneous formation of laterals. *Botanical Gazette* 143: 341–352.

Brechenmacher, L., Lee, J., Sachdev, S. et al. (2009). Establishment of a protein reference map for soybean root hair cells. *Plant Physiology* 149: 670–682.

Bridges, I.G., Hillman, J.R., and Wilkin, M.B. (1973). Identification and localisation of auxin in primary roots of *Zea mays* by mass spectrometry. *Planta* 115: 189–192.

Brown, L.K., George, T.S., Dupuy, L.X. et al. (2013). A conceptual model of root hair ideotypes for future agricultural environments: what combination of traits should be targeted to cope with limited P availability? *Annals of Botany* 112: 317–330.

Cai, H., Chen, F., Mi, G. et al. (2012). Mapping QTLs for root system architecture of maize (*Zea mays* L.) in the field at different developmental stages. *Theoretical and Applied Genetics* 125: 1313–1324.

Cannon, W.A. (1949). A tentative classification of root systems. *Ecology* 30: 542–548.

Casero, P.I., Rodriguez-Gallardo, L., Martin-Partigo, G. et al. (1993). Lateral root initiation by means of asymmetrical transversal divisions of the pericycle cells in adventitious roots of *Allium cepa*. *Protoplasma* 176: 138–144.

Celenza, J.L., Grisafi, P.L., and Fink, G.R. (1995). A pathway for lateral root formation in *Arabidopsis thaliana*. *Genes & Development* 9: 2131–2142.

Chapman, E.J. and Estelle, M. (2009). Mechanism of auxin-regulated gene expression in plants. *Annual Review of Genetics* 43: 265–285.

Chen, R., Hilson, P., Sedbrook, J. et al. (1998). The *Arabidopsis thaliana* AGRAVITROPIC 1 gene encodes a component of the polar-auxin-transport efflux carrier. *Proceedings of the National Academy of Sciences of the United States of America* 95: 15112–15117.

Chen, Y.L., Palta, J., Clements, J. et al. (2014). Root architecture alteration of narrow-leafed lupin and wheat in response to soil compaction. *Field Crops Research* 165: 61–70.

Chiou, T.J., Liu, H., and Harrison, M.J. (2001). The spatial expression patterns of a phosphate transporter (MtPT1) from *Medicago truncatula* indicate a role in phosphate transport at the root/soil interface. *The Plant Journal* 25: 281–293.

Cho, M., Lee, S.H., and Cho, H.T. (2007). P-Glycoprotein displays auxin efflux transporter–like action in *Arabidopsis* root hair cells and tobacco cells. *Plant Cell* 19: 3930–3943.

Christopher, J., Christopher, M., Jennings, R. et al. (2013). QTL for root angle and number in a population developed from bread wheats (*Triticum aestivum*) with contrasting adaptation to water-limited environments. *Theoretical and Applied Genetics* 126: 1563–1574.

Chu, G., Chen, T., Wang, Z. et al. (2014). Morphological and physiological traits of roots and their relationships with water productivity in water-saving and drought-resistant rice. *Field Crops Research* 162: 108–119.

Clark, L.J., Whalley, W.R., and Barraclough, P.B. (2003). How do roots penetrate strong soil? *Plant and Soil* 255: 93–104.

Clowes, F.A.L. (1961). Apical Meristems. Oxford, UK: Blackwell Science.

Comas, L., Becker, S., Cruz, V.M. et al. (2013). Root traits contributing to plant productivity under drought. *Frontiers in Plant Science* 4: 442.

Costa, S. and Shaw, P. (2006). Chromatin organization and cell fate switch respond to positional information in *Arabidopsis*. *Nature* 439: 493–496.

Coudert, Y., Périn, C., Courtois, B. et al. (2010). Genetic control of root development in rice, the model cereal. *Trends in Plant Science* 15: 219–226.

Craine, J.M., Wedin, D.A., Chapin, F.S. et al. (2003). Relationship between the structure of root systems and resource use for 11 North American grassland plants. *Plant Ecology* 165: 85–100.

De Rybel, B., Möller, B., Yoshida, S. et al. (2013). A bHLH complex controls embryonic vascular tissue establishment and indeterminate growth in *Arabidopsis*. *Developmental Cell* 24 (4): 426–437.

De Smet, I., Signora, L., Beeckman, T. et al. (2003). An abscisic acid-sensitive checkpoint in lateral root development of *Arabidopsis*. *The Plant Journal* 33: 543–555.

De Smet, I., Vassileva, V., De Rybel, B. et al. (2008). Receptor-like kinase ACR4 restricts formative cell divisions in the *Arabidopsis* root. *Science* 322: 594–597.

De Smet, I., Lau, S., Mayer, U. et al. (2010a). Embryogenesis: the humble beginnings of plant life. *The Plant Journal* 61: 959–970.

De Smet, I., Lau, S., Voß, U. et al. (2010b). Bimodular auxin response controls organogenesis in *Arabidopsis*. *Proceedings of the National Academy of Sciences of the United States of America* 107: 2705–2710.

De Smet, I. and Beeckman, T. (2011). Asymmetric cell division in land plants and algae: the driving force for differentiation. *Nature Reviews Molecular Cell Biology* 12: 177–188.

De Smet, I. (2012). Lateral root initiation: one step at a time. *New Phytologist* 193: 867–873.

Demchenko, K.N. and Demchenko, N.P. (1996). Lateral root initiation in *Triticum aestivum* (Poaceae) seedlings. *Botanicheskii Zhurnal* 81: 47–51.

Di Cristina, M., Sessa, G., Dolan, L. et al. (1996). The *Arabidopsis* ATHB10 (GLABRA2) is a HD-ZIP protein required for repression of ectopic root hair formation. *The Plant Journal* 10: 393–402.

Donald, C.M. (1968). The breeding of crop ideotypes. *Euphytica* 17 (3): 385–403.

Drew, M.C., Saker, L.R., and Ashley, T.W. (1973). Nutrient supply and the growth of the seminal root system in barley. I. The effect of nitrate concentration on the growth of axes and laterals. *Journal of Experimental Botany* 24: 1189–1202. https://doi.org/10.1093/jxb/24.6.1189.

Drew, M.C. (1975). Comparison of the effects of a localized supply of phosphate, nitrate, ammonium and potassium on the growth of the seminal root system, and the shoot in barley. *New Phytologist* 75: 479–490. https://doi.org/10.1111/j.1469-8137.1975.tb01409.x.

Dunbabin, V., Diggle, A., and Rengel, Z. (2003). Is there any optimal root architecture for nitrate capture in leaching environments? *Plant, Cell & Environment* 26: 835–844.

El-Antably, H.M.M. and Larsen, P. (1974). Distribution of gibberellins and abscisic acid in geotropically stimulated *Vicia faba* roots. *Physiologia Plantarum* 32: 322–329.

Elliott, M.C. and Greenwood, M.S. (1974). Indol-3yl-acetic acid in roots of *Zea mays*. *Phytochemistry* 13: 239–241.

Esau, K. (1965). Plant Anatomy, 2e. New York: Wiley.

Esau, K. (1977). Anatomy of Seed Plants. New York: Wiley.

Fageria, N.K., Gheyi, H.R., Carvalho, M.C.S. et al. (2015). Root growth, nutrient uptake and use efficiency by roots of tropical legume cover crops as influenced by phosphorus fertilization. *Journal of Plant Nutrition* 39 (6): 781–792. https://doi.org/10.1080/01904167.2015.1088020.

Farr, W.K. (1927). The elongation of root hairs in solutions of single calcium compounds. A comparison of three species, *Brassica oleracea*, *Avena sativa*, *Oryza sativa*. *American Journal of Botany* 14: 627.

Fitter, A. (1987). An architectural approach to the comparative ecology of plant root systems. *New Phytologist* 106: 61–77.

Foehse, D. and Jungk, A. (1983). Influence of phosphate and nitrate supply on root hair formation of rape, spinach and tomato plants. *Plant and Soil* 74: 359–368.

Forde, B.G. and Lorenzo, H. (2001). The nutritional control of root development. *Plant and Soil* 232: 51–68.

Forde, B.G. (2009). Is it good noise? The role of developmental instability in the shaping of a root system. *Journal of Experimental of Botany* 60: 3989–4002.

Gahoonia, T.S., Care, D., and Nielsen, N.E. (1997). Root hairs and phosphorus acquisition of wheat and barley cultivars. *Plant and Soil* 191: 181–188.

Galway, M.E., Masucci, J.D., Lloyd, A.M. et al. (1994). The TTG gene is required to specify epidermal cell fate and cell patterning in the Arabidopsis root. *Developmental Biology* 166: 740–754.

Gilroy, S. and Jones, D.L. (2000). Through form to function: root hair development and nutrient uptake. *Trends in Plant Science* 5: 56–60.

Goebel, K. (1928). Organographie der Pflanzen Insbesondere der Archegoniaten und Samenpflanzen. Erster Teil: Allgemeine Organographie. Jena: Gustav Fischer Ferlag.

Gregory, P.J. and Kirkegaard, J.A. (2017). Growth and function of root systems. In: Encyclopedia of Applied Plant Sciences, 2e (eds. B. Thomas, B.G. Murray and D.J. Murphy), 230–237. Academic Press.

Gulyaev, V.A. (1964). Initiation and formation of lateral radicles in several species within the Cucurbitaceae family. *Botanicheskii Zhurnal* 49: 1482–1485.

Hackett, C. (1972). A method of applying nutrients locally to roots under controlled conditions and some morphological effects of locally applied nitrate on the branching of wheat roots. *Australian Journal of Biological Sciences* 1972: 1169–1180.

Haines, B.J. (2002). Zincophilic root foraging in *Thlaspi caerulescens*. *New Phytologist* 155: 363–372.

Hanstein, J. (1870). Die Entwicklung des Keimes der Monokotylen und Dikotylen. Bonn: A. Marcus.

Hassan, H., Scheres, B., and Blilou, I. (2010). JACKDAW controls epidermal patterning in the Arabidopsis root meristem through a non-cell-autonomous mechanism. *Development* 137: 1523–1529.

Havlin, J., Tisdale, S., Nelson, W. et al. (2004). Soil Fertility and Fertilizers: An Introduction to Nutrient Management. Upper Saddle River, NJ: Prentice Hall.

He, J., Jin, Y., Du, Y.L. et al. (2017). Genotypic variation in yield, yield components, root morphology and architecture, in soybean in relation to water and phosphorus supply. *Frontiers in Plant Science* 81: 1499. https://doi.org/10.3389/flps.2017.01499.

Hochholdinger, F., Woll, K., Sauer, M. et al. (2005). Functional genomic tools in support of the genetic analysis of root development in maize (*Zea mays* L.). *Maydica* 50: 437.

Hochholdinger, F., Wen, T.J., Zimmermann, R. et al. (2008). The maize (*Zea mays* L.) *roothairless3* gene encodes a putative GPI-anchored, monocot-specific, COBRA-like protein that significantly affects grain yield. *The Plant Journal* 54: 888–898.

Hochholdinger, F. and Tuberosa, R. (2009). Genetic and genomic dissection of maize root development and architecture. *Current Opinion in Plant Biology* 12: 172–177.

Hodge, A., Berta, G., Doussan, C. et al. (2009). Plant root growth, architecture and function. *Plant and Soil* 321: 153–187.

Hu, B., Zhu, C., Li, F. et al. (2011). LEAF TIP NECROSIS1 plays a pivotal role in the regulation of multiple phosphate starvation responses in rice. *Plant Physiology* 156: 1101–1115.

Ilina, E.L., Kiryushkin, A.S., Semenova, V.A. et al. (2018). Lateral root initiation and formation within the parental root meristem of *Cucurbita pepo*: is auxin a key player? *Annals of Botany* 122 (5): 873–888.

Imin, N., Nizamidin, M., Wu, T. et al. (2006). Factors involved in root formation in *Medicago truncatula*. *Journal of Experimental Botany* 58: 439–451.

Ingram, P.A., Zhu, J., and Shariff, A. (2012). High-throughput imaging and analysis of root system architecture in *Brachypodium distachyon* under differential nutrient availability. *Philosophical Transactions of the Royal Society B: Biological Sciences* 367: 1559–1569.

Jamieson, T. (1893). On root hairs. *Science* 22: 354–356.

Johnson, J.F., Vance, C.P., and Allan, D.L. (1996). Phosphorus deficiency in *Lupinus albus*. Altered lateral root development and enhanced expression of phosphoenolpyruvate carboxylase. *Plant Physiology* 112: 657–665.

Ju, C., Buresh, R.J., Wang, Z. et al. (2015). Root and shoot traits for rice varieties with higher grain yield and higher nitrogen use efficiency at lower nitrogen rates application. *Field Crops Research* 175: 47–55. https://doi.org/10.1016/j.fcr.2015.02.00.

Jung, J.K.H.M. and McCouch, S.R.M. (2013). Getting to the roots of it: genetic and hormonal control of root architecture. *Frontiers in Plant Science* 4: 186.

Kamoshita, A., Zhang, J., Sarkarung, S. et al. (2002). Mapping QTLs for root morphology of a rice population adapted to rainfed lowland conditions. *Theoretical and Applied Genetics* 104: 880–893.

Kaneyasu, T., Kobayashi, A., Nakayama, M. et al. (2007). Auxin response, but not its polar transport, plays a role in hydrotropism of Arabidopsis roots. *Journal of Experimental Botany* 58: 1143–1150.

Kang, L., Yue, S., and Li, S. (2014). Effects of phosphorus application in different soil layers on root growth, yield, and water-use efficiency of winter wheat grown under semi-arid conditions. *Journal of Integrative Agriculture* 13: 2028–2039.

Kawahara, R., Komamine, A., and Fukuda, H. (1995). Isolation and characterization of homeobox-containing genes of carrot. *Plant Molecular Biology* 27: 155–164.

Khan, A.A. (1969). Cytokinin-inhibitor antagonism in the hormonal control of α-amylase synthesis and growth in barley seed. *Plant Physiology* 22: 94–103.

Kim, C.M., Park, S.H., Je, B.I. et al. (2007). OsCSLD1, a cellulose synthase-like D1 gene, is required for root hair morphogenesis in rice. *Plant Physiology* 143: 1220–1230.

Kiss, J.Z., Mullen, J.L., Correll, M.J., and Hangarter, R.P. (2003). Phytochromes A and B mediate red-light-induced positive phototropism in roots. *Plant Physiology* 131: 1411–1417.

Kitomi, Y., Ito, H., Hobo, T. et al. (2011). The auxin responsive AP2/ERF transcription factor CROWN ROOTLESS5 is involved in crown root initiation in rice through the induction of OsRR1, a type-A response regulator of cytokinin signaling. *The Plant Journal* 67: 472–484.

Kojima, S., Bohner, A., Gassert, B. et al. (2007). AtDUR3 represents the major transporter for high-affinity urea transport across the plasma membrane of nitrogen-deficient *Arabidopsis* roots. *The Plant Journal* 52: 30–40.

Kolek, J. (1974). Structure and Function of Primary Root Tissues. Veda, Slovak Academy of Sciences Bratislava: Bratislava.

Krouk, G., Lacombe, B., Bielach, A. et al. (2010). Nitrate-regulated auxin transport by NRT1.1 defines a mechanism for nutrient sensing in plants. *Developmental Cell* 18: 927–937.

Kutschera, L. (1960). Wurzelatlas Mitteleuropaeischer Ackerunkraeuter und Kulturpflanzen, Frankfurt am Main. Germany: DLG Verlag.

Kutz, A., Müller, A., Hennig, P. et al. (2002). A role for nitrilase 3 in the regulation of root morphology in sulphur-starving *Arabidopsis thaliana*. *The Plant Journal* 30: 95–106.

Kwak, S.H. and Schiefelbein, J. (2008). Regulated accumulation of the SCRAMBLED receptor and position-dependent cell type patterning in *Arabidopsis*. *Current Biology* 18: 1949–1954.

Lamont, B. (1974). The biology of dauciform roots in the sedge *Cyathochaete avenacea*. *New Phytologist* 73: 985–996.

Landi, P., Giuliani, S., Salvi, S. et al. (2010). Characterization of root-yield-1.06, a major constitutive QTL for root and agronomic traits in maize across water regimes. *Journal of Experimental Botany* 61: 3553–3562.

Lavenus, J., Goh, T., Roberts, I. et al. (2013). Lateral root development in *Arabidopsis*: fifty shades of auxin. *Trends in Plant Science* 18: 450–458.

Lelandais-Brière, C., Jovanovic, M., Torres, G.A. et al. (2007). Disruption of AtOCT1, an organic cation transporter gene, affects root development and carnitine-related responses in *Arabidopsis*. *The Plant Journal* 51: 154–164.

Li, Z., Gao, Y., Wang, X. et al. (2005). QTL mapping of root traits in a doubled haploid population from a cross between upland and lowland japonica rice in three environments. *Theoretical and Applied Genetics* 110: 1244–1252.

Li, G.W., Zhang, M.H., Cai, W.M. et al. (2008). Characterization of OsPIP2; 7, a water channel protein in rice. *Plant and Cell Physiology* 49: 1851–1858.

Li, L., Hey, S., Liu, S. et al. (2016). Characterization of maize roothairless6 which encodes a D-type cellulose synthase and controls the switch from bulge formation to tip growth. *Scientific Reports* 6: 34395.

Liao, H., Rubio, G., Yan, X. et al. (2001). Effect of phosphorus availability on basal root shallowness in common bean. *Plant and Soil* 232: 69–79.

Libault, M., Brechenmacher, L., Cheng, J. et al. (2010). Root hair systems biology. *Trends in Plant Science* 15: 641–650.

Linkohr, B.I., Williamson, L.C., Fitter, A.H. et al. (2002). Nitrate and phosphate availability and distribution have different effects on root system architecture of *Arabidopsis*. *The Plant Journal* 29: 751–760.

Liscum, E. and Briggs, W.R. (1995). Mutations in the NPH1 locus of *Arabidopsis* disrupt the perception of phototropic stimuli. *Plant Cell* 7: 473–485.

Lloret, P.G., Casero, P.J., Pulgarin, A. et al. (1989). The behaviour of two cell populations in the pericycle of *Allium cepa, Pisum sativum* and *Daucus carota* during early lateral root development. *Annals of Botany* 63: 465–475.

López-Bucio, J., Cruz-Ramırez, A., and Herrera-Estrella, L. (2003). The role of nutrient availability in regulating root architecture. *Current Opinion in Plant Biology* 6: 280–287.

Lynch, J.P. (2011). Root phenes for enhanced soil exploration and phosphorus acquisition: tools for future crops. *Plant Physiology* 156: 1041–1049.

Lynch, J.P. (2013). Steep, cheap and deep: an ideotype to optimize water and N acquisition by maize root systems. *Annals of Botany* 112: 347–357.

Mace, E.S., Singh, V., Van Oosterom, E.J. et al. (2012). QTL for nodal root angle in sorghum (*Sorghum bicolor* L. Moench) co-locate with QTL for traits associated with drought adaptation. *Theoretical and Applied Genetics* 124: 97–109.

Mähönen, A.P., Bishopp, A., Higuchi, M. et al. (2006). Cytokinin signaling and its inhibitor AHP6 regulate cell fate during vascular development. *Science* 311 (5757): 94–98.

Majer, C., Xu, C., Berendzen, K.W. et al. (2012). Molecular interactions of ROOTLESS CONCERNING CROWN AND SEMINAL ROOTS, a LOB domain protein regulating shoot-borne root initiation in maize (*Zea mays* L.). *Philosophical Transactions of the Royal Society B: Biological Sciences* 367: 1542–1551.

Malamy, J.E. (2005). Intrinsic and environmental response pathways that regulate root system architecture. *Plant, Cell & Environment* 28: 67–77.

Maruyama-Nakashita, A., Nakamura, Y., Yamaya, T. et al. (2004). Regulation of high-affinity sulphate transporters in plants: towards systematic analysis of sulphur signalling and regulation. *Journal of Experimental Botany* 55: 1843–1849.

Masucci, J.D., Rerie, W.G., Foreman, D.R. et al. (1996). The homeobox gene GLABRA2 is required for position dependent cell differentiation in the root epidermis in *Arabidopsis thaliana*. *Development* 122: 1253–1260.

Materechera, S.A., Dexter, A.R., and Alston, A.M. (1991). Penetration of very strong soils by seedling roots of different plant species. *Plant and Soil* 135: 31–41.

McClean, P.E., Burridge, J., Beebe, S. et al. (2011). Crop improvement in the era of climate change: an integrated, multi-disciplinary approach for common bean (*Phaseolus vulgaris*). *Functional Plant Biology* 38: 927–933.

McCully, M.E. (1975). The development of lateral roots. In: The Development and Function of Roots (eds. J.G. Torrey and D.T. Clarkson), 105–124. Academic Press.

McMichael, B.L. and Burke, J.J. (2002). Temperature effects on root growth. In: Plant Roots, 1120–1138. Boca Raton, FL: CRC Press.

Meijer, A.H., Scarpella, E., van Dijk, E.L. et al. (1997). Transcriptional repression by Oshox1, a novel homeodomain leucine zipper protein from rice. *The Plant Journal* 11: 263–276.

Meister, R., Rajani, M.S., Ruzicka, D. et al. (2014). Challenges of modifying root traits in crops for agriculture. *Trends in Plant Science* 19: 779–788.

Mergemann, H. and Sauter, M. (2000). Ethylene induces epidermal cell death at the site of adventitious root emergence in rice. *Plant Physiology* 124: 609–614.

Meyer, R.S., DuVal, A.E., and Jensen, H.R. (2012). Patterns and processes in crop domestication: an historical review and quantitative analysis of 203 global food crops. *New Phytologist* 196: 29–48.

Miller, E.C. (1938). Plant Physiology, 2e, 1201p. New York: Academic Press.

Misra, R.K., Alston, A.M., and Dexter, A.R. (1988). Role of root hairs in phosphorus depletion from a macrostructured soil. *Plant and Soil* 107: 11–18.

Mock, J.J. and Pearce, R.B. (1975). An ideotype of maize. *Euphytica* 24: 613–623.

Mouchel, C.F., Briggs, G.C., and Hardtke, C.S. (2004). Natural genetic variation in *Arabidopsis* identifies BREVIS RADIX, a novel regulator of cell proliferation and elongation in the root. *Genes & Development* 18: 700–714.

Mouchel, C.F., Osmont, K.S., and Hardtke, C.S. (2006). BRX mediates feedback between brassinosteroid levels and auxin signalling in root growth. *Nature* 443: 458–461.

Mylona, P., Linstead, P., Martienssen, R. et al. (2002). SCHIZORIZA controls an asymmetric cell division and restricts epidermal identity in the *Arabidopsis* root. *Development* 129: 4327–4334.

Nestler, J., Liu, S., Wen, T.J. et al. (2014). Roothairless5, which functions in maize (*Zea mays* L.) root hair initiation and elongation encodes a monocot-specific NADPH oxidase. *The Plant Journal* 79: 729–740.

Nye, P.H. (1966). The effect of the nutrient intensity and buffering power of a soil, and the absorbing power, size and root hairs of a root, on nutrient absorption by diffusion. *Plant and Soil* 25: 81–105.

O'Dell, D.H. and Foard, D.E. (1969). Presence of lateral root primordia in the radicle of buckwheat embryos. *Bulletin of the Torrey Botanical Club* 96: 1–3.

Ono, F., Frommer, W.B., and Von Wiren, N. (2000). Coordinated diurnal regulation of low-and high-affinity nitrate transporters in tomato. *Plant Biology* 2: 17–23.

Paez-Garcia, A., Motes, C.M., Scheible, W.R. et al. (2015). Root traits and phenotyping strategies for plant improvement. *Plants* 4: 334–355.

Parizot, B., Laplaze, L., Ricaud, L. et al. (2008). Diarch symmetry of the vascular bundle in *Arabidopsis* root encompasses the pericycle and is reflected in distich lateral root initiation. *Plant Physiology* 146: 140–148.

Parniske, M. (2008). Arbuscular mycorrhiza: the mother of plant root endosymbioses. *Nature Reviews Microbiology* 6: 763–775.

Peremyslov, V.V., Prokhnevsky, A.I., Avisar, D. et al. (2008). Two class XI myosins function in organelle trafficking and root hair development in *Arabidopsis*. *Plant Physiology* 146: 1109–1116.

Péret, B., De Rybel, B., Casimiro, I. et al. (2009). *Arabidopsis* lateral root development: an emerging story. *Trends in Plant Science* 14: 399–408.

Pérez-Torres, C.A., López-Bucio, J., Cruz-Ramírez, A. et al. (2008). Phosphate availability alters lateral root development in *Arabidopsis* by modulating auxin sensitivity via a mechanism involving the TIR1 auxin receptor. *Plant Cell* 20: 3258–3272.

Pernas, M., Ryan, E., and Dolan, L. (2010). SCHIZORIZA controls tissue system complexity in plants. *Current Biology* 20: 818–823.

Perrot-Rechenmann, C. (2010). Cellular responses to auxin: division versus expansion. *Cold Spring Harbor Perspectives in Biology* 2: a001446.

Pfeffer, W. (1894). Geotropic sensitiveness of the root-tip. *Annals of Botany* 8: 317–320.

Purnell, H.M. (1960). Studies of the family Proteaceae. I. Anatomy and morphology of the roots of some Victorian species. *Australian Journal of Botany* 8: 38–50.

Raghothama, K.G. (2005). Phosphorus and plant nutrition: an overview. In: Phosphorus: Agriculture and the Environment (eds. J.T. Sims and A.N. Sharpley), 355–378. Madison, WI: American Society of Agronomy.

Rebouillat, J., Dievart, A., Verdeil, J.L. et al. (2009). Molecular genetics of rice root development. *Rice* 2: 15–34.

Reynolds, M.P., Acevedo, E., Sayre, K.D. et al. (1994). Yield potential in modern wheat varieties: its association with a less competitive ideotype. *Field Crops Research* 37: 149–160.

Riou-Khamlichi, C., Huntley, R., Jacqmard, A. et al. (1999). Cytokinin activation of *Arabidopsis* cell division through a D-type cyclin. *Science* 283: 1541–1544.

Robinson, D. and Rorison, I.H. (1987). Root hairs and plant growth at low nitrogen availabilities. *New Phytologist* 107: 681–693.

Rogers, E.D. and Benfey, P.N. (2015). Regulation of plant root system architecture: implications for crop advancement. *Current Opinion in Biotechnology* 32: 93–98.

Ryan, P.R., Delhaize, E., and Jones, D.L. (2001). Function and mechanism of organic anion exudation from plant roots. *Annual Review of Plant Physiology and Plant Molecular Biology* 52: 527–560.

Sarquis, J.I., Jordon, W.R., and Morgan, P.W. (1991). Ethylene evolution from maize (*Zea mays* L.) seedling roots and shoots in response to mechanical impedance. *Plant Physiology* 96: 1171–1177.

Scacchi, E., Osmont, K.S., Beuchat, J. et al. (2009). Dynamic, auxin responsive plasma membrane-to-nucleus movement of *Arabidopsis* BRX. *Development* 136: 2059–2067.

Schmidt, J.E. and Gaudin, A.C. (2017). Toward an integrated root ideotype for irrigated systems. *Trends in Plant Science* 22: 433–443.

Seago, J.L. (1973). Developmental anatomy in roots of *Ipomoea purpurea*. 2. Initiation and development of secondary roots. *American Journal of Botany* 60: 607–618.

Seago, J.L. Jr. and Fernando, D.D. (2013). Anatomical aspects of angiosperm root evolution. *Annals of Botany* 112 (2): 223–238.

Shane, M.W., Dixon, K.W., and Lambers, H. (2005). The occurrence of dauciform roots amongst Western Australian reeds, rushes and sedges, and the impact of phosphorus supply on dauciform-root development in *Schoenus unispiculatus* (Cyperaceae). *New Phytologist* 165: 887–898.

Shaw, S. and Wilkins, M.B. (1974). Auxin transport in roots: x. relative movement of radioactivity from IAA in the stele and cortex of zea root segments. *Journal of Experimental Botany* 25: 199–207.

Shen, C., Wang, S., Zhang, S. et al. (2013). OsARF16, a transcription factor, is required for auxin and phosphate starvation response in rice (*Oryza sativa* L.). *Plant, Cell & Environment* 36: 607–620.

Sinnot, E.W. (1960). Plant Morphogenesis. New York: McGraw-Hill Book Company.

Snow, L.M. (1905). The development of root hairs. *Botanical Gazette* 40: 12–48.

St John, T.V. (1980). Root size, root hairs and mycorrhizal infection: a reexamination of Baylis's hypothesis with tropical trees. *New Phytologist* 84: 483–487.

Street, H.E. (1969). Factors influencing the initiation and activity of meristems in roots. *Root Growth* 235: 20–41.

Subbarao, G.V., Sahrawat, K.L., Nakahara, K. et al. (2012). A paradigm shift towards low-nitrifying production systems: the role of biological nitrification inhibition (BNI). *Annals of Botany* 112 (2): 297–316.

Suzuki, M., Sato, Y., Wu, S. et al. (2015). Conserved functions of the *MATE* transporter BIG EMBRYO1 in regulation of lateral organ size and initiation rate. *Plant Cell* 27: 2288–2300.

Takano, J., Wada, M., Ludewig, U. et al. (2006). The *Arabidopsis* major intrinsic protein NIP5; 1 is essential for efficient boron uptake and plant development under boron limitation. *Plant Cell* 18: 1498–1509.

Taramino, G., Sauer, M., Stauffer, J. et al. (2007). The *RTCS* gene in maize (*Zea mays* L.) encodes a LOB domain protein that is required for postembryonic shoot-borne and embryonic seminal root initiation. *The Plant Journal* 50: 649–659.

Thimann, K.V. (1936). Auxins and the growth of roots. *American Journal of Botany* 23: 561–569.

Thorup-Kristensen, K. and Kirkegaard, J. (2016). Root system based limits to agricultural productivity and efficiency: the farming systems context. *Annals of Botany* 118: 573–592.

Tian, Q. and Reed, J.W. (1999). Control of auxin-regulated root development by the *Arabidopsis thaliana SHY2/IAA3* gene. *Development* 126: 711–721.

Tominaga-Wada, R., Iwata, M., Sugiyama, J. et al. (2009). The GLABRA2 homeodomain protein directly regulates CESA5 and XTH17 gene expression in *Arabidopsis* roots. *The Plant Journal* 60: 564–574.

Torrey, J.G. (1965). Physiological bases of organization and development in the root. In: Differentiation and Development/Differenzierung und Entwicklung (eds. F.T. Addicott, A. Lang, W. Ruhland, et al.), 1256–1327. Encyclopedia of Plant Physiology. Berlin, Heidelberg: Springer.

Tsay, Y.F., Schroeder, J.I., Feldmann, A. et al. (1993). The herbicide sensitivity gene CM1 of Arabidopsis encodes a nitrate inducible nitrate transporter. *Cell* 72: 705–713.

Uga, Y., Sugimoto, K., Ogawa, S. et al. (2013). Control of root system architecture by deeper rooting 1 increases rice yield under drought conditions. *Nature Genetics* 45: 1097–1102.

van Overbeek, J. (1966). Plant hormones and regulators. *Science* 152 (72): 1–31.

Vanneste, S. and Friml, J. (2009). Auxin: a trigger of change in plant development. *Cell* 136: 1005–1016.

Vines, S.H. (1888). On the relation between the formation of tubercles on the roots of Leguminosae and the presence of nitrogen in the soil. *Annals of Botany* 2: 386–389.

von Behrens, I., Komatsu, M., Zhang, Y. et al. (2011). Rootless with undetectable meristem 1 encodes a monocot-specific AUX/IAA protein that controls embryonic seminal and post-embryonic lateral root initiation in maize. *The Plant Journal* 66: 341–353.

von Nägeli, C. (1858). Beiträge zur Wissenschaftlichen Botanik. Leipzig, Germany: von Wilhelm Engelmann.

Voronin, N.S. (1964). Evolution of the primary structures in plant roots. *Uch Zap Kaluzhskogo Gos Pedagog Inst* 13: 3–179.

Wada, T., Tachibana, T., Shimura, Y. et al. (1997). Epidermal cell differentiation in *Arabidopsis* is determined by a Myb homolog, CPC. *Science* 227: 1113–1116.

Wahid, P.A. (2000). A system of classification of woody perennials based on their root activity patterns. *Agroforestry Systems* 49: 123–130.

Waisel, T., Eshel, A., Beeckman, T. et al. (2002). Plant Roots: The Hidden Half, 3e. New York/ Basel: Marcel Dekker, Inc.

Walker, J.C. and Key, J.L. (1982). Isolation of cloned cDNAs to auxin-responsive poly(A)+ RNAs of elongating soybean hypocotyl. *Proceedings of the National Academy of Sciences of the United States of America* 79: 7185–7189.

Wang, Y., Zhu, Y., Ling, Y. et al. (2010). Disruption of actin filaments induces mitochondrial Ca^{2+} release to the cytoplasm and $[Ca^{2+}]$c changes in *Arabidopsis* root hairs. *BMC Plant Biology* 10: 53.

Wang, X.F., He, F.F., Ma, X.X. et al. (2011). OsCAND1 is required for crown root emergence in rice. *Molecular Plant* 4: 289–299.

Wasson, A.P., Richards, R.A., Chatrath, R. et al. (2012). Traits and selection strategies to improve root systems and water uptake in water-limited wheat crops. *Journal of Experimental Botany* 63: 3485–3498.

Watt, M. and Evans, J.R. (1999). Proteoid roots. Physiology and development. *Plant Physiology* 121: 317–323.

Weaver, J.E. and Clements, F.E. (1938). Plant Ecology, 2e. New York: Mc-Graw Hill.

Wen, T.J., Hochholdinger, F., Sauer, M. et al. (2005). The roothairless1 gene of maize encodes a homolog of sec3, which is involved in polar exocytosis. *Plant Physiology* 138: 1637–1643.

Whaley, W.G. and Kephart, J. (1957). Effect of gibberellic acid on growth of maize roots. *Science* 125 (3241): 125–234.

White, P.J. and Brown, P.H. (2010). Plant nutrition for sustainable development and global health. *Annals of Botany* 105: 1073–1080.

White, P.J., George, T.S., Gregory, P.J. et al. (2013). Matching roots to their environment. *Annals of Botany* 112: 207–222.

White, P.J. and Greenwood, D.J. (2013). Properties and management of cationic elements for crop growth. In: Soil Conditions and Plant Growth (eds. P.J. Gregory and S. Nortcliff), 160–194. Oxford: Blackwell Publishing.

Whiting, S.N., Leake, J.R., McGrath, S.P. et al. (2000). Positive responses to Zn and Cd by roots of the Zn and Cd hyperaccumulator *Thlaspi caerulescens*. *New Phytologist* 145: 199–210.

Wiersum, L.K. (1958). Density of root branching as affected by substrate and separate ions. *Acta Botanica Neerlandica* 7: 174–190.

Willemsen, V., Bauch, M., Bennett, T. et al. (2008). The NAC domain transcription factors FEZ and SOMBRERO control the orientation of cell division plane in *Arabidopsis* root stem cells. *Developmental Cell* 15: 913–922.

Wiśniewska, J., Xu, J., Seifertová, D. et al. (2006). Polar PIN localization directs auxin flow in plants. *Science* 312: 883.

Xu, C.R., Liu, C., Wang, Y.L. et al. (2005). Histone acetylation affects expression of cellular patterning genes in the *Arabidopsis* root epidermis. *Proceedings of the National Academy of Sciences of the United States of America* 102: 14469–14474.

Xu, J., Hofhuis, H., Heidstra, R. et al. (2006). A molecular framework for plant regeneration. *Science* 311: 385–388.

Xu, C., Tai, H., Saleem, M. et al. (2015). Cooperative action of the paralogous maize lateral organ boundaries (LOB) domain proteins RTCS and RTCL in shoot-borne root formation. *New Phytologist* 207: 1123–1133.

Yamamoto, Y., Kamiya, N., Morinaka, Y. et al. (2007). Auxin biosynthesis by the YUCCA genes in rice. *Plant Physiology* 143: 1362–1371.

Yang, C.M., Yang, L.Z., Yang, Y.X. et al. (2004). Rice root growth and nutrient uptake as influenced by organic manure in continuously and alternately flooded paddy soils. *Agricultural Water Management* 70: 67–81.

Yang, Z., Culvenor, R.A., Haling, R.E. et al. (2017). Variation in root traits associated with nutrient foraging among temperate pasture legumes and grasses. *Grass Forage Science* 72: 93–103.

Yoo, S.C., Cho, S.H., and Paek, N.C. (2013). Rice WUSCHEL-related homeobox 3A (OsWOX3A) modulates auxin-transport gene expression in lateral root and root hair development. *Plant Signaling and Behavior* 8: e25929.

Zhang, H. and Forde, B.G. (1998). An *Arabidopsis* MADS box gene that controls nutrient-induced changes in root architecture. *Science* 279: 407–409.

Zhang, H., Jennings, A., Barlow, P.W. et al. (1999). Dual pathways for regulation of root branching by nitrate. *Proceedings of the National Academy of Sciences of the United States of America* 96: 6529–6534.

Zhang, J.W., Xu, L., Wu, Y.R. et al. (2012). OsGLU3, a putative membrane-bound endo-1, 4-beta-glucanase, is required for root cell elongation and division in rice (*Oryza sativa* L.). *Molecular Plant* 5: 176–186.

Zhu, J.M., Zhang, C.C., and Lynch, J.P. (2010). The utility of phenotypic plasticity of root hair length for phosphorus acquisition. *Functional Plant Biology* 37: 313–322.

Zou, Y., Liu, X., Wang, Q. et al. (2014). OsRPK1, a novel leucine-rich repeat receptor-like kinase, negatively regulates polar auxin transport and root development in rice. *Biochimica et Biophysica Acta* 1840: 1676–1685.

5

Interplay Between Root Structure and Function in Enhancing Efficiency of Nitrogen and Phosphorus Acquisition

Jianbo Shen, Liyang Wang, Xin Wang, Kemo Jin, and Chuanyong Xiong

Department of Plant Nutrition, National Academy of Agriculture Green Development, College of Resources and Environmental Sciences, Key Laboratory of Plant-Soil Interactions, Ministry of Education, China Agricultural University, Beijing, P.R. China

5.1 Root Structure and Function in Structured Soils

Roots acquire water and nutrients from soil (Eshel and Beeckman 2013; Shen et al. 2013). Root structure and function determine soil exploration and exploitation and have significant impacts on nutrient and water uptake, stress tolerance, soil health and resilience, plant–soil interactions, rhizosphere dynamics, soil microbiome, and thus crop production and its sustainability. Root structure, the spatial distribution and characteristics of root systems, is fundamentally important to plant roots capturing soil resources (Lynch and Brown 2012). Root function is closely relevant to the capacity, activity, and efficiency of water and nutrient acquisition (McCully 1999). A better understanding of root structure, root function, and their interplay is essential for improving crop productivity and enhancing nutrient-use efficiency through maximizing root/rhizosphere efficiency (Shen et al. 2013; Jiao et al. 2016; Wang and Shen 2019).

5.1.1 Root Structure

Root structure can be interpreted at various levels, such as root system architecture (RSA), root morphology, and root anatomy. RSA refers to the spatial pattern of the entire root system in the soil and has a profound effect on soil exploration and consequently uptake of water and nutrients through modifying spatial availability of these resources. RSA, which is under genetic control as well as regulated by environmental factors (Shahzad and Amtmann 2017), is mainly determined by four root structural traits: growth, branching, surface area, and angle (Morris et al. 2017). Modifications of RSA are important in capturing the nutrients such as phosphorus (P, immobile in soil). Phosphorus is relatively concentrated in the topsoil (Lynch 2011). Root system architectural traits associated with enhanced topsoil foraging are therefore important for P acquisition. Lynch (2011) has proposed the

The Root Systems in Sustainable Agricultural Intensification, First Edition.
Edited by Zed Rengel and Ivica Djalovic.

ideal RSA for efficient P acquisition, including shallower growth angles and greater numbers of axial roots, enhanced adventitious rooting, and greater dispersion of lateral roots. For plant acquisition of the mobile resources such as water and N that leach into the deeper soil layers when sufficient water is available, greater exploitation of subsoil by roots would afford substantial benefits (Lynch and Wojciechowski 2015). Evidence exists that rooting depth is positively related to water and nitrogen capture, and rooting depth has been proposed as a central element of ideotype for drought-tolerant and N-efficient crops (Lynch and Wojciechowski 2015).

Given that root distribution and function change along the soil profile, a new model of ideotype RSA in cropping system has been proposed to match the changes in water and nutrient availability from topsoil to subsoil, which is able to effectively integrate soil water capture via "hydraulic lift" from subsoil by deep roots (Richards and Caldwell 1987), N and P acquisition via "induced root proliferation" by shallow roots in the topsoil containing spatially and temporally heterogeneous N and P content. This RSA model can enhance the synergy between water and nutrient capture and thus enhance use efficiency of various soil resources and increase crop productivity (Shen et al. 2013; Jiao et al. 2016), which is crucial not only for root and soil management, but also for breeding of new improved varieties (Shen et al. 2013).

There are two types of root structure: taproot and fibrous (see also Chapter 4). The taproot system is common in dicotyledonous plants, such as *Arabidopsis thaliana* L. and legumes. The taproot system consists of one distinguished embryo-derived primary root and several orders of postembryonic lateral roots developed from the primary root (Smith and De Smet 2012). In contrast, the fibrous roots are universal in monocotyledonous plants such as maize (*Zea mays*) and wheat (*Triticum aestivum*), forming relatively complex root systems (Rich and Watt 2013). After germination, cereals develop embryonic root system that comprises a single primary root and several seminal roots (Yu et al. 2016b). At later stages of development, a special class of postembryonic adventitious crown roots (not present in dicotyledonous plants) emerges from shoot tissues, dominating the root system development (Rogers and Benfey 2015). The adventitious crown roots have the capacity to form several orders of lateral roots that can provide enhanced anchorage and resource transport capacity (Hochholdinger et al. 2018). The taproot and fibrous root systems, due to their strikingly different root structure and different development mechanisms (Smith and De Smet 2012), have different plasticity in foraging for soil resources (Fry et al. 2018).

Branching is one of the most fundamental characteristics of root system structure. Lateral roots emerging from the axial roots contribute to anchorage and enhance soil exploration. In monocotyledonous and dicotyledonous plants, lateral roots arise from the pericycle-derived progenitor (stem) cells (Smith and De Smet 2012). In *Arabidopsis*, lateral roots emerge from the outermost cell layer of the central vascular cylinder, i.e. the pericycle cells opposite to xylem poles (De Smet et al. 2012). *Arabidopsis* has two xylem poles, whereby lateral roots emerge in an alternating pattern at each pole (Moller et al. 2017; Van Norman et al. 2013). Once activated, the specialized xylem pole pericycle cells start with several anticlinal and periclinal cell divisions to form lateral root primordia (Laskowski and Ten Tusscher 2017). Lateral root primordia then grow through the endodermis, the cortex, and the epidermis to complete the lateral root emergence. Recent findings demonstrated that the feedback from endodermal cells is required for lateral root primordia to grow through this cell layer (Motte et al. 2019). In contrast, in cereals, lateral roots emerge from phloem poles (Yu et al. 2016b). In cereals such as maize and wheat, phloem poles are more numerous (10 or more) than in *Arabidopsis*, enabling lateral roots to simultaneously emerge at many different radial positions (Smith and De Smet 2012). Therefore, the radial branching pattern of lateral root emergence in cereals appears to be unpredictable (Yu et al. 2016b), thus causing some difficulties in investigating root development in cereal crops.

Endodermis may be crucial in lateral root formation in cereals. In maize (Yu et al. 2016a), both pericycle and endodermal cells give rise to the new lateral roots, but how pericycle and endodermis cells coordinate their contribution to emerging lateral roots remains elusive. Auxin plays a crucial role in many processes of lateral root development (Lavenus et al. 2013). A localized auxin response primes pericycle founder cells to undergo several rounds of cell division to form the lateral root primordia (Jansen et al. 2012; Motte et al. 2019). Many studies showed that auxin content and distribution in roots can be altered by various nutrient treatments (Beeckman and Friml 2010; Giehl et al. 2012; Krouk et al. 2010; Lima et al. 2010); our recent study reported that heterogeneous phosphate supply influenced maize lateral root proliferation by regulating auxin redistribution (Wang et al. 2020a), offering an opportunity to manipulate root structure via nutrient management, particularly for localized supply of nutrients in cropping systems (Jing et al. 2010; Ma et al. 2015; Wang and Shen 2019).

Root anatomy is an important determinant of root penetration, soil exploration, and resource acquisition. In the longitudinal direction, a single root consists of three developmental zones (apex and the elongation and differentiation zones). At the root apex, progenitor (stem) cells divide and differentiate in the meristematic zone. These differentiated cells increase their size in the elongation zone and become mature in the differentiation zone (Eshel and Beeckman 2013). The root apex, where the complex crosstalk of phytohormones and signals regulate root development, is responsible for sensing gravity and the rhizosphere physical structure (Massa and Gilroy 2003), leading root penetration in structured soil. In the root differentiation zone, epidermal cells form root hairs, which significantly enlarge the absorbing surface and extend the range of root–soil interactions.

In the radial orientation, roots comprise functionally distinct cell types and tissues (Eshel and Beeckman 2013; Hochholdinger et al. 2018). The stele, which consists of xylem and phloem, is responsible for the transport of water, nutrients, and photosynthates. Xylem vessel number and diameter directly influence root water conductance (Lynch et al. 2014). The stele influences root strength. Root tensile strength is correlated positively with stele diameter rather than root diameter (Chimungu et al. 2015). Surrounding the stele, the pericycle cells, and endodermal cells are involved in lateral root formation as mentioned above. The cortex forms the largest proportion of root tissues in most plant species (Lynch et al. 2014). Living cortical area was proposed as an indicator of root metabolic costs and water stress adaptation (Jaramillo et al. 2013). In maize, large root cortical cell size and reduced root cortical cell file number were suggested to improve drought tolerance (Lynch et al. 2014). Root cortical tissue also plays an important role in root penetration, whereby smaller outer cortical cells increase root penetration of hard layers by stabilizing the root against ovalization (Chimungu et al. 2015). The epidermis, from which root hairs emerge, connects the root system to the rhizosphere. In some species, root hairs contribute to forming rhizosheath by aggregating soil particles in the rhizosphere (McCully 1999). In addition, the physical role of root hairs is to assist root penetration and improve the root–soil contact strength and anchoring (Bengough et al. 2016).

5.1.2 Root Functions

Roots have three basic functions: growth, absorption, and exudation. Root growth provides anchorage to plants and enhances spatial exploration of soil (Eshel and Beeckman 2013). Root absorption maintains plant development by providing water and nutrients. Root exudation assists roots in mobilizing rhizosphere nutrients, lubricating root–soil interface contact, and providing carbon substrates for the activity of microorganisms in the rhizosphere, thus improving plant adaptation to the soil environment (Bais et al. 2006; Shen et al. 2011, 2013). In addition to these functions, roots have other roles such as sensing, responding to, and modifying the rhizosphere. Once anchored in the soil, plants are unable to move. To survive in a limited environment, plant roots must be highly responsive to soil conditions varying in space and time.

Roots are capable of sensing and responding to soil physical properties. Roots prefer to grow in the existing biopores with appropriate size (Bengough 2012) (see also Chapter 3). When root apex encounters hard objects, the root elongation rate decreases significantly (Goss and Russell 1980). This touch sensing interplays with gravity sensing, resulting in roots growing angularly along the surface of soil solid particles (Jin et al. 2013).

The soil resources that plants require are often distributed heterogeneously in space and time. Plant roots have the capacity to cope with the heterogeneous resource signals to aid foraging. In heterogeneous environments, roots are allocated preferentially to the resource-rich areas. When roots are in contact with dry and wet surfaces, lateral roots and root hairs preferentially grow towards the higher moisture content (Bao et al. 2014). It is widely reported that localized nutrient supply triggers root proliferation as well as enhanced root absorption rates in the nutrient-rich patches (Hodge 2006). In case of nitrate, a dual-affinity nitrate transporter NRT1.1, which can alter its affinity, is a nitrate sensor on the plasma membrane, triggering many physiological and developmental processes in response to changing external nitrate availability (Bouguyon et al. 2015). In spatially heterogeneous nitrate supply, NRT1.1 shows (i) a stimulatory role in root preferential proliferation in the high-nitrate patches and (ii) participates in root architecture alteration by upregulating ARABIDOPSIS NITRATE-REGULATED 1 (ANR1) to promote lateral root development (Mounier et al. 2014). Regarding phosphate, there is evidence that the sensing of phosphate occurs at the root apex in *Arabidopsis*, and a direct contact of root tip with the medium is required to sense external phosphate availability (Svistoonoff et al. 2007). In root apex, the interaction of cell wall-targeted LOW PHOSPHATE ROOT1 (LPR1) and PHOSPHATE DEFICIENCY RESPONSE2 (PDR2) is involved in sensing low external phosphate concentration (Abel 2017). These findings provide insights into root functions with respect to sensing and responding to rhizosphere conditions.

Roots modify rhizosphere environment. Root growth exerts the compressive and shear stresses on soil, leading to deformation of the surrounding soil (Koebernick et al. 2017). The real-time imaging with particle image velocimetry (PIV) clearly showed the displacement and deformation of the root-surrounding matrix (Vollsnes et al. 2010). In addition to the root elongation, the root hair development has been demonstrated to significantly improve the adhesion, formation, and stability of aggregates in the rhizosphere soil (Moreno-Espíndola et al. 2007). In situ high-resolution synchrotron radiation computed tomography (SRCT) showed the effect of root hairs on porosity and connectivity of the ≥5 μm pores in rhizosphere (Koebernick et al. 2017) (see also Chapter 3).

Root absorption activities result in alteration of rhizosphere processes. Water absorption by roots modifies the radial water flow in the rhizosphere (Carminati 2012). Uptake of nutrients by plant roots induces the depletion gradients in the rhizosphere, resulting in the driving force for nutrient diffusion towards the root surface (Tinker and Nye 2000).

Root exudates strongly mediate rhizosphere processes (Hinsinger et al. 2009; Shen et al. 2013). Plant roots exude a range of compounds, such as mucilage, enzymes, and organic anions, that greatly modify rhizosphere soil properties (Ryan et al. 2001). Root-secreted enzymes, organic anions, and protons can alter soil chemical properties through mobilizing insoluble nutrients, changing soil pH, chelating metal ions, and increasing the number of ion exchange sites (Ryan et al. 2001; Shen et al. 2013). In the intercropping system, root exudates provide interspecific benefits through mobilizing nutrients from a common pool (Zhang et al. 2016) or enhancing nodulation and N_2 fixation (Li et al. 2016). In addition, a large proportion of photosynthetically fixed carbon partitioned to roots is transferred to the rhizosphere through root exudates such as sugars and amino acids (Walker et al. 2003). These organic substances feed the soil microbiome and enrich the microbial community in the rhizosphere (Sasse et al. 2018).

Roots are essential for plants to communicate with soil organisms, whereby root exudates play an important role as signal messengers (Walker et al. 2003). Plants release signal compounds that deliver biological information and govern the interaction with organisms in the rhizosphere (Hu et al. 2018; Venturi and Keel 2016). Many root exudates are involved in rhizosphere signalling and the shaping of root-associated microbial communities (Venturi and Keel 2016). The two most impressive signal compounds secreted by plant roots, strigolactones and flavonoids, are specifically sensed by, respectively, mycorrhiza and rhizobia, triggering the symbiosis establishment to contribute to plant supply of phosphorus and nitrogen (Sasse et al. 2018). The association between plants and beneficial microbes promotes nutrient uptake, root development, and pathogen defence of the host plants (Hiruma et al. 2018). There is also evidence that root exudates are involved in governing root–root interactions (Mommer et al. 2016). For example, root exudates produced by *Centaurea maculosa* can significantly inhibit the development of roots of neighbouring species (Thorpe et al. 2009).

5.1.3 Root Structure and Function in Structured Soils

The soil is a heterogeneous but structured environment. Soil physical structure has a considerable impact on root structure and function, such as root development, the delivery of nutrients to the root by mass flow and diffusion, and the consequent root responses (Bengough et al. 2006; Chapman et al. 2012). Defined by the spatial arrangement of solid, liquid, and gaseous phases, the soil structural characteristics vary both temporally and spatially regarding soil texture, porosity, aggregation, and soil strength (Jin et al. 2013). The soil physical properties, such as soil porosity and mechanical impedance, significantly influence root spatial exploration and proliferation (Bengough et al. 2006). Plant roots encounter heterogeneous obstacles or macropores in the soil, and they are able to grow in existing pores with suitable diameters, thus avoiding obstacles (Bengough 2012). High soil mechanical impedance leads to root morphological modifications, such as a decrease in root size and root elongation rate (Bingham and Bengough 2003) and altered branching pattern depending on the plant species (Potocka and Szymanowska-Pulka 2018). In addition to morphological alterations, other properties such as increasing root exudation, especially of mucilage, are observed under increased soil mechanical impedance (Potocka et al. 2011). Soil pores determine water and gas movement and availability. Hence, diffusion and mass flow of nutrients to the root surface are highly dependent on the soil pore system structure (Chapman et al. 2012).

There have been many efforts to understand plant root responses to nutrient supply using simplified experimental systems (Chapman et al. 2012). However, soil-free experimental systems, such as hydroponics or agar/gel plates, cannot mimic the physical properties of soil. Indeed, plant roots grown in nonporous artificial media considerably differ from those grown in soil (Atkinson et al. 2019). Soil in the field provides a much more complex environment for crop roots regarding physical structure (Chapman et al. 2012). The pattern and mechanisms of plant root traits recorded in the simplified experimental systems are difficult to extrapolate to field conditions and may not contribute significantly to our understanding of root structure and function in the field (Chapman et al. 2012). More advanced approaches are needed to take soil physical structure into consideration. Nondestructive imaging techniques, such as X-ray computed tomography (X-ray CT), magnetic resonance imaging (MRI), and positron emission tomography (PET), enable in situ observation of detailed dynamic root system structure and function in real structured soils (Atkinson et al. 2019; Gao et al. 2019; van Veelen et al. 2020) (see Chapter 3). Integrating innovative approaches will contribute to uncovering the root structure and function in real root–soil systems (Atkinson et al. 2019).

5.2 Root Growth in Structured Soils

Boosting crop yields is critical for providing food security to increasing global population and requires greater insights into rooting traits and their interactions with soil. A role soil physics plays in root architecture has attracted research interest, but the knowledge of heterogeneous properties of soil surrounding the roots is limited; hence, how roots engineer and change soil structure needs to be fully explored. In this section, we detail typical heterogeneous soil conditions to indicate how roots explore pathways through the complex soil environment and its feedback to the structured soil. The understanding of root growth in structured soil, especially root architecture adaptation traits under variable soil conditions, contributes significantly to searching for, and establishing, an optimal combination of root ideotype and soil in the cropping systems.

5.2.1 Root Architecture Benefits

Sustaining the human population of 10 billion in a changing climate is a pre-eminent challenge of the twenty-first century (Lynch and Wojciechowski 2015; Jiao et al. 2016). However, yield gaps of 45–70% of the theoretically possible yield are observed for most crops globally (Lobell et al. 2009; Mueller et al. 2012). Thus, crops with better adaptation to soil conditions would directly improve food production and therefore food security worldwide (Rebetzke 2016).

RSA, the spatial configuration of a root system in soil, plays a major role in plant fitness, crop performance, and grain yield (Rogers and Benfey 2015). An understanding of the impact of soil conditions on root architecture is essential to addressing the issues of sustainable agricultural systems. RSA is influenced by the external environment (soil moisture, nutrients, and soil structure), and impacts on how plants detect and respond to their surroundings (Paez-Garcia et al. 2015). The heterogeneous nature of soil environment results in a highly plastic RSA (Rogers and Benfey 2015). The plasticity of RSA in response to changing moisture and nutrient status of the soil provides opportunities for exploring natural variation to identify beneficial root traits to enhance plant productivity in agricultural systems (Paez-Garcia et al. 2015; Jiao et al. 2016).

We need to develop and design efficient roots that can reshape, biomine, and manipulate soil via a better understanding of root architecture traits under heterogeneous soil conditions to provide a way of modifying root traits to improve plant capacity to acquire resources (Figure 5.1). The integrated knowledge, from the adaptive growth of RSA to the identification of ideal root phenotypes for crops growing in specific soil conditions, can enable breeders to select, breed or engineer improved cultivars ideally suited to difficult soil conditions to improve agricultural production under global climate change (Clark et al. 2003; Paez-Garcia et al. 2015).

5.2.2 Heterogeneity of Soil Surrounding Roots and Root Adaptive Responses

The common interactions between a penetrating root and soil include the following: (i) roots can penetrate the structured soil and then elongate along an existing pore (or crack); (ii) or negotiate a crack at some angle before penetrating the soil (Figure 5.1). In homogeneous soils, elongation depends on the capacity of roots to produce a growth pressure large enough to deform the soil. When roots attempt to penetrate soil, the stiffness of the root, the angle of incidence, and the strength of the soil being penetrated all interact to determine the outcome.

5.2.2.1 Soil Compaction

Soil compaction involves significant interrelationships among most recognized physical and biological properties of soils; it adversely affects crop production in many parts of the world (McCarthy 2007; Alakukku 2012). Some physical properties, such as soil water, nutrients, and soil mechanical resistance, can be modified intensely by soil compaction. Effects of soil strength and mechanical impedance on root growth and root architecture typically increase with decreasing soil moisture (Iijima et al. 2007; Whitmore and Whalley 2009).

Most field soils exhibit spatial variability in soil structure. Roots growing in soils typically encounter a mix of loose and compacted soil. The spatial variation in mechanical resistance affects the degree of root clustering. Preferential growth of roots occurs in less compacted areas with large macropores, lower mechanical resistance, sufficient aeration, or high

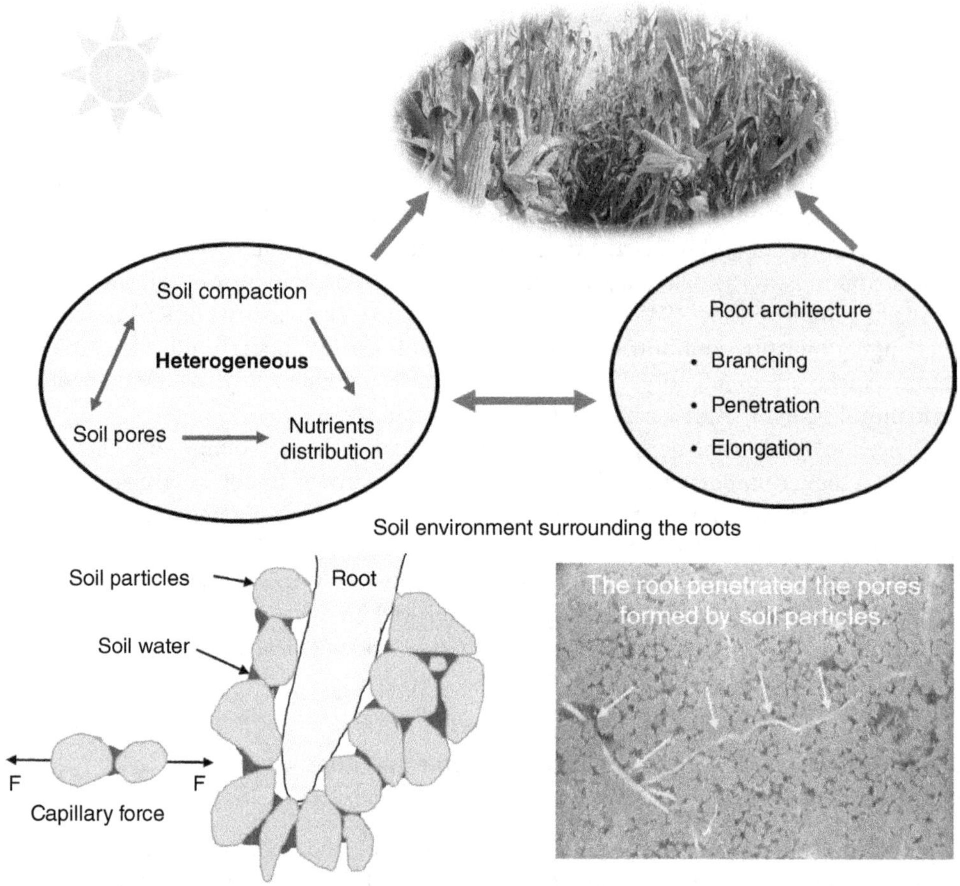

Figure 5.1 Interactions between root architecture and soil conditions. *Source:* Image of maize cobs by Liyang Wang (Department of Plant Nutrition, College of Resources and Environmental Sciences, China Agricultural University, P. R. China).

nutrient availability. When the scale of structural heterogeneity is larger than the extension of the root system of individual plants, the position of the plants relative to the most compacted parts of soil controls the opportunities for root growth (Boone 1988). Layered soil compaction (or vertical compaction discontinuity) is an important factor affecting root growth and plant water use. Vertical discontinuity usually occurs between aggregated seedbed and compacted soil below it after seedbed preparation or between a tilled layer and untilled subsoil (Håkansson 2005). Usually, the dense layers can be localized in the soil profile by measuring soil bulk density and strength (Lipiec and Hatano 2003; Bengough 2012; Nosalewicz and Lipiec 2014). Even in compacted soils, areas of lower mechanical impedance occur due to shrinkage cracks and channels/pores formed by earthworms or previous root growth (Tardieu 1988). Under these conditions, root systems encountering hard compacted zones of soil may tend to proliferate in zones of loose soil. Such plasticity in root system development (maintaining the total root length in response to heterogeneous soil conditions) has been reported in both pot (Garcia et al. 1988) and field experiments (Bennie and Botha 1986; Montagu et al. 1998).

Deep roots capable of penetrating hard soil horizons are important for improving crop productivity in compacted soils (Gao et al. 2016). Moreover, deep rooting is important for drought adaptation by improving utilization of water in deep soil layers, but the presence of compacted soil layers often precludes the exploitation of deep water and nutrients (Rogers and Benfey 2015). Thus, crop yields depend on optimum root growth and architecture in heterogeneous soils.

5.2.2.2 Location of Pores

Plant roots grow through the soil pores in search of water and nutrients (Figure 5.1). Soil pore system can be defined as the ensemble of the voids or spaces existing in a given volume of soil (Dexter et al. 2008). The basic character of the pore space influences, and is influenced by, almost everything that occurs in soil: the movement of air, water, and other fluids; the transport and the reactions of chemicals; and the residence of roots and other biota (Nimmo 2013). Soil pore system contains biopores and the network of interaggregate pores, cracks, and other structural voids; heterogeneous distribution of pores can be influenced by soil management.

Although a main gap in our current knowledge is in understanding the traits associated with heterogeneous location and distribution of pores, X-ray computer-aided tomography offers the possibility of 3-D measurements of the soil pore system. Cárdenas et al. (2010) presented a complex network model based on a heterogeneous preferential attachment scheme in order to quantify the structure of porous soils. Fractal analysis has been effective in characterizing heterogeneity and complexity of the soil pore attributes in microtomographic images (Gibson et al. 2006; Tarquis et al. 2009). Kravchenko et al. (2011) found that long-term tillage and land-use differences influenced intra-aggregate pore heterogeneity, and distribution of pore voxels belonging to large pores was most heterogeneous in the aggregates from soil under native succession vegetation, followed by no-till and conventional tillage aggregates. Bodner et al. (2013) studied wetting–drying effects on soil pore size distribution, and their model showed that wetting–drying enhanced macroporosity (higher $r_{\mathrm{m,\,Kosugi}}$) while decreasing pore heterogeneity (lower σ_{Kosugi}). This might be related to capillary-induced soil shrinkage reducing the median pore radii, whereas pore heterogeneity increased by formation of microfissures (Leij et al. 2002).

In soil, roots often exploit cracks, voids, and large pores, or enlarge smaller pores by displacing soil particles (Figure 5.1). Although roots are able to expand the diameter of a soil pore, the rate of root growth can be limited in a compacted soil with low porosity (Jin et al. 2015). Gao et al. (2016) reported that roots could grow deep only by exploiting the existing pore networks. Bodner et al. (2014) found that species with dense systems of thin roots exacerbated heterogeneity of the pore space and increased the micropore volume. Thus, it is crucial to understand (i) how roots sense and respond to soil physical heterogeneity to find a path through a network of pores and (ii) identify root architecture properties that are beneficial to soil exploration (Jin et al. 2015; Jiao et al. 2016).

5.2.2.3 Nutrient Distribution

Nutrient distribution in soil is spatially and temporally heterogeneous or "patchy" (Croft et al. 2015). Plants need to sense these changes in nutrient availability to optimize growth and biomass allocation by integrating root and shoot growth (Hodge 2004) as well as growth of various parts of root system (Bellini et al. 2014; Giehl and von Wirén 2014). Roots show considerable plasticity in their morphology and physiology in response to this variable nutrient distribution, particularly in their "foraging for resources" response, stimulating growth in the nutrient-rich zones (Kudoyarova et al. 2015; Hendriks et al. 2016).

In agriculture, there is interest in spatial nutrient variability at the single-plant scale. In such environments, Drew (1975) demonstrated that exposing parts of the main seminal roots to high concentrations of phosphorus caused a localized promotion of the initiation and subsequent extension of both first- and second-order laterals compared with the low-phosphorus zones. On encountering a soil patch enriched in nutrient (particularly N or P), the roots of many plant species increase their rate of growth and branching (Robinson 1994). Typically, root traits enabling "topsoil foraging" enhance Pi acquisition when this element is limiting (Lynch 2013). This topsoil-foraging ideotype was modelled in hydroponically and agar-grown *Arabidopsis* plants where roots of low-P-supplied plants were commonly shallower due to inhibition of primary root elongation and stimulation of lateral root formation (Giehl and von Wirén 2014). Studies with *Arabidopsis* showed that a localized supply of NO_3^- stimulated lateral root elongation by two- to threefold in the NO_3^- patch, and even 0.1 mM NO_3^- was sufficient to elicit the maximal response (Zhang and Forde 1998). However, water flux from the soil into the root determines growth of the primary root in *Arabidopsis* independently of NO_3^- concentration that varied from 0.1–10 mM (Chapman et al. 2011).

Changes in root morphology in response to spatially heterogeneous water and nutrient supply in soil are considered the primary way for plants to acquire soil resources (Forde and Lorenzo 2001). Understanding how root architecture and environmental conditions optimize crop performance and resource capture is important. However, general "rules" of how plants respond to their heterogeneous environment have proved hard to predict. It has recently been proposed that a root ideotype "steep, cheap, and deep" may maximize soil exploration for optimum water and nutrient acquisition (Lynch and Wojciechowski 2015).

5.2.3 Root-induced Changes in Soil Structure

Soil structure influences the growth and activity of roots living in soil; in addition, plant roots influence soil structure (Figure 5.2). Root-induced changes in soil structure influence plant growth mostly by modifying the root physical environment, and the water and nutrient cycles. Roots influence soil structure at different scales and through a wide range of mechanisms involving root activity, such as root penetration, root water extraction, root anchorage, and root exudation of compounds into the rhizosphere (Angers and Caron 1998).

Growing in existing pores or through the soil matrix, roots create compressive and shear stresses that can reach 2 MPa (Goss 1991), so root penetration could cause deformation of the soil structure. Radial pressure exerted by the growing roots compress the soil in the vicinity (Dexter 1987) and decrease the porosity in that zone (Guidi et al. 1985; Bruand et al. 1996). The compression of soil around roots results in the enlargement or shrinkage of the existing pores and the creation of new ones (Rich and Watt 2013). Bodner et al. (2014) found that root systems differing in morphology induced distinctive pore variation, with coarse root systems increasing macroporosity by 30%. The higher the plant biomass production was, the greater the area and the length of macropores (cracks) were.

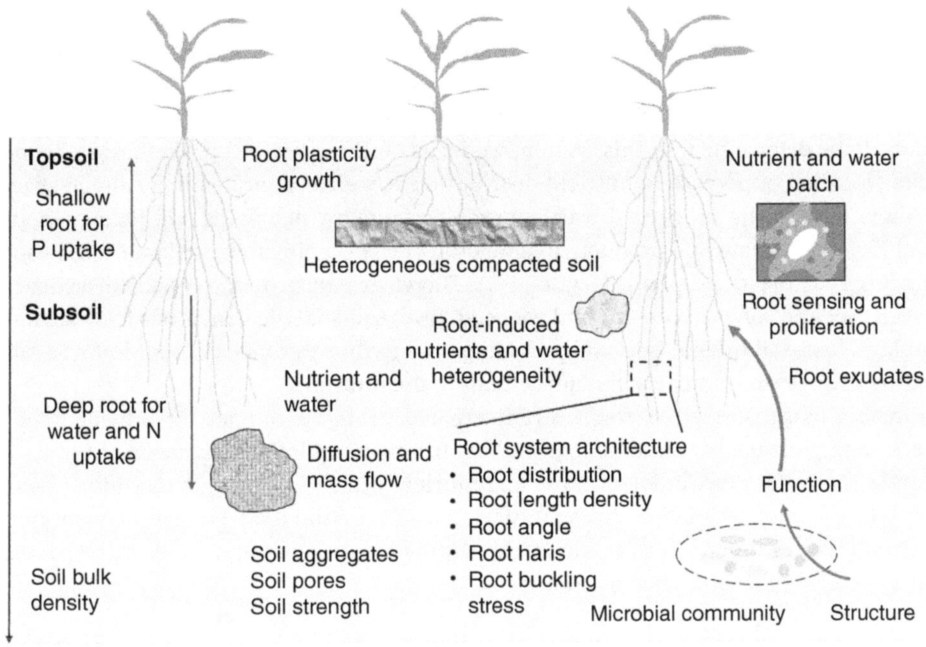

Figure 5.2 Root–soil feedbacks in cropping system based on temporal and spatial heterogeneity in soil.

The process of pore formation by roots is believed to be particularly important in undisturbed or no-till soils because tillage tends to disrupt the continuity of the pore system.

Root growth and architecture can also modify the soil water regime. Water uptake by roots usually results in an increase in soil strength (Horn and Dexter 1989). The heterogeneity of water extraction by roots gave rise to tensile stresses that led to the production of small aggregates, and compression resulting from water extraction by roots led to aggregates that were denser and of higher tensile strength than those in unplanted soils (Materechera et al. 1992, 1994). These findings were attributed partly to cracking of soil caused by heterogeneous water extraction giving rise to tensile stress patterns within the root grid.

Dorioz et al. (1993) observed that water uptake by roots altered the structure of clay soil and that this microenvironment was rich in root mucilage. Indeed, mucilage released by maize root tips significantly increased soil aggregate stability immediately upon addition to soil, i.e. independently of any microbial activity (Morel et al. 1991).

In addition to the immediate and short-term effects of roots on soil structure, plant roots also have a longer-term influence through their contribution to soil organic matter. A large proportion of carbon fixed by plants is allocated below-ground; hence, in many ecosystems, roots constitute the most important source of organic matter in soil.

Root-induced changes in soil structure strongly influence nutrient storage and turnover, and their eventual feedback impact on the root and plant growth is significant and needs to be studied further (Figure 5.2). Fully understanding root-induced soil structure changes will help us improve crop production through root traits identification and modification as well as specific soil management to underpin robust crop yield potential on a large scale.

Plants have adaptation mechanisms to deal with limitations during root establishment as well as late-stage development. The complicated root–soil interactions have a big effect on plant growth and yield formation, and the exact nature of the underlying mechanisms needs to be understood fully. In particular, the role of plant roots in modifying soil structure remains to be determined. In this section, we focus on three key soil properties: soil compaction, location of pores, and nutrient distribution as well as their heterogeneity in space and time, influencing crop yield. In the cropping systems, managing soil heterogeneity (variability) to increase soil productivity and crop yields is going to be reliant on root–soil interactions in the rhizosphere (Figure 5.2). Direct evidence of the dynamic feedback between root and soil is necessary, and we need novel soil analytical methods for assessing the role of root–soil interactions and providing quantitative estimates of root-induced soil structure changes to improve soil management in the future.

Climate change and more frequent/exacerbated droughts are felt worldwide. Deep understanding of root–soil interactions will contribute to optimized management of agricultural systems for more efficient water and nutrient utilization and enhanced food security for the growing population. In the future, root-specific soil management may improve soil structure and functions to optimize the supply of nutrients and water to crops and reduce nutrient losses into the environment.

5.3 Root Sensing of Soil Nutrients

5.3.1 Root Foraging in Heterogeneous Soil Environments

Soil environment is complex, with heterogeneously distributed water and mineral nutrients. In order to acquire essential resources for the successful completion of life cycle, plants have a capacity to sense soil nutrient and water availability by roots and respond accordingly, such as proliferating roots in nutrient-rich patches, and growing roots towards soil water. We can describe the act of acquiring resources as foraging. The foraging behaviour of plant roots is similar to that of animals, which means that plants explore soil for "food" through a series of root trait changes (Hutchings and De Kroon 1994). McNickle et al. (2009) described that plant behaviour as the phenotypic plasticity in response to a certain stimulus or signal, whereby the environment modifies expression of an individual genotype. The amount by which a phenotype can be modified is termed its plasticity. This plasticity comprises changes in both morphological and physiological traits. The plasticity is a general property of the whole genotype and is related to a general pattern of development (Bradshaw 1965). de Kroon et al. (2015) argue that the plasticity exhibited by plants in response to environmental changes is not a response of the plant as a whole, but rather of some part of it, such as blade or root tip sensing a local signal. Plants integrate the responses and interactions of all the parts influenced by the local signals. The fact an individual plant is composed of a series of repeating units ensures that the plants can perform different behaviours by different units simultaneously to adapt to spatially and temporally heterogeneous environments.

Plants adjust the morphological, structural, and physiological properties as a manifestation of foraging behaviour in response to heterogeneous distribution of nutrients in soil. The work on plant root foraging started with research on the response of barley (*Hordeum vulgare*) root system to heterogeneous supply of nutrients (Drew and Saker 1978). The foraging scale and precision, the two important components of the root foraging behaviour, should be considered. The scale of root foraging refers to the distribution of root system in 3-D space, which is relevant to the degree of resource supply in soil and is influenced mainly by the resources availability and the properties of a plant species or genotype (Campbell et al. 1991; Einsmann et al. 1999). The precision of plant foraging refers to a degree of root response to nutrient-rich patch (Campbell et al. 1991; Fransen et al. 1998; Einsmann et al. 1999; Hutchings and John 2004; Kembel and Cahill 2005). Kembel and Cahill (2005) found that the response of dicotyledonous roots to heterogeneous nutrient supply was more sensitive than that of monocotyledonous plants, indicating response differences among plant species. Grime et al. (1986) suggested that the "cost" and "benefit" of root plasticity were the main reasons for the specific foraging behaviour (selective growth) of roots. To analyze the optimal root foraging behaviour, McNickle and Cahill (2009) quantified the cost–benefit of root foraging by recording the dynamics of growth and distribution of plant roots in a nutrient-rich patch. With the changes in environmental factors, the plasticity response of roots caused the phenotype diversity among different individuals, thus giving the plant roots the capacity to adapt to surrounding soil and biological conditions (such as interacting with neighbouring plants, microorganisms, and even animals).

5.3.2 Plant Responses to Differential Availability of Mineral Nutrients

Plants count on a wide variety of metabolic, physiological, and developmental responses to adapt to variations in mineral nutrient availability (Rubio et al. 2009) because changes in nutrient availability, due to their heterogeneous distribution in soil, affect plant growth and development. It is necessary for plants to sense these changes and integrate root and shoot responses by optimizing biomass allocation for higher return in a "carbon-for-mineral nutrients" trade-off. In most cases under nutrient deficiency, the sensing and signalling cascade of phosphate and nitrate play an important role (Desnos 2008; Krouk et al. 2010; Gojon et al. 2011; Chien et al. 2017). Therefore, it is important to study the mechanisms of nutrient uptake from soil by plant roots. Growth of root system in response to variable external nutrient concentration is a combination of sensor or receptor proteins and cytoplasmic effector proteins. After nutritional sensing, sensor proteins initiate cellular relays involving transcription factors, kinases, ubiquitin ligases, and microRNAs (miRNAs) (Bhardwaj et al. 2015). In addition, phytohormones are involved in root-to-shoot signalling regarding availability of different nutrients and water, regulating physiological responses such as gene expression, root development, nutrient uptake, and stomatal and hydraulic conductance (Kudoyarova et al. 2015). Specific evidence can be obtained by measuring the physiological responses of genotypes with altered phytohormone responses or biosynthesis/translocation. Abscisic acid, indoleacetic acid, and cytokinin may regulate changes in shoot growth, assimilate partitioning to root, and root architecture caused by water, nitrate, and phosphorus availability (Kudoyarova et al. 2015). A better understanding of plant response mechanisms to availability of mineral nutrients may contribute not only to better crop management for efficient use of fertilizers, but also promote crop breeding for improved nutrient efficiency to underpin sustainable agriculture.

5.3.3 Nitrogen Sensing and Signalling

Nitrogen (N) is one of the main mineral nutrients in plants, and its availability has an important impact on the growth and development of plants. In most cases, nitrogen resources are restricted because soil supply cannot match plant requirement at key growth stages. Plants have evolved strategies to enhance N acquisition. For different N forms, the N uptake system combines high-affinity and low-affinity transporters belonging to the multigene families. The expression and activity of most uptake systems are regulated by the local signal (concentration of their substrates) and the systemic signal (the feedback of whole-plant N status), forming a complex composite signalling network (Nacry et al. 2013).

In addition to regulating the capacity of the transport system, plants can also regulate their growth and development to maintain N homeostasis. In particular, root structure has strong plasticity, and the change in root structure has a large impact on the acquisition of soil nitrogen. To optimize nutrition, plants need efficient sensing systems that can quickly sense and signal the local external concentration of individual nutrients. Nitrate is the important nitrogen source for most plants. In *A. thaliana*, the nitrate transporter NRT1.1 (CHL1) has been identified as the main sensor of nitrate, and it also promotes nitrate uptake by roots (Ho et al. 2009) (see also Chapter 4). In yeasts and animals, membrane transporters have dual nutritional transmission and signalling functions ("transceptors"). Hence, NRT1.1 is a transceptor (Gojon et al. 2011).

Nitrate stimulation of lateral root elongation is under the control of a nitrate-signalling pathway (Remans et al. 2006). The nutrient-sensing capacity of NRT1.1 is linked to the regulation of distribution of phytohormone auxin that controls the formation and emergence of lateral roots, with NRT1.1 promoting uptake of auxin (Krouk et al. 2010). However, nitrate inhibits NRT1.1-dependent auxin uptake. In *nrt1.1* mutants, lateral roots formation occurred even at low nitrate concentrations, which abrogates the capacity of roots to respond effectively to nitrate availability in soil and undermines the foraging behaviour of *nrt1.1* roots because of insufficient auxin uptake in the root tip cells (Beeckman and Friml 2010).

Studying complex regulatory networks that control nitrogen uptake, metabolism, and plant development will help understand plant adaptation to suboptimal nitrogen supply (thus increasing nitrogen-use efficiency) and temporally and spatially heterogeneous environments. It will also provide solutions to major problems threatening agricultural and ecological sustainability, such as nonpoint source pollution by nitrate, and greenhouse gas emissions and excessive economic costs associated with the overuse of nitrogen fertilizers.

5.3.4 Phosphorus Sensing and Signalling

Phosphorus is a structural component of nucleic acids and phospholipids and plays essential roles in energy metabolism, signal transduction, and enzyme activation. Plants often grow in soils that contain very low concentrations of available phosphorus due to chemical and biological fixation. To adapt and grow in phosphorus-deprived environments, plants must sense the balance of external and internal phosphorus concentrations and adjust growth and metabolism to match resource availability. The adaptive mechanisms include RSA optimization, exudation of organic compounds, remobilization of internal phosphorus, and beneficial mutualistic relationships with microorganisms, and they are linked with the upstream components of P signalling pathway involving membrane potential, voltage-dependent ion channels, intracellular Ca activity and pH, and transcription factors, as well as phytohormones and metabolites for systemic signalling.

One branch of the signal transduction cascade related to a phosphorus deprivation response involves a transcription factor regulated by the post-translational sumoylation of proteins (Rubio et al. 2003; Misson et al. 2005; Schachtman and Shin 2007). Phosphate starvation response 1 (PHR1) is a Myb transcription factor involved in certain aspects of response to phosphorus deprivation. Promoter analysis of genes responsive to phosphate deficiency showed the presence of PHR1-binding sites. Interestingly, the expression of PHR1 itself is not very responsive to phosphate deprivation, and therefore, it may be downstream in the phosphate signal transduction pathway. Although PHR1 is not regulated at the transcriptional level, the function of this factor is regulated by sumoylation (Miura et al. 2005).

Reactive oxygen species increase rapidly after mineral nutrient deprivation and may be an upstream mediator of nutrient signalling. Molecular analyses suggest that both short-term and long-term responses are important in understanding the progression of signalling events when the external, then internal, supplies of nutrients become depleted (Schachtman and Shin 2007).

Shoot–root communication is crucial for plant acclimation to phosphorus deficiency, requiring signals that could be transmitted from a sensor to a transponder. Both sugars and miRNAs have been implicated as potential long-distance signal molecules transported through phloem between shoot and roots for the regulation of gene expression, Pi uptake, and utilization. Plant miRNAs usually downregulate the abundance of their target mRNAs by post-transcriptional cleavage. The miR399 is upregulated during phosphate (Pi) deficiency. Upregulation of miR399 results in the downregulation of UBC24 encoding a ubiquitin-conjugating E2 enzyme. Plants overexpressing miR399 are defective in UBC24 and display Pi toxicity because of increased Pi uptake, enhanced root-to-shoot translocation, and retention of Pi in the old leaves. This observation suggests that the miR399-mediated regulation of UBC24 expression is critical in Pi homeostasis.

By studying the expression patterns of a serine/threonine phosphatase gene (PvHAD1) and miR399 in common bean (*Phaseolus vulgaris* L.), Liu et al. (2010) provided evidence for the interaction between light, phloem transport, and miR399 in the systemic regulation of gene expression under P deficiency. In contrast to systemic signalling, local sensing was the primary causal factor for rapid downregulation of PvHAD1 by Pi prior to a decrease in miR399 level in P-deficient roots. Furthermore, this initial response to Pi in P-deficient roots was also mimicked by the Pi analog phosphonate (Phi) that does not affect internal Pi status, but acts as a local signal. These findings suggest that plants have developed a highly coordinated dual regulatory pathway to integrate systemic signal from shoot to roots with local sensing of Pi in roots (Liu et al. 2010).

How local phosphorus signals are generated in roots is poorly understood; however, low Pi availability in soil often induces a reduction in primary root growth and an increase in the number and length of lateral roots. Hence, the primary root tip is the first promising organ to sense P availability in soil. Svistoonoff et al. (2007) showed that physical contact of the *A. thaliana* primary root tip with low Pi medium was necessary and sufficient to arrest root growth, even when leaves were in contact with a high-Pi medium. The loss-of-function mutations in *LPR1* and its close paralog *LPR2* strongly decreased this inhibition; indeed, LPR1 and LPR2 encode multicopper oxidases (MCOs) essential in plant development. Moreover, Reymond et al. (2006) identified *LPR1* as a major quantitative trait locus (QTL) controlling root growth response to phosphate starvation in *A. thaliana*; the molecular origin of this QTL was explained by the differential allelic expression of LPR1 in the root cap, suggesting the involvement of root cap in sensing nutrient deficiency and/or responding to it.

Despite many lines of evidence showing that roots can sense phosphorus availability in soil, information on phosphate receptor is lacking. Nevertheless, the knowledge on plant responses to local phosphorus availability has been applied in agricultural management practices to increase phosphorus-use efficiency and mitigate environmental risks in the agricultural systems (Jing et al. 2010; Shen et al. 2013).

5.4 Root/Rhizosphere Processes Influenced by Spatial and Temporal Heterogeneity of Soil Nutrients Supply

The soil is an uneven continuum with many different (placement, size, composition, and intensity) patches of nutrient enrichment (Figure 5.2). Root growth and rhizosphere processes are altered upon exposure to these nutrient patches. Plants can respond to these heterogeneously supplied nutrients by changing root morphology and/or physiology, including root proliferation and enhanced absorption capacity (Zhang et al. 2010; Shen et al. 2013; Wang and Shen 2019). Root morphological and physiological plasticity in response to heterogeneous nutrient supply is a kind of foraging behaviour of plants that can greatly increase nutrient acquisition in the local environment through changing root distribution and nutrient absorption rate to meet the needs of plant growth and development.

5.4.1 Root Morphology

Roots plasticity response to external environment is controlled by both genetic and environmental factors and is an important adaptive response to environmental conditions. Roots proliferate (elongate and branch to produce increases in total length, surface area, and/or biomass) when they encounter nutrient-rich patches in soil. A large increase in lateral root growth and branching in the nutrient patches offsets the impact on plant growth caused by the failure of roots outside the nutrient-rich area to obtain enough resources (Robinson 1994).

Nutrient supply intensity directly influences the morphological characteristics of plant roots, subsequently influencing uptake and utilization of nutrient resources. Low soil P concentration stimulated the formation of large number of cluster roots in white lupin (*Lupinus albus*) (Johnson and Biondini 2001; Shen et al. 2005), but in the split-root arrangement, the cluster root ratio in white lupin was similar on the two sides regardless of P supply (Shane et al. 2003). However, it has also been reported that the cluster roots formation in white lupin is not only regulated by the internal plant P status, but also greatly influenced by localized P supply. The cluster root percentage increased in the P-rich patches and decreased in nutrient-free zone (Shu et al. 2005, 2007a,b). Some studies also showed that the growth and distribution of wheat roots were determined by shoot P concentration and the supply of soil P (Ma and Rengel 2008; Ma et al. 2009). These cases suggest that plants can sense external nutrient resources and regulate root morphology to efficiently obtain soil nutrients.

The root morphology of fast-growing plants might be more plastic than that of slow-growing plants (Fransen et al. 1998). Indeed, Robinson and van Vuuren (1998) found that roots of fast-growing plants in nutrient-rich areas (N, P, and K) were much larger (number, length, and dry matter weight) than those of slow-growing varieties. It has also been reported that root morphological plasticity is influenced by nutrient specificity and species specificity, and cereal species (e.g. maize and wheat) exhibited greater morphological plasticity than leguminous species (e.g. faba bean and chickpea) (Li et al. 2014a). Lyu et al. (2016) compared the root morphological response of seven common crops to phosphorus and found that crops with a smaller root diameter (such as wheat and maize) had stronger morphological plasticity than those with a larger root diameter (such as chickpea and white lupin).

The size and concentration of nutrient patches in soil influence root growth and nutrient absorption. The P-uptake efficiency was influenced more by the size of nutrient patches than by the amount of P, indicating that small nutrient patches may be inconsequential because they can be missed by roots limited in growth (Yano and Kume 2005). Hence, plants have limits in adapting to, and utilizing, the heterogeneous supply of soil nutrients, if in small patches. Nutrient-rich patches alter the distribution of roots in soil. In uniform soils, lateral root elongation mainly occurs in the topsoil, whereas in soils with nutrient patches, lateral root elongation occurs mainly in these local nutrient supply areas (Linkohr et al. 2002).

5.4.2 Root Physiology

The effect of soil heterogeneity in nutrient supply on root physiological plasticity is reflected mainly in the change of nutrient absorption capacity, which plays an important role in uptake of soil-mobile nutrient (such as NO_3^-). Nutrient absorption per unit root length usually increases substantially (generally two to three times) when localized nutrients are applied to nutrient-deficient plants (Robinson 1994). Some reports found a fivefold increase or more, but the response was often short-lived rather than sustained (van Vuuren et al. 1996; Robinson 1994). An increase in ion absorption rate was regarded as a plant response to sensing the composition and intensity of nutrient-rich patches and may or may not be accompanied by a morphological response (root proliferation) as well (depending on the plant species, Dunbabin et al. 2001, 2003). The root physiological response is likely to be important in the short term, whereas morphological changes (that take longer to develop) enable the root system to obtain nutrients in the nutrient patch area more effectively in the long term (Dunbabin et al. 2003).

Due to the slow diffusion of nutrients with poor soil mobility, such as phosphate, the absorption of such nutrients is dependent on root growth. Indeed, Jackson et al. (1990) found that localized supply of P stimulated root proliferation, and uptake of P was 60% higher than that of the control. At the early stage of root growth, roots explore only a small volume in soil, and the localized P placement is of large significance for plants to obtain sufficient P by having high uptake rate in the localized area. How long this high absorption rate lasts depends on the intensity of local nutrient supply in soil (Caldwell et al. 1992).

5.4.3 Root/Rhizosphere Processes and Nutrient-use Efficiency under Heterogeneous Nutrient Supply

The response of roots to heterogeneous nutrient supply is a coordinated process balancing morphology and physiology to optimize soil exploration and foraging for nutrients in nutrient-rich patches (Hodge 2004, 2006). A range of morphological and physiological responses provide plant with the assured, yet flexible, capacity to obtain nutrients in heterogeneous soils. Plants constantly adjust the relationship between root morphology and physiology to meet their needs for resources (Lynch and Ho 2005; Lambers et al. 2006).

The distribution of nutrient-rich patches varies greatly in time and space (Hodge 2004). In order to obtain resources that are unevenly distributed in the soil, plant roots are preferentially distributed to nutrient-rich patches. However, recent studies show that when the soil background nutrient is low, plants put more roots to the area in which nutrient concentration varies with time, and when the soil background nutrient is high, plants tend to distribute roots preferentially to the areas with more stable nutrient concentrations, suggesting that plant roots show risk preferences related to the spatial and temporal heterogeneity of nutrient resources (Dener et al. 2016).

Phosphorus is easily adsorbed or fixed by soils, showing strong spatiotemporal heterogeneity (Shen et al. 2011, 2019). To obtain soil P, plant roots show intense plastic responses regarding morphology and physiology. Plants usually increase the growth of roots in P-rich patches, expand root surface area, and increase the uptake rate to enhance P acquisition. Different plant species showed various adaptive strategies for the acquisition of heterogeneously supplied soil P. Some crops (e.g. oilseed rape and wheat) responded to heterogeneous P supply by changing root morphological traits such as root elongation and biomass allocation, and thus, acquisition of P was based mainly on root morphological plasticity (Rose et al. 2009). White lupin can form numerous cluster roots under heterogeneous P environment, releasing a range of organic acid anions and acid phosphatases; hence, acquisition of P by white lupin has both morphological and physiological plasticity responses (Shen et al. 2005; Shu et al. 2007a,b). *Artemisia tridentata* improved P uptake by 80% under heterogeneous compared with the homogeneous P supply, indicating that P acquisition was dominated by the root physiological plasticity (Jackson et al. 1990).

The form of N supply largely controls the absorption ratio of cations and anions, influencing the root apoplastic pH and rhizosphere pH (Marschner 2011; Shen et al. 2013; Mpanga et al. 2019). Nitrate uptake occurs in cotransport with H^+, causing rhizosphere alkalinization, whereas ammonium uptake is accompanied by release of H^+, inducing rhizosphere acidification (Marschner 2011). In the field, the application of localized NH_4^+-N plus P significantly promoted the acquisition of P by maize in the calcareous soils (Jing et al. 2010). In such soils, the localized patch of ammonium plus phosphate resulted in rhizosphere acidification due to ammonium uptake (Jing et al. 2010), thus increasing P availability and improving uptake of P by maize roots (Jing et al. 2010; Ma et al. 2015). In addition, it also significantly stimulated root proliferation, especially of fine roots (diameter less than 0.2 mm) (Jing et al. 2010; Ma et al. 2014). However, recent studies have also shown that improper application of localized ammonium plus P can lead to increased interspecies competition and ultimately lower crop yields. For instance, the heterogeneous nutrient supply induced maize root competition at high planting density and thus decreased crop productivity (Li et al. 2019).

In addition to rhizosphere acidification induced by localized fertilization, intercropping of plant species with different nutrient acquisition strategies can create "biological nutrient patches." Intercropping systems based on species interactions are very popular in China, which is an effective way to improve crop productivity (Li et al. 2020) (see also Chapters 11 and 12). For instance, maize/faba bean intercropping is widely practised. Maize has large root system and its morphological responses to heterogeneous nutrients are dominant, whereas faba bean roots are smaller and depended mainly on physiological responses (exudation of organic acid anions, phosphatase, and protons) to obtain soil P (Li et al. 2007; Zhang et al. 2016). This morphological and physiological complementation is one of the important strategies for the efficient acquisition of soil P in this intercropping system. More importantly, the strong root exudation capacity of faba bean improve mobilization of sparingly soluble soil P (not available to maize), leading to the formation of "biological nutrient patches" along faba bean roots (Zhang et al. 2016). For maize, soluble P distributed along the roots of faba bean is readily available; therefore, maize roots tend to grow alongside the faba bean roots to gain access to soil P (Zhang et al. 2020).

5.5 Root/Rhizosphere Management in the Field to Enhance Nutrient-use Efficiency

High input, high yield, low nutrient-use efficiency, and high environmental pressure are typical characteristics of intensive agricultural systems in China (Zhang et al. 2010). Crop yields have not increased proportionally with increasing fertilizer inputs over the past 30 years, leading to nutrient inefficiency and increasing environmental risks (Shen et al. 2013). It is a big challenge for the sustainable development of intensive agriculture to achieve high nutrient-use efficiency and high crop productivity in China. Traditional nutrient management strategies heavily rely on external fertilizer inputs, but ignore the biological potential of plant rhizosphere processes for efficient mobilization and use of soil nutrients (Zhang et al. 2010; Shen et al. 2011, 2013; Jiao et al. 2016). Studies demonstrate that rhizosphere management provides a unique opportunity to coordinate crop yield, nutrient efficiency, and environmental impact. Rhizosphere management emphasizes maximizing the efficiency of root and rhizosphere processes to increase soil nutrient acquisition and use, rather than solely depending on excessive chemical fertilizer application (Shen et al. 2013; Wang and Shen 2019).

Plant roots absorb soil nutrients from the rhizosphere. The rhizosphere processes reflect the nature and the intensity of interactions among plants, soil, and microorganisms in both natural and agricultural ecosystems (Zhang et al. 2010; Shen et al. 2013). In intensive agricultural ecosystems, the interaction between different components determines the exchange of matter and energy between plants and soil, influencing nutrient efficiency, crop productivity, and ecosystem stability. Root growth and rhizosphere processes directly determine the transformation, mobilization, and efficient use of soil nutrients (Figure 5.3). During the evolution, plant roots developed the capacity to sense the surrounding environment and make a series of plasticity responses (Hodge 2004). Plants can strongly (i) regulate root morphological traits to adapt to the soil environment and (ii) change rhizosphere processes via exudation of organic acid anions, phosphatases, protons, and signalling substances, and by changing redox conditions (Shen et al. 2013). All these rhizosphere processes induced by roots determine the mobilization and acquisition of soil nutrients and nutrient-use efficiency by crops, thus profoundly influencing crop productivity and sustainability (Figure 5.3). Therefore, manipulating and managing root growth and rhizosphere processes are effective ways to improve nutrient-use efficiency and crop yield to ensure sustainable development of agriculture (Zhang et al. 2010; Wang and Shen 2019).

5.5.1 Manipulating Root-zone Nutrient Availability

Soil nutrient status directly influences plant health, root growth, and rhizosphere processes. The efficiency of the rhizosphere processes largely depends on the inherent soil fertility and the input of external nutrient resources (Shen et al. 2013; Jiao et al. 2016). In the low nutrient soils, the efficiency and impact of the rhizosphere processes are reduced because root growth and expansion are severely limited by inadequate external nutrient supply. In contrast, when the soil nutrient supply is at, or slightly below, the optimum, the root/rhizosphere efficiency can be maximized. Furthermore, when soil nutrient supply exceeds, the optimum level, root growth, and rhizosphere efficiency are limited (Shen et al. 2013; Wang and Shen 2019).

When plants such as maize are exposed to short-term N deficiency, they may develop longer axial roots, exploring larger soil volumes and thus increasing spatial availability of nitrogen (Marschner 2011). However, if N is not replenished in time, its limited supply eventually decreases root growth (Wang et al. 2003). Similarly, root growth is also limited under the conditions of N oversupply. Maize root growth was inhibited when the concentration of nitrate in the culture solution was greater than $5\,\text{mmol}\,\text{l}^{-1}$ (Tian et al. 2008). These reports suggested that root growth can be regulated or enhanced by optimizing the nutrient status in the root zone or the rhizosphere.

In intensive agricultural production systems, inorganic fertilizer application is frequently excessive. However, in most cases, over-applying fertilizers does not increase crop production further. Generally, shoot biomass and N uptake of maize were positively related to the root size during growth (Mackay and Barber 1986). Larger roots occupy a larger volume of soil and therefore can take up more soil nutrients, enhance the soil N use, and reducing the risk of nitrate leaching. For maize genotypes grown under the same nutrient supply, the nutrient-efficient genotypes had higher biomass, larger roots, and absorbed more nutrients than the inefficient ones (Peng et al. 2010). However, soil fertility in intensive agricultural systems is generally high due to a large (or even excessive) input of fertilizers. In that case, the amount of nutrients taken up by crop is driven by the demand of shoot growth rather than the root size (Peng et al. 2010; Ning et al. 2012). In other words, increasing fertilizer inputs beyond an optimum level does not increase yields. Compared with optimal N application, excessive N application can inhibit maize root growth in both the early growth period (V6) and the rapid growth period (before flowering stage), and fail to improve the final N content of shoots, thus reducing the final grain yield (Shen et al. 2013). For root-zone nutrient management, nutrient supply optimization not only controls the total amount of inorganic fertilizers used, but also synchronizes crop nutrient demand and soil nutrient supply (Peng et al. 2014).

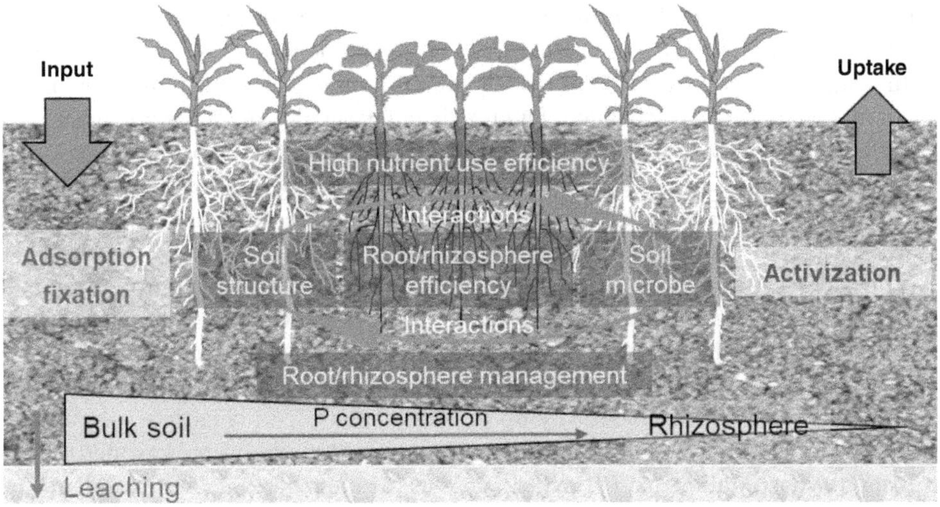

Figure 5.3 Root/rhizosphere management for improving nutrient-use efficiency and sustainable crop production.

Soil N shows high temporal and spatial variability, and crop N demand and soil N supply are difficult to synchronize. However, the root-zone N management for high-yielding crops must achieve that synchronization of root-zone soil N supply from soil, environment, and fertilizers with crop N demand in time and space (Chen et al. 2011, 2014). Based on this, the N requirements in the root-zone were determined according to the different crop growth stages based on the in-season monitoring of soil N in the root zone, and the N fertilizer application amounts were decided accordingly. Numerous field experiments proved that root-zone nutrient management could meet the crop demands, significantly decreasing the amount of N fertilizer input and losses of N to the environment while ensuring high yield (Zhao et al. 2006; Cui et al. 2008, 2010). Compared with conventional N management, in-season root-zone N management saved 40–60% of fertilizer input, increased crop yield by 4–5%, reduced N losses by 65–116 $kg\,ha^{-1}$, and greatly improved fertilizer-use efficiency. Nitrate in the soil profile (0–90 cm) can be controlled with a fertilization rate not exceeding 100 $kg\,ha^{-1}$, further reducing the risk of nitrate leaching to underground water (Cui et al. 2010).

5.5.2 Manipulating Plant Roots

Plants can change rhizosphere processes by changing root morphology and physiology to mobilize and obtain soil nutrients, so the manipulation of the roots includes the regulation of both root morphology and physiology (Shen et al. 2013; Wang and Shen 2019). In the early stage of crop growth, because roots are relatively short and only occupy a small volume in soil, the capacity to resist environmental stresses (such as low temperature, drought, nutrient deficiency) is poor, making targeted regulation of root growth very important. Localized application of starter fertilizer can significantly promote root growth, especially the proliferation of root branches and fine roots, thus improving the crop resistance to environmental stresses at the early stage of growth (Jing et al. 2010; Ma et al. 2015). For spring maize in northern China, in addition to short roots in the early growth stage, there is poor availability of P caused by low soil temperature, making it difficult for soil to meet the P demand for growth and development. Therefore, the supply of P fertilizer at maize seedling stage plays an important role in the shoot growth and yield formation (Shen et al. 2013; Wang and Shen 2019). In addition, localized nutrient supply can also meet the crop needs for growth by (i) increasing the nutrient concentration in localized areas and (ii) promoting strong uptake of nutrients by roots.

The physiological processes in roots can be altered by changing the forms and composition of fertilizers. For instance, the rational selection of ammonium-N, nitrate-N, and/or urea in different soil types can induce a desirable rhizosphere pH change and improve availability of soil nutrients. Applying ammonium nitrogen to calcareous soils in northern China significantly decreased (by up to 2 units) the pH of rhizosphere soil (Jing et al. 2010). In contrast, the absorption of nitrate by the roots is accompanied by the cotransport (consumption) of H^+, which may lead to an increase in rhizosphere pH (Marschner 2011; McClure et al. 1990). Hence, the physiological processes at the root/soil interface can be regulated by changing nutrient forms depending on the soil properties. In addition, nutrient supply intensity also significantly affects root growth and biological potential; as mentioned above, excess nutrient supply inhibits root growth. Hence, to ensure high yield and fertilizer-use efficiency of crops, the appropriate nutrient applications are needed to promote root growth and exudation for improving nutrient uptake and use efficiency.

5.5.3 Manipulating the Root-influenced Rhizosphere Environment

Rhizosphere processes such as root-induced rhizosphere acidification and root exudation play an important role in soil nutrient mobilization and absorption (Marschner 2011). The N supply forms and supply intensity significantly influence the rhizosphere chemical processes. Root uptake of ammonium-N is accompanied by H^+ release, leading to acidification of the rhizosphere (Marschner 2011). Calcareous soils in northern China have the soil pH generally higher than 8. Hence, the rhizosphere acidification induced by ammonium-N uptake can significantly promote uptake of nutrients, especially of water-insoluble P such as various apatite compounds. Localized supply of ammonium-N and P reduce the pH of rhizosphere from 8 to 5, and this change lasted for eight weeks (Jing et al. 2010), which effectively helped the small root system obtain enough P to meet maize demand in the early growth stage.

Localized supply of nutrients makes use of the root responses to heterogeneous nutrient supply, creating a concentration gradient with surrounding soil that can effectively alter the rhizosphere environment and manipulate root growth and physiology. Localized application of P and ammonium-N significantly promoted root growth and improved nutrient-use efficiency in early growth stage of maize. Compared with conventional nutrient management, the rhizosphere management (localized supply of nutrients) significantly (by 14.4%) increased maize grain yield and partial factor productivity (kilograms of grain produced per kilogram N applied) by 31.4% based on the average results during the 2010–2015 period (Jing et al. 2010; Ma et al. 2014; Jiao et al. 2016).

5.5.4 Manipulating Root/Rhizosphere Interactions

Root/rhizosphere interactions include root–soil interactions, root–root interactions (interspecific and intraspecific), and root–microorganism interactions (Zhang et al. 2010). In fact, the soil environment is so complex that root–soil–microbe interactions are so intertwined they cannot be distinguished completely. The roots interact with the surrounding soil and exchange mineral nutrients, organic compounds, and chemical signals (Figure 5.3). Soil characteristics that influence plant growth include water content (insufficient or excessive), compaction, low or excessive nutrient supply, mineral toxicity, and salinity. Roots respond both spatially and temporally to the heterogeneous soil environment by avoiding stress areas and proliferating in more favourable ones (Jin et al. 2013). Using root morphological and physiological plasticity in the soil–water–plant–microbe continuum, specific agricultural management measures (e.g. no-till, cover crops, diversification) to reshape/create an optimal rhizosphere soil environment may provide us with new opportunities to improve soil functions for enhanced agricultural production (Jin et al. 2017).

Intercropping system has a long history in China, which can improve the nutrient-use efficiency and crop productivity by regulating the rhizosphere interactions between different crop species (see Chapters 11 and 12). Judicious intercropping of different species is an effective way to reduce competition for nutrients by different crops. With an increasing research focus recently on the root–soil–microbe interactions in intercropping, we understand that underground interactions significantly change the spatial distribution of roots, and their morphological and physiological processes in the rhizosphere (Li et al. 2014b; Zhang et al. 2016).

Intercropping legumes with cereals plays an important role in maintaining N balance. However, recent studies found that leguminous crops, in addition to symbiosis with rhizobia for fixing N_2, also have strong capacity to secrete protons, organic acid anions, and phosphatases, which can significantly improve the availability of water-insoluble soil P in intercropping systems (Li et al. 2014b; Betencourt et al. 2012; Darch et al. 2018). In addition, the niche complementarity of root distribution in soil helped promote uptake of P and other nutrients by cereal crops (Zhang et al. 2010, 2016, 2017; Li et al. 2014b; Wang et al. 2020b). Many studies showed that intercropping not only increased crop productivity, land-use rate, and use efficiency of water and nutrient resources (Song et al. 2007; Mei et al. 2012; Brooker et al. 2015), but also increased farmland biodiversity, controlled the spread of diseases and pests, and reduced the use of pesticides and fertilizers (Zhu et al. 2003), thus decreasing the negative impact of agricultural production on the environment. Intercropping might also have a positive effect on soil health, with an increase in the organic carbon content in deep soil layers, thus enhancing carbon sequestration and reducing greenhouse gas emissions in agricultural production (Cong et al. 2015). Optimizing nutrient supply in intercropping systems can further strengthen rhizosphere effects and improve the efficiency of nutrient use and crop yield. For instance, localized supply of P and NH_4^+-N promoted uptake and utilization of P in intercropping of maize and faba bean by expanding the absorption range of target maize roots and increasing P uptake by maize (Zhang et al. 2016).

Soil microorganisms act as modifiers in the process of plant nutrient acquisition (see Chapter 6). The root–microbe interactions in the rhizosphere are important for plant growth, nutrition, and health (Zhang et al. 2017). Soil microorganisms are the driving force behind soil organic matter decomposition and soil nutrient cycling and transformation. Stable soil microbial community with appropriate abundance and appropriate population structure is the basis for promoting nutrient-use efficiency and beneficial nutrient cycling. Interactions between roots and microorganisms can be altered by inoculation with mycorrhiza or other beneficial microorganisms (Mueller and Sachs 2015). Arbuscular mycorrhizal fungi (AMF) help more than 80% of terrestrial plant species acquire P from soil. Rhizobia can fix N_2, and P-solubilizing bacteria can increase the amount of P available to plants.

Recently, many advances have been made in the development and application of plant-growth-promoting rhizobacteria (PGPR) in an attempt to enhance the biological potential of crops (see Chapter 6). PGPR colonize roots and promote plant growth and may play an important role in root growth and soil P mobilization (Dodd 2009). *Variovorax paradoxus* 5C2 can decrease the ethylene content in roots and promote root growth by decomposing the ethylene precursor aminocyclopropane carboxylic acid (ACC) (Belimov et al. 2009). In addition, root exudates play an important role in root–microbe interactions and even may act as signals to initiate/regulate root–microbe interactions (Werner et al. 2014).

The application of biotechnology has an important theoretical value and may have significance for improving soil nutrient availability and fertilizer-use efficiency. By regulating and optimizing the interaction of roots, soil, and microorganisms, and strengthening the modifying effects of microorganisms on the rhizosphere processes, the capacity to manipulate root/rhizosphere interactions can significantly increase nutrient-use efficiency and crop productivity with lower inputs, which contributes to the sustainable development of agriculture.

5.6 Concluding Remarks

Effective roots have an important impact on the exploration and exploitation of soil resources, crop production, and its sustainability. A better understanding of roots and their growth as well as the interactions with the surrounding environment is essential for improving crop productivity and enhancing nutrient-use efficiency through maximizing root/rhizosphere efficiency. The soil structure influences root growth, and the root growth also changes soil structure. The understanding of root growth in structured soils, especially root architecture adaptation to heterogeneous soils, is important to setting up an optimal root ideotype and managing soil appropriately in the agricultural systems. However, the knowledge of soil heterogeneity is limited, and how roots alter soil structure needs to be elucidated further.

The root/rhizosphere management based on understanding the relevant processes provides a new approach for high nutrient-use efficiency and productivity in intensive agricultural production systems. Traditionally, fertilizers are applied only to provide nutrients to crops, but more importantly they should be used as regulators of root growth and rhizosphere processes. Such management is gradually becoming a reality in Chinese agricultural production. However, the biological potential of roots is large, and how to further regulate it based on biological knowledge and the existing growth modifiers (such as auxin, ethylene, amino acids, and new fertilizers) is worthy of further study.

References

Abel, S. (2017). Phosphate scouting by root tips. *Current Opinion in Plant Biology* 39: 168–177. https://doi.org/10.1016/j.pbi.2017.04.016.

Alakukku, L. (2012). Soil compaction. In: Ecosystem Health and Sustainable Agriculture 1: Sustainable Agriculture (ed. C. Jakobsson). Uppsala: Uppsala University.

Angers, D.A. and Caron, J. (1998). Plant-induced changes in soil structure: processes and feedbacks. *Biogeochemistry* 42 (1–2): 55–72. https://doi.org/10.1023/A:1005944025343.

Atkinson, J.A., Pound, M.P., Bennett, M.J. et al. (2019). Uncovering the hidden half of plants using new advances in root phenotyping. *Current Opinion in Biotechnology* 55: 1–8. https://doi.org/10.1016/j.copbio.2018.06.002.

Bais, H.P., Weir, T.L., Perry, L.G. et al. (2006). The role of root exudates in rhizosphere interactions with plants and other organisms. *Annual Review of Plant Biology* 57: 233–266. https://doi.org/10.1146/annurev.arplant.57.032905.105159.

Bao, Y., Aggarwal, P., Robbins, N.E. et al. (2014). Plant roots use a patterning mechanism to position lateral root branches toward available water. *Proceedings of the National Academy of Sciences of the United States of America* 111 (25): 9319–9324. https://doi.org/10.1073/pnas.1400966111.

Beeckman, T. and Friml, J. (2010). Nitrate contra auxin: nutrient sensing by roots. *Development Cell* 18 (6): 877–878. https://doi.org/10.1016/j.devcel.2010.05.020.

Belimov, A.A., Dodd, I.C., Hontzeas, N. et al. (2009). Rhizosphere bacteria containing 1-aminocyclopropane-1-carboxylate deaminase increase yield of plants grown in drying soil via both local and systemic hormone signalling. *New Phytologist* 181 (2): 413–423. https://doi.org/10.1111/j.1469-8137.2008.02657.x.

Bellini, C., Pacurar, D.I., and Perrone, I. (2014). Adventitious roots and lateral roots: similarities and differences. *Annual Review of Plant Biology* 65: 639–666. https://doi.org/10.1146/annurev-arplant-050213-035645.

Bengough, A.G. (2012). Root elongation is restricted by axial but not by radial pressures: so what happens in field soil? *Plant and Soil* 360 (1–2): 15–18. https://doi.org/10.1007/s11104-012-1316-2.

Bengough, A.G., Bransby, M.F., Hans, J. et al. (2006). Root responses to soil physical conditions; growth dynamics from field to cell. *Journal of Experimental Botany* 57 (2): 437–447. https://doi.org/10.1093/jxb/erj003.

Bengough, A.G., Loades, K., and McKenzie, B.M. (2016). Root hairs aid soil penetration by anchoring the root surface to pore walls. *Journal of Experimental Botany* 67 (4): 1071–1078. https://doi.org/10.1093/jxb/erv560.

Bennie, A.T.P. and Botha, F.J.P. (1986). Effect of deep tillage and controlled traffic on root growth, water-use efficiency and yield of irrigated maize and wheat. *Soil and Tillage Research* 7 (1–2): 85–95. https://doi.org/10.1016/0167-1987(86)90010-3.

Betencourt, E., Duputel, M., Colomb, B. et al. (2012). Intercropping promotes the ability of durum wheat and chickpea to increase rhizosphere phosphorus availability in a low P soil. *Soil Biology and Biochemistry* 46: 181–190. https://doi.org/10.1016/j.soilbio.2011.11.015.

Bhardwaj, D., Medici, A., Gojon, A. et al. (2015). A new insight into root responses to external cues: paradigm shift in nutrient sensing. *Plant Signaling and Behavior* 10 (12): e1049791. https://doi.org/10.1080/15592324.2015.1049791.

Bingham, I.J. and Bengough, A.G. (2003). Morphological plasticity of wheat and barley roots in response to spatial variation in soil strength. *Plant and Soil* 250 (2): 273–282. https://doi.org/10.1023/A:1022891519039.

Bodner, G., Scholl, P., and Kaul, H.P. (2013). Field quantification of wetting–drying cycles to predict temporal changes of soil pore size distribution. *Soil and Tillage Research* 133: 1–9. https://doi.org/10.1016/j.still.2013.05.006.

Bodner, G., Leitner, D., and Kaul, H.P. (2014). Coarse and fine root plants affect pore size distributions differently. *Plant and Soil* 380 (1–2): 133–151. https://doi.org/10.1007/s11104-014-2079-8.

Boone, F.R. (1988). Weather and other environmental factors influencing crop responses to tillage and traffic. *Soil and Tillage Research* 11 (3–4): 283–324. https://doi.org/10.1016/0167-1987(88)90004-9.

Bouguyon, E., Brun, F., Meynard, D. et al. (2015). Multiple mechanisms of nitrate sensing by *Arabidopsis* nitrate transceptor NRT1.1. *Nature Plants* 1 (3): 15015. https://doi.org/10.1038/nplants.2015.15.

Bradshaw, A.D. (1965). Evolutionary significance of phenotypic plasticity in plants. *Advances in Genetics* 13 (1): 115–155. https://doi.org/10.1016/S0065-2660(08)60048-6.

Brooker, R.W., Bennett, A.E., Cong, W.F. et al. (2015). Improving intercropping: a synthesis of research in agronomy, plant physiology and ecology. *New Phytologist* 206 (1): 107–117. https://doi.org/10.1111/nph.13132.

Bruand, A., Cousin, I., Nicoullaud, B. et al. (1996). Backscattered electron scanning images of soil porosity for analyzing soil compaction around roots. *Soil Science Society of America Journal* 60 (3): 895–901. https://doi.org/10.2136/sssaj1996.03615995006000030031x.

Caldwell, M.M., Dudley, L.M., and Lilieholm, B. (1992). Soil solution phosphate, root uptake kinetics and nutrient acquisition: implications for a patchy soil environment. *Oecologia* 89 (3): 305–309. https://doi.org/10.1007/BF00317406.

Campbell, B.D., Grime, J.P., and Mackey, J.M. (1991). A trade-off between scale and precision in resource foraging. *Oecologia* 87 (4): 532–538. https://doi.org/10.1007/BF00320417.

Cárdenas, J.P., Santiago, A., Tarquis, A.M. et al. (2010). Soil porous system as heterogeneous complex network. *Geoderma* 160 (1): 13–21. https://doi.org/10.1016/j.geoderma.2010.04.024.

Carminati, A. (2012). A model of root water uptake coupled with rhizosphere dynamics. *Vadose Zone Journal* 11 (3): vzj2011.0106. https://doi.org/10.2136/vzj2011.0106.

Chapman, N., Whalley, W.R., Lindsey, K. et al. (2011). Water supply and not nitrate concentration determines primary root growth in *Arabidopsis*. *Plant, Cell and Environment* 34 (10): 1630–1638. https://doi.org/10.1111/j.13653040.2011.02358.x.

Chapman, N., Miller, A.J., Lindsey, K. et al. (2012). Roots, water, and nutrient acquisition: let's get physical. *Trends in Plant Science* 17 (12): 701–710. https://doi.org/10.1016/j.tplants.2012.08.001.

Chen, X.P., Cui, Z.L., Vitousek, P.M. et al. (2011). Integrated soil-crop system management for food security. *Proceedings of the National Academy of Sciences of the United States of America* 108 (16): 6399–6404. https://doi.org/10.1073/pnas.1101419108.

Chen, X., Cui, Z., Fan, M. et al. (2014). Producing more grain with lower environmental costs. *Nature* 514 (7523): 486–489. https://doi.org/10.1038/nature13609.

Chien, P.S., Chiang, C.B., Wang, Z., and Chiou, T.J. (2017). MicroRNA-mediated signaling and regulation of nutrient transport and utilization. *Current Opinion in Plant Biology* 39: 73–79. https://doi.org/10.1016/j.pbi.2017.06.007.

Chimungu, J.G., Loades, K.W., and Lynch, J.P. (2015). Root anatomical phenes predict root penetration ability and biomechanical properties in maize (*Zea mays*). *Journal of Experimental Botany* 66 (11): 3151–3162. https://doi.org/10.1093/jxb/erv121.

Clark, L.J., Whalley, W.R., and Barraclough, P.B. (2003). How do roots penetrate strong soil? *Plant and Soil* 255 (1): 93–104. https://doi.org/10.1023/A:1026140122848.

Cong, W.F., Hoffland, E., Li, L. et al. (2015). Intercropping enhances soil carbon and nitrogen. *Global Change Biology* 21 (4): 1715–1726. https://doi.org/10.1111/gcb.12738.

Croft, S.A., Pitchford, J.W., and Hodge, A. (2015). Fishing for nutrients in heterogeneous landscapes: modelling plant growth trade-offs in monocultures and mixed communities. *AoB Plants* 7: plv109. https://doi.org/10.1093/aobpla/plv109.

Cui, Z., Zhang, F., Chen, X. et al. (2008). On-farm evaluation of an in-season nitrogen management strategy based on soil N_{min} test. *Field Crops Research* 105 (1–2): 48–55. https://doi.org/10.1016/j.fcr.2007.07.008.

Cui, Z., Zhang, F., Chen, X. et al. (2010). In-season nitrogen management strategy for winter wheat: maximizing yields, minimizing environmental impact in an over-fertilization context. *Field Crops Research* 116 (1-2): 140–146. https://doi.org/10.1016/j.fcr.2009.12.004.

Darch, T., Giles, C.D., Blackwell, M.S. et al. (2018). Inter-and intra-species intercropping of barley cultivars and legume species, as affected by soil phosphorus availability. *Plant and Soil* 427 (1–2): 125–138. https://doi.org/10.1007/s11104-017-3365-z.

De Smet, I., White, P.J., Bengough, A.G. et al. (2012). Analyzing lateral root development: how to move forward. *The Plant Cell* 24 (1): 15–20. https://doi.org/10.1105/tpc.111.094292.

Dener, E., Kacelnik, A., and Shemesh, H. (2016). Pea plants show risk sensitivity. *Current Biology* 26 (13): 1763–1767. https://doi.org/10.1016/j.cub.2016.05.008.

Desnos, T. (2008). Root branching responses to phosphate and nitrate. *Current Opinion in Plant Biology* 11 (1): 82–87. https://doi.org/10.1016/j.pbi.2007.10.003.

Dexter, A.R. (1987). Compression of soil around roots. *Plant and Soil* 97 (3): 401–406. https://doi.org/10.1007/BF02383230.

Dexter, A.R., Richard, G., Arrouays, D. et al. (2008). Complexed organic matter controls soil physical properties. *Geoderma* 144 (3–4): 620–627. https://doi.org/10.1016/j.geoderma.2008.01.022.

Dodd, I.C. (2009). Rhizosphere manipulations to maximize 'crop per drop'during deficit irrigation. *Journal of Experimental Botany* 60 (9): 2454–2459. https://doi.org/10.1093/jxb/erp192.

Dorioz, J.M., Robert, M., and Chenu, C. (1993). The role of roots, fungi and bacteria on clay particle organization. An experimental approach. In: Soil Structure/Soil Biota Interrelationships (eds. L. Brussaard and M.J. Kooistra), 179–194. Elsevier.

Drew, M.C. (1975). Comparison of the effects of a localised supply of phosphate, nitrate, ammonium and potassium on the growth of the seminal root system, and the shoot, in barley. *New Phytologist* 75 (3): 479–490. https://doi.org/10.1111/j.1469-8137.1975.tb01409.x.

Drew, M.C. and Saker, L.R. (1978). Nutrient supply and the growth of the seminal root system in barley. III. Compensatory increases in growth of lateral roots, and in rates of phosphate uptake in response to a localised supply of phosphate. *Journal of Experimental Botany* 29 (2): 435–451. https://doi.org/10.1093/jxb/29.2.435.

Dunbabin, V., Rengel, Z., and Diggle, A. (2001). The root growth response to heterogeneous nitrate supply differs for *Lupinus angustifolius* and *Lupinus pilosus*. *Australian Journal of Agricultural Research* 52 (4): 495–503. https://doi.org/10.1071/AR00098.

Dunbabin, V., Diggle, A., and Rengel, Z. (2003). Is there an optimal root architecture for nitrate capture in leaching environments? *Plant. Cell and Environment* 26 (6): 835–844. https://doi.org/10.1046/j.1365-3040.2003.01015.x.

Einsmann, J., Jones, R., Pu, M., and Mitchell, R. (1999). Nutrient foraging traits in 10 co-occurring plant species of contrasting life forms. *Journal of Ecology* 87 (4): 609–619. https://doi.org/10.1046/j.1365-2745.1999.00376.x.

Eshel, A. and Beeckman, T. (2013). Plant Roots: The Hidden Half. Boca Raton: CRC Press.

Forde, B. and Lorenzo, H. (2001). The nutritional control of root development. *Plant and Soil* 232 (1–2): 51–68. https://doi.org/10.1023/A:1010329902165.

Fransen, B., de Kroon, H., and Berendse, F. (1998). Root morphological plasticity and nutrient acquisition of perennial grass species from habitats of different nutrient availability. *Oecologia* 115 (3): 351–358. https://doi.org/10.1007/s004420050527.

Fry, E.L., Evans, A.L., Sturrock, C.J. et al. (2018). Root architecture governs plasticity in response to drought. *Plant and Soil* 433 (1–2): 189–200. https://doi.org/10.1007/s11104-018-3824-1.

Gao, W., Hodgkinson, L., Jin, K. et al. (2016). Deep roots and soil structure. *Plant, Cell and Environment* 39 (8): 1662–1668. https://doi.org/10.1111/pce.12684.

Gao, W., Schlüter, S., Blaser, S.R.G.A. et al. (2019). A shape-based method for automatic and rapid segmentation of roots in soil from X-ray computed tomography images: rootine. *Plant and Soil* 441 (1–2): 643–655. https://doi.org/10.1007/s11104-019-04053-6.

Garcia, F., Cruse, R.M., and Blackmer, A.M. (1988). Compaction and nitrogen placement effect on root growth, water depletion, and nitrogen uptake. *Soil Science Society of America Journal* 52 (3): 792–798. https://doi.org/10.2136/sssaj1988.03615995005200030035x.

Gibson, J.R., Lin, H., and Bruns, M.A. (2006). A comparison of fractal analytical methods on 2-and 3 dimensional computed tomographic scans of soil aggregates. *Geoderma* 134: 335–348. https://doi.org/10.1016/j.geoderma.2006.03.052.

Giehl, R.F. and von Wirén, N. (2014). Root nutrient foraging. *Plant Physiology* 166 (2): 509–517. https://doi.org/10.1104/pp.114.245225.

Giehl, R.F., Lima, J.E., and von Wiren, N. (2012). Localized iron supply triggers lateral root elongation in *Arabidopsis* by altering the AUX1-mediated auxin distribution. *The Plant Cell* 24 (1): 33–49. https://doi.org/10.1105/tpc.111.092973.

Gojon, A., Krouk, G., Perrine-Walker, F. et al. (2011). Nitrate transceptor (s) in plants. *Journal of Experimental Botany* 62 (7): 2299–2308. https://doi.org/10.1093/jxb/erq419.

Goss, M.J. (1991). Consequences of the activity of roots on soil. In: Plant Root Growth (ed. D. Atkinson). Oxford: Blackwell Publishing Oxford.

Goss, M. and Russell, R.S. (1980). Effects of mechanical impedance on root growth in barley (*Hordeum vulgare* L.) III. Observations on the mechanism of response. *Journal of Experimental Botany* 31 (2): 577–588. https://doi.org/10.1093/jxb/31.2.577.

Grime, J.P., Crick, J.C., and Rincon, J.E. (1986). The ecological significance of plasticity. In: Plasticity in plants. The Company of Biologists (eds. D.H. Jennings and A.J. Trewavas). Cambridge, UK: Springer.

Guidi, G., Poggio, G., and Petruzzelli, G. (1985). The porosity of soil aggregates from bulk soil and from soil adhering to roots. *Plant and Soil* 87 (2): 311–314. https://doi.org/10.1007/BF02181871.

Håkansson, I. (2005). Machinery-induced compaction of arable soils. Technical Report. Uppsala. Rapporter från Jordbearbetningsavdelningen, Sveriges Lantbruksuniversitet, 109.

Hendriks, P.W., Kirkegaard, J.A., Lilley, J.M. et al. (2016). A tillering inhibition gene influences root–shoot carbon partitioning and pattern of water use to improve wheat productivity in rainfed environments. *Journal of Experimental Botany* 67 (1): 327–340. https://doi.org/10.1093/jxb/erv457.

Hinsinger, P., Bengough, A.G., Vetterlein, D. et al. (2009). Rhizosphere: biophysics, biogeochemistry and ecological relevance. *Plant and Soil* 321 (1–2): 117–152. https://doi.org/10.1007/s11104-008-9885-9.

Hiruma, K., Kobae, Y., and Toju, H. (2018). Beneficial associations between Brassicaceae plants and fungal endophytes under nutrient-limiting conditions: evolutionary origins and host-symbiont molecular mechanisms. *Current Opinion in Plant Biology* 44: 145–154. https://doi.org/10.1016/j.pbi.2018.04.009.

Ho, C.H., Lin, S.H., Hu, H.C., and Tsay, Y.F. (2009). CHL1 functions as a nitrate sensor in plants. *Cell* 138 (6): 1184–1194. https://doi.org/10.1016/j.cell.2009.07.004.

Hochholdinger, F., Yu, P., and Marcon, C. (2018). Genetic control of root system development in maize. *Trends in Plant Science* 23 (1): 79–88. https://doi.org/10.1016/j.tplants.2017.10.004.

Hodge, A. (2004). The plastic plant: root responses to heterogeneous supplies of nutrients. *New Phytologist* 162 (1): 9–24. https://doi.org/10.1111/j.1469-8137.2004.01015.x.

Hodge, A. (2006). Plastic plants and patchy soils. *Journal of Experimental Botany* 57 (2): 401–411. https://doi.org/10.1093/jxb/eri280.

Horn, R. and Dexter, A.R. (1989). Dynamics of soil aggregation in an irrigated desert loess. *Soil and Tillage Research* 13 (3): 253–266. https://doi.org/10.1016/0167-1987(89)90002-0.

Hu, L., Robert, C.A.M., Cadot, S. et al. (2018). Root exudate metabolites drive plant-soil feedbacks on growth and defense by shaping the rhizosphere microbiota. *Nature Communications* 9 (1): 2738. https://doi.org/10.1038/s41467-018-05122-7.

Hutchings, M.J. and de Kroon, H. (1994). Foraging in plants: the role of morphological plasticity in resource acquisition. *Advances in Ecological Research* 25: 159–238. https://doi.org/10.1016/S0065-2504(08)60215-9.

Hutchings, M.J. and John, E.A. (2004). The effects of environmental heterogeneity on root growth and root/shoot partitioning. *Annals of Botany* 94 (1): 1–8. https://doi.org/10.1093/aob/mch111.

Iijima, M., Kato, J., and Taniguchi, A. (2007). Combined soil physical stress of soil drying, anaerobiosis and mechanical impedance to seedling root growth of four crop species. *Plant Production Science* 10 (4): 451–459. https://doi.org/10.1626/pps.10.451.

Jackson, R.B., Manwaring, J.H., and Caldwell, M.M. (1990). Rapid physiological adjustment of roots to localized soil enrichment. *Nature* 344 (6261): 58–60. https://doi.org/10.1038/344058a0.

Jansen, L., Roberts, I., De Rycke, R., and Beeckman, T. (2012). Phloem-associated auxin response maxima determine radial positioning of lateral roots in maize. *Philosophical Transactions of the Royal Society B: Biological Sciences* 367 (1595): 1525–1533. https://doi.org/10.1098/rstb.2011.0239.

Jaramillo, R.E., Nord, E.A., Chimungu, J.G. et al. (2013). Root cortical burden influences drought tolerance in maize. *Annals of Botany* 112 (2): 429–437. https://doi.org/10.1093/aob/mct069.

Jiao, X., Lyu, Y., Wu, X. et al. (2016). Grain production versus resource and environmental costs: towards increasing sustainability of nutrient use in China. *Journal of Experimental Botany* 67 (17): 4935–4949. https://doi.org/10.1093/jxb/erw282.

Jin, K., Shen, J., Ashton, R.W. et al. (2013). How do roots elongate in a structured soil? *Journal of Experimental Botany* 64 (15): 4761–4777. https://doi.org/10.1093/jxb/ert286.

Jin, K.M., Shen, J.B., Ashton, R.W. et al. (2015). Wheat root growth responses to horizontal stratification of fertiliser in a water-limited environment. *Plant and Soil* 386 (1–2): 77–88. https://doi.org/10.1007/s11104-014-2249-8.

Jin, K., White, P.J., Whalley, W.R. et al. (2017). Shaping an optimal soil by root–soil interaction. *Trends in Plant Science* 22 (10): 823–829. https://doi.org/10.1016/j.tplants.2017.07.008.

Jing, J., Rui, Y., Zhang, F. et al. (2010). Localized application of phosphorus and ammonium improves growth of maize seedlings by stimulating root proliferation and rhizosphere acidification. *Field Crops Research* 119 (2–3): 355–364. https://doi.org/10.1016/j.fcr.2010.08.005.

Johnson, H.A. and Biondini, M.E. (2001). Root morphological plasticity and nitrogen uptake of 59 plant species from the Great Plains grasslands, USA. *Basic and Applied Ecology* 2 (2): 127–143. https://doi.org/10.1078/1439-1791-00044.

Kembel, S.W. and Cahill, J.F. Jr. (2005). Plant phenotypic plasticity belowground: a phylogenetic perspective on root foraging trade-offs. *The American Naturalist* 166 (2): 216–230. https://doi.org/10.1086/431287.

Koebernick, N., Daly, K.R., Keyes, S.D. et al. (2017). High-resolution synchrotron imaging shows that root hairs influence rhizosphere soil structure formation. *New Phytologist* 216 (1): 124–135. https://doi.org/10.1111/nph.14705.

Kravchenko, A.N., Wang, A.N.W., Smucker, A.J.M. et al. (2011). Long-term differences in tillage and land use affect intra-aggregate pore heterogeneity. *Soil Science Society of America Journal* 75 (5): 1658–1666. https://doi.org/10.2136/sssaj2011.0096.

de Kroon, H., Huber, H., Stuefer, J.F., and van Groenendael, J.M. (2015). A modular concept of phenotypic plasticity in plants. *New Phytologist* 166 (1): 73–82. https://doi.org/10.1111/j.1469-8137.2004.01310.x.

Krouk, G., Lacombe, B., Bielach, A. et al. (2010). Nitrate-regulated auxin transport by NRT1.1 defines a mechanism for nutrient sensing in plants. *Developmental Cell* 18 (6): 927–937. https://doi.org/10.1016/j.devcel.2010.05.008.

Kudoyarova, G.R., Dodd, I.C., Veselov, D.S. et al. (2015). Common and specific responses to availability of mineral nutrients and water. *Journal of Experimental Botany* 66 (8): 2133–3144. https://doi.org/10.1093/jxb/erv017.

Lambers, H., Shane, M.W., Cramer, M.D. et al. (2006). Root structure and functioning for efficient acquisition of phosphorus: matching morphological and physiological traits. *Annals of Botany* 98 (4): 693–713. https://doi.org/10.1093/aob/mcl114.

Laskowski, M. and Ten Tusscher, K.H. (2017). Periodic lateral root priming: what makes it tick? *The Plant Cell* 29 (3): 432–444. https://doi.org/10.1105/tpc.16.00638.

Lavenus, J., Goh, T., Roberts, I. et al. (2013). Lateral root development in *Arabidopsis*: fifty shades of auxin. *Trends in Plant Science* 18 (8): 450–458. https://doi.org/10.1016/j.tplants.2013.04.006.

Leij, F.J., Ghezzehei, T.A., and Or, D. (2002). Modeling the dynamics of the soil pore-size distribution. *Soil and Tillage Research* 64 (1–2): 61–78. https://doi.org/10.1016/S0167-1987(01)00257-4.

Li, L., Li, S.M., Sun, J.H. et al. (2007). Diversity enhances agricultural productivity via rhizosphere phosphorus facilitation on phosphorus-deficient soils. *Proceedings of the National Academy of Sciences of the United States of America* 104 (27): 11192–11196. https://doi.org/10.1073/pnas.0704591104.

Li, H., Ma, Q., Li, H. et al. (2014a). Root morphological responses to localized nutrient supply differ among crop species with contrasting root traits. *Plant and Soil* 376 (1–2): 151–163. https://doi.org/10.1007/s11104-013-1965-9.

Li, L., Tilman, D., Lambers, H. et al. (2014b). Plant diversity and overyielding: insights from belowground facilitation of intercropping in agriculture. *New Phytologist* 203 (1): 63–69. https://doi.org/10.1111/nph.12778.

Li, B., Li, Y.Y., Wu, H.M. et al. (2016). Root exudates drive interspecific facilitation by enhancing nodulation and N_2 fixation. *Proceedings of the National Academy of Sciences of the United States of America* 113 (23): 6496–6501. https://doi.org/10.1073/pnas.1523580113.

Li, H., Wang, X., Brooker, R.W. et al. (2019). Root competition resulting from spatial variation in nutrient distribution elicits decreasing maize yield at high planting density. *Plant and Soil* 439 (1–2): 219–232. https://doi.org/10.1007/s11104-018-3812-5.

Li, C., Hoffland, E., Kuyper, T.W. et al. (2020). Syndromes of production in intercropping impact yield gains. *Nature Plants* 6: 653–660. https://doi.org/10.1038/s41477-020-0680-9.

Lima, J.E., Kojima, S., Takahashi, H. et al. (2010). Ammonium triggers lateral root branching in *Arabidopsis* in an AMMONIUM TRANSPORTER1; 3-dependent manner. *The Plant Cell* 22 (11): 3621–3633. https://doi.org/10.1105/tpc.110.076216.

Linkohr, B.I., Williamson, L.C., Fitter, A.H. et al. (2002). Nitrate and phosphate availability and distribution have different effects on root system architecture of *Arabidopsis*. *The Plant Journal* 29 (6): 751–760. https://doi.org/10.1046/j.1365-313X.2002.01251.x.

Lipiec, J. and Hatano, R. (2003). Quantification of compaction effects on soil physical properties and crop growth. *Geoderma* 116 (1–2): 107–136. https://doi.org/10.1016/S0016-7061(03)00097-1.

Liu, J.Q., Allan, D.L., and Vance, C.P. (2010). Systemic signaling and local sensing of phosphate in common bean: cross-talk between photosynthate and microRNA399. *Molecular Plant* 3 (2): 428–437. https://doi.org/10.1093/mp/ssq008.

Lobell, D.B., Cassman, K.G., and Field, C.B. (2009). Crop yield gaps: their importance, magnitudes, and causes. *Annual Review of Environment and Resources* 34: 179–204. https://doi.org/10.1146/annurev.environ.041008.093740.

Lynch, J.P. (2011). Root phenes for enhanced soil exploration and phosphorus acquisition: tools for future crops. *Plant Physiology* 156 (3): 1041–1049. https://doi.org/10.1104/pp.111.175414.

Lynch, J.P. (2013). Steep, cheap and deep: an ideotype to optimize water and N acquisition by maize root systems. *Annals of Botany* 112 (2): 347–357. https://doi.org/10.1093/aob/mcs293.

Lynch, J.P. and Brown, K.M. (2012). New roots for agriculture: exploiting the root phenome. *Philosophical Transactions of the Royal Society B: Biological Sciences* 367 (1595): 1598–1604. https://doi.org/10.1098/rstb.2011.0243.

Lynch, J.P. and Ho, M.D. (2005). Rhizoeconomics: carbon costs of phosphorus acquisition. *Plant and Soil* 269 (1-2): 45–56. https://doi.org/10.1007/s11104-004-1096-4.

Lynch, J.P. and Wojciechowski, T. (2015). Opportunities and challenges in the subsoil: pathways to deeper rooted crops. *Journal of Experimental Botany* 66 (8): 2199–2210. https://doi.org/10.1093/jxb/eru508.

Lynch, J.P., Chimungu, J.G., and Brown, K.M. (2014). Root anatomical phenes associated with water acquisition from drying soil: targets for crop improvement. *Journal of Experimental Botany* 65 (21): 6155–6166. https://doi.org/10.1093/jxb/eru162.

Lyu, Y., Tang, H., Li, H. et al. (2016). Major crop species show differential balance between root morphological and physiological responses to variable phosphorus supply. *Frontiers in Plant Science* 7: 1939. https://doi.org/10.3389/fpls.2016.01939.

Ma, Q. and Rengel, Z. (2008). Phosphorus acquisition and wheat growth are influenced by shoot phosphorus status and soil phosphorus distribution in a split-root system. *Journal of Plant Nutrition and Soil Science* 171 (2): 266–271. https://doi.org/10.1002/jpln.200700183.

Ma, Q., Rengel, Z., and Palta, J. (2009). Changes in soil-plant P under heterogeneous P supply influence C allocation between the shoot and roots. *Functional Plant Biology* 36 (9): 826–831. https://doi.org/10.1071/FP08259.

Ma, Q., Wang, X., Li, H. et al. (2014). Localized application of NH_4^+-N plus P enhances zinc and iron accumulation in maize via modifying root traits and rhizosphere processes. *Field Crops Research* 164: 107–116. https://doi.org/10.1016/j.fcr.2014.05.017.

Ma, Q., Wang, X., Li, H. et al. (2015). Comparing localized application of different N fertilizer species on maize grain yield and agronomic N-use efficiency on a calcareous soil. *Field Crops Research* 180: 72–79. https://doi.org/10.1016/j.fcr.2015.05.011.

Mackay, A.D. and Barber, S.A. (1986). Effect of nitrogen on root growth of two corn genotypes in the field. *Agronomy Journal* 78 (4): 699–703. https://doi.org/10.2134/agronj1986.00021962007800040028x.

Marschner, P. (2011). Marschner's Mineral Nutrition of Higher Plants. Elsevier.

Massa, G.D. and Gilroy, S. (2003). Touch modulates gravity sensing to regulate the growth of primary roots of *Arabidopsis thaliana*. *The Plant Journal* 33 (3): 435–445. https://doi.org/10.1046/j.1365-313X.2003.01637.x.

Materechera, S.A., Alston, A.M., Kirby, J.M., and Dexter, A.R. (1992). Influence of root diameter on the penetration of seminal roots into a compacted subsoil. *Plant and Soil* 144 (2): 297–303. https://doi.org/10.1007/BF00012888.

Materechera, S.A., Kirby, J.M., Alston, A.M., and Dexter, A.R. (1994). Modification of soil aggregation by watering regime and roots growing through beds of large aggregates. *Plant and Soil* 160 (1): 57–66. https://doi.org/10.1007/BF00150346.

McCarthy, D.F. (2007). Essentials of Soil Mechanics and Foundations. Upper Saddle River, NJ: Pearson Prentice Hall.

McClure, P.R., Kochian, L.V., Spanswick, R.M. et al. (1990). Evidence for cotransport of nitrate and protons in maize roots: II. Measurement of NO_3^- and H^+ fluxes with ion-selective microelectrodes. *Plant Physiology* 93 (1): 290–294. https://doi.org/10.1104/pp.93.1.290.

McCully, M.E. (1999). Roots in soil: unearthing the complexities of roots and their rhizospheres. *Annual Review of Plant Physiology and Plant Molecular Biology* 50 (1): 695–718. https://doi.org/10.1146/annurev.arplant.50.1.695.

McNickle, G.G. and Cahill, J.F. Jr. (2009). Plant root growth and the marginal value theorem. *Proceedings of the National Academy of Sciences of the United States of America* 106 (12): 4747–4751. https://doi.org/10.1073/pnas.0807971106.

McNickle, G.G., St Clair, C.C., and Cahill, J.F. Jr. (2009). Focusing the metaphor: plant root foraging behaviour. *Trends in Ecology and Evolution* 24 (8): 419–426. https://doi.org/10.1016/j.tree.2009.03.004.

Mei, P.P., Gui, L.G., Wang, P. et al. (2012). Maize/faba bean intercropping with rhizobia inoculation enhances productivity and recovery of fertilizer P in a reclaimed desert soil. *Field Crops Research* 130: 19–27. https://doi.org/10.1016/j.fcr.2012.02.007.

Misson, J., Raghothama, K.G., Jain, A. et al. (2005). A genome-wide transcriptional analysis using *Arabidopsis thaliana* Affymetrix gene chips determined plant responses to phosphate deprivation. *Proceedings of the National Academy of Sciences of the United States of America* 102 (33): 11934–11939. https://doi.org/10.1073/pnas.0505266102.

Miura, K., Rus, A., Sharkhuu, A. et al. (2005). The *Arabidopsis* SUMO E3 ligase SIZ1 controls phosphate deficiency responses. *Proceedings of the National Academy of Sciences of the United States of America* 102 (21): 7760–7765. https://doi.org/10.1073/pnas.0500778102.

Moller, B.K., Xuan, W., and Beeckman, T. (2017). Dynamic control of lateral root positioning. *Current Opinion in Plant Biology* 35: 1–7. https://doi.org/10.1016/j.pbi.2016.09.001.

Mommer, L., Kirkegaard, J., and van Ruijven, J. (2016). Root-root interactions: towards a rhizosphere framework. *Trends in Plant Science* 21 (3): 209–217. https://doi.org/10.1016/j.tplants.2016.01.009.

Montagu, K.D., Conroy, J.P., and Francis, G.S. (1998). Root and shoot response of field-grown lettuce and broccoli to a compact subsoil. *Australian Journal of Agricultural Research* 49 (1): 89–98. https://doi.org/10.1071/A97051.

Morel, J.L., Habib, L., Plantureux, S., and Guckert, A. (1991). Influence of maize root mucilage on soil aggregate stability. *Plant and Soil* 136 (1): 111–119. https://doi.org/10.1007/BF02465226.

Moreno-Espíndola, I.P., Rivera-Becerril, F., de Jesús Ferrara-Guerrero, M. et al. (2007). Role of root-hairs and hyphae in adhesion of sand particles. *Soil Biology and Biochemistry* 39 (10): 2520–2526. https://doi.org/10.1016/j.soilbio.2007.04.021.

Morris, E.C., Griffiths, M., Golebiowska, A. et al. (2017). Shaping 3D root system architecture. *Current Biology* 27 (17): R919–R930. https://doi.org/10.1016/j.cub.2017.06.043.

Motte, H., Vanneste, S., and Beeckman, T. (2019). Molecular and environmental regulation of root development. *Annual Review of Plant Biology* 70: 465–488. https://doi.org/10.1146/annurev-arplant-050718-100423.

Mounier, E., Pervent, M., Ljung, K. et al. (2014). Auxin-mediated nitrate signalling by NRT1.1 participates in the adaptive response of *Arabidopsis* root architecture to the spatial heterogeneity of nitrate availability. *Plant, Cell and Environment* 37 (1): 162–174. https://doi.org/10.1111/pce.12143.

Mpanga, I.K., Gomez-Genao, N., Moradtalab, N. et al. (2019). The role of N form supply for PGPM-host plant interactions in maize. *Journal of Plant Nutrition and Soil Science* 182 (6): 908–920. https://doi.org/10.1002/jpln.201900133.

Mueller, U.G. and Sachs, J.L. (2015). Engineering microbiomes to improve plant and animal health. *Trends in Microbiology* 23 (10): 606–617. https://doi.org/10.1016/j.tim.2015.07.009.

Mueller, N.D., Gerber, J.S., Johnston, M. et al. (2012). Closing yield gaps through nutrient and water management. *Nature* 490 (7419): 254–257. https://doi.org/10.1038/nature11420.

Nacry, P., Bouguyon, E., and Gojon, A. (2013). Nitrogen acquisition by roots: physiological and developmental mechanisms ensuring plant adaptation to a fluctuating resource. *Plant and Soil* 370 (1/2): 1–29. https://doi.org/10.1007/s11104-013-1645-9.

Nimmo, J.R. (2013). Porosity and Pore Size Distribution, Reference Module in Earth Systems and Environmental Sciences. Elsevier.

Ning, P., Liao, C., Li, S. et al. (2012). Maize cob plus husks mimics the grain sink to stimulate nutrient uptake by roots. *Field Crops Research* 130: 38–45. https://doi.org/10.1016/j.fcr.2012.02.010.

Nosalewicz, A. and Lipiec, J. (2014). The effect of compacted soil layers on vertical root distribution and water uptake by wheat. *Plant and Soil* 375 (1–2): 229–240. https://doi.org/10.1007/s11104-013-1961-0.

Paez-Garcia, A., Motes, C.M., Scheible, W.R. et al. (2015). Root traits and phenotyping strategies for plant improvement. *Plants* 4 (2): 334–355. https://doi.org/10.3390/plants4020334.

Peng, Y., Niu, J., Peng, Z. et al. (2010). Shoot growth potential drives N uptake in maize plants and correlates with root growth in the soil. *Field Crops Research* 115 (1): 85–93. https://doi.org/10.1016/j.fcr.2009.10.006.

Peng, Y., Li, C., and Fritschi, F.B. (2014). Diurnal dynamics of maize leaf photosynthesis and carbohydrate concentrations in response to differential N availability. *Environmental and Experimental Botany* 99: 18–27. https://doi.org/10.1016/j.envexpbot.2013.10.013.

Potocka, I. and Szymanowska-Pulka, J. (2018). Morphological responses of plant roots to mechanical stress. *Annals of Botany* 122 (5): 711–723. https://doi.org/10.1093/aob/mcy010.

Potocka, I., Szymanowska-Pułka, J., Karczewski, J. et al. (2011). Effect of mechanical stress on Zea root apex. I. Mechanical stress leads to the switch from closed to open meristem organization. *Journal of Experimental Botany* 62 (13): 4583–4593. https://doi.org/10.1093/jxb/err169.

Rebetzke, G.J. (2016). From inspiration to impact: delivering value from global root research. *Journal of Experimental Botany* 67 (12): 3601–3603. https://doi.org/10.1093/jxb/erw215.

Remans, T., Nacry, P., Pervent, M. et al. (2006). The *Arabidopsis* NRT1. 1 transporter participates in the signaling pathway triggering root colonization of nitrate-rich patches. *Proceedings of the National Academy of Sciences of the United States of America* 103 (50): 19206–19211. https://doi.org/10.1073/pnas.0605275103.

Reymond, M., Svistoonoff, S., Loudet, O. et al. (2006). Identification of QTL controlling root growth response to phosphate starvation in *Arabidopsis thaliana*. *Plant, Cell and Environment* 29 (1): 115–125. https://doi.org/10.1111/j.1365-3040.2005.01405.x.

Rich, S.M. and Watt, M. (2013). Soil conditions and cereal root system architecture: review and considerations for linking Darwin and Weaver. *Journal of Experimental Botany* 64 (5): 1193–1208. https://doi.org/10.1093/jxb/ert043.

Richards, J.H. and Caldwell, M.M. (1987). Hydraulic lift: substantial nocturnal water transport between soil layers by *Artemisia tridentata* roots. *Oecologia* 73 (4): 486–489. https://doi.org/10.1007/BF00379405.

Robinson, D. (1994). The responses of plants to non-uniform supplies of nutrients. *New Phytologist* 127 (4): 635–674. https://doi.org/10.1111/j.1469-8137.1994.tb02969.x.

Robinson, D. and van Vuuren, M.M.I. (1998). Responses of wild plants to nutrient patches in relation to growth rate and life-form. In: Variation in Plant Growth (eds. H. Lambers, H. Poorter and M.M.I. van Vuuren), 237–257. Leiden, The Netherlands: Backhuys.

Rogers, E.D. and Benfey, P.N. (2015). Regulation of plant root system architecture: implications for crop advancement. *Current Opinion in Biotechnology* 32: 93–98. https://doi.org/10.1016/j.copbio.2014.11.015.

Rose, T.J., Rengel, Z., Ma, Q.F. et al. (2009). Crop species differ in root plasticity response to localised P supply. *Journal of Plant Nutrition and Soil Science* 172 (3): 360–368. https://doi.org/10.1002/jpln.200800031.

Rubio, G., Liao, H., Yan, X., and Lynch, J.P. (2003). Topsoil foraging and its role in plant competitiveness for phosphorus in common bean. *Crop Science* 43 (2): 598–607. https://doi.org/10.2135/cropsci2003.0598.

Rubio, V., Bustos, R., Irigoyen, M.L. et al. (2009). Plant hormones and nutrient signaling. *Plant Molecular Biology* 69 (4): 361-373. https://doi.org/10.1007/s11103-008-9380-y.

Ryan, P.R., Delhaize, E., and Jones, D.L. (2001). Function and mechanism of organic anion exudation from plant roots. *Annual Review of Plant Biology* 52 (1): 527–560. https://doi.org/10.1146/annurev.arplant.52.1.527.

Sasse, J., Martinoia, E., and Northen, T. (2018). Feed your friends: do plant exudates shape the root microbiome? *Trends in Plant Science* 23 (1): 25–41. https://doi.org/10.1016/j.tplants.2017.09.003.

Schachtman, D.P. and Shin, R. (2007). Nutrient sensing and signaling: NPKS. *Annual Review of Plant Biology* 58: 47–69. https://doi.org/10.1146/annurev.arplant.58.032806.103750.

Shahzad, Z. and Amtmann, A. (2017). Food for thought: how nutrients regulate root system architecture. *Current Opinion in Plant Biology* 39: 80–87. https://doi.org/10.1016/j.pbi.2017.06.008.

Shane, M.W., De Vos, M., De Roock, S. et al. (2003). Shoot P status regulates cluster-root growth and citrate exudation in *Lupinus albus* grown with a divided root system. *Plant, Cell and Environment* 26 (2): 265–273. https://doi.org/10.1046/j.1365-3040.2003.00957.x.

Shen, J., Li, H., Neumann, G. et al. (2005). Nutrient uptake, cluster root formation and exudation of protons and citrate in *Lupinus albus* as affected by localized supply of phosphorus in a split-root system. *Plant Science* 168 (3): 837–845. https://doi.org/10.1016/j.plantsci.2004.10.017.

Shen, J., Yuan, L., Zhang, J. et al. (2011). Phosphorus dynamics: from soil to plant. *Plant Physiology* 156 (3): 997–1005. https://doi.org/10.1104/pp.111.175232.

Shen, J., Li, C., Mi, G. et al. (2013). Maximizing root/rhizosphere efficiency to improve crop productivity and nutrient use efficiency in intensive agriculture of China. *Journal of Experimental Botany* 64 (5): 1181–1192. https://doi.org/10.1093/jxb/ers342.

Shen, J., Wang, L., Jiao, X. et al. (2019). Innovations of phosphorus sustainability: implications for the whole chain. *Frontiers of Agricultural Science and Engineering* 6 (4): 321–331. https://doi.org/10.15302/J-FASE-2019283.

Shu, L., Shen, J., Rengel, Z. et al. (2005). Growth medium and phosphorus supply affect cluster root formation and citrate exudation by *Lupinus albus* grown in a sand/solution split-root system. *Plant and Soil* 276 (1-2): 85–94. https://doi.org/10.1007/s11104-005-3105-7.

Shu, L., Shen, J., Rengel, Z. et al. (2007a). Cluster root formation by *Lupinus albus* is modified by stratified application of phosphorus in a split-root system. *Journal of Plant Nutrition* 30 (2): 271–288. https://doi.org/10.1080/01904160601118075.

Shu, L., Shen, J., Rengel, Z. et al. (2007b). Formation of cluster roots and citrate exudation by *Lupinus albus* in response to localized application of different phosphorus sources. *Plant Science* 172 (5): 1017–1024. https://doi.org/10.1016/j.plantsci.2007.02.006.

Smith, S. and De Smet, I. (2012). Root system architecture: insights from *Arabidopsis* and cereal crops. *Philosophical Transactions of the Royal Society of London B: Biological Sciences* 367 (1595): 1441–1452. https://doi.org/10.1098/rstb.2011.0234.

Song, Y.N., Zhang, F.S., Marschner, P. et al. (2007). Effect of intercropping on crop yield and chemical and microbiological properties in rhizosphere of wheat (*Triticum aestivum* L.), maize (*Zea mays* L.), and faba bean (*Vicia faba* L.). *Biology and Fertility of Soils* 43 (5): 565–574. https://doi.org/10.1007/s00374-006-0139-9.

Svistoonoff, S., Creff, A., Reymond, M. et al. (2007). Root tip contact with low-phosphate media reprograms plant root architecture. *Nature Genetics* 39 (6): 792–796. https://doi.org/10.1038/ng2041.

Tardieu, F. (1988). Analysis of the spatial variability of maize root density: II. Distances between roots. *Plant and Soil* 107: 267–272.

Tarquis, A.M., Heck, R.J., Andina, D. et al. (2009). Pore network complexity and thresholding of 3D soil images. *Ecological Complexity* 6 (3): 230–239. https://doi.org/10.1016/j.ecocom.2009.05.010.

Thorpe, A.S., Thelen, G.C., Diaconu, A. et al. (2009). Root exudate is allelopathic in invaded community but not in native community: field evidence for the novel weapons hypothesis. *Journal of Ecology* 97 (4): 641–645. https://doi.org/10.1111/j.1365-2745.2009.01520.x.

Tian, Q., Chen, F., Liu, J. et al. (2008). Inhibition of maize root growth by high nitrate supply is correlated with reduced IAA levels in roots. *Journal of Plant Physiology* 165 (9): 942–951. https://doi.org/10.1016/j.jplph.2007.02.011.

Tinker, P.B. and Nye, P.H. (2000). Solute Movement in the Rhizosphere. Oxford University Press.

Van Norman, J.M., Xuan, W., Beeckman, T. et al. (2013). To branch or not to branch: the role of pre-patterning in lateral root formation. *Development* 140 (21): 4301–4310. https://doi.org/10.1242/dev.090548.

Van Veelen, A., Koebernick, N., Scotson, C.S. et al. (2020). Root-induced soil deformation influences Fe, S and P: rhizosphere chemistry investigated using synchrotron XRF and XANES. *New Phytologist* 225 (4): 1476–1490. https://doi.org/10.1111/nph.16242.

Van Vuuren, M.M.I., Robinson, D., and Griffiths, B.S. (1996). Nutrient inflow and root proliferation during the exploitation of a temporally and spatially discrete source of nitrogen in soil. *Plant and Soil* 178 (2): 185–192. https://doi.org/10.1007/BF00011582.

Venturi, V. and Keel, C. (2016). Signaling in the rhizosphere. *Trends in Plant Science* 21 (3): 187–198. https://doi.org/10.1016/j.tplants.2016.01.005.

Vollsnes, A.V., Futsaether, C.M., and Bengough, A.G. (2010). Quantifying rhizosphere particle movement around mutant maize roots using time-lapse imaging and particle image velocimetry. *European Journal of Soil Science* 61 (6): 926–939. https://doi.org/10.1111/j.1365-2389.2010.01297.x.

Walker, T.S., Bais, H.P., Grotewold, E. et al. (2003). Root exudation and rhizosphere biology. *Plant Physiology* 132 (1): 44–51. https://doi.org/10.1104/pp.102.019661.

Wang, L. and Shen, J. (2019). Root/rhizosphere management for improving phosphorus use efficiency and crop productivity. *Better Crops with Plant Food* 103 (1): 36–39. https://doi.org/10.24047/bc103136.

Wang, Y., Mi, G., Chen, F. et al. (2003). Genotypic differences in nitrogen uptake by maize inbred lines its relation to root morphology. *Acta Ecologica Sinica* 23 (2): 297–302.

Wang, X., Feng, J., White, P.J. et al. (2020a). Heterogeneous phosphate supply influences maize lateral root proliferation by regulating auxin redistribution. *Annals of Botany* 125 (1): 119–130. https://doi.org/10.1093/aob/mcz154.

Wang, L., Hou, B., Zhang, D. et al. (2020b). The niche complementarity driven by rhizosphere interactions enhances phosphorus-use efficiency in maize/alfalfa mixture. *Food and Energy Security* 9: e252. https://doi.org/10.1002/fes3.252.

Werner, G.D., Strassmann, J.E., Ivens, A.B. et al. (2014). Evolution of microbial markets. *Proceedings of the National Academy of Sciences of the United States of America* 111 (4): 1237–1244. https://doi.org/10.1073/pnas.1315980111.

Whitmore, A.P. and Whalley, W.R. (2009). Physical effects of soil drying on roots and crop growth. *Journal of Experimental Botany* 60 (10): 2845–2857. https://doi.org/10.1093/Jxb/Erp200.

Yano, K. and Kume, T. (2005). Root morphological plasticity for heterogeneous phosphorus supply in *Zea mays* L. *Plant Production Science* 8 (4): 427–432. https://doi.org/10.1626/pps.8.427.

Yu, P., Baldauf, J.A., Lithio, A. et al. (2016a). Root type-specific reprogramming of maize pericycle transcriptomes by local high nitrate results in disparate lateral root branching patterns. *Plant Physiology* 170 (3): 1783–1798. https://doi.org/10.1104/pp.15.01885.

Yu, P., Gutjahr, C., Li, C. et al. (2016b). Genetic control of lateral root formation in cereals. *Trends in Plant Science* 21 (11): 951–961. https://doi.org/10.1016/j.tplants.2016.07.011.

Zhang, H. and Forde, B.G. (1998). An *Arabidopsis* MADS box gene that controls nutrient-induced changes in root architecture. *Science* 279 (5349): 407–409. https://doi.org/10.1126/science.279.5349.407.

Zhang, F., Shen, J., Zhang, J. et al. (2010). Rhizosphere processes and management for improving nutrient use efficiency and crop productivity: implications for China. *Advances in Agronomy* 107: 1–32. https://doi.org/10.1016/S0065-2113(10)07001-X.

Zhang, D., Zhang, C., Tang, X. et al. (2016). Increased soil phosphorus availability induced by faba bean root exudation stimulates root growth and phosphorus uptake in neighbouring maize. *New Phytologist* 209 (2): 823–831. https://doi.org/10.1111/nph.13613.

Zhang, R., Vivanco, J.M., and Shen, Q. (2017). The unseen rhizosphere root-soil-microbe interactions for crop production. *Current Opinion in Microbiology* 37: 8–14. https://doi.org/10.1016/j.mib.2017.03.008.

Zhang, W.P., Liu, G.C., Sun, J.H. et al. (2017). Temporal dynamics of nutrient uptake by neighbouring plant species: evidence from intercropping. *Functional Ecology* 31 (2): 469–479. https://doi.org/10.1111/1365-2435.12732.

Zhang, D., Lyu, Y., Li, H. et al. (2020). Neighbouring plants modify maize-root foraging for phosphorus: coupling nutrients and neighbours for improved nutrient-use efficiency. *New Phytologist* 226 (1): 244–253. https://doi.org/10.1111/nph.16206.

Zhao, R.F., Chen, X.P., Zhang, F.S. et al. (2006). Fertilization and nitrogen balance in a wheat-maize rotation system in North China. *Agronomy Journal* 98 (4): 938–945. https://doi.org/10.2134/agronj2005.0157.

Zhu, Y.Y., Wang, Y.Y., Chen, H.R. et al. (2003). Conserving traditional rice through biodiversity management. *BioScience* 53: 158–162.

6

Root–microbe Interactions Influencing Water and Nutrient Acquisition Efficiency

Teofilo Vamerali[1], Anna Panozzo[1], Giovanna Visioli[2], and Cristian Dal Cortivo[1]

[1] *Department of Agronomy, Food, Natural Resources, Animals and the Environment, University of Padua, Legnaro – Padua, Italy*
[2] *Department of Chemistry, Life Sciences and Environmental Sustainability, University of Parma, Parma, Italy*

6.1 Introduction

Agriculture currently is having to cope with a rapid growth in the world's population and demand for food while dealing with critical issues such as the scarcity of resources (e.g. water and P fertilizers) and environmental pollution, which require the adoption of more sustainable approaches. In the context of the need to move from intensive to environmentally friendly practices, research on plant–microbe interactions can open up new opportunities for agriculture.

As water and nutrients are the major factors limiting crop yield, more attention should be focused on the below-ground compartment. Besides anchorage, plant roots provide essential resources, and much of successful crop production is contingent on the growth and efficiency of root system, which is largely dependent on its interaction with the soil microbiome. The rhizosphere is the soil volume under the influence of plant roots and is where microbial activity reaches its peak compared with the surrounding bulk soil (Dobbelaere et al. 2003). A variety of compounds released into the rhizosphere as root exudates, which include polysaccharides and proteins, depending on the plant species and its physiological activity (Kang et al. 2010), attract microbial populations to colonize roots (Somers et al. 2004; Rodriguez-Navarro et al. 2007). The rhizosphere microbial community associated with plant roots is known as the rhizo-microbiome. The principal components are bacteria that are more abundant (10–1000 times) (Bulgarelli et al. 2013) and more diverse in the rhizosphere than the surrounding bulk soil (Visioli et al. 2018).

Some bacterial species and strains play key beneficial roles in their associated plants through various mechanisms and are therefore known as plant growth-promoting rhizobacteria (PGPR). PGPR act mainly at the soil/root interface because of their capacity to colonize different root microhabitats (Bishnoi 2015). They are classified into two main types: (i) extracellular plant growth-promoting rhizobacteria (ePGPR), which inhabit the rhizoplane or the

The Root Systems in Sustainable Agricultural Intensification, First Edition.
Edited by Zed Rengel and Ivica Djalovic.

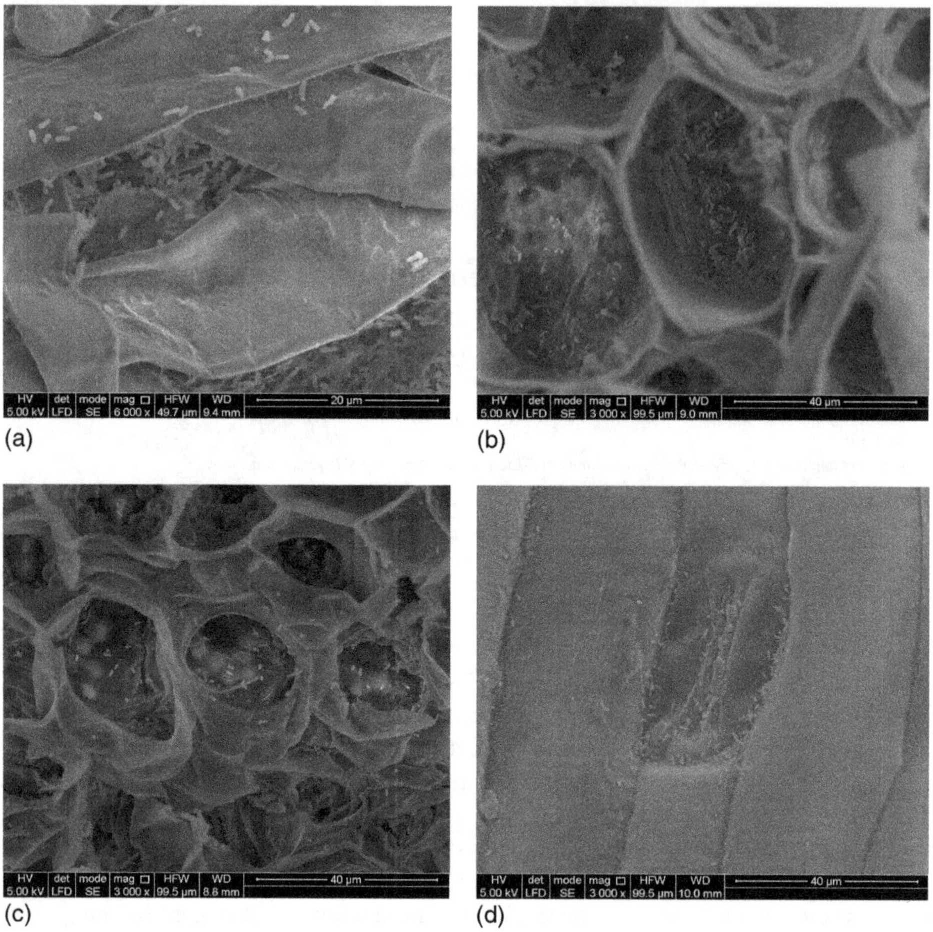

Figure 6.1 ESEM micrographs of root surface (a), microdissected transversal root section (b), leaf transversal section (c), and surface around closed stomata (d) of seven-day-old wheat seedlings inoculated with a consortium of *Azospirillum* spp., *Azoarcus* spp., and *Azorhizobium* spp. in sterile conditions. *Source:* Monica Mattarozzi (University of Parma, Italy).

intercellular spaces of the root cortex, and (ii) intracellular plant growth-promoting rhizobacteria (iPGPR), which live within specialized nodular structures of the root cells (Viveros et al. 2010). The main ePGPR bacterial genera include *Azotobacter*, *Serratia*, *Azospirillum*, *Bacillus*, *Caulobacter*, *Chromobacterium*, *Agrobacterium*, *Erwinia*, *Flavobacterium*, *Arthrobacter*, *Micrococcus*, *Pseudomonas*, and *Burkholderia*; in contrast, iPGPR include the root-nodulating endophyte genera, such as *Allorhizobium*, *Bradyrhizobium*, *Mesorhizobium*, and *Rhizobium*, as well as *Frankia* (Gouda et al. 2018). It has been found that ePGPR can pass through the endodermis barrier to move from the cortex to the vascular system, where they subsequently thrive as endophytes in stems, leaves, and other plant organs (Compant et al. 2005a), as we recently demonstrated in wheat seedlings inoculated with a PGPR consortium in aseptic conditions (Dal Cortivo et al. 2017) (Figure 6.1).

PGPR enhance plant growth directly by improving the plant physiological processes and tolerance to adverse conditions (Vessey 2003; Zakry et al. 2012), and they also protect plants indirectly by producing pathogen-suppressing substances (Singh and Jha 2015) such as hydrolytic enzymes (e.g. chitinases, cellulases, proteases, and 1-aminocyclopropane-1-carboxylate [ACC]-deaminase), antibiotics in response to pathogen attack, siderophores, rhizobitoxines, exopolysaccharides, and volatile organic compounds (Sandhya et al. 2009; Gupta et al. 2014). There is wide variation in the activity and effectiveness of PGPR due to plant species (Garcia-Fraile et al. 2015) and some biotic (plant genotype and growth stage, interaction with other members of the microbial community) and abiotic factors (soil structure and management, nutrient abundance, and climatic conditions) (Vacheron et al. 2013).

PGPR improve plant growth and development through several mechanisms and processes, such as nitrogen fixation, mineralization of organic matter, solubilization of mineral nutrients in the rhizosphere, and the production of auxins, that promote root growth (Vessey 2003; Bhardwaj et al. 2014).

6.2 PGPR and Nutrient-use Efficiency

6.2.1 Biological Nitrogen Fixation (BNF)

Nitrogen (N) is the major plant growth-limiting nutrient; plants have high requirements but low use efficiency of nitrogen in chemical fertilizers (approximately 50%) (Zhang et al. 2018; Bhattacharjee et al. 2008; Zhu and Chen 2002). This means that a large portion of fertilizers may be lost through run-off, erosion, denitrification, volatilization, and leaching. PGPR greatly contribute to improving soil fertility and nitrogen-use efficiency through biological N_2 fixation (BNF), which is carried out by various symbiotic and nonsymbiotic bacteria (Shridhar 2012) through the activity of nitrogenase enzyme (Kim and Rees 1994). Symbiotic PGPR N_2-fixers, collectively known as rhizobia, are most frequently reported as fixing atmospheric N_2 in root nodules of leguminous species, although there are others, such as the filamentous bacteria of the genus *Frankia* in actinorhizal plant species from families *Betulaceae and Elaeagnaceae* (Rashid et al. 2015). The rhizobia genera involved are *Rhizobium*, *Sinorhizobium*, *Bradyrhizobium*, *Mesorhizobium*, and *Azorhizobium*. Sometimes, there is no real symbiotic structure, as in the case of *Azorhizobium caulinodans* that enters the root system between the rhizodermal cells and colonizes the plant xylem, providing a non-nodular niche for endosymbiotic nitrogen fixation in wheat, rice, maize, and other nonlegume crops (Cocking 2003). Nonsymbiotic N_2 fixation is basically carried out by free-living diazotroph bacteria belonging to the genera *Azoarcus*, *Azospirillum*, *Azotobacter*, *Burkholderia*, *Gluconacetobacter*, and *Pseudomonas* (Bhattacharyya and Jha 2012).

The potential for exploiting N_2-fixing PGPR bacteria as commercial biofertilizers has been investigated in various crop species in recent years, with *Azospirillum* and *Pseudomonas* seemingly the genera receiving most attention. For instance, microbe-mediated mineralization and N_2 fixation are essential for increasing N availability and uptake in rice fields (Richardson et al. 2009; Lu et al. 2009). Plant inoculation with *Azospirillum brasilense*

(Pedraza et al. 2009), *Pseudomonas fluorescens*, and *Pseudomonas putida* (Ghaffari et al. 2018), or co-inoculation with *A. brasilense* and *P. fluorescens* (Zhang et al. 2018) have been shown to improve greatly nitrogenase activity in the soil and the mineralization rate in rice fields, which translated into the growth and yield improvements. *P. putida* (Cakmakçi et al. 2006), *Pseudomonas aeruginosa* (Arif et al. 2017), and various *Azotobacter* and *Azospirillum* species (Shoghi-Kalkhoran et al. 2013) increased N-uptake efficiency in sunflower and sugar beet.

An essential area to explore is the interaction between nitrogen fertilizers (type and rate) and bacteria inoculation, with the aim of optimizing fertilization management for each crop, particularly cereals such as maize and wheat (Calvo et al. 2017; Kumar et al. 2014). In this regard, it seems that higher N fertilizer doses do not interact negatively with rooting enhancement in wheat as a result of inoculation with *Azospirillum* together with other bacteria, so the negative effect of fertilizer N is countered, while N accumulation by the crop is improved (Dal Cortivo et al. 2017). Nitrogen fixation is sensitive to external factors, including N availability; N_2 fixation can be modulated rapidly through expression of the *nif* genes that are involved in activating the iron protein and synthesizing the iron–molybdenum cofactor, as well as many other essential genes regulating nitrogenase synthesis and activity (Reed et al. 2011).

6.2.2 Phosphorus-solubilizing Bacteria

Phosphorus (P) is one of the major essential macronutrients for plant growth and development. However, 95–99% of the world's soil phosphorus is present in insoluble, immobilized, or precipitated forms. A large part of the soluble inorganic phosphate applied to soils as chemical fertilizer is rapidly immobilized after application due to phosphate fixation by calcium, aluminium, iron, and soil colloids, depending on soil pH (Ekin 2010). It is estimated that suboptimal P availability reduces crop yield in 30–40% of arable land (Vance et al. 2003).

Several PGPR are known as phosphate-solubilizing bacteria (PSB), and their capacity to convert insoluble P to soluble forms is considered fundamental in plant nutrition (Chen 2006). These bacteria can increase P availability through local rhizosphere soil acidification by the secretion of protons and/or organic anions (e.g. acetate, lactate, oxalate, tartarate, succinate, citrate, gluconate, ketogluconate, and glycolate) that lowers the rhizosphere pH, which in turn allows $Ca_3(PO_4)_2$ to dissociate in calcareous soils (Afzal and Bano 2008) and facilitates chelation and exchange reactions (Richardson et al. 2009).

Another important source of P for plants is mineralization of organic P by soil bacteria, which takes place through different phosphatases that catalyze the hydrolysis of phosphoric esters (Glick 2012). Phosphate solubilization and mineralization can be performed by various bacterial species and strains, such as *Azotobacter*, *Bacillus*, *Burkholderia*, *Erwinia*, *Flavobacterium*, *Pseudomonas*, *Rhizobium*, and *Serratia* (Bhattacharyya and Jha 2012), some of which also enhance the availability of essential micronutrients (e.g. iron and zinc). Phosphate solubilization by PGPR directly promotes root growth (Zaidi et al. 2009; Collavino et al. 2010; Bashan et al. 2013). *Bacillus*, *Rhizobia*, and *Pseudomonas* are currently the most frequently studied P-solubilizing bacteria. The rapid colonization and

P-solubilizing activities of *Bacillus* species in the rhizosphere of maize (Breedt et al. 2017), sunflower (Ekin 2010), sugar beet, wheat, and oilseed rape (De Freitas et al. 1997; Cakmakçi et al. 2006) have been extensively investigated, with particularly good results obtained with *Bacillus megaterium*. Among the *Pseudomonas* species, *P. putida* has been reported to be an efficient P-solubilizer (Ghaffari et al. 2018).

6.2.3 Potassium-solubilizing Bacteria

Like P, more than 90% of soil potassium (K) is present as insoluble rock and silicate minerals, whereas soluble forms are present usually in very low concentrations (Parmar and Sindhu 2013). Plant use efficiency of K is also relatively low, estimated at 40% (Zou and Lu 2010), highlighting the importance of improving uptake of this mineral by crop plants. Potassium deficiency results in poorly developed root systems, weak fruits, low seed production, and slow growth rate, thus compromising yield. K-solubilizing bacteria (KSB), such as the genera *Acidithiobacillus*, *Bacillus*, *Pseudomonas*, *Burkholderia*, *Paenibacillus*, *Pantoea*, and *Frateuria* can release K from insoluble minerals (Nath et al. 2017; Ramarethinam and Chandra 2005). The release of protons and organic anions by these bacteria acidifies the rhizosphere environment, with consequent enhancement of mineral K solubilization (Meena et al. 2014).

The use of microorganisms appears to be a feasible strategy for addressing deficiencies in poorly mobile nutrients and thus improving productivity, particularly in degraded or marginal soils (Mohammadi and Sohrabi 2012). Attempts have been made to isolate bacterial strains from different rhizosphere soils in order to identify and characterize those that efficiently solubilize K (e.g. Nath et al. 2017; Meena et al. 2016), with many of these screenings carried out in rice fields (Khanghahi et al. 2018; Bakhshandeh et al. 2017). Although only a few investigations have been undertaken so far, increased K uptake after field application of KSB has been reported in sorghum (Badr 2006; Sheng and Huang 2002), and in maize and wheat (Meena et al. 2018; Singh et al. 2010).

6.2.4 Zinc and Iron Solubilization by Rhizosphere Bacteria

Zinc (Zn) deficiency in crops is linked to its low solubility compared with the total soil amount. Zinc is essential for plant growth and reproduction as a cofactor in many enzymatic complexes.

PGPR that can solubilize water-insoluble zinc compounds (e.g. ZnO, ZnS, $Zn_3(PO_4)_2$, and $ZnCO_3$) include species of *Arthrobacter*, *Achromobacter*, *Bacillus*, *Bordetella*, *Exiguobacterium*, *Flavobacterium*, *Kocuria*, *Pantoea*, *Pseudomonas*, and *Staphylococcus*. Of the various mechanisms involved, pH reduction through excretion of protons appears to be the most important, even though synthesis and exudation of specific chelating agents is considered fundamental (Ramesh et al. 2014).

Iron (Fe) is an essential microelement involved as a cofactor in many enzymes and numerous physiological processes, such as N_2 fixation, photosynthesis, and respiration. Even though the roots of strategy I plants (dicots and nongraminaceous monocots) are adapted to absorbing iron in its reduced form (Fe^{2+}), it is the oxidized ferric ion (Fe^{3+}) that prevails in soils (Violante et al. 2003). To overcome this problem, microorganisms and plants have evolved specific mechanisms to exude soluble organic compounds (chelators and phytosiderophores) that prevent precipitation of Fe^{3+} into insoluble compounds (Saha et al. 2016). Chelators exuded by strategy I plants carry the ferric ion to the root surface where it is absorbed after reduction to Fe^{2+}. Bacteria exude siderophores to complex and take up Fe^{3+}. The siderophore-Fe^{3+} complex can be absorbed also by roots via specific receptor proteins in the plasma membrane, although bacteria are more competitive than plants (Sharma and Johri 2003; Marschner et al. 2011). Siderophores include hydroxamates, phenol-catecholes, and carboxylates (Podile and Kishore 2006); their synthesis is influenced by environmental conditions (Duffy and Défago 1999), and their effectiveness in providing iron to plants is enhanced in the presence of other metals, such as nickel and cadmium (Beneduzi et al. 2012). Various studies have identified siderophore-producing bacteria belonging to the genera *Azospirillum*, *Bradyrhizobium*, *Pseudomonas*, *Rhizobium*, *Serratia*, and *Streptomyces* (e.g. Kuffner et al. 2008). The majority of bacteria can take up only self-produced siderophores, but *Pseudomonas* spp. can exploit those produced by other genera or species (Loper and Henkels 1999; Rathore 2015).

Bacterial siderophores have biocontrol properties against plant pathogens (Compant et al. 2005b) and also have direct effects on plant growth, particularly when iron is not a limiting factor for plants (Carrillo-Castañeda et al. 2002).

6.3 Role of PGPR in Promoting Plant Growth

Plant growth and development, and many biochemical and physiological processes are regulated by phytohormones, organic substances acting in very low concentrations (<1 mM) (Damam et al. 2016; Davière and Achard 2013). There is considerable evidence that PGPR stimulate plant growth through direct production of plant growth-promoting substances (PGPS), including several phytohormones, such as auxins (indole acetic acid [IAA]), gibberellins, cytokinins, abscisic acid (ABA), ethylene, jasmonates, and brassinosteroids, and/or by regulating their endogenous concentrations in plant tissues (Spaepen et al. 2007; Sureshbabu et al. 2016). This leads to larger root systems and improvements in resource acquisition, enabling plants to cope better with adverse conditions (Jung et al. 2013; Ahemad and Kibret 2014); in addition, bacteria can provide also N_2 fixation (BNF) and/or mineral solubilization at various plant growth stages (De Souza et al. 2013).

6.3.1 Plant Growth-promoting Substances (PGPS)

About 80% of root-associated bacteria can synthesize auxins as secondary metabolites (Patten and Glick 2002; Spaepen et al. 2007). Among auxin-producing microorganisms are *Azospirillum*, *Pseudomonas*, *Rhizobium*, and *Bacillus* (Khalid et al. 2011; Khakipour et al. 2008; Etesami et al. 2009; Lim and Kim 2009). IAA is the main bacterial auxin promoting plant growth, which it does by stimulating root cell proliferation, lateral root initiation, and overproduction of root hairs, thereby increasing the active area for absorption of mineral nutrients and water (Morel et al. 2012; Vessey 2003; Fibach-Paldi et al. 2012). In screening for IAA-overproducing bacterial strains, several studies found a strong correlation between exogenous IAA concentration and primary and lateral root elongation (Raheem et al. 2018; Breedt et al. 2017; Allam et al. 2018). Basaglia et al. (2003) observed large improvements (+44%) in root length density of sorghum only when seeds were inoculated with the IAA-overproducing *Azospirillum brasilense* strain Sp6-IAA^{++}. As IAA does not function as a hormone in the bacterial cells, it is thought that the capacity to produce IAA evolved with plant association (Hallmann et al. 1997). IAA also regulates the expression of bacterial genes involved in adhesion, adaptation, and virulence, and hence influences the capacity of bacteria to colonize plants (Spaepen et al. 2007).

Gibberellins regulate plant growth and development from germination to senescence. The most evident physiological effect of gibberellins in plants is shoot elongation (overcoming dwarfism), although they govern many other physiological processes, such as seed and bud dormancy, flowering induction, and fruit ripening (Argüeso et al. 2012).

Several PGPR species, such as *Achromobacter xylosoxidans*, *Acinetobacter calcoaceticus*, *Azospirillum* spp., *Azotobacter* spp., *Bacillus* spp., *Herbaspirillum seropedicae*, and *Gluconobacter diazotrophicus* (Spaepen and Vanderleyden 2011; Ivanova et al. 2008; Masciarelli et al. 2014), as well as many rhizobia (Gutierrez-Manero et al. 2001; Bottini et al. 2004; Dodd et al. 2010), are known to synthesize and excrete gibberellins and/or gibberellic acid-like substances.

Cytokinins also affect many aspects of plant growth and development, including stimulation of cell division, and control of leaf senescence and apical dormancy. They promote root meristem differentiation and root hair formation by concurrently inhibiting lateral root formation and primary root elongation (Riefler et al. 2006).

Many pathogenic and plant-aiding bacteria can synthesize cytokinins and therefore have a direct effect on shoot growth stimulation (Ivanova et al. 2001; Arkhipova et al. 2007; Pallai et al. 2012). Symbiotic N_2-fixing bacteria, such as *Rhizobium* and *Bradyrhizobium*, and free-living N_2-fixing bacteria, such as *Azospirillum, Azotobacter*, and *Bacillus*, are recognized producers of cytokinins and other growth-promoting substances in the rhizosphere (Perrig et al. 2007; Ahemad and Kibret 2014; García de Salamone et al. 2001).

Plant beneficial bacteria do not synthesize ethylene, but they regulate its endogenous concentration in plants, and possibly suppress its negative effects (Shaharoona et al. 2008). High concentration of ethylene is responsible for plant senescence and growth inhibition leading to defoliation, as well as inhibition of shoot and root elongation (Saleem et al. 2007; Salisbury 1994). In response to some abiotic stresses, such as drought, waterlogging, salinity, high temperatures, altered pH, and metal contamination, plants synthesize ACC, the immediate precursor for ethylene in higher plants (Glick 2012). Some PGPR facilitate plant growth by lowering internal ethylene concentration in plants, particularly in roots, where ACC is exuded into the rhizosphere, then re-absorbed and converted into ethylene (Glick 1995). As ethylene synthesis is related directly to the concentration of ACC, bacteria clearly play a role by producing the enzyme ACC deaminase that cleaves ACC into α-ketobutyrate and ammonium, thereby reducing the production of ethylene in plants (Goswami et al. 2016; Glick et al. 1998; Shaharoona et al. 2011). This improves plant growth and slows down senescence, a beneficial effect particularly under stress conditions, including metal contamination (Verma et al. 2014a, 2014b, 2015; Lally et al. 2017; Masciarelli et al. 2014; Visioli et al. 2014).

Rhizosphere PGP bacteria with ACC deaminase activity belong to a wide range of genera, such as *Arthrobacter*, *Achromobacter*, *Acinetobacter*, *Azospirillum*, *Burkholderia*,

Enterobacter, *Pseudomonas*, *Ralstonia*, *Rhizobium*, *Serratia*, and *Bacillus*, as well as *Bacillus*-related genera (Glick 1995; Khalid et al. 2006; Xu et al. 2014; Jacobson et al. 1994; Penrose et al. 2001).

Other phytohormones are involved in the plant root–PGPR interaction, although further work is required. *A. brasilense* has been reported to aid plant growth under drought stress by producing ABA (Cohen et al. 2008), a phytohormone involved in regulating stomatal opening and leaf senescence (Pieterse et al. 2009). Similarly, Shahzad et al. (2017) reported that some *Bacillus amyloliquefaciens* strains isolated from the rice rhizosphere could produce high amounts of ABA, and inoculating sensitive crop species might help them tolerate salinity better.

6.4 Plant Growth Enhancement

Changing the phytohormone status of host plants and improving nutrition by PGPR application often results in increased crop yields (Prathap and Ranjitha 2015; Gholami et al. 2012; Pérez-Montaño et al. 2014), although there are exceptions (Dobbelaere et al. 2001). Phytohormones produced by PGPR play an important role in root growth (Ahemad and Kibret 2014), and most of the existing literature agrees that PGPR influence root more than shoot growth (De Souza et al. 2013; Sandhya et al. 2010; Díaz-Zorita and Fernández-Canigia 2009; Dobbelaere et al. 2003). In most cases, PGPR increase root length, area, and branching (Molla et al. 2001; Salamone et al. 2005; Ma et al. 2016). Root length has been found to be considerably enhanced in field crops following plant/seed/soil inoculation with *Azotobacter chroococcum* (Allam et al. 2018), *Azotobacter vinelandii* (Dal Cortivo et al. 2017), *Bacillus simplex, Bacillus thuringiensis*, *Pseudomonas stutzeri* (Raheem et al. 2018), *Bacillus caryophylli*, and *P. fluorescens* (Shaharoona et al. 2008). Root morphological responses vary according to plant growth stage, with some studies revealing the best impact in the early stages after inoculation (Cakmakçi et al. 2006) or in mature plants (Calvo et al. 2017).

PGPR can increase the number of root hairs, as observed by Fallik et al. (1994) in maize inoculated with *Azospirillum brasilense*, although stimulation of lateral root initiation and root length appears to be the most frequent effect (Esitken et al. 2003; Barazani and Friedman 1999). Improvements in root length, as well as the number of root tips and branches, were also our main finding after applying PGPR and a "PGPR + mycorrhizal fungus" consortium to wheat as a seed inoculum in a controlled environment or by foliar + soil spraying at the tillering stage in the open field (Dal Cortivo et al. 2017, 2018) (Figure 6.2).

Seed treatment with PGPR has been found to increase root length and number in several crop species, such as oilseed rape, lettuce, tomato, barley, and oats. This accelerates root establishment and allows plants to explore greater soil volumes, thereby gaining access to greater amounts of nutrients and water (Hall et al. 1996; Qiaosi et al. 2005). Seed germination and seedling vigour have also been found to be enhanced in various species (Gholami et al. 2009; Ashrafuzzaman et al. 2009; Cassan et al. 2009) because of phytohormone production and root respiration (Vedder-Weiss et al. 1999).

A further appreciable benefit from PGPR activity is related to root nodulation because the application of endophytic bacteria isolated from legume root nodules and screened

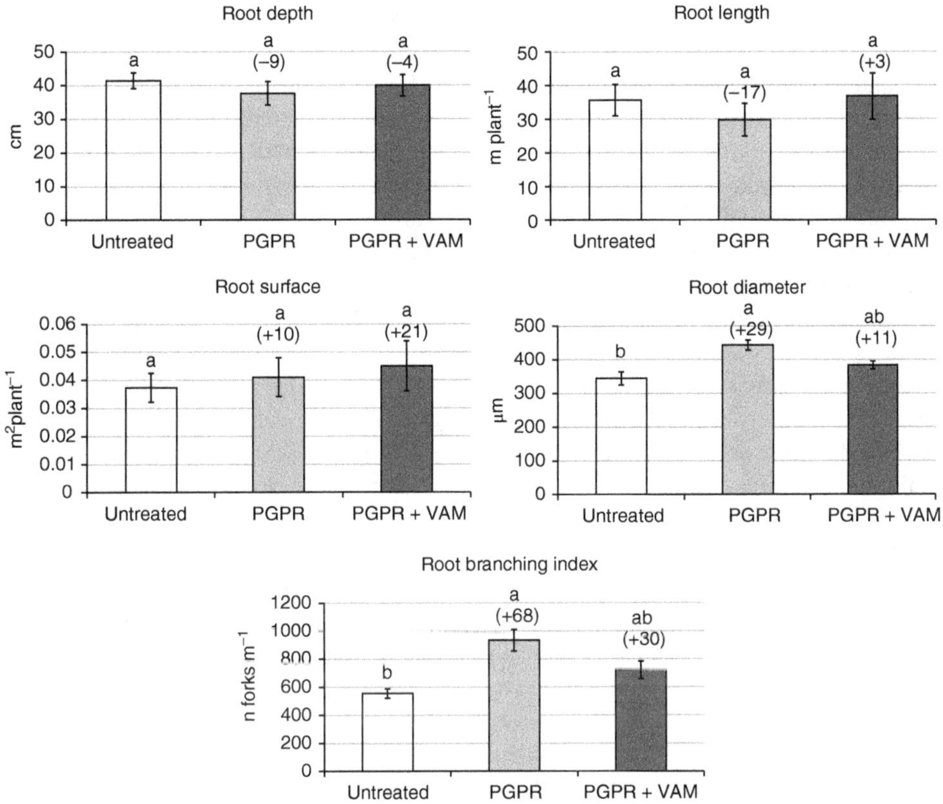

Figure 6.2 Root parameters in seedlings of *Triticum aestivum* L. following seed inoculation with PGPR (*Azospirillum* spp.+*Azorhizobum* spp.+*Azoarcus* spp.) and PGPR+arbuscular mycorrhizal (VAM) fungus consortium (*Azotobacter vinelandii* +*Rhizophagus irregularis*) vs. noninoculated controls after 50 days of growth in sterilized soil in rhizoboxes. Different letters indicate significant differences among treatments (Newman–Keuls test, $P \leq 0.05$). Parentheses indicate % variation in inoculated plants vs. noninoculated controls. *Source:* Dal Cortivo et al. (2017, 2018).

for their capacity to improve nodulation has a positive impact on the *Rhizobium*–legume interaction. Co-inoculation with *Bradyrhizobium japonicum* and *Bacillus* species has been found to favour root nodulation in soybean, leading to greater plant biomass compared with plants inoculated with *B. japonicum* alone (Bai et al. 2002; Masciarelli et al. 2014).

Despite the promising effects observed in controlled conditions, in the open field a huge variation in responses can be expected from inoculation with specific bacteria. Many biotic and abiotic factors influence the role played by PGPR, such as indigenous microbiome, plant genotype and age, climatic conditions, soil properties, and agronomic management (Roesti et al. 2006). Many effects recorded in the lab cannot be reproduced in the field (Zehnder et al. 1999; Smyth et al. 2011; Nadeem et al. 2013; Dal Cortivo et al. 2017), suggesting the need to develop more efficient inocula and appropriate application strategies.

6.5 Role of PGPR in Plant Tolerance to Stresses

Biotic and abiotic stressors commonly encountered by plants, including nutrient deficiency, drought, salinity, extreme temperatures, and metal contamination, as well as insect and pathogen attacks, can significantly limit crop yield potential (Nadeem et al. 2015; Vejan et al. 2016). For instance, drought causes plant dehydration, alters membrane permeability, and reduces nutrient diffusion in soil, root uptake, and translocation to the shoot as a result of reduced transpiration rates (Sardans et al. 2008). Similarly, under extreme salinity, increased Na^+ uptake and decreased K^+ and Ca^{2+} concentrations cause dehydration and cell osmotic alterations; in addition, under stress conditions plants produce significant quantities of ethylene, which inhibits root growth and causes leaf epinasty and early senescence and abscission (Morgan and Drew 1997).

PGPR can play an active role in helping plants grow under adverse environmental conditions by preventing or neutralizing the negative effects of abiotic stressors. In unfavourable conditions, PGPR may enhance plant growth through the overexpression of stress-responsive genes and accumulation of osmoprotectants and antioxidant enzymes (Nadeem et al. 2013, 2014). For instance, a high degree of adaptability to osmotic stress of the genera *Burkholderia, Enterobacter, Rhizobium,* and *Pseudomonas* has led to their widespread use in managing water stress in various crops (Ahmad et al. 2013; Naveed et al. 2014a; Zheng et al. 2018), including soybean, chickpea, and wheat (Ngumbi and Kloepper 2016).

In a stress environment, photosynthesis is commonly reduced due to decreased leaf expansion and premature leaf senescence, as well as decreased photosystem efficiency (Wahid and Rasul 2005). *Bacillus* and *Pseudomonas* species enable plants to maintain high growth by regulating stomatal closure, thereby significantly enhancing photosynthesis rate (Golpayegani and Tilebeni 2011).

6.5.1 Drought Stress

The severity and frequency of drought events have risen rapidly in recent decades (Vinocur and Altman 2005). Low water availability can seriously compromise plant growth and productivity due to reduced cell turgor, chlorophyll content, and nutrient uptake and translocation, as well as increased oxidative stress and ethylene synthesis (Rahdari and Hoseini 2012; Caravaca et al. 2005; Ali et al. 2014). Drought can also negatively affect soil bacteria because of changes in osmotic pressure and biochemical modifications that may cause nucleic acid damage and the accumulation of free oxygen radicals (Bérard et al. 2015). On the other hand, there are mechanisms that help bacteria cope with a lack of water, including accumulation of solutes, production of exopolysaccharides and extracellular polymeric substances (EPS), and even sporulation (Schimel et al. 2007; Naseem and Bano 2014). PGPR can improve drought tolerance in plants through the phytohormone and ACC deaminase activities, and production of osmolytes and exopolysaccharides (Shaharoona et al. 2007; Lally et al. 2017; Vurukonda et al. 2016). Some EPS-producing bacterial species form a biofilm on the root surface that minimizes dehydration while enhancing soil aggregate stability and root exudation (Sandhya et al. 2010; Alami et al. 2000; Raheem et al. 2018). Improved drought tolerance after inoculation with EPS-producing bacteria can be observed in both monocotyledonous and dicotyledonous species, often combined with tolerance to other abiotic stresses (Ngumbi and Kloepper 2016; Coleman-Derr and Tringe 2014).

Specific root traits associated with crop productivity under drought conditions include (i) increases in the numbers of roots with thin diameter and (ii) deep rooting (Comas et al. 2013). Plant inoculation with *Bacillus megaterium* and *Pseudomonas* species increases shoot and root biomasses and water content under drought conditions, with improved root length and architecture correlating with increased concentration of IAA produced by rhizobacteria (Marulanda et al. 2009).

Plants inoculated with selected *Bacillus* sp. strains maintained a near-normal shoot growth rate under water shortage, allowing crop productivity to be preserved or ameliorated (Ngumbi and Kloepper 2016; Vardharajula et al. 2011), probably through plant physiological changes that maintain the relative water content at a higher level compared with noninoculated plants (Grover et al. 2014; Naveed et al. 2014b). The release of ABA by bacteria improves plant stomatal closure (Dodd et al. 2010), and the release of organic solutes enhances osmotic adjustment. Some studies have reported increased concentrations of proline, free amino acids (e.g. aspartate, glutamate, alanine, and threonine), and soluble sugars in plants inoculated with some *Azospirillum* and *Pseudomonas* strains, resulting in improved tolerance to drought (Bano et al. 2013; Cohen et al. 2015; Sandhya et al. 2010; García et al. 2017). The most important challenges now are (i) the selection of strains tolerant to osmotic stress isolated from plant roots growing in arid soils, and (ii) assessment of their effectiveness when applied to field crops (García et al. 2017).

6.5.2 Salt Stress

Salinity is a critical environmental stress hampering crop cultivation in many areas of the world. Excessive soil Na^+ concentrations cause irreversible nutritional imbalances, suppress growth, and can even lead to plant death, with the extent of damage dependent on the salt concentration and specific plant tolerance (Shrivastava and Kumar 2015). Decreased ethylene accumulation in plants due to the ACC deaminase activity of rhizosphere bacteria can alleviate salt stress (Ahmad et al. 2011; Nadeem et al. 2010; Allam et al. 2018). Strains of *Pseudomonas* (*P. aurantiaca* TSAU22 and *P. extremorientalis* TSAU6 and TSAU20) applied as seed inocula were found to stimulate plant growth and prevent the effects of salinity, at least during the seedling stage (Egamberdieva 2009).

Germination is commonly associated with decreased concentrations of endogenous ABA, a phytohormone that has been found recently to play a role in improving stress tolerance (Shahzad et al. 2017). Under high soil salinity, beneficial bacteria with ACC deaminase activity were reported to significantly increase plant biomass and water-use efficiency in tomato (Mayak et al. 2004), and the chlorophyll content, shoot nutrient concentrations, and yield in maize (Hamdia et al. 2004; Nadeem et al. 2009). Many of these positive effects have been suggested to be due to improved root elongation rate and decreased Na^+ uptake following, for instance, *Rhizobium* and *Pseudomonas* co-inoculation (Ahmad et al. 2012; Dobbelaere and Okon 2007).

Table 6.1 List of beneficial bacteria investigated in recent studies in controlled conditions and field trials with various crop species.

Investigated effects		Bacterial genera	Bacterial species and strains	Crop	Experimental conditions	References
Nutrient uptake efficiency (biofertilizers)	Phosphorus-solubilizing bacteria (PSB)	*Pseudomonas*	*P. aeruginosa* 54RB	Wheat	Pot exp. – Soil inoc.	Afzal and Bano (2008)
		Rhizobium	*R. leguminosarum* Thal-8/SK			
		Azotobacter	*A. chroococcum*	Rice	Field and pot exp. – Soil inoc.	Ghaffari et al. (2018)
		Bacillus	*B. safensis* S7, *B. pumilus* A26	Maize	Field exp. – Seed inoc.	Breedt et al. (2017)
			B. megaterium RC07, *B. licheniformis* RC08	Sugar beet	Field and pot exp. – Seed inoc.	Cakmakçi et al. (2006)
			B. megaterium, *B. polymyxa*, *B. sphaericus*, *B. thuringiensis*	Oilseed rape	Pot in growth chamber exp. – Seed inoc.	De Freitas et al. (1997)
			Bacillus M-13	Sunflower	Field exp. – Seed inoc.	Ekin (2010)
		Bradyrhizobium	*B. japonicum*	Wheat	Field exp. – Seed inoc.	Mahanta et al. (2018)
	Nitrogen-use efficiency	*Pseudomonas*	*P. aeruginosa*	Sunflower	Field exp. – Seed inoc.	Arif et al. (2017)
			P. putida (Rc6)	Sugar beet	Field and pot exp. – Seed inoc.	Cakmakçi et al. (2006):
			P. putida-1, *P. putida*-2, *P. fluorescens*	Rice	Field and pot exp. – Soil inoc.	Ghaffari et al. (2018)
					Field and pot exp. – Seedling root inoc.	Zhang et al. (2018)
			P. fluorescens	Maize	Field exp. – Seed inoc.	Di Salvo et al. (2018)
		Azotobacter	*Azotobacter* spp.	Sunflower	Field exp. – Seed inoc.	Shoghi-Kalkhoran et al. (2013)

		Azospirillum	*A. brasilense*	Maize	Field exp. – Seed inoc.	Di Salvo et al. (2018)
			A. brasilense	Rice	Field and pot exp. – Seedling root inoc.	Zhang et al. (2018)
					Field and pot exp. – Seed inoc.	Pedraza et al. (2009)
			A. brasilense	Sunflower	Field exp. – Seed inoc.	Shoghi-Kalkhoran et al. (2013)
		Bacillus	*B. safensis* T4, *B. pumilus* INR7, *B. subtilis*, *B. sphaericus* SE56	Maize	Containers in greenhouse exp. – Soil inoc.	Calvo et al. (2017)
			Bacillus spp.		Pot exp. – Soil inoc.	Wu et al. (2005)
			B. megaterium	Wheat	Field and pot exp. – Seed inoc.	Kumar et al. (2014)
		Azorhizobium	*A. caulinodans*	Cereals	Pot exp. – Soil inoc.	Cocking (2003)
		Azotobacter	*Azotobacter* spp.	Rice	Field exp. – Seedling root inoc.	Alam et al. (2001)
Plant growth promotion	Effects on plant growth	*Pseudomonas*	*P. cepacia* R55, R85, *P. aeruginosa* R80, *P. fluorescens* R92, *P. putida* R104	Wheat	Field exp. – Seed inoc.	Germida and Walley (1996)
						Roesti et al. (2006)
						Shaharoona et al. (2007)
			P. gessardii, *P. fluorescens*	Sunflower	Pot exp. – Seed inoc.	Saleem et al. (2018)
			P. putida Rc6	Sugar beet	Field and pot exp. – Seed inoc.	Cakmakçi et al. (2006)

(*Continued*)

Table 6.1 (Continued)

Investigated effects	Bacterial genera	Bacterial species and strains	Crop	Experimental conditions	References
		P. putida-1, *P. putida*-2, *P. fluorescens*	Rice	Field and pot exp. – Seed inoc.	De Souza et al. (2013)
				Field and pot exp. – Soil inoc.	Ghaffari et al. (2018)
		Pseudomonas spp.		Pot exp. – Seed inoc.	Mirza et al. (2006)
		P. fluorescens strain R-93, *P. fluorescens* DSM 50090, *P. putida* DSM291	Maize	Pot exp. – Seed inoc.	Gholami et al. (2009)
		P. entomophila BV-P13, *P. stutzeri* GRFHAP-P14, *P. putida* GAP-P45, *P. syringae* GRFHYTP52, *P. monteilli* WAPP53		Pot exp. – Seed inoc.	Sandhya et al. (2010)
		P. putida-A Q7, *P. fluorescens* Q14, *P. fluorescens*-G N3		Field exp. – Soil inoc.	Naveed et al. (2008)
		P. fluorescens F113	Oilseed rape	Field and pot exp. – Seed inoc.	Lally et al. (2017)
	Rhizobium	*Rhizobium* spp.	Rice	Field and pot exp. – Seed inoc.	Ashrafuzzaman et al. (2009)
	Azotobacter	*A. vinelandii*	Wheat	Rhizoboxes and field exp. – Seed, leaves and soil inoc.	Dal Cortivo et al. (2018)
		A. chroococcum	Rice	Field and pot exp. – Soil inoc.	Ghaffari et al. (2018)
		Azotobacter spp.	Sunflower	Field exp. – Seed inoc.	Shoghi-Kalkhoran et al. (2013)
	Azospirillum	*A. lipoferum* DSM 1691, *A. brasilense*	Maize	Pot exp. - Seed inoc.	Gholami et al. (2009)

		Azospirillum spp.	Rice	Field exp. – Seed inoc.	De Souza et al. (2013)
		A. brasilense		Pot exp. – Seedling root inoc.	Zhang et al. (2018)
			Wheat	Field exp. – Seed inoc.	Díaz-Zorita and Fernández-Canigia (2009)
			Sunflower	Field exp. – Seed inoc.	Shoghi-Kalkhoran et al. (2013)
	Bacillus	*B. safensis* T4, *B. pumilus* INR7, *B. subtilis*, *B. sphaericus* SE56	Maize	Containers in greenhouse exp. – Soil inoc.	Calvo et al. (2017)
		B. amiloliquefaciens, *B. cereus*, *B. methylotrophicus*		Containers in greenhouse exp. – Seed inoc.	Lin et al. (2018)
		Bacillus spp.		Field exp. – Soil inoc.	Wu et al. (2005)
		B. megaterium RC07, *B. licheniformis* RC08	Sugar beet	Field and pot exp. – Seed inoc.	Cakmakçi et al. (2006)
		B. megaterium	Wheat	Pot and field exp. – Seed inoc.	Kumar et al. (2014)
		B. amyloliquefaciens NBRISN13	Rice	Pot exp. – Soil inoc.	Nautiyal et al. (2013)
		Bacillus spp.	Soybean	Pot exp. – Seedling inoc.	Bai et al. (2002)
	Agrobacterium	*Agrobacterium* spp.	Rice	Field exp. – Seed inoc.	De Souza et al. (2013)
	Bradyrhizobium	*B. japonicum*	Soybean	Pot exp. – Seed inoc.	Masciarelli et al. (2014)
	Azotobacter	*Azotobacter* spp.	Rice	Field exp. – Seedling root inoc.	Alam et al. (2001)

(*Continued*)

Table 6.1 (Continued)

Investigated effects	Bacterial genera	Bacterial species and strains	Crop	Experimental conditions	References
Phytohormone production	*Pseudomonas*	*P. aurantiaca* TSAU22, *P. extremorientalis* TSAU6, *P. extremorientalis* TSAU20	Wheat	Petri in growth chamber exp. – Seed inoc.	Egamberdieva (2009)
		P. fluorescens AK1, *P. aeruginosa* AK2	Rice	Pot exp. – Seed inoc.	Karnwal (2009)
	Rhizobium	*Rhizobium* spp.	Rice	Field and pot exp. – Seed inoc.	Ashrafuzzaman et al. (2009)
			Maize	Pot exp. – Seed and soil inoc.	Hossain and Mårtensson (2008)
	Azotobacter	*A. chroococcum*, *A. vinelandii*	Wheat	Field and pot exp. – Soil inoc.	Allam et al. (2018)
	Azospirillum	*A. brasilense* Az39	Maize	Pot in growth chamber exp. – Seed inoc.	Cassan et al. (2009)
		Azospirillum spp.		Pot in growth chamber exp. – Seedling inoc.	García et al. (2017)
	Bacillus	*B. safensis* (S7), *B. pumilus* (A26)	Maize	Field exp. – Seed inoc.	Breedt et al. (2017)
		B. thuringiensis S-26, D-2, *B. muralis* D-5, *B. amyloliquefaciens* S-134	Wheat	Pot exp. – Seed inoc.	Raheem et al. (2018)
		B. amyloliquefaciens RWL-1	Rice	Pot in growth chamber – Soil inoc.	Shahzad et al. (2017)
	Bradyrhizobium	*B. japonicum* E109	Soybean	Pot in growth chamber exp. – Seed inoc.	Cassan et al. (2009)

Tolerance to stresses	Tolerance to pathogens (biocontrol)	*Pseudomonas*	*P. protegens* CHA0, *P. chlororaphis* PCL1391	Wheat	Field exp. – Soil inoc.	Imperiali et al. (2017)
			P. fluorescens	Rice	Field exp. – Seed and leaves inoc.	Lucas et al. (2009)
			P. aeruginosa, *P. fluorescens*	Maize	Pot exp. – Seed inoc.	Akhtar et al. (2018)
		Bacillus	*Bacillus* spp.	Rice	Field exp. – Seedling root inoc.	Rais et al. (2018)
	Tolerance to abiotic stress	*Pseudomonas*	*P. aurantiaca* TSAU22, *P. extremorientalis* TSAU6, *P. extremorientalis* TSAU20	Wheat	Petri in growth chamber exp. – Seed inoc.	Egamberdieva (2009)
			P. gessardii, *P. fluorescens*	Sunflower	Pot exp. – Seed inoc.	Saleem et al. (2018)
			P. fluorescens	Rice	Field exp. – Seed and leaves inoc.	Lucas et al. (2009)
			P. entomophila BV-P13, *P. stutzeri* GRFHAP-P14, *P. putida* GAP-P45, *P. syringae* GRFHYTP52, *P. monteilli* WAPP53	Maize	Pot exp. – Seed inoc.	Sandhya et al. (2010)
		Azotobacter	*A. chroococcum*, *A. vinelandii*	Wheat	Field and pot exp. – Seed inoc.	Allam et al. (2018)
		Azospirillum	*A. brasilense* SP-7	Maize	Trays in greenhouse exp. – Seed inoc.	Curá et al. (2017)
			Azospirillum spp.		Pot in growth chamber exp. – Seedling inoc.	García et al. (2017)
		Bacillus	*B. subtilis*	Wheat	Pot exp. – Soil inoc.	Upadhyay et al. (2012)
			B. thuringiensis S-26, D-2, *B. muralis* D-5, *B. amyloliquefaciens* S-134		Pot exp. – Seed inoc.	Raheem et al. (2018)

(*Continued*)

Table 6.1 (Continued)

Investigated effects	Bacterial genera	Bacterial species and strains	Crop	Experimental conditions	References
		B. amyloliquefaciens	Rice	Pot exp. – Soil inoc.	Nautiyal et al. (2013)
				Pot in growth chamber – Soil inoc.	Shahzad et al. (2017)
Modification of field rhizosphere microbial communities after bacteria inoculation	*Pseudomonas*	*P. fluorescens*	Wheat	Field exp.	Roesti et al. (2006)
			Maize		Di Salvo et al. (2018)
	Rhizobium	*Rhizobium* spp.	Soybean		Sugiyama et al. (2014)
	Azotobacter	*Azotobacter* spp.	Maize		Piromyou et al. (2011)
	Azospirillum	*A. lipoferum*	Maize		Baudoin et al. (2009)
		A. brasilense			Di Salvo et al. (2018)
	Bacillus	*Bacillus* spp.	Soybean		Sugiyama et al. (2014)
	Bradyrhizobium	*Bradyrhizobium* spp.	Soybean		Sugiyama et al. (2014)
Isolation, identification, and characterization of PGPR strains associated with field crops	*Pseudomonas*	*Pseudomonas* spp.	Wheat		Shaharoona et al. (2007)
			Rice		De Souza et al. (2013)

	P. chlororaphis	Oilseed rape		García de Salamone et al. (2001)
Rhizobium	*Rhizobium* spp.	Rice		Ashrafuzzaman et al. (2009)
Azotobacter	*Azotobacter* spp.	Maize		Piromyou et al. (2011)
Azospirillum	*Azospirillum* spp.	Rice		De Souza et al. (2013)
Bacillus	*B. amyloliquefaciens*, *B. firmus*	Maize		Mendis et al. (2018)
	B. megaterium, *B. polymyxa*, *B. sphaericus*, *B. thuringiensis*	Oilseed rape		De Freitas et al. (1997)
	Bacillus spp.	Sunflower		Forchetti et al. (2007)
	B. megaterium	Wheat		Kumar et al. (2014)
	Bacillus spp.	Soybean		Wahyudi et al. (2011)
				Bai et al. (2002)
Agrobacterium	*Agrobacterium* spp.	Rice		De Souza et al. (2013)

6.5.3 Other Abiotic Stresses

Bacterial modulation of high ethylene concentration in plants under stress conditions may also intensify temperature and metal stress tolerance. Bacterial ACC deaminase has been shown to protect plants from low and high temperature stresses that affect plant growth by increasing ethylene production (Nadeem et al. 2015). In addition, plant–microbe interactions in the rhizosphere have the potential to improve plant biomass production and heavy metal stress tolerance through siderophore production, and the synthesis of phytohormones and enzymes (Kumar et al. 2008; Belimov and Wenzel 2009). Several *Pseudomonas* species have been reported to improve sunflower growth and antioxidant activities when applied to lead-contaminated soil (Saleem et al. 2018). This can be exploited in phytoremediation of soils polluted with metals (Macek et al. 2000; Vamerali et al. 2010), although the selection of suitable bacterial species is crucial. Visioli et al. (2015) found that combining the endophytic strains of *Arthrobacter* and *Microbacterium* improved shoot and root growth and nickel accumulation in *Noccaea caerulescens* and *Arabidopsis thaliana* better than each strain alone. It has been suggested that improved metal uptake is related to the capacity of these rhizobacteria to lower soil pH, thereby increasing metal solubilization and uptake (Abou-Shanab et al. 2006). Amending soils with suitable PGPR could be useful for establishing plants in degraded/contaminated areas (Schoebitz et al. 2014).

6.6 Conclusions

A large body of literature confirms the essential role played by the abundant and diverse rhizosphere bacteria in plant growth and adaptation to extreme conditions, mainly through modifications of root architecture and activity (Table 6.1). This opens up opportunities to improve stress resistance and crop productivity by developing appropriate biofertilizers and applying them to soil/seeds/plants, although in formulating them attention must also be paid to the positive interaction between rhizobacteria and mycorrhizal fungi. Biofertilizers have been acknowledged as offering a new sustainable tool for agriculture, mainly for enhancing plant capacity to cope with adverse environmental conditions, which may or may not translate into yield gains. Future research should focus on selecting suitable microorganisms for specific crops and environments, and developing appropriate application methods to reduce the variability in plant responses.

References

Abou-Shanab, R.A.I., Angle, J.S., and Chaney, R.L. (2006). Bacterial inoculants affecting nickel uptake by *Alyssum murale* from low, moderate and high Ni soils. *Soil Biology and Biochemistry* 38 (9): 2882–2889.

Afzal, A. and Bano, A. (2008). Rhizobium and phosphate solubilizing bacteria improve the yield and phosphorus uptake in wheat (*Triticum aestivum*). *International Journal of Agriculture and Biology* 10 (1): 85–88.

Ahemad, M. and Kibret, M. (2014). Mechanisms and applications of plant growth promoting rhizobacteria: current perspective. *Journal of King Saud University – Science* 26: 1–20.

Ahmad, M., Zahir, Z.A., Asghar, H.N. et al. (2011). Inducing salt tolerance in mung bean through co-inoculation with *Rhizobium* and PGPR containing ACC deaminase. *Canadian Journal of Microbiology* 57: 578–589.

Ahmad, M., Zahir, Z.A., Asghar, H.N. et al. (2012). The combined application of rhizobial strains and plant growth promoting rhizobacteria improves growth and productivity of mung bean (*Vigna radiata* L.) under salt-stressed conditions. *Annals of Microbiology* 62: 1321–1330.

Ahmad, M., Zahir, Z.A., and Khalid, M. (2013). Efficacy of *Rhizobium* and *Pseudomonas* strains to improve physiology, ionic balance and quality of mung bean under salt-affected conditions on farmer's fields. *Plant Physiology and Biochemistry* 63: 170–176.

Akhtar, N., Naveed, M., Khalid, M. et al. (2018). Effect of bacterial consortia on growth and yield of maize grown in *Fusarium* infested soil. *Soil and Environment* 37 (1): 35–44.

Alam, M.S., Cui, Z.J., Yamagishi, T. et al. (2001). Grain yield and related physiological characteristics of rice plants (*Oryza sativa* L.) inoculated with free-living rhizobacteria. *Plant Production Science* 4 (2): 126–130.

Alami, Y., Achouak, W., Marol, C. et al. (2000). Rhizosphere soil aggregation and plant growth promotion of sunflowers by an exopolysaccharide-producing *Rhizobium* sp. strain isolated from sunflower roots. *Applied and Environmental Microbiology* 66 (8): 3393–3398.

Ali, S.K.Z., Sandhya, V., and Rao, L.V. (2014). Isolation and characterization of drought-tolerant ACC deaminase and exopolysaccharide-producing fluorescent *Pseudomonas* sp. *Annals of Microbiology* 64: 493–502.

Allam, N.G., Kinany, R., El-Refai, E. et al. (2018). Potential use of beneficial salt tolerant bacteria for improving wheat productivity grown in salinized soil. *Journal of Microbiology Research* 8 (2): 43–53.

Argüeso, C.T., Ferreira, F.J., Epple, P. et al. (2012). Two-component elements mediate interactions between cytokinin and salicylic acid in plant immunity. *PLoS Genetics* 8: e1002448.

Arif, M.S., Shahzad, S.M., Riaz, M. et al. (2017). Nitrogen-enriched compost application combined with plant growth-promoting rhizobacteria (PGPR) improves seed quality and nutrient use efficiency of sunflower. *Journal of Plant Nutrition and Soil Science* 180 (4): 464–473.

Arkhipova, T.N., Prinsen, E., Veselov, S.U. et al. (2007). Cytokinin producing bacteria enhance plant growth in drying soil. *Plant and Soil* 292: 305–315.

Ashrafuzzaman, M., Hossen, F.A., Ismail, M.R. et al. (2009). Efficiency of plant growth-promoting rhizobacteria (PGPR) for the enhancement of rice growth. *African Journal of Biotechnology* 8 (7): 1247–1252.

Badr, M.A. (2006). Efficiency of K-feldspar combined with organic materials and silicate dissolving bacteria on tomato yield. *Journal of Applied Sciences Research* 2: 1191–1198.

Bai, Y., D'Aoust, F., Smith, D.L. et al. (2002). Isolation of plant-growth-promoting *Bacillus* strains from soybean root nodules. *Canadian Journal of Microbiology* 48 (3): 230–238.

Bakhshandeh, E., Pirdashti, H., and Lendeh, K.S. (2017). Phosphate and potassium-solubilizing bacteria effect on the growth of rice. *Ecological Engineering* 103: 164–169.

Bano, Q., Ilyas, N., Bano, A. et al. (2013). Effect of *Azospirillum* inoculation on maize (*Zea mays* L.) under drought stress. *Pakistan Journal of Botany* 45: 13–20.

Barazani, O. and Friedman, J. (1999). Is IAA the major root growth factor secreted from plant-growth-mediating bacteria? *Journal of Chemical Ecology* 25 (10): 2397–2406.

Basaglia, M., Casella, S., Peruch, U. et al. (2003). Field release of genetically marked *Azospirillum brasilense* in association with *Sorghum bicolor* L. *Plant and Soil* 256 (2): 281–290.

Bashan, Y., Kamnev, A.A., and de Bashan, L.E. (2013). A proposal for isolating and testing phosphate- solubilizing bacteria that enhance plant growth. *Biology and Fertility of Soils* 49: 1–2.

Baudoin, E., Nazaret, S., Mougel, C. et al. (2009). Impact of inoculation with the phytostimulatory PGPR *Azospirillum lipoferum* CRT1 on the genetic structure of the rhizobacterial community of field-grown maize. *Soil Biology and Biochemistry* 41 (2): 409–413.

Belimov, A.A. and Wenzel, W.W. (2009). The role of rhizosphere microorganisms in heavy metal tolerance of higher plants. *Aspects of Applied Biology* 98: 81–90.

Beneduzi, A., Ambrosini, A., and Passaglia, L.M.P. (2012). Plant growth-promoting rhizobacteria: their potential as antagonists and biocontrol agents. *Genetics and Molecular Biology* 35 (4): 1044–1051.

Bérard, A., Ben Sassi, M., Kaisermann, A. et al. (2015). Soil microbial community responses to heat wave components: drought and high temperature. *Climate Research* 66: 243–264.

Bhardwaj, D., Ansari, M.W., Sahoo, R.K. et al. (2014). Biofertilizers function as key player in sustainable agriculture by improving soil fertility, plant tolerance and crop productivity. *Microbial Cell Factories* 13 (66): 1–10.

Bhattacharjee, R.B., Singh, A., and Mukhopadhyay, S.N. (2008). Use of nitrogen-fixing bacteria as biofertiliser for non-legumes: prospects and challenges. *Applied Microbiology and Biotechnology* 80 (2): 199–209.

Bhattacharyya, P.N. and Jha, D.K. (2012). Plant growth-promoting rhizobacteria (PGPR): emergence in agriculture. *World Journal of Microbiology and Biotechnology* 28: 1327–1350.

Bishnoi, U. (2015). PGPR Interaction: an ecofriendly approach promoting the sustainable agriculture system. *Advances in Botanical Research* 75: 81–113.

Bottini, R., Cassán, F., and Piccoli, P. (2004). Gibberellin production by bacteria and its involvement in plant growth promotion and yield increase. *Applied Microbiology and Biotechnology* 65: 497–503.

Breedt, G., Labuschagne, N., and Coutinho, T.A. (2017). Seed treatment with selected plant growth-promoting rhizobacteria increases maize yield in the field. *Annals of Applied Biology* 171 (2): 229–236.

Bulgarelli, D., Schlaeppi, K., Spaepen, S. et al. (2013). Structure and functions of the bacterial microbiota of plants. *Annual Review of Plant Biology* 64: 807–838.

Cakmakçi, R., Dönmez, F., Aydın, A. et al. (2006). Growth promotion of plants by plant growth-promoting rhizobacteria under greenhouse and two different field soil conditions. *Soil Biology and Biochemistry* 38 (6): 1482–1487.

Calvo, P., Watts, D.B., Kloepper, J.W. et al. (2017). Effect of microbial-based inoculants on nutrient concentrations and early root morphology of corn (*Zea mays*). *Journal of Plant Nutrition and Soil Science* 180 (1): 56–70.

Caravaca, F., Alguacil, M.M., Hernández, J.A. et al. (2005). Involvement of antioxidant enzyme and nitrate reductase activities during water stress and recovery of mycorrhizal *Myrtus communis* and *Phillyrea angustifolia* plants. *Plant Science* 169 (1): 191–197.

Carrillo-Castañeda, G., Juàrez Muños, J., Peralta-Videa, J.R. et al. (2002). Alfalfa growth promotion by bacteria grown under iron limiting conditions. *Advances in Environmental Research* 6: 391–399.

Cassan, F., Perrig, D., Sgroy, V. et al. (2009). *Azospirillum brasilense* Az39 and *Bradyrhizobium japonicum* E109, inoculated singly or in combination, promote seed germination and early seedling growth in corn (*Zea mays* L.) and soybean (*Glycine max* L.). *European Journal of Soil Biology* 45 (1): 28–35.

Chen, J.H. (2006). The combined use of chemical and organic fertilizers and/or biofertilizer for crop growth and soil fertility. *International Workshop on Sustained Management of the Soil-Rhizosphere System for Efficient Crop Production and Fertilizer Use*, 16, 20. Thailand: Land Development Department Bangkok.

Cocking, E.C. (2003). Endophytic colonization of plant roots by nitrogen-fixing bacteria. *Plant and Soil* 252 (1): 169–175.

Cohen, A.C., Bottini, R., and Piccoli, P.N. (2008). *Azospirillum brasilense* Sp 245 produces ABA in chemically defined culture medium and increases ABA content in arabidopsis plants. *Plant Growth Regulation* 54: 97–103.

Cohen, A., Bottini, R., Pontin, M. et al. (2015). *Azospirillum brasilense* ameliorates the response of *Arabidopsis thaliana* to drought mainly via enhancement of ABA levels. *Physiologia Plantarum* 153: 79–90.

Coleman-Derr, D. and Tringe, S.G. (2014). Building the crops of tomorrow: advantages of symbiont-based approaches to improving abiotic stress tolerance. *Frontiers in Microbiology* 5: 283.

Collavino, M.M., Sansberro, P.A., Mroginski, L.A. et al. (2010). Comparison of *in vitro* solubilization activity of diverse phosphate-solubilizing bacteria native to acid soil and their ability to promote *Phaseolus vulgaris* growth. *Biology and Fertility of Soils* 46: 727–738.

Comas, L.H., Becker, S.R., Cruz, V.V. et al. (2013). Root traits contributing to plant productivity under drought. *Frontiers in Plant Science* 4: 442.

Compant, S., Duffy, B., Nowak, J. et al. (2005a). Use of plant growth-promoting bacteria for biocontrol of plant diseases: principles, mechanisms of action, and future prospects. *Applied and Environmental Microbiology* 71 (9): 4951–4959.

Compant, S., Reiter, B., Sessitsch, A. et al. (2005b). Endophytic colonization of *Vitis vinifera* L. by plant growth promoting bacterium *Burkholderia* sp strain PsJN. *Applied and Environmental Microbiology* 71: 1685–1693.

Curá, J.A., Franz, D.R., Filosofía, J.E. et al. (2017). Inoculation with *Azospirillum* sp. and *Herbaspirillum* sp. bacteria increases the tolerance of maize to drought stress. *Microorganisms* 5 (3): 41.

Dal Cortivo, C., Barion, G., Visioli, G. et al. (2017). Increased root growth and nitrogen accumulation in common wheat following PGPR inoculation: assessment of plant-microbe interactions by ESEM. *Agriculture, Ecosystems and Environment* 247: 396–408.

Dal Cortivo, C., Barion, G., Ferrari, M. et al. (2018). Effects of field inoculation with VAM and bacteria consortia on root growth and nutrients uptake in common wheat. *Sustainability* 10: 3286.

Damam, M., Kaloori, K., Gaddam, B. et al. (2016). Plant growth promoting substances (phytohormones) produced by rhizobacterial strains isolated from the rhizosphere of medicinal plants. *International Journal of Pharmaceutical Sciences Review and Research* 37 (1): 130–136.

Davière, J.M. and Achard, P. (2013). Gibberellin signaling in plants. *Development* 140: 1147–1151.

De Freitas, J.R., Banerjee, M.R., and Germida, J.J. (1997). Phosphate-solubilizing rhizobacteria enhance the growth and yield but not phosphorus uptake of canola (*Brassica napus* L.). *Biology and Fertility of Soils* 24 (4): 358–364.

De Souza, R., Beneduzi, A., Ambrosini, A. et al. (2013). The effect of plant growth-promoting rhizobacteria on the growth of rice (*Oryza sativa* L.) cropped in southern Brazilian fields. *Plant and Soil* 366 (1–2): 585–603.

Díaz-Zorita, M. and Fernández-Canigia, M.V. (2009). Field performance of a liquid formulation of *Azospirillum brasilense* on dryland wheat productivity. *European Journal of Soil Biology* 45 (1): 3–11.

Di Salvo, L.P., Cellucci, G.C., Carlino, M.E. et al. (2018). Plant growth-promoting rhizobacteria inoculation and nitrogen fertilization increase maize (*Zea mays* L.) grain yield and modified rhizosphere microbial communities. *Applied Soil Ecology* 126: 113–120.

Dobbelaere, S., Croonenborghs, A., Thys, A. et al. (2001). Responses of agronomically important crops to inoculation with *Azospirillum*. *Functional Plant Biology* 28 (9): 871–879.

Dobbelaere, S., Venderleyden, J., and Okon, Y. (2003). Plant growth promoting effects of diazotrophs in the rhizosphere. *Critical Reviews in Plant Sciences* 22: 107–149.

Dobbelaere, S. and Okon, Y. (2007). The plant growth promoting effect and plant responses. In: Associative and Endophytic Nitrogen-Fixing Bacteria and Cyanobacterial Associations (eds. C. Elmerich and W.E. Newton), 1–26. Dordrecht: Kluwer Academic Publishers.

Dodd, I.C., Zinovkina, N.Y., Safronova, V.I. et al. (2010). Rhizobacterial mediation of plant hormone status. *Annals of Applied Biology* 157: 361–379.

Duffy, B.K. and Défago, G. (1999). Environmental factors modulating antibiotic and siderophore biosynthesis by *Pseudomonas fluorescens* biocontrol strains. *Applied and Environmental Microbiology* 65: 2429–2438.

Egamberdieva, D. (2009). Alleviation of salt stress by plant growth regulators and IAA producing bacteria in wheat. *Acta Physiologiae Plantarum* 31 (4): 861–864.

Ekin, Z. (2010). Performance of phosphate solubilizing bacteria for improving growth and yield of sunflower (*Helianthus annuus* L.) in the presence of phosphorus fertilizer. *African Journal of Biotechnology* 9 (25): 3794–3800.

Esitken, A., Ercisli, S., Sevik, I. et al. (2003). Effect of indole-3-butyric acid and different strains of *Agrobacterium rubi* on adventive root formation from softwood and semi-hardwood wild sour cherry cuttings. *Turkish Journal of Agriculture and Forestry* 27: 37–42.

Etesami, H., Alikhani, H.A., and Akbari, A.A. (2009). Evaluation of plant growth hormones production (IAA) ability by Iranian soils rhizobial strains and effects of superior strains application on wheat growth indexes. *World Applied Sciences Journal* 6: 1576–1584.

Fallik, E., Sarig, S., and Okon, Y. (1994). Morphology and physiology of plant roots associated with *Azospirillum*. In: Azospirillum/Plant Associations (ed. Y. Okon), 77–86. Boca Raton, FL: CRC Press.

Fibach-Paldi, S., Burdman, S., and Okon, Y. (2012). Key physiological properties contributing to rhizosphere adaptation and plant growth promotion abilities of *Azospirillum brasilense*. *FEMS Microbiology Letters* 326 (2): 99–108.

Forchetti, G., Masciarelli, O., Alemano, S. et al. (2007). Endophytic bacteria in sunflower (*Helianthus annuus* L.): isolation, characterization, and production of jasmonates and abscisic acid in culture medium. *Applied Microbiology and Biotechnology* 76 (5): 1145–1152.

García de Salamone, I.E., Hynes, R.K., and Nelson, L.M. (2001). Cytokinin production by plant growth promoting rhizobacteria and selected mutants. *Canadian Journal of Microbiology* 47 (5): 404–411.

Garcia-Fraile, P., Menendez, E., and Rivas, R. (2015). Role of bacterial bio fertilizers in agriculture and forestry. *AIMS Bioengineering* 2: 183–205.

García, J.E., Maroniche, G., Creus, C. et al. (2017). *in vitro* PGPR properties and osmotic tolerance of different *Azospirillum* native strains and their effects on growth of maize under drought stress. *Microbiological Research* 202: 21–29.

Germida, J.J. and Walley, F.L. (1996). Plant growth-promoting rhizobacteria alter rooting patterns and arbuscular mycorrhizal fungi colonization of field-grown spring wheat. *Biology and Fertility of Soils* 23 (2): 113–120.

Ghaffari, H., Gholizadeh, A., Biabani, A. et al. (2018). Plant growth promoting rhizobacteria (PGPR) application with different nitrogen fertilizer levels in rice (*Oryza sativa* L.). *Pertanika Journal of Tropical Agricultural Science* 41 (2): 715–728.

Gholami, A., Shahsavani, S., and Nezarat, S. (2009). The effect of plant growth promoting rhizobacteria (PGPR) on germination, seedling growth and yield of maize. *International Journal of Biological Life Science* 5 (1): 35–40.

Gholami, A., Biyari, A., Gholipoor, M. et al. (2012). Growth promotion of maize (*Zea mays* L.) by plant-growth-promoting rhizobacteria under field conditions. *Communications in Soil Science and Plant Analysis* 43: 1263–1272.

Glick, B.R. (1995). The enhancement of plant growth by free-living bacteria. *Canadian Journal of Microbiology* 41 (2): 109–117.

Glick, B.R., Penrose, D.M., and Li, J.P. (1998). A model for the lowering of plant ethylene concentrations by plant growth-promoting bacteria. *Journal of Theoretical Biology* 190 (1): 63–68.

Glick, B.R. (2012). Plant growth-promoting bacteria: mechanisms and applications. *Scientifica* 2012: 963401.

Golpayegani, A. and Tilebeni, H.G. (2011). Effect of biological fertilizers on biochemical and physiological parameters of basil (*Ociumum basilicum* L.) medicine plant. *American-Eurasian Journal of Agricultural and Environmental Sciences* 11: 411–416.

Goswami, D., Thakker, J.N., and Dhandhukia, P.C. (2016). Portraying mechanics of plant growth promoting rhizobacteria (PGPR): a review. *Cogent Food and Agriculture* 2: 1–19.

Gouda, S., Kerry, R.G., Das, G. et al. (2018). Revitalization of plant growth promoting rhizobacteria for sustainable development in agriculture. *Microbiological Research* 206: 131–140.

Grover, M., Madhubala, R., Ali, S.Z. et al. (2014). Influence of *Bacillus* spp. strains on seedling growth and physiological parameters of sorghum under moisture stress conditions. *Journal of Basic Microbiology* 54: 951–961.

Gupta, S., Meena, M.K., and Datta, S. (2014). Isolation, characterization of plant growth promoting bacteria from the plant *Chlorophytum borivilianum* and *in-vitro* screening for activity of nitrogen fixation, phosphate solubilization and IAA production. *International Journal of Current Microbiology and Applied Sciences* 3: 1082–1090.

Gutierrez-Manero, F.J., Ramos-Solano, B., Probanza, A. et al. (2001). The plant-growth promoting rhizobacteria *Bacillus pumilus* and *Bacillus licheniformis* produce high amounts of physiologically active gibberellins. *Physiologia Plantarum* 111 (2): 206–211.

Hall, J.A., Peirson, D., Ghosh, S. et al. (1996). Root elongation in various agronomic crops by the plant growth promoting rhizobacterium *Pseudomonas putida* GR12-2. *Israel Journal of Plant Sciences* 44: 37–42.

Hallmann, J., Quadt-Hallmann, A., Mahaffee, W. et al. (1997). Bacterial endophytes in agricultural crops. *Canadian Journal of Microbiology* 43 (10): 895–914.

Hamdia, M.A., Shaddad, M.A.K., and Doaa, M.M. (2004). Mechanism of salt tolerance and interactive effect of *Azospirillum brasilense* inoculation on maize cultivars grown under salt stress conditions. *Plant Growth Regulation* 44: 165–174.

Hossain, S. and Mårtensson, A. (2008). Potential use of *Rhizobium* spp. to improve fitness of non-nitrogen-fixing plants. *Acta Agriculturae Scandinavica Section B – Soil and Plant Science* 58 (4): 352–358.

Imperiali, N., Chiriboga, X., Schlaeppi, K. et al. (2017). Combined field inoculations of *Pseudomonas* bacteria, arbuscular mycorrhizal fungi, and entomopathogenic nematodes and their effects on wheat performance. *Frontiers in Plant Science* 8: 1809.

Ivanova, E., Doronina, N., and Trotsenko, Y.A. (2001). Aerobic methylobacteria are capable of synthesizing auxins. *Microbiology* 70 (4): 392–397.

Ivanova, E., Pirttilä, A., Fedorov, D. et al. (2008). Association of methylotrophic bacteria with plants: metabolic aspects. In: Prospects and Applications for Plant Associated Microbes a Laboratory Manual, Part A: Bacteria (eds. S. Sorvari and A.M. Pirttilä), 225–231. Turku: Biobien Innovations.

Jacobson, C.B., Pasternak, J.J., and Glick, B.R. (1994). Partial purification and characterization of 1-aminocyclopropane-1-carboxylate deaminase from the plant growth promoting rhizobacterium *Pseudomonas putida* GR12-2. *Canadian Journal of Microbiology* 40 (12): 1019–1025.

Jung, H., Janelle, K., and McCouch, S. (2013). Getting to the roots of it: genetic and hormonal control of root architecture. *Frontiers in Plant Science* 4 (186): 1–32.

Kang, B.G., Kim, W.T., Yun, H.S. et al. (2010). Use of plant growth-promoting rhizobacteria to control stress responses of plant roots. *Plant Biotechnology Reports* 4: 179–183.

Karnwal, A. (2009). Production of indole acetic acid by fluorescent *Pseudomonas* in the presence of l-tryptophan and rice root exudates. *Journal of Plant Pathology* 91 (1): 61–63.

Khakipour, N.K., Khavazi Mojallali, H., Pazira, E. et al. (2008). Production of auxin hormone by fluorescent pseudomonads. *American-Eurasian Journal of Agricultural and Environmental Sciences* 4: 687–692.

Khalid, A., Akhtar, M., Mahmood, M. et al. (2006). Effect of substrate-dependent microbial ethylene production on plant growth. *Microbiology* 75 (2): 231–236.

Khalid, A., Sultana, S., Arshad, M. et al. (2011). Performance of auxin producing rhizobacteria for improving growth and yield of wheat and rice grown in rotation under field conditions. *International Journal of Agriculture and Applied Sciences* 3: 44–50.

Khanghahi, M.Y., Pirdashti, H., Rahimian, H. et al. (2018). Potassium solubilising bacteria (KSB) isolated from rice paddy soil: from isolation, identification to K use efficiency. *Symbiosis* 76 (1): 13–23.

Kim, J. and Rees, D.C. (1994). Nitrogenase and biological nitrogen fixation. *Biochemistry* 33: 389–397.

Kuffner, M., Puschenreiter, M., Wieshammer, G. et al. (2008). Rhizosphere bacteria affect growth and metal uptake of heavy metal accumulating willows. *Plant and Soil* 304: 35–44.

Kumar, K.V., Singh, N., Behl, H.M. et al. (2008). Influence of plant growth promoting bacteria and its mutant on heavy metal toxicity in *Brassica juncea* grown in fly ash amended soil. *Chemosphere* 72: 678–683.

Kumar, A., Maurya, B.R., and Raghuwanshi, R. (2014). Isolation and characterization of PGPR and their effect on growth, yield and nutrient content in wheat (*Triticum aestivum* L.). *Biocatalysis and Agricultural Biotechnology* 3 (4): 121–128.

Lally, R.D., Galbally, P., Moreira, A.S. et al. (2017). Application of endophytic *Pseudomonas fluorescens* and a bacterial consortium to *Brassica napus* can increase plant height and biomass under greenhouse and field conditions. *Frontiers in Plant Science* 8: 2193.

Lim, J.H. and Kim, S.D. (2009). Synergistic plant growth promotion by the indigenous auxins-producing PGPR *Bacillus subtilis* AH18 and *Bacillus licheniformis* K11. *Journal of the Korean Society for Applied Biological Chemistry* 52: 531–538.

Lin, Y., Watts, D.B., Kloepper, J.W. et al. (2018). Influence of plant growth-promoting rhizobacteria on corn growth under different fertility sources. *Communications in Soil Science and Plant Analysis* 49 (10): 1239–1255.

Loper, J.E. and Henkels, M.D. (1999). Utilization of heterologous siderophores enhances levels of iron available to *Pseudomonas putida* in the rhizosphere. *Applied and Environmental Microbiology* 65: 5357–5363.

Lu, J., Hu, Z., Xu, Z. et al. (2009). Effects of rice cropping intensity on soil nitrogen mineralization rate and potential in buried ancient paddy soils from the Neolithic Age in China's Yangtze river delta. *Journal of Soils and Sediments* 9 (6): 526–536.

Lucas, J.A., Solano, B.R., Montes, F. et al. (2009). Use of two PGPR strains in the integrated management of blast disease in rice (*Oryza sativa*) in Southern Spain. *Field Crops Research* 114 (3): 404–410.

Ma, Y., Rajkumar, M., Zhang, C. et al. (2016). Beneficial role of bacterial endophytes in heavy metal phytoremediation. *Journal of Environmental Management* 174: 14–25.

Macek, T., Mackova, M., and Káš, J. (2000). Exploitation of plants for the removal of organics in environmental remediation. *Biotechnology Advances* 18 (1): 23–34.

Mahanta, D., Rai, R.K., Dhar, S. et al. (2018). Modification of root properties with phosphate solubilizing bacteria and arbuscular mycorrhiza to reduce rock phosphate application in soybean-wheat cropping system. *Ecological Engineering* 111: 31–43.

Marschner, P., Crowley, D.E., and Rengel, Z. (2011). Rhizosphere interactions between microorganisms and plants govern iron and phosphorus acquisition along the root axis – model and research methods. *Soil Biology and Biochemistry* 43: 883–894.

Marulanda, A., Barea, J.-M., and Azcón, R. (2009). Stimulation of plant growth and drought tolerance by native microorganisms (AM fungi and bacteria) from dry environments: mechanisms related to bacterial effectiveness. *Journal of Plant Growth Regulation* 28: 115–124.

Masciarelli, O., Llanes, A., and Luna, V. (2014). A new PGPR co-inoculated with *Bradyrhizobium japonicum* enhances soybean nodulation. *Microbiological Research* 169 (7–8): 609–615.

Mayak, S., Tirosh, T., and Glick, B.R. (2004). Plant growth-promoting bacteria confer resistance in tomato plants to salt stress. *Plant Physiology and Biochemistry* 42: 565–572.

Meena, V.S., Maurya, B.R., and Verma, J.P. (2014). Does a rhizospheric microorganism enhance K^+ availability in agricultural soils? *Microbiological Research* 169: 337–347.

Meena, V.S., Maurya, B.R., Verma, J.P. et al. (2016). Potassium Solubilizing Microorganisms for Sustainable Agriculture. India: Springer.

Meena, V.S., Zaid, A., Maurya, B.R. et al. (2018). Evaluation of potassium solubilizing rhizobacteria (KSR): enhancing K-bioavailability and optimizing K-fertilization of maize plants under Indo-Gangetic Plains of India. *Environmental Science and Pollution Research* 25: 36412–36424.

Mendis, H.C., Thomas, V.P., Schwientek, P. et al. (2018). Strain-specific quantification of root colonization by plant growth promoting rhizobacteria *Bacillus firmus* I-1582 and *Bacillus amyloliquefaciens* QST713 in non-sterile soil and field conditions. *PLoS One* 13 (2): e0193119.

Mirza, M.S., Mehnaz, S., Normand, P. et al. (2006). Molecular characterization and PCR detection of a nitrogen-fixing *Pseudomonas* strain promoting rice growth. *Biology and Fertility of Soils* 43 (2): 163–170.

Mohammadi, K. and Sohrabi, Y. (2012). Bacterial biofertilizers for sustainable crop production: a review. *ARPN Journal of Agricultural and Biological Science* 7: 307–316.

Molla, A.H., Shamsuddin, Z.H., Halimi, M.S. et al. (2001). Potential for enhancement of root growth and nodulation of soybean co-inoculated with *Azospirillum* and *Bradyrhizobium* in laboratory systems. *Soil Biology and Biochemistry* 33: 457–463.

Morel, M.A., Braña, V., and Castro-Sowinski, S. (2012). Legume crops, importance and use of bacterial inoculation to increase the production. In: Crop Plant (ed. A. Goyal), 217–240. Rijeka: InTech.

Morgan, P.W. and Drew, M.C. (1997). Ethylene and plant responses to stress. *Physiologia Plantarum* 100: 620–630.

Nadeem, S.M., Zahir, Z.A., Naveed, M. et al. (2009). Rhizobacteria containing ACC deaminase confer salt tolerance in maize grown on salt affected soils. *Canadian Journal of Microbiology* 55: 1302–1309.

Nadeem, S.M., Zahir, Z.A., Naveed, M. et al. (2010). Rhizobacteria capable of producing ACC-deaminase may mitigate salt stress in wheat. *Soil Science Society of America Journal* 74: 533–542.

Nadeem, S.M., Naveed, M., Zahir, Z.A. et al. (2013). Plant-Microbe interactions for sustainable agriculture: fundamentals and recent advances. In: Plant Microbe Symbiosis: Fundamentals and Advances (ed. N. Arora), 51–103. New Delhi: Springer.

Nadeem, S.M., Ahmad, M., Zahir, Z.A. et al. (2014). The role of mycorrhizae and plant growth promoting rhizobacteria (PGPR) in improving crop productivity under stressful environments. *Biotechnology Advances* 32: 429–448.

Nadeem, S.M., Naveed, M., Ahmad, M. et al. (2015). Rhizosphere bacteria for crop production and improvement of stress tolerance: mechanisms of action, applications, and future prospects. In: Plant Microbes Symbiosis: Applied Facets (ed. N.K. Arora), 1–36. New Delhi: Springer.

Naseem, H. and Bano, A. (2014). Role of plant growth-promoting rhizobacteria and their exopolysaccharide in drought tolerance of maize. *Journal of Plant Interactions* 9: 689–701.

Nath, D., Maurya, B.R., and Meena, V.S. (2017). Documentation of five potassium-and phosphorus-solubilizing bacteria for their K and P-solubilization ability from various minerals. *Biocatalysis and Agricultural Biotechnology* 10: 174–181.

Nautiyal, C.S., Srivastava, S., Chauhan, P.S. et al. (2013). Plant growth-promoting bacteria *Bacillus amyloliquefaciens* NBRISN13 modulates gene expression profile of leaf and rhizosphere community in rice during salt stress. *Plant Physiology and Biochemistry* 66: 1–9.

Naveed, M., Khalid, M., Jones, D.L. et al. (2008). Relative efficacy of *Pseudomonas* spp., containing ACC-deaminase for improving growth and yield of maize (*Zea mays* L.) in the presence of organic fertilizer. *Pakistan Journal of Botany* 40 (3): 1243–1251.

Naveed, M., Hussain, M.B., Zahir, Z.A. et al. (2014a). Drought stress amelioration in wheat through inoculation with *Burkholderia phytofirmans* strain PsJN. *Plant Growth Regulation* 73: 121–131.

Naveed, M., Mitter, B., Reichenauer, T.G. et al. (2014b). Increased drought stress resilience of maize through endophytic colonization by *Burkholderia phytofirmans* PsJN and *Enterobacter* sp. FD 17. *Environmental and Experimental Botany* 97: 30–39.

Ngumbi, E. and Kloepper, J. (2016). Bacterial-mediated drought tolerance: current and future prospects. *Applied Soil Ecology* 105: 109–125.

Pallai, R., Hynes, R.K., Verma, B. et al. (2012). Phytohormone production and colonization of canola (*Brassica napus* L.) roots by *Pseudomonas fluorescens* 6-8 under gnotobiotic conditions. *Canadian Journal of Microbiology* 58 (2): 170–178.

Parmar, P. and Sindhu, S.S. (2013). Potassium solubilisation by rhizosphere bacteria: influence of nutritional and environmental conditions. *Journal of Microbiological Research* 3: 25–31.

Patten, C.L. and Glick, B.R. (2002). Role of *Pseudomonas putida* indole acetic acid in development of the host plant root system. *Applied and Environmental Microbiology* 68: 3795–3801.

Pedraza, R.O., Bellone, C.H., De Bellone, S.C. et al. (2009). *Azospirillum* inoculation and nitrogen fertilization effect on grain yield and on the diversity of endophytic bacteria in the phyllosphere of rice rainfed crop. *European Journal of Soil Biology* 45 (1): 36–43.

Penrose, D.M., Moffatt, B.A., and Glick, B.R. (2001). Determination of 1-aminocycopropane-1-carboxylic acid (ACC) to assess the effects of ACC deaminase-containing bacteria on roots of canola seedlings. *Canadian Journal of Microbiology* 47 (1): 77–80.

Pérez-Montaño, F., Alías-Villegas, C., Bellogín, R.A. et al. (2014). Plant growth promotion in cereal and leguminous agricultural important plants: from microorganism capacities to crop production. *Microbiological Research* 169: 325–336.

Perrig, D., Boiero, M.L., Masciarelli, O.A. et al. (2007). Plant-growth-promoting compounds produced by two agronomically important strains of *Azospirillum brasilense*, and implications for inoculant formulation. *Applied Microbiology and Biotechnology* 75: 1143–1150.

Pieterse, C., Leon-Reyes, A., Van der Ent, S. et al. (2009). Networking by small-molecule hormones in plant immunity. *Nature Chemical Biology* 5: 308–317.

Piromyou, P., Buranabanyat, B., Tantasawat, P. et al. (2011). Effect of plant growth promoting rhizobacteria (PGPR) inoculation on microbial community structure in rhizosphere of forage corn cultivated in Thailand. *European Journal of Soil Biology* 47 (1): 44–54.

Podile, A.R. and Kishore, G.K. (2006). Plant growth-promoting rhizobacteria. In: Plant-Associated Bacteria (ed. S.S. Gnanamanickam), 195–230. Netherlands: Springer.

Prathap, M. and Ranjitha, K.B.D. (2015). A critical review on plant growth promoting rhizobacteria. *Journal of Plant Pathology and Microbiology* 6 (4): 1–4.

Qiaosi, L., Saleh-Lakha, S., and Glick, B.R. (2005). The effect of native and ACC deaminase containing *Azospirillum brasilense* Cd1843 on the rooting of carnation cuttings. *Canadian Journal of Microbiology* 51: 511–514.

Rahdari, P. and Hoseini, S.M. (2012). Drought stress, a review. *International Journal of Agronomy and Plant Production* 3: 443–446.

Raheem, A., Shaposhnikov, A., Belimov, A.A. et al. (2018). Auxin production by rhizobacteria was associated with improved yield of wheat (*Triticum aestivum* L.) under drought stress. *Archives of Agronomy and Soil Science* 64 (4): 574–587.

Rais, A., Shakeel, M., Malik, K. et al. (2018). Antagonistic *Bacillus* spp. reduce blast incidence on rice and increase grain yield under field conditions. *Microbiological Research* 208: 54–62.

Ramarethinam, S. and Chandra, K. (2005). Studies on the effect of potash solubilizing/mobilizing bacteria *Frateuria aurantia* on brinjal growth and yield. *Pestology* 30 (11): 35–39.

Ramesh, A., Sharma, S.K., Sharma, M.P. et al. (2014). Inoculation of zinc solubilizing *Bacillus aryabhattai* strains for improved growth, mobilization and biofortification of zinc in soybean and wheat cultivated in vertisols of central India. *Applied Soil Ecology* 73: 87–96.

Rashid, M.H., Krehenbrink, M., and Akhtar, M.S. (2015). Nitrogen-fixing plant-microbe symbioses. In: Sustainable Agriculture Reviews, vol. 15 (ed. E. Lichtfouse), 193–234. Cham: Springer International Publishing.

Rathore, P. (2015). A review on approaches to develop plant growth promoting rhizobacteria. *Internatonal Journal of Recent Scientific Research* 5 (2): 403–407.

Reed, S.C., Cleveland, C.C., and Townsend, A.R. (2011). Functional ecology of free-living nitrogen fixation: a contemporary perspective. *Annual Review of Ecology and Systematics* 42: 489–512.

Richardson, A.E., Barea, J.M., McNeill, A.M. et al. (2009). Acquisition of phosphorus and nitrogen in the rhizosphere and plant growth promotion by microorganisms. *Plant and Soil* 321 (1-2): 305–339.

Riefler, M., Novak, O., Strnad, M. et al. (2006). Arabidopsis cytokinin receptor mutants reveal functions in shoot growth, leaf senescence, seed size, germination, root development, and cytokinin metabolism. *Plant Cell* 18: 40–54.

Rodriguez-Navarro, D.N., Dardanelli, M.S., and Ruiz-Sainz, J.E. (2007). Attachment of bacteria to the roots of higher plants. *FEMS Microbiology Letters* 272: 127–136.

Roesti, D., Gaur, R., Johri, B.N. et al. (2006). Plant growth stage, fertiliser management and bio-inoculation of arbuscular mycorrhizal fungi and plant growth promoting rhizobacteria affect the rhizobacterial community structure in rain-fed wheat fields. *Soil Biology and Biochemistry* 38: 1111–1120.

Saha, M., Sarkar, S., Sarkar, B. et al. (2016). Microbial siderophores and their potential applications: a review. *Environmental Science and Pollution Research* 23 (5): 3984–3999.

Salamone, I.E.G., Hynes, R.K., and Nelson, L.M. (2005). Role of cytokinins in plant growth promotion by rhizosphere bacteria. In: PGPR: Biocontrol and Biofertilization (ed. Z.A. Siddiqui), 173–195. The Netherlands: Springer.

Saleem, M., Arshad, M., Hussain, S. et al. (2007). Perspective of plant growth promoting rhizobacteria (PGPR) containing ACC-deaminase in stress agriculture. *Journal of Industrial Microbiology and Biotechnology* 34: 635–648.

Saleem, M., Asghar, H.N., Zahir, Z.A. et al. (2018). Impact of lead tolerant plant growth promoting rhizobacteria on growth, physiology, antioxidant activities, yield and lead content in sunflower in lead contaminated soil. *Chemosphere* 195: 606–614.

Salisbury, F.B. (1994). The role of plant hormones. In: Plant-Environment Interactions (ed. R.E. Wilkinson), 39–81. New York: Marcel Dekker.

Sandhya, V.S.K.Z., Ali, S.K.Z., Grover, M. et al. (2009). Alleviation of drought stress effects in sunflower seedlings by the exopolysaccharides producing *Pseudomonas putida* strain GAP-P45. *Biology and Fertility of Soils* 46: 17–26.

Sandhya, V.S.K.Z., Ali, S.Z., Grover, M. et al. (2010). Effect of plant growth promoting *Pseudomonas* spp. on compatible solutes, antioxidant status and plant growth of maize under drought stress. *Plant Growth Regulation* 62 (1): 21–30.

Sardans, J., Penuelas, J., and Ogaya, R. (2008). Drought-induced changes in C and N stoichiometry in a *Quercus ilex* Mediterranean forest. *Forest Science* 54 (5): 513–522.

Schimel, J.P., Balser, T.C., and Wallenstein, M. (2007). Microbial stress response physiology and its implications for ecosystem function. *Ecology* 88: 1386–1394.

Schoebitz, M., Mengual, C., and Roldán, A. (2014). Combined effects of clay immobilized *Azospirillum brasilense* and *Pantoea dispersa* and organic olive residue on plant performance and soil properties in the revegetation of a semiarid area. *Science of the Total Environment* 466–467: 67–73.

Shaharoona, B., Jamro, G.M., Zahir, Z.A. et al. (2007). Effectiveness of various *Pseudomonas* spp. and *Burkholderia caryophylli* containing ACC-deaminase for improving growth and yield of wheat (*Triticum aestivum* L.). *Journal of Microbiology and Biotechnology* 17 (8): 1300–1307.

Shaharoona, B., Naveed, M., Arshad, M. et al. (2008). Fertilizer-dependent efficiency of *Pseudomonads* for improving growth, yield, and nutrient use efficiency of wheat (*Triticum aestivum* L.). *Applied Microbiology and Biotechnology* 79: 147–155.

Shaharoona, B., Imran, M., Arshad, M. et al. (2011). Manipulation of ethylene synthesis in roots through bacterial ACC deaminase for improving nodulation in legumes. *Critical Reviews in Plant Science* 30: 279–291.

Shahzad, R., Khan, A.L., Bilal, S. et al. (2017). Inoculation of abscisic acid-producing endophytic bacteria enhances salinity stress tolerance in *Oryza sativa*. *Environmental and Experimental Botany* 136: 68–77.

Sharma, A. and Johri, B.N. (2003). Growth promoting influence of siderophore-producing *Pseudomonas* strains GRP3A and PRS9 in maize (*Zea mays* L.) under iron depriving conditions. *Microbiological Research* 158: 243–248.

Sheng, X.F. and Huang, W.Y. (2002). Study on the conditions of potassium release by strain NBT of silicate bacteria. *Scientia Agricultura Sinica* 35: 673–677.

Shoghi-Kalkhoran, S., Ghalavand, A., Modarres-Sanavy, S.A.M. et al. (2013). Integrated fertilization systems enhance quality and yield of sunflower (*Helianthus annuus* L.). *Journal of Agricultural Science and Technology* 15: 1343–1352.

Shridhar, B.S. (2012). Review: nitrogen fixing microorganisms. *International Journal of Microbiological Research* 3 (1): 46–52.

Shrivastava, P. and Kumar, R. (2015). Soil salinity: a serious environmental issue and plant growth promoting bacteria as one of the tools for its alleviation. *Saudi Journal of Biological Sciences* 22 (2): 123–131.

Singh, G., Biswas, D.R., and Marwaha, T.S. (2010). Mobilization of potassium from waste mica by plant growth promoting rhizobacteria and its assimilation by maize (*Zea mays*) and wheat (*Triticum aestivum* L.): a hydroponics study under phytotron growth chamber. *Journal of Plant Nutrition* 33: 1236–1251.

Singh, R.P. and Jha, P.N. (2015). Molecular identification and characterization of rhizospheric bacteria for plant growth promoting ability. *International Journal of Current Biotechnology* 3: 12–18.

Smyth, E.M., McCarthy, J., Nevin, R. et al. (2011). in vitro analyses are not reliable predictors of the plant growth promotion capability of bacteria; a *Pseudomonas fluorescens* strain that promotes the growth and yield of wheat. *Journal of Applied Microbiology* 111: 683–692.

Somers, E., Vanderleyden, J., and Srinivasan, M. (2004). Rhizosphere bacterial signalling: a love parade beneath our feet. *Critical Reviews in Microbiology* 30: 205–240.

Spaepen, S., Vanderleyden, J., and Remans, R. (2007). Indole-3-acetic acid in microbial and microorganism-plant signaling. *FEMS Microbiology Reviews* 31: 425–448.

Spaepen, S. and Vanderleyden, J. (2011). Auxin and plant-microbe interactions. *Cold Spring Harbor Perspectives in Biology* 3 (4): a001438.

Sugiyama, A., Ueda, Y., Zushi, T. et al. (2014). Changes in the bacterial community of soybean rhizospheres during growth in the field. *PLoS One* 9 (6): e100709.

Sureshbabu, K., Amaresan, N., and Kumar, K. (2016). Amazing multiple function properties of plant growth promoting rhizobacteria in the rhizosphere soil. *International Journal of Current Microbiology and Applied Sciences* 5 (2): 661–683.

Upadhyay, S.K., Singh, J.S., Saxena, A.K. et al. (2012). Impact of PGPR inoculation on growth and antioxidant status of wheat under saline conditions. *Plant Biology* 14 (4): 605–611.

Vacheron, J., Desbrosses, G., Bouffaud, M.L. et al. (2013). Plant growth promoting rhizobacteria and root system functioning. *Frontiers in Plant Science* 4 (356): 1–19.

Vamerali, T., Bandiera, M., and Mosca, G. (2010). Field crops for phytoremediation of metal-contaminated land. A review. *Environmental Chemistry Letters* 8 (1): 1–17.

Vance, C.P., Uhde-Stone, C., and Allan, D.L. (2003). Phosphorus acquisition and use: critical adaptations by plants for securing a nonrenewable resource. *New Phytologist* 157 (3): 423–447.

Vardharajula, S., Ali, S.Z., Grover, M. et al. (2011). Drought-tolerant plant growth promoting *Bacillus* spp.: effect on growth osmolytes, and antioxidant status of maize under drought stress. *Journal of Plant Interactions* 6: 1–14.

Vedder-Weiss, D., Jurkevitch, E., Burdman, S. et al. (1999). Root growth, respiration and beta-glucosidase activity in maize (*Zea mays*) and common bean (*Phaseolus vulgaris*) inoculated with *Azospirillum brasilense*. *Symbiosis* 26: 363–377.

Vejan, P., Abdullah, R., Khadiran, T. et al. (2016). Role of plant growth promoting rhizobacteria in agricultural sustainability-a review. *Molecules* 21 (573): 1–17.

Verma, J.P., Yadav, J., Tiwari, K.N. et al. (2014a). Evaluation of plant growth promoting activities of microbial strains and their effect on growth and yield of chickpea (*Cicer arietinum* L.) in India. *Soil Biology and Biochemistry* 70: 33–37.

Verma, P., Yadav, A.N., Kazy, S.K. et al. (2014b). Evaluating the diversity and phylogeny of plant growth promoting bacteria associated with wheat (*Triticum aestivum*) growing in central zone of India. *International Journal of Current Microbiology and Applied Sciences* 3 (5): 432–447.

Verma, P., Yadav, A.N., Khannam, K.S. et al. (2015). Assessment of genetic diversity and plant growth promoting attributes of psychrotolerant bacteria allied with wheat (*Triticum aestivum*) from the northern hills zone of India. *Annals of Microbiology* 65 (4): 1885–1899.

Vessey, J.K. (2003). Plant growth promoting rhizobacteria as biofertilizers. *Plant and Soil* 255 (2): 571–586.

Vinocur, B. and Altman, A. (2005). Recent advances in engineering plant tolerance to abiotic stress: achievements and limitations. *Current Opinion in Biotechnology* 16: 123–132.

Violante, A., Barberis, E., Pigna, M. et al. (2003). Factors affecting the formation, nature, and properties of iron precipitation products at the soil-root interface. *Journal of Plant Nutrition* 26 (10-11): 1889–1908.

Visioli, G., D' Egidio, S., Vamerali, T. et al. (2014). Culturable endophytic bacteria enhance Ni translocation in the hyperaccumulator *Noccaea caerulescens*. *Chemosphere* 117 (1): 538–544.

Visioli, G., Vamerali, T., Mattarozzi, M. et al. (2015). Combined endophytic inoculants enhance nickel phytoextraction from serpentine soil in the hyperaccumulator *Noccaea caerulescens*. *Frontiers in Plant Science* 6: 638.

Visioli, G., Sanangelantoni, A.M., Vamerali, T. et al. (2018). 16S rDNA profiling to reveal the influence of seed-applied biostimulants on the rhizosphere of young maize plants. *Molecules* 23: 1–19.

Viveros, O.M., Jorquera, M.A., Crowley, D.E. et al. (2010). Mechanisms and practical considerations involved in plant growth promotion by rhizobacteria. *Journal of Soil Science and Plant Nutrition* 10: 293–319.

Vurukonda, S.S.K.P., Vardharajula, S., Shrivastava, M. et al. (2016). Enhancement of drought stress tolerance in crops by plant growth promoting rhizobacteria. *Microbiological Research* 184: 13–24.

Wahid, A. and Rasul, E. (2005). Photosynthesis in leaf, stem, flower and fruit. In: Handbook of Photosynthesis, 3e (ed. M. Pessarakli), 479–497. Boca Raton, FL: CRC Press.

Wahyudi, A.T., Astuti, R.P., Widyawati, A. et al. (2011). Characterization of *Bacillus* sp. strains isolated from rhizosphere of soybean plants for their use as potential plant growth for promoting rhizobacteria. *Journal of Microbiology and Antimicrobials* 3 (2): 34–40.

Wu, S.C., Cao, Z.H., Li, Z.G. et al. (2005). Effects of biofertilizer containing N-fixer, P and K solubilizers and AM fungi on maize growth: a greenhouse trial. *Geoderma* 125 (1–2): 155–166.

Xu, M., Sheng, J., Chen, L. et al. (2014). Bacterial community compositions of tomato (*Lycopersicum esculentum* Mill.) seeds and plant growth promoting activity of ACC deaminase producing *Bacillus subtilis* (HYT-12-1) on tomato seedlings. *World Journal of Microbiology and Biotechnology* 30 (3): 835–845.

Zaidi, A., Khan, M.S., Ahemad, M. et al. (2009). Recent advances in plant growth promotion by phosphate-solubilizing microbes. In: Microbial Strategies for Crop Improvement (eds. M.S. Khan, A. Zaidi and J. Musarrat), 23–50. Berlin: Springer.

Zakry, F.A.A., Shamsuddin, Z.H., Khairuddin, A.R. et al. (2012). Inoculation of *Bacillus sphaericus* UPMB-10 to young oil palm and measurement of its uptake of fixed nitrogen using the N isotope dilution technique. *Microbes and Environment* 27: 257–262.

Zehnder, G.W., Yao, C., Murphy, J.F. et al. (1999). Microbe induced resistance against pathogens and herbivores: evidence of effectiveness in agriculture. In: Induced Plant Defenses Against Pathogens and Herbivores: Biochemistry, Ecology and Agriculture (eds. A.A. Agarwal, S. Tuzun and E. Bent), 335–355. St. Paul, MN: APS Press.

Zhang, J., Hussain, S., Zhao, F. et al. (2018). Effects of *Azospirillum brasilense* and *Pseudomonas fluorescens* on nitrogen transformation and enzyme activity in the rice rhizosphere. *Journal of Soils and Sediments* 18 (4): 1453–1465.

Zheng, W., Zeng, S., Bais, H. et al. (2018). Plant growth-promoting rhizobacteria (PGPR) reduce evaporation and increase soil water retention. *Water Resources Research* 54: 3673–3687.

Zhu, Z.L. and Chen, D.L. (2002). Nitrogen fertilizer use in China–contributions to food production, impacts on the environment and best management strategies. *Nutrient Cycling in Agroecosystems* 63 (2–3): 117–127.

Zou, J. and Lu, J.W. (2010). Yield response of winter rapeseed to potassium fertilization, use efficiency and soil's potassium critical level in the Yangtze River Valley. *Journal of Agronomy* 3: 45–54.

7

The Interplay Between Roots and Arbuscular Mycorrhizal Fungi Influencing Water and Nutrient Acquisition and Use Efficiency

Thomas W. Kuyper[1], *Xinxin Wang*[2], *and Mary N. Muchane*[3]

[1] *Soil Biology Group, Wageningen University & Research, Wageningen, The Netherlands*
[2] *Mountain Area Research Institute, Agricultural University of Hebei, Baoding, P.R. China*
[3] *Botany Department, National Museums of Kenya, Nairobi, Kenya*

7.1 Introduction

Since plants colonized the terrestrial environment, they have made a living through an intimate and usually mutualistic partnership with mycorrhizal and other root-inhabiting fungi. In fact, the mycorrhizal symbiosis enabled plants to conquer the land 450–500 million years ago. Plant roots, specialized organs with a gravitropic response, a protective root cap, and root hairs, evolved much later, around 415 million years ago (Kenrick and Strullu-Derrien 2014). Plant roots therefore (co-)evolved within the constraints imposed by mycorrhizal fungi, and some plant species still cannot make it on their own (without mycorrhizal fungi, Rodriguez and Redman 2008). The mycorrhizal symbiosis has been a great success, as the overwhelming majority of land plants form mycorrhizal associations. Even though nonmycorrhizal plants are known, a complete loss of capacity to be colonized by mycorrhizal fungi is exceptional and occurs in plants that have evolved the alternative resource acquisition strategies (Maherali et al. 2016; Werner et al. 2018). Examples from this category of nonmycorrhizal plants are carnivorous plants (Droseraceae), parasitic plants (Orobanchaceae, Santalaceae), and plants that form cluster roots or dauciform roots (Proteaceae and Cyperaceae). A few other families are also largely nonmycorrhizal, including Caryophyllaceae and agronomically important Brassicaceae.

Compared to the long history of the mycorrhizal symbiosis, agriculture is a recent phenomenon, starting around 12 000 years ago. Agriculture has allowed the species *Homo sapiens* to construct and expand its own niche, in terms of appropriation of carbon and nutrients. It allowed the human population to grow to more than 7.8 billion people. The need to ensure sufficient food of sufficient quality to all humans requires further intensification of agriculture; however, we have to recognize that we need *sustainable or ecological intensification* to achieve this (Struik and Kuyper 2017). The challenge is how knowledge of this permanent partnership between crops and arbuscular mycorrhizal fungi

The Root Systems in Sustainable Agricultural Intensification, First Edition.
Edited by Zed Rengel and Ivica Djalovic.

(AMF) can be harnessed to contribute to ecological intensification. Addressing that question forces us to sail between the Scylla of a phytocentric perspective (where mycorrhizal fungi are seen as just extensions of the root system of the plant, neglecting the fact that fungal fitness matters as much as plant fitness) and the Charybdis of looking at mycorrhizal fungi with disregard for roots and forgetting how mycorrhizal fungi lead a dual life in both soil and roots. In that middle ground, we have to understand how the mycorrhizal symbiosis affects the structure and functioning of roots. Phrased from the opposite perspective, we need to understand how evolution (due to the selection of favourable crop traits) has modified mycorrhizal functioning. As Bonfante and Genre (2015) eloquently worded it: in the middle ground, we have to understand both fungish and plantish in our study of the interplay between both organisms.

To give an example: the Green Revolution that allowed spectacular yield increases of major (mycorrhizal) cereal crops was enabled by the introduction of dwarfing genes into these cereals, together with enhanced resistance against or tolerance of pathogens and pests. Both types of genetic modification could unintentionally have affected mycorrhizal functioning because of underlying changes in the hormonal regulatory pathways. Strigolactones, a class of plant hormones, started as internal plant hormones in early (still aquatic) plants regulating simple developmental processes. Later, they were also exuded and played an additional role in the communication with AMF (and subsequently also in the interaction with parasitic plants of the family Orobanchaceae, Andreo-Jimenez et al. 2015). Other plant hormones (auxin, gibberellin, and abscisic acid) also modify mycorrhizal functioning (Pozo et al. 2015; Lanfranco et al. 2018).

Modern crop breeding has not resulted in novel crops that have abandoned the mycorrhizal symbiosis. However, over ecological times, our current (unsustainable) agricultural practices could have made the symbiosis less important or less efficient (Verbruggen and Kiers 2010). Will such ecological changes ultimately translate into evolutionary changes that make ecological intensification more difficult? Answers to that question have ranged from a decisive "no" to a somewhat cautious "yes". Ryan and Graham (2002) claimed that AMF do not play a crucial role in the nutrient acquisition and growth of plants in production-oriented agricultural systems, due to management that reduces mycorrhizal benefits (high levels of fertilizers, use of fungicides, disturbance through tillage, rotations with nonmycorrhizal plants, wrong cultivar selection by including less responsive cultivars, etc.). Observations of negative growth responses to mycorrhiza in the major crops under controlled conditions have strengthened this opinion (Smith and Smith 2011). However, a growth depression of (facultatively) mycorrhizal plants can be effective in weed suppression (Veiga et al. 2011). Ryan and Graham (2002) also referred to exceptions to their general conclusion, viz. when the mycorrhizal symbiosis makes a major contribution to soil structure (not treated here, but see the review by Rillig and Mummey (2006), and also Chapter 3) or when enhancing micronutrient concentrations (see Section 7.5). In a recent review, when Ryan and Graham looked at the literature published in the intervening 16 years, they came to a similar conclusion (Ryan and Graham, 2018). They concluded that the yield benefits of maintaining high abundance and diversity of AMF in agro-ecosystems have often been overstated because of an unwarranted view that high colonization levels and/or high species richness are needed for high yields. On the other side of the spectrum, Thirkell et al. (2017) expressed cautious optimism when they referred to mycorrhizal fungi as our sustainable saviours. Other claims that mycorrhizal fungi act as biofertilizer, bioregulator, or biocontrol agent (Rouphael et al. 2015; Berruti et al. 2016; Bitterlich et al. 2018) are also expressions of the prime importance of the arbuscular mycorrhizal symbiosis for ecological intensification. However, we wish to state from the outset that the term biofertilizer is misleading. Contrary to N_2-fixing rhizobia that add nitrogen to the agro-ecosystem, AMF scavenge, but do not mine for, nutrients and therefore have no effect on the total pool of nutrients (Lambers et al. 2008). In fact, under conditions of very low P availability, a scavenging strategy is not successful and a mining strategy through the exudation of carboxylates by nonmycorrhizal plants contributes to P acquisition much better (Parfitt 1979). However, in cases of a legacy of high fertilizer application, the superior scavenging activities of mycorrhizal fungi allow them to use residual soil P, giving the false impression that there is no need for P replacement.

In this chapter, we evaluate arbuscular mycorrhizal functioning in agriculture (especially cropland) and look at the challenges ahead for intensification based on ecological principles. We will consider mycorrhizal function in relation to plant roots. We assume a basic background knowledge of the arbuscular mycorrhizal symbiosis among the readers of this chapter. Our focus is on nutrients and water, and the efficiency of both acquisition and subsequent use is discussed. We will not deal with the role of mycorrhizal fungi in creating and maintaining soil structure (but see Chapter 3) or conferring tolerance/resistance to pathogens and herbivores. In the latter case, both nutritional and hormonal effects are important.

7.2 A Trait-based Approach towards Mycorrhizal Root Systems

Trait ecology is currently a very active and promising field of research, both for ecologists and agronomists (Garnier et al. 2016). The strength of an above-ground trait approach has been shown in the general validity of the leaf economics spectrum, a gradient along which plants can be described from having acquisitive to conservative traits (Reich 2014). Reich hypothesized substantial trait coordination above-ground and below-ground, and this would allow a generalized plant economics spectrum. However, the current literature has shown that trait coordination is not universal (Weemstra et al. 2016; Laliberté 2017). A clear example is provided by specific root length (SRL; root length per unit root biomass). SRL was hypothesized to be equivalent to specific leaf area (SLA; leaf area per unit leaf biomass). Theoretical arguments showed that a high SRL would be beneficial under conditions of both high nutrient availability (as predicted by comparison with SLA) and limited nutrient and water availability, where exploration growth is more efficient than exploitation growth. Two major explanations can be forwarded why trait correlation did not occur. Firstly, roots fulfil a diversity of functions – they need to acquire nutrients of very different mobility. They also need to acquire nutrients and water, and transport them to above-ground tissues. Water availability is often not highest at sites where nutrient concentrations are highest. Secondly, and more importantly for our arguments, plant roots evolved in a mycorrhizal world, and the mycorrhizal symbiosis may impose constraints on economic optimality.

Table 7.1 Shifts in plant trait values due to arbuscular mycorrhizal symbiosis.

Traits	Shift due to mycorrhizal symbiosis	References
Leaf and root nutrient status	Increases, as mycorrhizal plants often have higher nutrient (especially phosphorus) concentrations; shift towards acquisitive side	See Section 7.4
Specific Leaf Area (SLA)	Possibly positive	Zhang et al. (2018)
Root : shoot ratio	Declines; shift towards the acquisitive side; however, after allometric correction (larger plants generally have lower R : S ratios), the effect is unlikely to be significant	Veresoglou et al. (2012)
Decomposability	Increases; shift towards acquisitive side	Schädler et al. (2010)
Specific root length (SRL)	Generally decreases; shift towards conservative side	Berta et al. (1993); see below
Root diameter	Increases; shift towards conservative side	See below
Root tissue density (RTD)	Increases; shift towards conservative side?	See below
Root hairs	Fewer and shorter root hairs; shift towards conservative side	Shao et al. (2018); see below
Root branching	Enhanced branching, shift from herringbone to dichotomous branching with increased fractal dimension; shift towards acquisitive side	Berta et al. (1993) and Yang et al. (2014); see below
Foraging precision	Declines; shift towards conservative side	Johnson and Biondini (2001) and Li et al. (2014); see below

Even though several authors have indicated how plants (notably trees) with different mycorrhizal associations fit into the hypothetical plant economics spectrum, there has been less attention paid to how the mycorrhizal symbiosis has an effect on the expression of root traits (compared to a nonmycorrhizal plant of the same species). Friesen et al. (2011) pointed out that microbes mediate plant functional traits. Table 7.1 provides examples on how the arbuscular mycorrhizal symbiosis modifies plant traits and how such changes could fit in a general economics spectrum.

These changes (Table 7.1) indicate that above-ground traits shift towards the acquisitive side when plants are mycorrhizal, whereas below-ground traits generally shift towards more conservative trait expression in the mycorrhizal plants. The shift in root traits towards the more conservative side benefits the mycorrhizal fungus as it maintains control over the plant. Fungal control has been shown in studies where mycorrhizal fungi dominated P uptake even in the absence of a growth benefit because, under the mycorrhizal condition, the plant P transporters were downregulated (Smith et al. 2011). From a mycocentric perspective, the shift towards a more branched root system may equally benefit the mycorrhizal fungus because a higher branching frequency decreases the chance that the plant root outgrows the growth of the mycorrhizal fungus inside the root. At the same time, the increased nutrient status equally benefits the fungus, as a higher nutrient status translates into higher photosynthesis rates and these would increase carbon flux to, and alleviate carbon limitation of, the fungus.

Bayliss (1970) proposed that plants with few and short root hairs and plants with thick root systems (low SRL) derived more benefit from the mycorrhizal symbiosis than plants with abundant and long root hairs and with thin root systems (high SRL). Data by Schweiger et al. (1995) provided confirmation of that hypothesis. From a fungal fitness perspective, a decline in abundance and length of root hairs would be advantageous, as this reduction maintains mycorrhizal control over nutrient uptake. Over the evolutionary time-scale, there has been a tendency of plants to develop thinner roots (higher SRL) and this has reduced both dependency on, and benefits derived from, mycorrhizal fungi (Ma et al. 2018). Thinner roots provide less cortical space for colonization by mycorrhizal fungi. As thinner roots reduce the cortical volume more than the diameter of the stele (Valverde-Barrantes et al. 2016), this evolutionary trend should result in increases in root tissue density (RTD), as the density of the lignified cells of the stele is higher than that of the cortex. Stele rather than cortex was a major determinant of variation in RTD in the study by Hummel et al. (2007). There may, however, be limits to reduction of root diameter (and hence options for plants to forego the mycorrhizal symbiosis). Transport through roots (or mycorrhizal hyphae) obeys the Hagen–Poiseuille law, which states that flow rates scale with the fourth power of diameter. However, uptake rates scale with surface, so with the second power of diameter. Thinner roots would thereby face constraints in balancing water acquisition and subsequent transport through the roots towards aboveground tissues (Kong et al. 2017).

Mycorrhizal fitness would be enhanced if cortical volume increases more than the volume of the stele, as mycorrhizal colonization should increase with an increase in the relative cortical volume. There are only few data to test this hypothesis. Both Kong et al. (2014) and Valverde-Barrantes et al. (2016) noted a positive correlation between cortex-to-stele ratio and mycorrhizal colonization. However, Berta et al. (1993) studied the effects of AMF on root traits of leek (*Allium porrum*) and noted similar increases in cortex and stele.

A final trait may be foraging precision of (mycorrhizal) roots, usually expressed as the relative increase in root proliferation in nutrient-patches compared to nonenriched patches. Chen et al. (2018b) posited that high precision is associated with an acquisitive strategy (and in line with that, they found higher precision among arbuscular mycorrhizal trees than among ectomycorrhizal ones, and higher precision among arbuscular mycorrhizal herb species than among arbuscular mycorrhizal trees). They also suggested that roots with lower RTD may be more precise than roots with higher RTD. That hypothesis would seemingly contradict the earlier observation that root diameter is a major determinant of RTD, with thin roots having higher RTD because of lower cortical volume. However, in addition to the cortex-to-stele ratio, RTD is also determined by the degree of lignification, and more acquisitive plants contain less lignin and hence have lower RTD. The balance between the two opposing processes needs further study. Data by Johnson and Biondini (2001) showed about twice as high precision by dicotyledonous herbs as by grasses in the nonmycorrhizal condition. In an intercropping field experiment, where plants were likely mycorrhizal, Li et al. (2014) noted that foraging precision was lower by two legumes (chickpea [*Cicer arietinum*] and faba bean [*Vicia faba*]) than by two cereals (common wheat [*Triticum aestivum*] and maize [*Zea mays*]). The discrepancy has not been explained.

Foraging precision by AMF has also been studied. Because these fungi react negatively to increased P availability, foraging precision is likely more important in case of organic than inorganic patches containing N and P (Wang et al. 2016).

Trait approaches to AMF are in their infancy. A trait that currently receives limited attention is architecture of the mycelium, notably the difference between thin hyphae and runner hyphae. Hyphal architecture (the balance between thin, short-lived hyphae/hyphal networks, and much thicker, unbranched, and long-lived runner hyphae) is both a species trait (Friese and Allen 1991) and also dependent on soil fertility (Olsson et al. 2014). Under high P and high C availability, AMF reduce the amount of runner hyphae (and hence become more explorative and less acquisitive). There are only very few publications of the fractal geometry of the mycelium. Juge et al. (2009) noted, *in vitro*, a final fractal dimension of 1.62, similar to the dimension of highly branched root systems. Mycorrhizal plants exhibited a higher fractal dimension of the root system than nonmycorrhizal plants, and the effect was more pronounced during drought (Yang et al. 2014). The interplay between changes in the root and hyphal branching is likely caused by phytohormonal changes (Pozo et al. 2015).

A potentially interesting fungal trait is colonization rate. Graham and Abbott (2000) showed that trait expression was independent of soil fertility and that fungal species that showed high colonization rate (called aggressive colonizers) at low (limiting) P supply also showed high rates at high (nonlimiting) P supply. There was also some evidence that aggressive colonizers generally resulted in a lower and often negative growth response.

7.3 Acquisition and Use of Water by Arbuscular Mycorrhizal Plants

Many studies have shown that under drought mycorrhizal benefit tends to increase compared to the conditions of normal water supply (Augé et al. 2015, 2016). These studies showed increased photosynthesis, stomatal conductance, and transpiration by mycorrhizal compared to nonmycorrhizal plants. The meta-analysis by Augé et al. (2016) showed a 49% increase in carbon exchange rate, a 28% rise in stomatal conductance, and a 26% increase in transpiration, compared with nonmycorrhizal plants. Because the increases in photosynthesis were larger than the increases in transpiration, the net effect was positive. Under less favourable conditions, the effect on photosynthesis was larger than under more favourable conditions (72% versus 30%), whereas the effects on stomatal conductance and transpiration did not change. The authors noted the somewhat troubling effect of changes in photosynthesis over time; in contrast to the initial studies (during the 1980s and 1990s) that reported a mycorrhizal effect on photosynthesis of more than 70%, this effect declined to 30–40% in the most recent decade. Stomatal conductance and transpiration rate did not show a consistent trend over time. It could be possible, therefore, that beneficial mycorrhizal effects on photosynthesis decline over time, if these results were not due to research bias (the use of experimental conditions, treatments, or plant and mycorrhizal fungal species that are conducive to a positive outcome) or publication bias (the tendency to publish significant and positive results rather than nonsignificant outcomes).

Understanding the mechanisms by which the mycorrhizal symbiosis enhances plant performance under drought has lagged behind. Often, the intuitive argument is used that mycorrhizal hyphae, by virtue of their small diameter, have both better access to soil water in small pores and provide better hyphal contact with the water film. However, water transport through mycorrhizal hyphae may be relatively slow due to the physical constraints of small hyphal diameter according to the Hagen–Poiseuille law. Even though most nutrients are taken up through the mycorrhizal mycelium (up to 100%; Smith et al. 2011), Ruth et al. (2011) estimated that the combined direct and indirect hyphal contribution to total plant water uptake was around 20%.

Augé et al. (2001, 2004) argued that to understand relations between soil, water, and the mycorrhizal association we should look at pathways through which the symbiosis improved hydraulic properties of soil as well as plant-related pathways. They demonstrated that the first pathway, whereby mycorrhizal fungi improve soil structure, is as important as the plant-related pathway. They noted that a nonmycorrhizal mutant of bean (*Phaseolus vulgaris*) showed equally increased stomatal conductance as the mycorrhizal wild type. Mycorrhizal effects on soil hydraulic properties have been reviewed by Bitterlich et al. (2018) who pointed out that the mycorrhizal enhancement of soil structure improved water retention and also mobility of water, i.e. hydraulic conductivity. These changes influence the way plants with and without mycorrhiza experience drought, even under equal drought treatments. A further soil-related mycorrhizal effect could be hydraulic redistribution, which may help sustain hyphal activity during drought (Allen 2007).

Plant-related mechanisms are associated with both nutritional (enhanced uptake of K and P) and phytohormonal effects. As the mycorrhizal enhancement of stomatal conductance occurred both with and without a positive mycorrhizal growth response, it is likely that phytohormonal changes play a major role (Augé et al. 2015). Not only are mycorrhizal plants better able to withstand drought, the recovery from drought is also faster in mycorrhizal than nonmycorrhizal rice (*Oryza sativa*) (Ruiz-Sánchez et al. 2010).

Water limitation (physiological drought) can also be caused by high concentration of salt (salinity stress). The mechanisms by which AMF can alleviate growth reduction due to excess salinity are similar to the mechanisms by which mycorrhizal plants cope with drought (Evelin et al. 2009; Chandrasekaran et al., 2014). Like in the case of water shortage, mycorrhizal plants show morphological and physiological effects in roots. Mycorrhizal plants then maintain photosynthesis and stomatal conductance for a longer time, show higher water-use efficiency (the ratio between photosynthesis and transpiration), and production of osmoprotectants (Santander et al. 2017). Some of these effects are regulated by phytohormones. However, shifts in the balance between K uptake and Na uptake (and hence a more favourable K : Na balance in shoots and roots) are likely the major mechanism of coping with high salinity.

7.4 Acquisition and Use of Nutrients by Arbuscular Mycorrhizal Plants

The arbuscular mycorrhizal symbiosis is first and foremost known for its beneficial effects on uptake of immobile (diffusion-transported) nutrients. The hyphal network extends beyond the depletion zone of plant roots (and root hairs), thereby increasing access to these nutrients. The beneficial effects are especially noted with P, but also with diffusion-limited cations Zn, Cu, Fe, Mn, K, and NH_4^+. Beneficial effects on uptake of other plant nutrients e.g. (NO_3^-, Ca, and Mg) have been reported, but these play a minor role. Other beneficial effects of the mycorrhizal symbiosis have been reported, and sometimes they are a consequence of the improved plant nutritional status. Because of the mechanism of extension of the hyphal network beyond the depletion zone, one could expect there is a positive correlation between the extent of the extraradical network and uptake of immobile nutrients. This relationship was found by Munkvold et al. (2004), who compared a number of different strains of the same mycorrhizal fungus, and noted a significant linear correlation between the hyphal length density and P acquisition.

The extent to which plants benefit from the mycorrhizal symbiosis depends on the nature of fungus, properties of root system, environmental context, and management. These factors do not act in isolation, and the interactions (including the most complex one, fungus × plant × environment × management) should receive more attention.

Root system characteristics (root hair density and length and root diameter) largely determine the plant benefits arising from the symbiosis (but see Maherali (2014) for a meta-analysis that did not find support for this hypothesis). Plants can therefore be placed in different categories, ranging from those with only minor growth increments under conditions of normal nutrient supply to those with large benefits, in addition to those where plant growth in the nonmycorrhizal condition is hardly possible (Habte and Manjunath 1991; Tawaraya 2003). The latter category includes the members of Alliaceae (onion [*Allium cepa*] and leek) and Solanaceae (bell pepper [*Capsicum annuum*]), as well as cassava [*Manihot esculenta*]. In cases where the mycorrhizal growth response is very low, the mycorrhizal fungi can still be responsible for almost all P uptake (Smith et al. 2011).

Environmental context has an important role; with increasing nutrient availability, the mycorrhizal benefits decline, whereas at declining water availability (i.e. increasing tortuosity – the length of nutrient flow path) mycorrhizal benefits increase. Finally, management of agro-ecosystems is of major importance. Sustainable management should therefore avoid practices that reduce the amount or activity of mycorrhiza, such as excessive disturbance, unbalanced fertilization, overfertilization, use of fungicides, use of nonmycorrhizal crops in a rotation, etc. The latter topic may be particularly relevant considering the increasing area of rapeseed/canola (*Brassica napus*) or fodder radish (*Raphanus sativus*) as (winter) cover crop or as part of a wider crop rotation. Several studies reported decreased mycorrhizal colonization and poorer plant performance of crops after canola, and noted that a crop responsive to mycorrhiza should not be grown after canola (Gavito and Miller 1998; McGonigle et al. 2011).

Normally, benefits of the mycorrhizal symbiosis are shown by comparing plants with and without mycorrhiza at the same soil nutrient status. Because of the very poor performance of some nonmycorrhizal plants, or even the impossibility of growing nonmycorrhizal plants in the field, an alternative approach would be to compare plants whose colonization level has been experimentally reduced, e.g. through rotation cores. Comparisons between mycorrhizal and nonmycorrhizal plants, or between plants with high ("normal") and low ("experimentally reduced") mycorrhizal colonization, indicate how much primary production can be increased if we were to manage the mycorrhizal symbiosis well. It is also possible to compare mycorrhizal and nonmycorrhizal plants at the same size, and then assess the attainable fertilizer savings. Such savings are increasingly important, considering that P is a finite, nonrenewable resource. Wang et al. (2020a) observed that under adequate mycorrhizal management, mycorrhizal maize plants fertilized by 30 kg P ha^{-1} performed equally well as plants with reduced mycorrhizal abundance but with a double dose of fertilizer. Assuming that fertilizer savings of 50% were possible for maize globally, we can calculate the economic value of this ecosystem service (Gianinazzi et al. 2010; Chen et al. 2018a) by the mycorrhizal symbiosis: US$1.5 billion annually. Based on yield increases, Kuyper and Giller (2011) calculated that assuming AMF increase P crop uptake by 10%, the ecosystem service provided by AM fungi would be US$2 billion per year in additional crop yield. However, they did not calculate the economic benefits of fertilizer savings at a constant yield.

Even though Ryan and Graham (2002, 2018) questioned the use of mycorrhiza in production agriculture, they did explicitly refer to cases of micronutrient deficiencies where the mycorrhizal symbiosis may be crucial. Awareness of the mycorrhizal role in remedying micronutrient deficiencies has been increasing, with micronutrient deficiencies, notably that of Zn, in human populations being a major health hazard in many (sub-) tropical areas. Awareness of low micronutrient content in crops increased when it became clear that P fertilization increased cereal yield, but decreased plant and especially grain Zn concentrations. This effect was first shown by Lambert et al. (1979) and was due to P fertilization increasing plant size and grain yield without additional Zn uptake (e.g. in maize), resulting in a dilution effect. In soybean (*Glycine max*), the results were even bleaker: Zn concentration was much higher in mycorrhizal compared with nonmycorrhizal plants. However, increased P fertilization of nonmycorrhizal plants had a small beneficial effect on total Zn uptake, but the same P fertilization of mycorrhizal plants resulted in a decline of mycorrhizal colonization of roots and a decrease in Zn uptake, resulting in a more severe negative effect than what could be explained by dilution only. It is likely therefore that the use of P fertilizer in order to achieve yield increases without management of micronutrients and mycorrhizal fungi can result in declining food quality. The study by Ryan et al. (2008) showed a 30–40% decrease in wheat grain Zn concentration due to either P fertilization, or cropping wheat after fallow or in a rotation with nonmycorrhizal crop, implying the crucial role of AMF in Zn acquisition. Moreover, high P fertilization increased concentration of phytic acid in grains, which adversely affected Zn bioavailability in flour.

In addition to biofortification (enhanced content of micronutrients, for instance Zn), the mycorrhizal symbiosis has been implicated in higher production of bio-active compounds with positive health effects for humans (Avio et al. 2018). The same applies to anti-oxidant enzymes – suggesting the specific use of AMF in production of functional foods. However, underlying mechanisms have yet to be elucidated.

Mycorrhiza-related increases in micronutrient content in crops confronts us with a mycorrhizal conundrum that has not yet been explained convincingly. Even though mycorrhizal plants show both increased water-acquisition efficiency and enhanced water-use efficiency (see Section 7.3), there seems to be a disconnect between nutrient-acquisition efficiency and nutrient-use efficiency (the amount of biomass produced per unit of nutrient acquired). Many researchers have noted that the mycorrhizal P responsiveness (relative increases in P uptake) was larger than the mycorrhizal responsiveness based on biomass. An empirical relationship between both parameters was proposed by Van Der Heijden (2002), but without an underlying physiological mechanism. This effect (an increase in P concentration or enhanced P uptake that is not translated into additional biomass) has usually been explained by referring to carbon costs.

Cardoso et al. (2004) noted an apparent paradox that the mycorrhizal symbiosis in many cases increases nutrient-acquisition efficiency but reduces nutrient-use efficiency; however, they disagreed with the conventional explanation that referred to carbon costs. They noted problems with the comparison per se because in some cases plants of very different sizes were compared. In order to correct for differences in plant size, Wang et al. (2020a) compared plants with high and low colonization along a P-availability gradient. They then plotted biomass against the natural logarithm of P content. The slope in this graph can be used as a proxy for nutrient-use efficiency, with a slope = 0 indicating luxury uptake (whereby additional P acquired cannot be used to increase biomass) and a positive slope indicating that additional P acquired is partly converted into biomass. A higher slope for mycorrhizal than nonmycorrhizal plants indicates that the mycorrhizal symbiosis improves nutrient-use efficiency (Elser et al. 2010). In the study of Wang et al. (2020a), a higher slope was associated with plants that had high colonization compared to plants with reduced colonization, implying that nutrient-use efficiency of plants with high (or normal) colonization was actually increased.

Cardoso et al. (2004) applied the double-pot technique and noted that under conditions where P was limiting for both mycorrhizal and nonmycorrhizal plants, P-use efficiency was similar. They therefore suggested that reduced P-use efficiency of mycorrhizal plants could be due to the fact that in certain experiments nonmycorrhizal plants were P-limited, whereas mycorrhizal plants were limited by an unknown factor. However, Stribley et al. (1980a,b) observed that shoots of mycorrhizal leek contained higher P concentration than nonmycorrhizal plants of equal size. Under the assumption that an increase in P concentration was due to carbon invested in the mycorrhizal symbiosis, they calculated carbon costs of around 40–60%, which is much higher than the carbon flux to mycorrhizal fungi that has been commonly reported (around 10%). Physiological studies are recommended to understand the underlying basis for changes in nutrient-use efficiency of mycorrhizal plants. However, as noted above: even if nutrients taken up are not converted to biomass, food quality may still be improved by increased nutrient content.

Theories on carbon costs of the mycorrhizal symbiosis have been regularly invoked to explain conditions in which nonmycorrhizal plants outperform mycorrhizal plants. In several cases, even the term *mycorrhizal parasitism* has been coined to refer to this negative mycorrhizal growth response (Johnson et al. 1997). However, the debate on carbon costs of the mycorrhizal symbiosis has been complicated; in most cases, the term has been used to explain the cases of lower mycorrhizal performance, without any measurement of the actual carbon costs. Carbon costs have sometimes been defined as carbon fluxes to the fungus, but this measurement is biased because larger plants (due to the mycorrhizal benefits) allow a larger flux of carbon below-ground. The issue of carbon costs assumes that carbon is a costly or even limiting element, which is unlikely to be the case in many conditions, with carbon often being available in surplus amounts in the plant. Carbon costs should be evaluated in relation to the carbon gained through enhanced photosynthesis due to alleviating N and/or P limitation. One should therefore compare plants of the same size with and without mycorrhiza. Kaschuk et al. (2009) demonstrated that plants, due to mycorrhizal fungi resulting in enhanced photosynthesis because of sink stimulation, had photosynthesis rates 5–10% higher than equally sized plants without root symbionts. The magnitude of the effect is similar to the amount of carbon (10%) that is invested in the root symbiont.

If carbon costs do not explain negative mycorrhizal responsiveness, there are several other plausible hypotheses that may explain this negative effect. Jin et al. (2017) divided possible explanations in phytocentric and mycocentric. Their list is not complete, and an additional mechanism underlying negative responses is the potential for nutrient immobilization in the mycorrhizal mycelium (Püschel et al. 2016; Wang et al. 2018). This mycorrhizal trap (Kuyper and Kiers 2014) may also explain why the mycorrhizal symbiosis persists and why it is unlikely that facultative mycorrhizal plant species (that can complete their life cycle in the absence of mycorrhiza) would evolve towards becoming nonmycorrhizal. Finally, cases where the mycorrhizal symbiosis impacts on the root traits (see Table 7.1) could also result in a reduced capacity to acquire nutrients and hence in a negative growth response.

7.5 Breeding to Enhance Mycorrhizal Benefits

Crop species differ in the degree to which they depend on, and are responsive to, the mycorrhizal symbiosis (Janos 2007). There is not only variation among different plant species, but also intraspecific variation in responsiveness of wild plants and crops. Table 7.2 provides a list of major crops for which genetic variation in mycorrhizal responsiveness has been demonstrated. Because of the likely genetic basis of this intraspecific variation, (natural) selection has apparently directly or indirectly favoured traits that are associated with the mycorrhizal symbiosis. The existence of variation in responsiveness among crops raises questions about whether it would be desirable to consider plant breeding for mycorrhizal benefits as a component of ecological intensification.

Variation in responsiveness was first described and analysed by Manske (1989), who compared 22 landraces and 22 high-yielding varieties of wheat. Under nutrient-poor conditions, the landraces showed a higher mycorrhizal responsiveness than the modern high-yielding cultivars. The author noted that high responsiveness was correlated with poor plant growth in the nonmycorrhizal condition. The author also made reciprocal crosses between an efficient and a nonefficient cultivar and noted that mycorrhizal responsiveness of the hybrid was different in the reciprocal crosses, suggesting a role of both chromosomal and cytoplasmatic effects.

The question of a genetic basis of mycorrhizal responsiveness gained increased prominence when Hetrick et al. (1992) reported that ancestors and landraces of wheat were more responsive to mycorrhizal fungi than modern cultivars. The authors suggested that differential responsiveness was due to breeding history and stated that selection under fertilized conditions could have reduced the frequency of genes that foster mycorrhizal associations. The authors proposed that the conditions under which cultivar selection took place, i.e. at adequate to high soil fertility, had inadvertently selected against mycorrhizal responsiveness. Their intuitively plausible explanation has initially gained general acceptance, and several papers made a plea for breeding for (enhanced) mycorrhizal

Table 7.2 Major crops for which genetic variation in mycorrhizal responsiveness has been demonstrated.

Plant species	Reference
Allium cepa	Taylor et al. (2015)
Allium fistulosum	Tawaraya et al. (2001)
Allium trihybrid	Galván et al. (2011)
Arachis hypogaea	Quilambo et al. (2005) and Daft (1991)
Avena sativa	Koide and Lu (1992)
Bactris gasipaes	Clement and Habte (1995)
Capsicum annuum	Sensoy et al. (2007)
Carica papaya	Trindade et al. (2001)
Cicer arietinum	Daft (1991)
Cucumis sativus	Rouphael et al. (2010)
Dioscorea rotundata	Dare et al. (2014)
Dioscorea spp.	Dare et al. (2008) and Lu et al. (2015)
Elaeis guineensis	Blal and Gianinazzi-Pearson (1989)
Fragaria vesca	Mark and Cassells (1996)
Glycine max	Khalil et al. (1994) and Salloum et al. (2016)
Helianthus annuus	Turrini et al. (2016)
Hordeum vulgare	Hetrick et al. (1992) and Baon et al. (1993)
Ipomoea batatas	Yooyongwech et al. (2016)
Lactuca sativa	Jackson et al. (2002)
Musa acuminata	Declerck et al. (1995) and Elsen et al. (2003)
Nicotiana tabacum	Janoušková et al. (2007)
Olea europaea	Citernesi et al. (1998)
Oryza sativa	Dhillion (1992) and Gao et al. (2007)
Pennisetum glaucum	Krishna et al. (1985)
Phaseolus vulgaris	Izaguirre-Mayoral et al. (2000)
Pisum sativum	Mårtensson and Rydberg (1994) and Estaún et al. (1987)
Solanum lycopersicum	Bryla and Koide (1990)
Sorghum bicolor	Lendzemo et al. (2006)
Trifolium repens	Hall et al. (1977) and Eason et al. (2001)
Triticum durum	Kapulnik and Kushnir (1991)
Triticum aestivum	Azcón and Ocampo (1981) and Hetrick et al. (1992, 1993, 1995)
Vigna unguiculata	Mercy et al. (1990) and Oruru et al. (2018)
Zea mays	Hall (1978) and Kaeppler et al. (2000)

responsiveness in order to harness mycorrhizal benefits in sustainable agriculture (Gosling et al. 2006; Bennett et al. 2013).

Kaeppler et al. (2000) investigated the genetic basis for variation in mycorrhizal responsiveness of inbred lines of maize by characterizing quantitative trait loci (QTL) for mycorrhizal responsiveness. Their study, based on a hypothesized relationship between plant breeding under nutrient-rich conditions and decreased responsiveness to AMF, showed a negative correlation between mycorrhizal responsiveness and performance of nonmycorrhizal maize cultivars under nutrient-poor conditions. The implications of the study were twofold: (i) if this pattern was generally valid, any selection for enhanced mycorrhizal responsiveness could entail selecting plants that perform more poorly under nonmycorrhizal conditions, and (ii) a quest for genetic markers (e.g. QTL) for responsiveness might ultimately pinpoint genes associated with poor performance under nonmycorrhizal conditions. The same conclusions were reached in a genetic analysis of mycorrhizal responsiveness in the trihybrid *A. cepa* × (*Allium fistulosum* × *Allium roylei*), which is used to introgress genes of Welsh onion (*A. fistulosum*) into onion. Galván et al. (2011) observed that (relative) mycorrhizal responsiveness was negatively correlated with growth under the nonmycorrhizal conditions and noted that this relation was almost universal.

In contrast to the studies listed above, Lehmann et al. (2012) performed a meta-analysis and found that new cultivars were more responsive than ancestral cultivars. They suggested that the effect could be due to the fact that most experiments were executed under conditions of low P supply, whereas the new cultivars have usually been bred under conditions of P sufficiency. Under P-sufficient conditions, the plants may have lost the capacity to perform well when nonmycorrhizal, which translates into increased responsiveness. The study also found a strong relation between the year the study was published and mycorrhizal responsiveness, suggesting that research and/or publication bias could impact on the generality of these research findings. Martín-Robles et al. (2018) did not find an impact of domestication on mycorrhizal responsiveness. However, modern crops benefitted only under P-limiting conditions, whereas their ancestors benefitted under a wider range of P availability.

Sawers et al. (2008, 2010) critically reflected on the issue of the genetic basis of mycorrhizal responsiveness and supported claims that a negative correlation between performance of nonmycorrhizal plants and mycorrhizal responsiveness is inevitable and unrelated to the actual mechanisms by which mycorrhizal fungi increase acquisition of nutrients. However, by separating dependence-based and nondependence-based variation in mycorrhizal responsiveness, they tried to save the idea that plant breeding can be based on genetic variation in crops for nondependence-based variation in responsiveness. Apparently, they refrained from drawing the inevitable conclusion that the attempt to breed for mycorrhizal responsiveness is misguided.

As an alternative for breeding for mycorrhizal responsiveness, we could breed plants for enhanced mycorrhizal colonization. The underlying rationale would be that there are generally good correlations between partial colonization and plant performance (Lekberg and Koide 2005). Several studies showed genetic variation in mycorrhizal colonization (maize: An et al. 2010; common wheat and durum wheat: Lehnert et al. 2017; De Vita et al. 2018; subterranean clover [*Trifolium subterraneum*]: Ryan et al. 2016). However, the traits associated with high colonization in sorghum (*Sorghum bicolor*) had low heritability;

hence, the breeding for enhanced mycorrhizal colonization may not be very effective (Leiser et al. 2016). A further caveat might be in order: breeding efforts and plant selection that result in thinner roots, usually caused by a decreased cortex-to-stele ratio, result in lower mycorrhizal colonization. However, thinner plant roots also facilitate increased P uptake.

What has not been attempted in the case of the mycorrhizal symbiosis is whether we can breed for enhanced plant selectivity, whereby plants associate only with a specific subset of mycorrhizal fungi that provide largest benefits. Such breeding has been successfully used in legumes, and soybean (*Glycine max*) has been bred for (i) promiscuity because that would guarantee nodulation and some N_2-fixation without the need for artificial inoculation, which would benefit resource-poor farmers in sub-Saharan Africa, and (ii) high selectivity in order to achieve maximal yield gains under conditions of efficient inoculation, as in Brazil. Even though mycorrhizal fungi are generally considered to be nonspecific (i.e. able to form mycorrhizal symbiosis with a large range of plant species), there are many studies that suggest high selectivity, in terms of either the plant × fungus combination or the benefit that a certain fungus provides to a plant. It would be an interesting set of experiments to test to which extent the breeding of soybean for promiscuity vs. selectivity would produce a variable degree of specificity with respect to AMF. Such breeding, however, is not trade-off free. Enhanced selectivity of crop cultivars for certain mycorrhizal fungi would make crop rotations more complex. A related option that has been suggested, but never tested, is breeding for enhanced discriminatory powers and sanctioning capacity of plants (i.e. to selectively withhold carbon from less beneficial or less cooperative symbionts) (Duhamel and Vandenkoornhuyse 2013).

It is still debated whether plant selectivity and fungal cooperativeness have a hereditary basis. If they have and if plants have sufficient capacity to discriminate, it is hard to understand why many plant individuals are colonized by 10–25 different fungal species rather than harbour a mycorrhizal fungal monoculture in/on their roots, because it is unlikely there would be beneficial effects of fungal species richness.

Plant breeding for mycorrhizal responsiveness could also have unexpected effects. Breeding for pathogen resistance negatively impacted on mycorrhizal colonization in maize (Toth et al. 1990), which may also have an impact on mycorrhizal functioning. Jacott et al. (2017) reiterated the warning that there could be a trade-off between disease resistance and mycorrhizal symbiosis. However, Hohmann and Messmer (2017) were more positive, and suggested beneficial effects through targeted breeding for pathogen suppression, possibly through priority effects or enhanced competition for root space, enhanced nutritional status (a case of tolerance rather than resistance) or because of the capacity of mycorrhizal fungi to modify phytohormone biosynthesis, notably of jasmonic acid and salicylic acid.

Even though it is unlikely plant breeding for mycorrhizal responsiveness would make a major contribution to sustainable agriculture, one might wonder whether we could breed or genetically improve AMF, resulting in enhanced plant benefit. Angelard et al. (2010) showed a fivefold enhancement of growth of rice when inoculated with segregated, but not with crossed lines. However, such very large effects in a crop poorly responsive to the mycorrhizal symbiosis raises a question of why plants did not select such beneficial strains during evolution. Rather than direct genetic improvement of AMF, we could train them to provide enhanced benefits to the desired host(s) and fewer/lesser benefits to the alternate hosts. Observations, however, suggest that this is unlikely to work. The host shifts can generate rapid (genetic or epigenetic) changes in mycorrhizal fungi (Angelard et al. 2014; Rua et al. 2016; Koyama et al. 2017) that would hinder selectivity in terms of host preference and/or benefit.

We may have unintentionally exerted selective pressure on mycorrhizal fungi; by increasing fertilizer doses or by tillage and other disruptive practices, we may have selected more ruderal species that are not necessarily the most beneficial (Verbruggen and Kiers 2010). In many cases, such changes happen over ecological rather than evolutionary times because these disruptive agricultural practices are relatively recent.

A final caveat appears to be in order. Plants have coevolved with their root symbionts for millions of years before the advent of agriculture slightly more than 10 000 years ago. Many agriculturally important crops are now grown outside their area of origin, and this movement of crops has intensified during the last century. Hence, it may be worth asking why natural selection has not enhanced benefits to both plants and microbes beyond the current level, considering that fitness alignment is likely the default state. Denison (2012, 2015) argued that it is extremely unlikely that natural selection would not have favoured genetic improvements free from trade-offs. Denison also pointed out the exception to this general rule, viz. cases where we can deliberately select for the collective traits that are traded-off against the individual fitness benefits. In that case, genetic improvement should start with a primary focus on the selection arena (the current forms of agriculture and their potential for ecological intensification) rather than on the individual crop or microbe species.

7.6 Managing Diversity and Abundance of Mycorrhizal Fungi

It is generally accepted that diversity or species richness, due to niche differentiation, increases ecosystem functioning. Several studies have intended to show that higher mycorrhizal species richness increases plant productivity. However, there is one major issue that needs to be addressed before such a relation can be demonstrated. We should be aware that the consensus pertains to the relationship within one trophic level (mycorrhizal fungal diversity and fungal productivity), not between two trophic levels (mycorrhizal fungal diversity and plant productivity). Powell and Rillig (2018) made the point that we should independently assess the relation between fungal diversity and productivity (a mycocentric question) and between fungal productivity and plant productivity (a phytocentric question). The latter relation would only pertain if the fitness of both partners is aligned and if there are excess resources (both carbon and nutrients) in both partners to be shared.

Several studies have shown a direct relation between fungal diversity and plant productivity of either plant mixtures in grasslands or a single crop. Most of these studies reported the number of fungal species inoculated, not the number of species that survived. Gosling et al. (2016) tested the relationship for a range of one to seven species of AMF and the growth of onion. For most parameters, the relationship saturated at low diversity, i.e.

two or three species. Such numbers are very low indeed, considering that most roots harbour between 10 and 25 species and that numbers of up to 66 species per root of an individual plant have been reported by Davison et al. (2015). Gosling et al. (2016) questioned the importance of agricultural management contributing to the maintenance of high fungal species richness. In other words, should we worry that current management has created impoverished arbuscular mycorrhizal communities? They suggested the conclusion could be premature, and that a lack of a relationship between fungal species diversity and onion biomass was due to experimental conditions. Their study took place in a glasshouse with very little environmental variation, whereby only fungal niche differentiation with respect to overcoming P-limitation would drive the relationship. Under multiple limiting factors, it could still be beneficial for plants to maintain high fungal species richness. Even though their suggestion is relevant (how suitable are our experiments under controlled conditions for understanding mycorrhizal functioning in agro-ecosystems?), it has the disadvantage of shielding the theory against falsification, as the number of factors that allow niche differentiation by AMF can in principle be infinite. We would therefore argue that large fungal redundancy is to be expected, even though Gosling et al. (2016) observed that most plant benefits could actually accrue by inoculation with one species (*Funneliformis caledonium*).

When the biomass of individual fungal species can be determined, e.g. through qPCR, results of experiments can be used to evaluate the importance of complementarity effects (niche differentiation between different fungal species) and selection effects (the increasing likelihood of including species in mixtures that cause above-average productivity). Wagg et al. (2011) grew a legume and a grass in all possible combination of one to four species of AMF in two soils (N-rich and P-poor "low-sand" soil vs. N-poor and P-rich "high-sand" soil). They observed overyielding of both plant species, but with a major role of the selection effect in the high-sand soil, and a major role of complementarity in the low-sand soil. The study by Gosling et al. (2016) provided support for, but did not test, the hypothesis that the selection effect rather than the complementarity effect would drive the diversity–productivity relationship.

Maherali and Klironomos (2007) showed that the relationship was only evident in the case of species of different evolutionary history, likely a proxy for functional differentiation (Powell et al. 2009). When species that belong to the same fungal genus were added, there was no evidence for niche differentiation and yield increase. Their results were confirmed in a meta-analysis by Yang et al. (2017) who showed a lack of relationship between species richness and plant productivity; however, they also noted (i) a positive relationship between family richness and plant productivity, and (ii) a counterintuitive result that phylogenetic diversity correlated negatively with plant response (family richness should be a proxy for phylogenetic diversity). They suggested that the latter negative relationship was due to the nature of available datasets and hence was likely an artefact.

An alternative to using phylogenetic diversity would be to use functional diversity. A link between phylogeny and functional differentiation was shown by Hart and Reader (2002a,b), who noted that members of Glomeraceae are better root colonizers, whereas members of Gigasporaceae are better soil colonizers. The latter group was also less efficient in the carbon-for-phosphorus trade than the first group. The meta-analysis by Yang et al. (2017) confirmed that members of Glomeraceae are better in enhancing P uptake. Powell et al.

(2009), however, showed that the capacity to colonize soils and roots were positively correlated rather than traded-off.

A further issue that needs consideration is that of intraspecific diversity (Hazard and Johnson 2018). For reasons poorly understood, some traits are phylogenetically conserved between species but variable within species. Large intraspecific variation in hyphal length (which is directly proportional to enhanced plant P content) was shown by Munkvold et al. (2004) for *Funneliformis mosseae*. Growth responses by test plants to different strains of the same fungus also showed high variability (Koch et al. 2017). Roger et al. (2013) tested the effect of genetic relatedness within one species of AMF, *Rhizophagus irregularis*, on fungal coexistence in the root system and plant performance. They observed that greater relatedness actually increased plant performance compared to more distantly related strains. Considering the large intraspecific diversity of AMF and also the large diversity of arbuscular mycorrhizal fungal species in the same agro-ecosystem, it is unclear how we can generalize the mycorrhizal fungal diversity effects and identify and subsequently manage agro-ecological conditions under which species richness of mycorrhizal fungi is or is not redundant.

Similarly, the effect of arbuscular mycorrhizal fungal networks (connections between different plant species by the same mycorrhizal fungal individual) on productivity of plant mixtures has resulted in contradictory outcomes. Klironomos et al. (2000) showed that the relationship between plant species richness and productivity saturated at lower plant species diversity when plants are mycorrhizal than nonmycorrhizal. They hypothesized that this effect was due to an increased niche overlap (reduced complementarity) of plant species interconnected in a mycorrhizal network. However, Wang et al. (2020b) actually found overyielding when two maize varieties were grown together with the same mycorrhizal fungus, whereas there was no overyielding in the nonmycorrhizal condition. Analysis of their data showed that this finding was due to complementarity effects.

To manage mycorrhizal diversity and abundance in the field, we need to know cases of inoculum limitation, or inefficiency of field inoculum. We also need to understand the extent to which plants function better with a species-rich mycorrhizal inoculum; and whether plants perform better with indigenous or autochthonous inoculum. Most commercial inoculum suppliers provide a few strains of a small number of fungi (especially fungi that can be multiplied easily, and therefore have the typical characteristics of r-selected or ruderal strategies) without too much concern about potential invasiveness and ecosystem effects (replacement of endemic species, or preferential benefit to exotic plants, with a risk of starting a cascade of subsequent plant invasions) in case these fungi "escape". However, recent concerns about this topic have surfaced (Hart et al. 2018; Thomsen and Hart 2018). These authors concluded that the (ruderal) traits selected for successful inocula might also be the traits that enhance invasion success (e.g. high number of propagules produced, strong competitive behaviour towards indigenous species, etc.). Fast colonization, as was shown for so-called aggressive mycorrhizal fungi (Graham and Abbott 2000), may be a further trait of these r-selected species.

Apart from possible risks that exotic AMF escape from managed fields, there is the need to evaluate if indigenous (autochthonous) or exotic (allochthonous) fungi generally function better. In a scenario of plant sanctions or rewards (towards less or more beneficial fungi), it is expected that indigenous strains would perform better. Fitness alignment (a positive correlation between the fitness of the microbial symbiont [microbial biomass] and the fitness of the plant [biomass used as proxy]), was demonstrated for the rhizobium–legumes symbiosis (Friesen 2012). The author concluded that fitness alignment is the default state, and that cases of fitness conflict mainly involve non-coevolved associations. That conclusion is particularly relevant for agriculture, which is only slightly over 10 000 years old, and especially for modern agriculture with exotic crops (crops are commonly grown outside their area of origin in the last one or two centuries), where time for coevolution has been limited. For mycorrhizal symbiosis, evidence is mixed (Rua et al. 2016). Koch et al. (2017) found weak evidence for fitness alignment at best. They referred to evolutionary asymmetry whereby fungal species have their traits highly conserved, but plant responses are highly plastic.

In many cases it appears there is no inherent inoculum limitation in well-managed agro-ecosystems, which raises questions about the efficacy of applying commercial inocula. If agro-ecosystems are poorly managed, it is likely the commercial inocula would deteriorate as well. A meta-analysis by Bowles et al. (2017) showed that less intensive tillage and winter cover cropping similarly increased (by around 30%) AMF colonization of roots of summer annual cash crops. Their data indicate the farmers can manage their agro-ecosystems in ways that enhance or maintain the mycorrhizal symbiosis. However, there could be specific crop production systems where the use of commercial inoculants is likely to be beneficial, such as greenhouse horticulture, soilless cultivation, or the initial growth of plantlets generated *in vitro*. In such systems, the mycorrhizal symbiosis may be more important for protection against pathogens rather than nutritional benefits.

7.7 Conclusions and Outlook

Even though the mycorrhizal symbiosis can still be considered primarily from the nutritional perspective, this chapter also points out the importance of the hormonal crosstalk between plants and mycorrhizal fungi resulting in plant trait modifications that enhance fungal fitness and often also enhance plant fitness. However, the interplay is poorly understood. The differential impact of drought and nutrient scarcity on that crosstalk needs to be further disentangled. Additional physiological research in this area is likely to yield novel insights into the evolution of mycorrhizas and their current ecological functioning.

Several authors have made a plea for a rethinking of our plant breeding activities as a contribution to ecological intensification of agriculture (Lammerts van Bueren et al. 2018). It is likely that this rethinking will also include breeding for traits that relate to more beneficial root symbioses. However, we should be aware of Denison's (2012) plea to take a Darwinian perspective – breeding for mycorrhizal benefits is unlikely to be trade-off free.

It has been noted that elevated atmospheric CO_2 directly impacts on food quality by reducing nutrient content of our foods (Loladze 2014). In this chapter, we have indicated the beneficial effects of arbuscular mycorrhiza not just on the amount of biomass produced but also on food quality, especially micronutrient content. There is some evidence about regulation of nutrient content through stoichiometric homeostasis under a range of environmental conditions. It may therefore be important to investigate whether the mycorrhizal symbiosis would counteract the deterioration of food quality. The topic has been clearly underinvestigated; we found only one paper dealing with that topic (Goicoechea et al. 2016). The authors showed that elevated CO_2 reduced concentrations of almost all macro- and micronutrients in durum wheat (*Triticum durum*), with mycorrhiza making no difference in most cases. Moreover, for copper, iron, manganese, and zinc, mycorrhizal plants even had lower concentrations than nonmycorrhizal plants under elevated CO_2. The role of mycorrhiza in enhancing nutritional quality of the major crops under elevated CO_2 is therefore an important topic for further research that should likely focus on stoichiometry (Elser et al. 2010). Uptake of limiting nutrients to an extent greater than what can be converted to biomass represents luxury consumption, but also contributes to enhanced nutritional quality of food. A degree to which plants exhibit stoichiometric homeostasis (unchanged nutrient concentrations and ratios despite changes in nutrient supply) likely varies for different plant species and different nutrients. It is also unclear to what extent mycorrhizal fungi impact on stoichiometric homeostasis.

Questions about the relation between intraspecific/interspecific diversity of mycorrhizal fungi and plant performance (biomass, yield, and/or nutritional quality) will likely remain in the near future. However, with the current molecular tools, it is possible to determine the persistence and relative abundance of different fungal species in roots, and hence extend our mechanistic knowledge on differential importance of the complementarity vs. selection effects.

References

Allen, M.F. (2007). Mycorrhizal fungi: highways for water and nutrients in arid soils. *Vadose Zone Journal* 6: 291–297.

An, G.H., Kobayashi, S., Enoki, H. et al. (2010). How does arbuscular mycorrhizal colonization vary with host plant genotype? An example based on maize (*Zea mays*) germplasms. *Plant and Soil* 327: 441–453.

Andreo-Jimenez, B., Ruyter-Spira, C., Bouwmeester, H.J. et al. (2015). Ecological relevance of strigolactones in nutrient uptake and other biotic stresses, and in plant-microbe interactions below–ground. *Plant and Soil* 394: 1–19.

Angelard, C., Colard, A., Niculita-Hirzel, H. et al. (2010). Segregation in a mycorrhizal fungus alters rice growth and symbiosis-specific gene transcription. *Current Biology* 20: 1216–1221.

Angelard, C., Tanner, C.J., Fontanillas, P. et al. (2014). Rapid genotypic change and plasticity in arbuscular mycorrhizal fungi is caused by a host shift and enhanced by segregation. *ISME Journal* 8: 284–294.

Augé, R.M., Stodola, A.J.W., Tims, J.E. et al. (2001). Moisture retention properties of a mycorrhizal soil. *Plant and Soil* 230: 87–97.

Augé, R.M., Sylvia, D.M., Park, S. et al. (2004). Partitioning mycorrhizal influence on water relations of *Phaseolus vulgaris* into soil and plant components. *Canadian Journal of Botany* 82: 503–514.

Augé, R.M., Toler, H.D., and Saxton, A.M. (2015). Arbuscular mycorrhizal symbiosis alters stomatal conductance of host plants more under drought than under amply water conditions: a meta-analysis. *Mycorrhiza* 25: 13–24.

Augé, R.M., Toler, H.D., and Saxton, A.M. (2016). Mycorrhizal stimulation of leaf gas exchange in relation to root colonization, shoot size, leaf phosphorus and nitrogen: a quantitative analysis of the literature using meta-regression. *Frontiers in Plant Science* 7: 1084.

Avio, L., Turrini, A., Giovanetti, M. et al. (2018). Designing the ideotype mycorrhizal symbionts for the production of healthy food. *Frontiers in Plant Science* 9: 0189.

Azcón, R. and Ocampo, J.A. (1981). Factors affecting the vesicular arbuscular infection and mycorrhizal dependency of thriteen wheat cultivars. *New Phytologist* 87: 677–685.

Baon, J.B., Smith, S.E., and Alston, A.M. (1993). Mycorrhizal responses of barley cultivars differing in P efficiency. *Plant and Soil* 157: 97–105.

Bayliss, G.T.S. (1970). Root hairs and phycomycetous mycorrhizas in phosphorous deficient soil. *Plant and Soil* 33: 713–716.

Bennett, A.E., Daniel, T.J., and White, P.J. (2013). Benefits of breeding crops for yield response to soil organisms. In: Molecular Microbial Ecology of the Rhizosphere, 1e, vol. 1 (ed. F.J. de Bruijn), 17–27. John Wiley & Sons, Inc.

Berruti, A., Lumini, E., Balestrini, R. et al. (2016). Arbuscular mycorrhizal fungi as natural biofertilizer: let's benefit from past successes. *Frontiers in Microbiology* 6: 1559.

Berta, G., Fusconi, A., and Trotta, A. (1993). VA mycorrhizal infection and the morphology and function of root systems. *Environmental and Experimental Botany* 33: 159–173.

Bitterlich, M., Rouphael, Y., Graefe, J. et al. (2018). Arbuscular mycorrhizas: a promising component of plant production systems provided favorable conditions for their growth. *Frontiers in Plant Science* 9: 1329.

Blal, B. and Gianinazzi-Pearson, V. (1989). Interest of mycorrhiza for the production of micropropagated oil palm clones. *Agriculture, Ecosystems & Environment* 29: 39–43.

Bonfante, P. and Genre, A. (2015). Arbuscular mycorrhizal dialogues: do you speak "plantish" or "fungish"? *Trends in Plant Science* 20: 150–154.

Bowles, T.M., Jackson, L.E., Loeher, M. et al. (2017). Ecological intensification and arbuscular mycorrhizas: a meta-analysis of tillage and cover crop effects. *Journal of Applied Ecology* 54: 1785–1793.

Bryla, D.R. and Koide, R.T. (1990). Role of mycorrhizal infection in the growth and reproduction of wild vs. cultivated plants. *Oecologia* 84: 82–92.

Cardoso, I.M., Boddington, C.L., Janssen, B.H. et al. (2004). Double pot and double compartment: Integrating two approaches to study nutrient uptake by arbuscular mycorrhizal fungi. *Plant and Soil* 260: 301–310.

Chandrasekaran, M., Boughatta, S., Hu, S.J. et al. (2014). A meta-analysis of arbuscular mycorrhizal effects on plants grown under salt stress. *Mycorrhiza* 24: 611–625.

Chen, M., Arato, M., Borghi, L. et al. (2018a). Beneficial services of arbuscular mycorrhizal fungi – from ecology to application. *Frontiers in Plant Science* 9: 1270.

Chen, W., Koide, R.T., and Eissenstat, D.M. (2018b). Nutrient foraging by mycorrhizas: from species functional traits to ecosystem processes. *Functional Ecology* 32: 858–869.

Citernesi, A.S., Vitagliano, C., and Giovannetti, M. (1998). Plant growth and root system morphology of *Olea europaea* L. rooted cutting as influenced by arbuscular mycorrhiza. *Journal of Horticultural Science and Biotechnology* 75: 647–654.

Clement, C.R. and Habte, M. (1995). Genotypic variation in vesicular-arbuscular mycorrhizal dependence of the pejibaye palm. *Journal of Plant Nutrition* 18 (9): 1907–1916.

Daft, M.J. (1991). Influence of genotypes, rock phosphate and plant densities on mycorrhizal development and the growth responses of five different crops. *Agriculture, Ecosystems & Environment* 35: 151–169.

Dare, M.O., Abaidoo, R.C., Fagbola, O. et al. (2008). Genetic variation and genotype × environment interaction in yams (*Dioscorea* spp.) for root colonization by arbuscular mycorrhiza. *Journal of Food Agriculture and Environment* 6: 227–233.

Dare, M.O., Fagbola, O., Abaidoo, R.C. et al. (2014). Evaluation of white yam (*Dioscorea rotundata*) genotypes for arbuscular mycorrhizal colonization, leaf nutrient concentrations and tuber yield under NPK fertilizer application. *Journal of Plant Nutrition* 37: 658–673.

Davison, J., Moore, M., Öpik, M. et al. (2015). Global assessment of arbuscular mycorrhizal fungal diversity reveals very low endemism. *Science* 349: 970–973.

De Vita, P., Svio, L., Sbrana, C. et al. (2018). Genetic markers associated to arbuscular mycorrhizal colonization in durum wheat. *Scientific Reports* 8: 10612.

Declerck, S., Plenchette, C., and Strullu, D.G. (1995). Mycorrhizal dependency of banana (*Musa acuminata*, AAA group) cultivar. *Plant and Soil* 176: 183–187.

Denison, R.F. (2012). Darwinian Agriculture: How Understanding Evolution can Improve Agriculture. Princeton: Princeton University Press.

Denison, R.F. (2015). A Darwinian perspective on improving nitrogen-fixation efficiency of legume crops and forages. In: Crop Physiology, 2nde (eds. V.O. Sadras and D.F. Calderini), 207–222. Amsterdam: Elsevier.

Dhillion, S.S. (1992). Host–endophyte specificity of vesicular-arbuscular mycorrhizal colonization of *Oryza sativa* L. at the pre-transplant stage in low or high phosphorus soil. *Soil Biology & Biochemistry* 24: 405–411.

Duhamel, M. and Vandenkoornhuyse, P. (2013). Sustainable agriculture: possible trajectories from mutualistic symbiosis and plant neodomestication. *Trends in Plant Sciences* 18: 597–600.

Eason, W.R., Webb, K.J., Michaelson-Yeates, T.P.T. et al. (2001). Effect of genotype of *Trifolium repens* on mycorrhizal symbiosis with *Glomus mosseae*. *Journal of Agricultural Science* 137: 27–36.

Elsen, A., Baimey, H., Swennen, R. et al. (2003). Relative mycorrhizal dependency and mycorrhiza-nematode interaction in banana cultivars (*Musa* spp.) differing in nematode susceptibility. *Plant and Soil* 256: 303–313.

Elser, J.J., Fagan, W.F., Kerkhoff, A.J. et al. (2010). Biological stoichiometry of plant production: metabolism, scaling and ecological response to global change. *New Phytologist* 186: 593–608.

Estaún, V., Calvet, C., and Hayman, D.S. (1987). Influence of plant genotype on mycorrhizal infection: Response of three pea cultivars. *Plant and Soil* 103: 295–298.

Evelin, H., Kapoor, R., and Giri, B. (2009). Arbuscular mycorrhizal fungi in alleviation of salt stress: a review. *Annals of Botany* 104: 1263–1280.

Friese, C.F. and Allen, M.F. (1991). The spread of VA mycorrhizal fungal hyphae in the soil – inoculum types and external hyphal architecture. *Mycologia* 83: 409–418.

Friesen, M.L. (2012). Widespread fitness alignment in the legume–rhizobium symbiosis. *New Phytologist* 194: 1096–1111.

Friesen, M.L., Porter, S.S., Stark, S.C. et al. (2011). Microbially mediated plant functional traits. *Annual Review of Ecology, Evolution, and Systematics* 42: 23–46.

Galván, G.A., Kuyper, T.W., Burger, K. et al. (2011). Genetic analysis of the interaction between *Allium* species and arbuscular mycorrhizal fungi. *Theoretical and Applied Genetics* 122: 947–960.

Gao, X., Kuyper, T.W., Zou, C. et al. (2007). Mycorrhizal responsiveness of aerobic rice genotypes is negatively correlated with their zinc uptake when nonmycorrhizal. *Plant and Soil* 290: 283–291.

Garnier, E., Navas, M.L., and Grigulis, K. (2016). Plant Functional Diversity: Organism Traits, Community Structure, and Ecosystem Properties. Oxford: Oxford University Press.

Gavito, M.E. and Miller, M.H. (1998). Changes in mycorrhiza development in maize induced by crop management practices. *Plant and Soil* 198: 185–192.

Gianinazzi, S., Golotte, A., Binet, M.N. et al. (2010). Agroecology: the key role of arbuscular mycorrhizas in ecosystem services. *Mycorrhiza* 20: 519–530.

Goicoechea, N., Bettoni, M.M., Fuentes-Mandizábal, T. et al. (2016). Durum wheat quality traits affected by mycorrhizal inoculation, water availability and atmospheric CO_2 concentration. *Crop & Pasture Science* 67: 147–155.

Gosling, P., Hodge, A., Goodlass, G. et al. (2006). Arbuscular mycorrhizal fungi and organic farming. *Agriculture, Ecosystems & Environment* 113: 17–35.

Gosling, P., Jones, J., and Bending, G.D. (2016). Evidence for functional redundancy in arbuscular mycorrhizal fungi and implications for agroecosystem management. *Mycorrhiza* 26: 77–83.

Graham, J.H. and Abbott, L.K. (2000). Wheat responses to aggressive and non-aggressive arbuscular mycorrhizal fungi. *Plant and Soil* 220: 207–218.

Habte, M. and Manjunath, A. (1991). Categories of vesicular-arbuscular mycorrhizal dependency of host species. *Mycorrhiza* 1: 3–12.

Hall, I.R. (1978). Effect of vesicular-arbuscular mycorrhizas on two varieties of maize and one of sweetcorn. *New Zealand Journal of Agricultural Research* 21: 517–519.

Hall, I.R., Scott, R.S., and Johnstone, P.D. (1977). Effect of vesicular-arbuscular mycorrhizas on response to 'Grasslands Huia' and 'Tamar' white clovers to phosphorus. *New Zealand Journal of Agricultural Research* 20: 349–355.

Hart, M.M. and Reader, R.J. (2002a). Host plant benefit from association with arbuscular mycorrhizal fungi: Variation due to differences in size of mycelium. *Biology and Fertility of Soils* 36: 357–366.

Hart, M.M. and Reader, R.J. (2002b). Taxonomic basis for variation in the colonization strategy of arbuscular mycorrhizal fungi. *New Phytologist* 153: 335–344.

Hart, M.M., Antunes, P.M., Chaudhary, V.B. et al. (2018). Fungal inoculants in the field: is the reward greater than the risk? *Functional Ecology* 32: 126–135.

Hazard, C. and Johnson, D. (2018). Does genotypic and species diversity of mycorrhizal plants and fungi affect ecosystem function? *New Phytologist* 220: 1122–1128.

Hetrick, B.A.D., Wilson, G.W.T., and Cox, T.S. (1992). Mycorrhizal dependence of modern wheat varieties, landraces, and ancestors. *Canadian Journal of Botany* 70: 2032–2040.

Hetrick, B.A.D., Wilson, G.W.T., and Cox, T.S. (1993). Mycorrhizal dependence of modern wheat cultivars and ancestors – a synthesis. *Canadian Journal of Botany* 71: 512–518.

Hetrick, B.A.D., Wilson, G.W.T., Gill, B.S. et al. (1995). Chromosome location of mycorrhizal responsive genes in wheat. *Canadian Journal of Botany* 73: 891–897.

Hohmann, P. and Messmer, M.M. (2017). Breeding for mycorrhizal symbiosis: focus on disease resistance. *Euphytica* 213: 113.

Hummel, I., Vile, D., Violle, C. et al. (2007). Relating root structure and anatomy to whole-plant functioning in 14 herbaceous Mediterranean species. *New Phytologist* 173: 313–321.

Izaguirre-Mayoral, M.L., Carballo, O., Carreno, L. et al. (2000). Effects of arbuscular mycorrhizal inoculation on growth, yield, nitrogen, and phosphorus nutrition of nodulating bean varieties in two soil substrates of contrasting fertility. *Journal of Plant Nutrition* 23: 1117–1133.

Jackson, L.E., Miller, D., and Smith, S.E. (2002). Arbuscular mycorrhizal colonization and growth of wild and cultivated lettuce in response to nitrogen and phosphorus. *Scientia Horticulturae* 94: 205–218.

Jacott, C.N., Murray, J.D., and Ridout, C.J. (2017). Trade-offs in arbuscular mycorrhizal symbiosis: disease resistance, growth responses and perspectives for crop breeding. *Agronomy* 7: 75.

Janos, D.P. (2007). Plant responsiveness to mycorrhizas differs from dependence upon mycorrhizas. *Mycorrhiza* 17: 75–91.

Janoušková, M., Vosatka, M., Rossi, L. et al. (2007). Effects of arbuscular mycorrhizal inoculation on cadmium accumulation by different tobacco (*Nicotiana tabacum* L.) types. *Applied Soil Ecology* 35 (3): 502–510.

Jin, L., Wang, Q., Wang, Q. et al. (2017). Mycorrhizal-induced growth depression in plants. *Symbiosis* 72: 81–88.

Johnson, H.A. and Biondini, M.E. (2001). Root morphological plasticity and nitrogen uptake of 59 plant species from the Great Plains grasslands, USA. *Basic and Applied Ecology* 2: 127–143.

Johnson, N.C., Graham, J.H., and Smith, F.A. (1997). Functioning of mycorrhizal associations along the mutualism-parasitism continuum. *New Phytologist* 135: 575–585.

Juge, C., Champagne, A., Coughlan, A.P. et al. (2009). Quantifying the growth of arbuscular mycorrhizal fungi: usefulness of the fractal dimensions. *Botany* 87: 387–400.

Kaeppler, S.M., Parke, J.L., Mueller, S.M. et al. (2000). Variation among maize inbred lines and detection of quantitative trait loci for growth at low phosphorus and responsiveness to arbuscular mycorrhizal fungi. *Crop Science* 40: 358–364.

Kapulnik, Y. and Kushnir, U. (1991). Growth dependency of wild, primitive and modern cultivated wheat lines on vesicular-arbuscular mycorrhiza fungi. *Euphytica* 56: 27–36.

Kaschuk, G., Kuyper, T.W., Leffelaar, P.A. et al. (2009). Are the rates of photosynthesis stimulated by the carbon sink strength of rhizobial and arbuscular mycorrhizal symbioses? *Soil Biology and Biochemistry* 41: 1233–1244.

Kenrick, P. and Strullu-Derrien, C. (2014). The origin and early evolution of roots. *Plant Physiology* 166: 570–580.

Khalil, S., Loynachan, T.E., and Tabatabai, M.A. (1994). Mycorrhizal dependency and nutrient uptake by improved and unimproved corn and soybean cultivars. *Agronomy Journal* 86: 949–958.

Klironomos, J.N., McCune, J., Hart, M. et al. (2000). The influence of arbuscular mycorrhizae on the relationship between plant diversity and productivity. *Ecology Letters* 3: 137–141.

Koch, A.M., Antunes, P.M., Maherali, H. et al. (2017). Evolutionary asymmetry in the arbuscular mycorrhizal symbiosis: conservatism in fungal morphology does not predict host plant growth. *New Phytologist* 214: 1330–1337.

Koide, R.T. and Lu, X. (1992). Mycorrhizal infection of wild oats: maternal effects on offspring growth and reproduction. *Oecologia* 90: 218–226.

Kong, D., Ma, C., Zhang, Q. et al. (2014). Leading dimensions in absorptive root trait variation across 96 subtropical species. *New Phytologist* 203: 863–872.

Kong, D., Wang, J., Zeng, H. et al. (2017). The nutrient absorption-transpiration hypothesis: optimizing structural traits in absorptive roots. *New Phytologist* 213: 1569–1572.

Koyama, A., Pietrangelo, O., Sanderson, L. et al. (2017). An empirical investigation of the possibility of adaptability of arbuscular mycorrhizal fungi to new hosts. *Mycorrhiza* 27: 553–563.

Krishna, K.R., Shetty, K.G., Dart, P.J. et al. (1985). Genotype dependent variation in mycorrhizal colonization and response to inoculation of pearl millet. *Plant and Soil* 86: 113–125.

Kuyper, T.W. and Giller, K.E. (2011). Biodiversity and ecosystem functioning below-ground. In: Agrobiodiversity Management for Food Security (eds. J.M. Lenné and D. Wood), 134–149. Wallingford: CABI.

Kuyper, T.W. and Kiers, E.T. (2014). The danger of mycorrhizal traps? *New Phytologist* 203: 352–354.

Laliberté, E. (2017). Below-ground frontiers in trait-based plant ecology. *New Phytologist* 213: 1597–1603.

Lambers, H., Raven, J.A., Shaver, G.R. et al. (2008). Plant nutrient-acquisition strategies change with soil age. *Trends in Ecology and Evolution* 23: 95–103.

Lambert, D.H., Baker, D.E., and Cole, J.H. (1979). The role of mycorrhizae in the interactions of phosphorus with zinc, copper and other elements. *Journal of the Soil Science Society of America* 43: 976–980.

Lammerts van Bueren, E.T., Struik, P.C., Van Eekeren, N. et al. (2018). Towards resilience through systems–based plant breeding. a review. *Agronomy for Sustainable Development* 38: 42.

Lanfranco, L., Fiorilli, V., Venice, F. et al. (2018). Strigolactones cross the kingdoms: plants, fungi, and bacteria in the arbuscular mycorrhizal symbiosis. *Journal of Experimental Botany* 69: 2175–2188.

Lehmann, A., Barto, E.K., Powell, J.R. et al. (2012). Mycorrhizal responsiveness trends in annual crop plants and their wild relatives – a meta-analysis on studies from 1981 to 2010. *Plant and Soil* 355: 231–250.

Lehnert, H., Serfling, A., Enders, M. et al. (2017). Genetics of mycorrhizal symbiosis in winter wheat (*Triticum aestivum*). *New Phytologist* 215: 779–791.

Leiser, W.L., Olatoye, M.O., Rattunde, H.F.W. et al. (2016). No need to breed for enhanced colonization by arbuscular mycorrhizal fungi to improve low-P adaptation of West African sorghums. *Plant and Soil* 405: 51–64.

Lekberg, Y. and Koide, R.T. (2005). Is plant performance limited by abundance of arbuscular mycorrhizal fungi? A meta-analysis of studies published between 1988 and 2003. *New Phytologist* 168: 189–204.

Lendzemo, V.W., van Ast, A., and Kuyper, T.W. (2006). Can arbuscular mycorrhizal fungi contribute to *Striga* management on cereals in Africa? *Outlook on Agriculture* 35: 307–311.

Li, H.B., Ma, Q.H., Li, H.G. et al. (2014). Root morphological responses to localized nutrient supply differ among crop species with contrasting root traits. *Plant and Soil* 376: 151–163.

Loladze, I. (2014). Hidden shift of the ionome of plants exposed to elevated CO_2 depletes minerals at the base of human nutrition. *eLife* 3: 02245.

Lu, F., Lee, C., and Wang, C. (2015). The influence of arbuscular mycorrhizal fungi inoculation on yam (*Dioscorea* spp.) tuber weights and secondary metabolite content. *PeerJ* 3: e1266.

Ma, Z., Guo, D., Xu, X. et al. (2018). Evolutionary history resolves global organization of root functional traits. *Nature* 555: 94–97.

Maherali, H. (2014). Is there an association between root architecture and mycorrhizal growth response? *New Phytologist* 204: 192–200.

Maherali, H. and Klironomos, J.N. (2007). Influence of phylogeny on fungal community assembly and ecosystem functioning. *Science* 316: 1746–1748.

Maherali, H., Oberle, B., Stevens, P.F. et al. (2016). Mutualism persistence and abandonment during the evolution of the mycorrhizal symbiosis. *American Naturalist* 188: E113–E125.

Manske, G.G.B. (1989). Genetical analysis of the efficiency of VA mycorrhiza with spring wheat. *Agriculture, Ecosystems & Environment* 29: 273–280.

Mark, G.L. and Cassells, A.C. (1996). Genotype-dependence in the interaction between *Glomus fistulosum*, *Phytophthora fragariae* and the wild strawberry (*Fragaria vesca*). *Plant and Soil* 185: 233–239.

Mårtensson, A. and Rydberg, M. (1994). Variability among pea varieties for infection with arbuscular mycorrhizal fungi. *Swedish Journal of Agricultural Research* 24: 13–19.

Martín-Robles, N., Lehmann, A., Seco, E. et al. (2018). Impacts of domestication on the arbuscular mycorrhizal symbiosis of 27 crop species. *New Phytologist* 218: 322–324.

McGonigle, T.P., Hutton, M., Greenley, A. et al. (2011). Role of mycorrhiza in wheat-flax versus canola-flax rotation: a case study. *Communications in Soil Science and Plant Analysis* 42: 2134–2142.

Mercy, M.A., Shivashankar, G., and Bagyaraj, D.J. (1990). Mycorrhizal colonization in cowpea is host dependent and heritable. *Plant and Soil* 121: 292–294.

Munkvold, L., Kjøller, R., Vestberg, M. et al. (2004). High functional diversity within species of arbuscular mycorrhizal fungi. *New Phytologist* 164: 357–364.

Olsson, O., Olsson, P.A., and Hammer, E.C. (2014). Phosphorus and carbon availability regulate structural composition and complexity of AM fungal mycelium. *Mycorrhiza* 24: 443–451.

Oruru, M.B., Njeru, E.M., Pasquet, R. et al. (2018). Response of a wild-type and modern cowpea cultivars to arbuscular mycorrhizal inoculation in sterilized and non-sterilized soil. *Journal of Plant Nutrition* 41 (1): 90–101.

Parfitt, R.L. (1979). The availability of P from phosphate-goethite bridging complexes. Desorption and uptake by ryegrass. *Plant and Soil* 53: 55–65.

Powell, J.R. and Rillig, M.C. (2018). Biodiversity of arbuscular mycorrhizal fungi and ecosystem function. *New Phytologist* 220: 1059–1075.

Powell, J.R., Parrent, J.L., Hart, M.M. et al. (2009). Phylogenetic trait conservatism and the evolution of functional trade-offs in arbuscular mycorrhizal fungi. *Proceedings of the Royal Society B: Biological Sciences* 276: 4237–4245.

Pozo, M.J., López-Ráez, J.A., Azcón-Aguilar, C. et al. (2015). Phytohormones as integrators of environmental signals in the regulation of mycorrhizal symbioses. *New Phytologist* 205: 1431–1436.

Püschel, D., Janoušková, M., Hujslová, M. et al. (2016). Plant–fungus competition for nitrogen erases mycorrhizal growth benefits of *Andropogon gerardii* under limited nitrogen supply. *Ecology and Evolution* 6: 4332–4346.

Quilambo, O.A., Weissenhorn, I., Doddema, H. et al. (2005). Arbuscular mycorrhizal inoculation of peanut in low-fertile tropical soil II. Alleviation of drought stress. *Journal of Plant Nutrition* 28: 1645–1662.

Reich, P.B. (2014). The world-wide 'fast-slow' plant economics spectrum: a traits manifesto. *Journal of Ecology* 102: 275–301.

Rillig, M.C. and Mummey, D.L. (2006). Mycorrhizas and soil structure. *New Phytologist* 171: 41–53.

Rodriguez, R. and Redman, R. (2008). More than 400 million years of evolution and some plants still can't make it on their own: plant stress tolerance via fungal symbiosis. *Journal of Experimental Botany* 59: 1109–1114.

Roger, A., Colard, A., Angelard, C. et al. (2013). Relatedness among arbuscular mycorrhizal fungi drives plant growth and intraspecific fungal coexistence. *ISME Journal* 7: 2137–2146.

Rouphael, Y., Cardarelli, M., and Di Mattia, E. (2010). Enhancement of alkalinity tolerance in two cucumber genotypes inoculated with an arbuscular mycorrhizal biofertilizer containing *Glomus intraradices*. *Biology and Fertility of Soils* 46: 499–509.

Rouphael, Y., Franken, P., Schneider, C. et al. (2015). Arbuscular mycorrhizal fungi act as biostimulants in horticultural crops. *Scientia Horticulturae* 196: 91–108.

Rua, M.A., Antoninka, A., Antunes, P.M. et al. (2016). Home-field advantage? Evidence of local adaptation among plants, soil and arbuscular mycorrhizal fungi through meta-analysis. *BMC Evolutionary Biology* 16: 122.

Ruiz-Sánchez, M., Aroca, R., Muñoz, Y. et al. (2010). The arbuscular mycorrhizal symbiosis enhances the photosynthetic efficiency and the antioxidative response of rice plants subjected to drought stress. *Journal of Plant Physiology* 167: 862–869.

Ruth, B., Khalvati, M., and Schmidthalter, U. (2011). Quantification of mycorrhizal water uptake via high-resolution on-line water content sensors. *Plant and Soil* 342: 459–468.

Ryan, M.H. and Graham, J.H. (2002). Is there a role for arbuscular mycorrhizal fungi in production agriculture? *Plant and Soil* 244: 263–271.

Ryan, M.H. and Graham, J.H. (2018). Little evidence that farmers should consider abundance or diversity of arbuscular mycorrhizal fungi when managing crops. *New Phytologist* 220: 1092–1097.

Ryan, M.H., McInerney, J.K., Record, I.R. et al. (2008). Zinc bioavailability in wheat grain in relation to phosphorus fertiliser, crop sequence and mycorrhizal fungi. *Journal of the Science of Food and Agriculture* 88: 1208–1216.

Ryan, M.H., Kidd, D.R., Sandral, G.A. et al. (2016). High variation in the percentage of root length colonised by arbuscular mycorrhizal fungi among 139 lines representing the species subterranean clover (*Trifolium subterraneum*). *Applied Soil Ecology* 98: 221–232.

Salloum, M.S., Guzzo, M.C., Velazquez, M.S. et al. (2016). Variability in colonization of arbuscular mycorrhizal fungi and its effect on mycorrhizal dependency of improved and unimproved soybean cultivars. *Canadian Journal of Microbiology* 62 (12): 1034–1040.

Santander, C., Aroca, R., Ruiz-Lozano, J.M. et al. (2017). Arbuscular mycorrhiza effects on plant performance under osmotic stress. *Mycorrhiza* 27: 639–657.

Sawers, R.J.H., Gutjahr, C., and Paszkowski, U. (2008). Cereal mycorrhiza: an ancient symbiosis in modern agriculture. *Trends in Plant Sciences* 13: 93–97.

Sawers, R.J.H., Gebreselassie, M.N., Janos, D.P. et al. (2010). Characterizing mycorrhiza effect among diverse plant varieties. *Theoretical and Applied Genetics* 120: 1029–1039.

Schädler, M., Brandl, L., and Kempel, A. (2010). "Afterlife" effects of mycorrhization on the decomposition of plant residues. *Soil Biology & Biochemistry* 42: 521–523.

Schweiger, P.F., Robson, A.D., and Barrow, N.J. (1995). Root hair length determines beneficial effects of a *Glomus* species on shoot growth of some pasture species. *New Phytologist* 131: 247–254.

Sensoy, S., Demir, S., Turkmen, O. et al. (2007). Responses of some different pepper (*Capsicum annuum* L.) genotypes to inoculation with two different arbuscular mycorrhizal fungi. *Scientia Horticulturae* 113: 92–95.

Shao, Y.D., Zhang, D.J., Hu, X.C. et al. (2018). Mycorrhiza-induced changes in root growth and nutrient absorption on tea plants. *Plant Soil and Environment* 64: 283–289.

Smith, F.A. and Smith, S.E. (2011). What is the significance of the arbuscular mycorrhizal colonisation of many economically important crop plants? *Plant and Soil* 348: 63–79.

Smith, S.E., Jakobsen, I., Grønlund, M. et al. (2011). Roles of arbuscular mycorrhizas in plant phosphorus nutrition: Interactions between pathways of phosphorus uptake in arbuscular mycorrhizal roots have important implications for understanding and manipulating plant phosphorus acquisition. *Plant Physiology* 156: 1050–1057.

Stribley, D.P., Tinker, P.B., and Rayner, J.H. (1980a). Relation of internal phosphorus concentration and pant weight in plants infected by vesicular-arbuscular mycorrhizas. *New Phytologist* 86: 261–266.

Stribley, D.P., Tinker, P.B., and Snellgrove, R.C. (1980b). Effect of vesicular-arbuscular mycorrhizal fungi on the relations of plant growth, internal phosphorus concentration and soil phosphate analyses. *Journal of Soil Science* 31: 655–672.

Struik, P.C. and Kuyper, T.W. (2017). Sustainable intensification in agriculture: the darker shade of green. *Agronomy for Sustainable Development* 37: 39.

Tawaraya, K. (2003). Arbuscular mycorrhizal dependency of different plant species and cultivars. *Soil Science and Plant Nutrition* 49: 655–668.

Tawaraya, K., Tokairin, K., and Wagatsuma, T. (2001). Dependence of *Allium fistulosum* cultivars on the arbuscular mycorrhizal fungus, *Glomus fasciculatum*. *Applied Soil Ecology* 17: 119–124.

Taylor, A., Pereira, N., Thomas, B. et al. (2015). Growth and nutritional responses to arbuscular mycorrhizal fungi are dependent on onion genotype and fungal species. *Biology and Fertility of Soils* 51: 801–813.

Thirkell, T.J., Charters, M.D., Elliott, A.J. et al. (2017). Are mycorrhizal fungi our sustainable saviours? Considerations for achieving food security. *Journal of Ecology* 105: 921–929.

Thomsen, C.N. and Hart, M.M. (2018). Using invasion theory to predict the fate of arbuscular mycorrhizal fungal inoculants. *Biological Invasions* 20: 2695–2706.

Toth, R., Toth, D., Starke, D. et al. (1990). Vesicular arbuscular mycorrhizal colonization in *Zea mays* affected by breeding for resistance to fungal pathogens. *Canadian Journal of Botany* 68: 1039–1044.

Trindade, A.V., Siqueira, J.O., and Almeida, F.P. (2001). Dependência micorrízica de variedades comerciais de mamoeiro. *Pesquisa Agropecuária Brasileira* 36: 1485–1494.

Turrini, A., Giordani, T., Avio, L. et al. (2016). Large variation in mycorrhizal colonization among wild accessions, cultivars, and inbreds of sunflower (*Helianthus annuus* L.). *Euphytica* 207: 331–342.

Valverde-Barrantes, O.J., Horning, A.L., Smemo, K.A. et al. (2016). Phylogenetically structured traits in root systems influence arbuscular mycorrhizal colonization in woody angiosperms. *Plant and Soil* 404: 1–12.

Van der Heijden, M.G.A. (2002). Arbuscular mycorrhizal fungi as a determinant of plant diversity: In search of underlying mechanisms and general principles. In: Mycorrhizal Ecology, Ecological Studies, vol. 157 (eds. M.G.A. Van der Heijden and I. Sanders), 243–265. Berlin: Springer-Verlag.

Veiga, R.S.L., Jansa, J., Frossard, E. et al. (2011). Can arbuscular mycorrhizal fungi reduce the growth of agricultural weeds? *PLoS One* 6: e27825.

Verbruggen, E. and Kiers, E.T. (2010). Evolutionary ecology of mycorrhizal functional diversity in agricultural systems. *Evolutionary Applications* 3: 547–560.

Veresoglou, S.D., Menexes, G., and Rillig, M.C. (2012). Do arbuscular mycorrhizal fungi affect the allometric partition of host plant biomass to shoots and roots? A meta-analysis of studies from 1990 to 2010. *Mycorrhiza* 22: 227–235.

Wagg, C., Jansa, J., Schmidt, B. et al. (2011). Belowground biodiversity effects of plant symbionts support aboveground productivity. *Ecology Letters* 14: 100–1009.

Wang, G.Z., Li, X., Christie, P. et al. (2016). Response of arbuscular mycorrhizal fungi to soil phosphorus depends on context. *Crop & Pasture Science* 67: 1116–1125.

Wang, X.X., Van der Werf, W, Yang, Y. et al. (2020a). Field performance of different maize varieties in growth cores at natural and reduced mycorrhizal colonization: yield gains and possible fertilizer savings in relation to phosphorus application. *Plant and Soil* 450: 613–624.

Wang, X.X., Hoffland, E., Feng, G. et al. (2020b). Arbuscular mycorrhizal symbiosis increases phosphorus uptake and productivity of mixtures of maize varieties compared to monocultures. *Journal of Applied Ecology* 57: 2203–2211.

Wang, X.X., Wang, X., Sun, Y. et al. (2018). Arbuscular mycorrhizal fungi negatively affect nitrogen acquisition and grain yield of maize in a N-deficient soil. *Frontiers in Microbiology* 9: 418.

Weemstra, M., Mommer, L., Visser, E.J.W. et al. (2016). Towards a multidimensional root trait framework: a tree root review. *New Phytologist* 211: 1159–1169.

Werner, G.D.A., Cornelissen, J.H.C., Cornwell, W.K. et al. (2018). Symbiont switching and alternative resource acquisition strategies drive mutualism breakdown. *Proceedings of the National Academy of Sciences of the United States of America* 115: 5229–5234.

Yang, Y.R., Tang, M., Sulpice, R. et al. (2014). Arbuscular mycorrhizal fungi alter fractal dimensions characteristics of *Robinia pseudoacacia* L. seedlings through regulating plant growth, leaf water status, photosynthesis, and nutrient concentration under drought stress. *Journal of Plant Growth Regulation* 33: 612–625.

Yang, H., Zhang, Q., Koide, R.T. et al. (2017). Taxonomic resolution is a determinant of biodiversity effects in arbuscular mycorrhizal fungal communities. *Journal of Ecology* 105: 219–228.

Yooyongwech, S., Samphumphuang, T., Tisarum, R. et al. (2016). Arbuscular mycorrhizal fungi (AMF) improved water deficit tolerance in two different sweet potato genotypes involves osmotic adjustments via soluble sugar and free proline. *Scientia Horticulturae* 198: 107–117.

Zhang, B.B., Zhang, H., Wang, H. et al. (2018). Effect of phosphorus additions and arbuscular mycorrhizal fungal inoculation on the growth, physiology, and phosphorus uptake of wheat under two water regimes. *Communications in Soil Science and Plant Analysis* 49: 862–874.

8

Role of Seed Priming in Root Development and Crop Production

Abdul Rehman[1], Faisal Nadeem[2], and Muhammad Farooq[3,4]

[1] *Department of Crop Sciences and Biotechnology, Dankook University, Cheonan-si, South Korea*
[2] *Department of Agronomy, The University of Haripur, Haripur, Khyber Pakhtunkhwa, Pakistan*
[3] *Department of Plant Sciences, College of Agricultural and Marine Sciences, Sultan Qaboos University, Al-Khoud, Oman*
[4] *Department of Agronomy, University of Agriculture, Faisalabad, Pakistan*

8.1 Introduction

Poor seedling emergence and stand establishment are the major constraints to crop production, but many farmers lack resources for suitable seedbed preparation and crop production practices (Farooq et al. 2019). Poor quality seed, delayed sowing, dry soil, and adverse edaphic conditions interfere with good stand establishment (Farooq et al. 2009, 2019; Carrillo-Reche et al. 2018). Seed priming is a cost-effective technology that improves crop performance under diverse climatic conditions. It is a controlled-hydration treatment that allows pre-germination metabolic events to take place without actual germination (radicle protrusion). During priming, seeds are soaked in water (with aeration [hydropriming] or without aeration [on-farm priming]) or solution with low water potential (osmopriming) (Bradford 1986). Seed priming agents include various chemicals (such as plant growth regulators: phytohormones, polyamines, salicylic acid, etc.), nutrients (e.g. Zn, B, Mn, Fe, etc.), and inorganic salts (e.g. $CaCl_2$, KCl, and NaCl) (Farooq et al. 2019).

Primed seeds exhibit early and uniform seedling emergence, early flowering, and higher crop yields (Harris et al. 2007; Farooq et al. 2019). Early and vigorous establishment contributes to the acquisition of more water and nutrient resources due to a better root system, which results in greater leaf areas with enhanced photosynthesis and consequently increased crop yield (Farooq et al. 2012, 2019; Jafar et al. 2012; Imran et al. 2013). Roots play a crucial role in crop establishment and growth and grain yield because they provide anchorage to plants and help in nutrient and water uptake (White and Brown 2010; Figure 8.1) and act as the site for synthesis of organic acids (Rehman et al. 2018a), amino acids, and plant hormones.

Healthy crop growth in the early stage is due to the efficient remobilization of nutrients to the growing embryonic axis (Kathiresan et al. 1984). For instance, seed priming increases the volume and surface area of embryo and perisperm, with distinctive morphological

The Root Systems in Sustainable Agricultural Intensification, First Edition.
Edited by Zed Rengel and Ivica Djalovic.

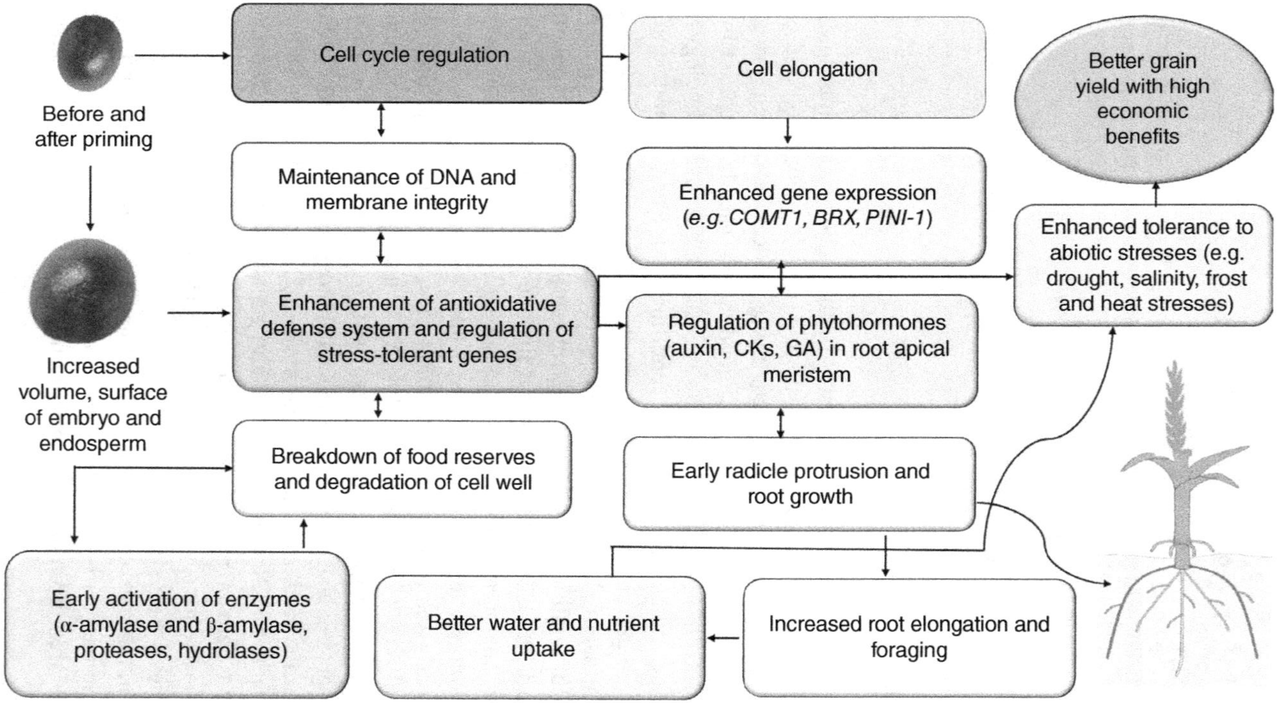

Figure 8.1 Seed-priming-mediated root growth and abiotic stress tolerance. Primed seed has early and synchronized germination due to activation of the starch metabolism enzymes (α- and β-amylases), increased cell division and cell elongation, and enhanced gene expression. Primed seed has a larger embryonic surface area with increased nutrient concentrations that enhance nutrient availability to a developing radicle. Seed priming also regulated the phytohormone concentrations (e.g. cytokinins, auxins, and GA_3), which help in establishment of the better root system architecture. Primed seeds have low oxidative damage due to enhanced signalling and activation of anti-oxidant defence system, imparting abiotic stress tolerance. Furthermore, increased root growth due to priming helps plants extract water and nutrients more efficiently to maintain active growth under optimal as well as suboptimal growth conditions. Better root growth enhances translocation of nutrients and water to developing leaves, thus increasing the leaf surface area, enhancing photosynthesis, and ultimately improving crop yield.

changes between primed and nonprimed seeds during germination (Blunk et al. 2019). Early and uniform emergence and seedling establishment are associated with a shorter lag phase for water imbibition, enzyme activation, accumulation of germination-aiding metabolites, and osmotic adjustment (Bradford 1986; Farooq et al. 2006b; Ullah et al. 2020). Moreover, the enhanced early plant growth in primed seeds is linked to rapid root development, which increases the volume of soil explored for nutrients (Blunk et al. 2019), resulting in better seedling emergence and plant growth, earlier flowering, and ultimately higher crop yields (Farooq et al. 2008, 2019; Ullah et al. 2020; Figure 8.1). Seed priming can help in reducing the adverse effects of agriculture intensification on environmental, social, and economic sustainability. Seed priming contributes to achieving sustainable yields of many crops under optimal and stress environments (Jafar et al. 2012; Farooq et al. 2007, 2019), with reduced inputs (e.g. fertilizers and agrochemicals), high resource-use efficiencies, and profitable economic returns without endangering the environment. In this chapter, the role of seed priming in root development and acquisition of water and nutrients under optimal and suboptimal conditions is critically appraised. Moreover, the role of seed priming in sustainable agricultural intensification is discussed.

8.2 Physiology of Root Development During Germination

Seed germination starts with imbibition and ends after radicle protrusion from the seed coat. Coleorhiza (protective cover of radicle) and radicle are the first structures that protrude from the seed in monocots and dicots, respectively (Nieuwland et al. 2016). Seedling growth from the embryo is driven by cell expansion in the embryonic axis. Primary root development starts during embryogenesis after asymmetric cell divisions that give rise to the quiescent centre and consequently generate root apical meristem (De Smet et al. 2010; Figure 8.2; see also Chapters 2 and 4). Activation of the embryo root meristem is crucial for the initiation of root growth and is a vital part of seedling establishment (Nieuwland et al. 2016; Figure 8.2). Lateral root formation is initiated from pericycle via highly coordinated asymmetric divisions giving rise to lateral root primordium that is transformed into postembryonic lateral root emerging from the primary root (Péret et al. 2009; Figure 8.1).

The embryo size in a dry seed is very small (Wolny et al. 2018), which may affect the imbibition and embryonic root growth. A higher proportion of 2C than 4C DNA in the primary root and shoot shows that majority of cells are in the G1 stage. In contrast, seed priming increases the size of perisperm and embryo, suggesting the improved embryonic axis growth during germination compared with nonprimed seeds (Blunk et al. 2019). Primed seeds exhibit an increase in the nuclear 4C DNA content, resulting in cell cycle advancement from the G1 to G2 phase and the cell cycle arrest in the G2 phase to allow cell synchronization during priming (Farooq et al. 2019) and to improve primary root growth. Cell division and other mitotic activities take place in primed seeds upon imbibition. This pre-activation is one of the mechanisms that improve the performance of primed seeds. Moreover, seed priming enhances the activity of α-amylase and β-amylase, which results in increased energy supply from starch metabolism fuelling root growth and seedling emergence (Li et al. 2014; Farooq et al. 2019; Figure 8.1).

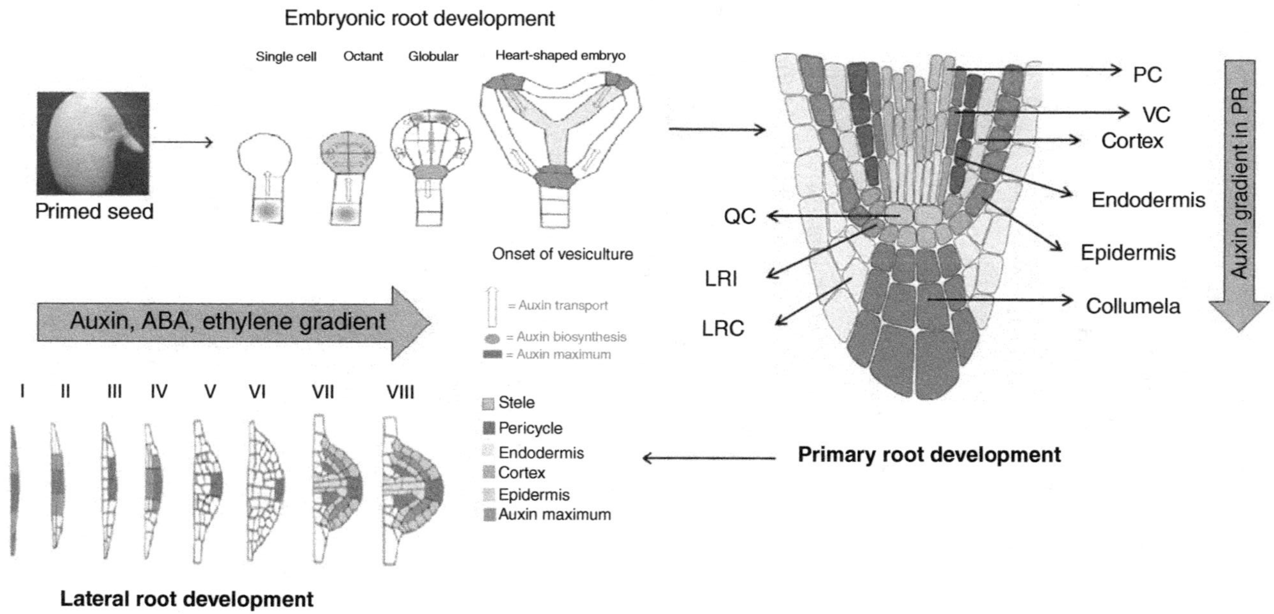

Figure 8.2 Physiology of root development in primed seeds during germination. PC, pericycle; VC, vascular cells; QC, Quiescent centre; LRI, lateral root initiation; LRC, Lateral root cap. *Source:* Modified from De Smet et al. (2015), Olatunji et al. (2017), and Aichinger et al. (2012).

Phytohormones play a key role in seed germination and regulation of root growth. The mitotic cell cycle activity in the root and shoot apical meristems during later stages of germination is influenced by gibberellic acid (GA), and many scientists have identified the vital roles of GA and abscisic acid (ABA) during germination and root growth. Phytohormones and reactive oxygen species have been found to modulate the radicle growth during germination (Oracz and Karpinski 2016). The external supplies of the phytohormones through seed priming can help in regulating the germination and root growth of plants. For instance, seed priming with cytokinin (CK), indole acetic acid (IAA), GA_3, or ascorbate increased the lateral root growth and biomass production of bromegrass and wheat (Eisvand et al. 2010; Khan et al. 2011).

Nutrient supply also affects germination and root growth. For instance, Zn deficiency limits the root growth of rice and wheat (Rehman et al. 2015); importantly, Zn is present in high concentration in germinating embryo. Seed priming with micronutrients has been effective in improving germination, root elongation, lateral root growth, root biomass, and organic acid production in roots and their translocation to shoots and seeds (Imran et al. 2013, 2015; Rehman et al. 2015, 2018a,b).

Embryonic root development during germination and seedling establishment influences the root system architecture. Seed priming modulates the germination-related processes that facilitate germination and root development. Moreover, the application of phytohormones and plant nutrients through seed priming improves the root system architecture by regulating the concentration and availability of these growth regulators and nutrients to the developing roots.

8.3 Seed Priming and Root Development

Roots provide anchorage, acquire water and nutrient, and synthesize phytohormones and organic acids. Root morphology and physiology are closely associated with shoot growth and development. Seed priming plays a vital role in root development as discussed in the section below.

8.3.1 Seed Priming and Root Initiation

Roots are vital organs of vascular plants, having a multicellular axis and highly distinctive structure with root cap, lateral roots, pericycle, and endodermis, which help in nutrient and water uptake (Seago Jr and Fernando 2013). Radicle (primary root) is the first structure that emerges from the seed of most plants (Clowes 1961). Seed priming improves the radicle emergence and root initiation in plants. For instance, seed priming with auxin increased the germination percentage and radicle growth of wheat seedlings (Akbari et al. 2007; Figure 8.2) because auxin is a key regulator of cell division and elongation to underpin plant growth. Auxin travels acropetally in roots and aids in the regulation of root cell elongation in the proximal axis of root apical meristem.

Early root initiation and radicle development were observed in maize seeds primed with melatonin compared to nonprimed seeds (Cao et al. 2019). The external supply of melatonin increased IAA concentration in roots, which contributed to early root initiation and radicle growth (Chen et al. 2009). Similarly, seed priming with phytohormones enhanced early root initiation and radicle development. For example, Dotto and Silva (2017) found that seed priming with GA_3 accelerated the root initiation and primary root development as GA promotes growth through cell expansion by initiating the destruction of growth repressing DELLA proteins; kinematic analysis showed that these proteins play a role in restraining the cell production in the root meristem. Furthermore, GA modulates the cell cycle activity via DELLA-dependent mechanism in the root meristem (Achard et al. 2009; Figure 8.2).

8.3.2 Root Elongation

Root length and distribution in soil are influenced by the edaphic, environmental, and genetic factors (Rogers and Benfey 2015). Phytohormones also influence root elongation (Table 8.1). Seed priming with phytohormones improves the emergence and root growth of aged seeds. For instance, aged seeds of bromegrass after priming with IAA showed increased germination and longer seedling root length. Seed priming with chitosan (0.5% and 0.25% w/v) substantially increased root elongation and root dry weight of maize (Guan et al. 2009). Seed priming of maize with melatonin substantially increased the radicle and hypocotyl length, root elongation, and seed vigour index compared with non-primed control seeds (Cao et al. 2019). Similarly, the exogenous supply of melatonin had a positive effect on root growth in *Brassica juncea* due to the increased concentration of IAA in roots (Chen et al. 2009).

Seed priming with nutrients helps in improving root development on low-nutrient soils (Table 8.1). Seed priming of rice improved root growth during the early seedling stage, leading to the establishment of the better root system with an increased capacity for acquisition of nutrients, particularly of Zn on low-Zn soils. Seed priming with 2.5 mM Zn contributed to improved seedling emergence and root growth of rice in soil with low or adequate Zn supply (Prom-u-thai et al. 2012). Likewise, seed priming with Zn (0.5 M $ZnSO_4$) improved the seedling emergence and root elongation of wheat. A similar observation was made for wheat grown in field conditions of Central Anatolia where Zn deficiency is the major constraint in wheat production (Cakmak et al. 1996). Seed priming with Zn facilitated root elongation during the early stages of plant growth due to increased Zn supply to the embryonic axis because Zn was present at higher concentration in the developing coleoptile and radicle (Ozturk et al. 2006; Rehman et al. 2018a,b). Similarly, Fe and Zn + Mn seed priming of maize improved the root length, biomass production, and nutrient uptake. Seed priming with a low concentration of B (0.001% w/v) improved the seedling root length in wheat (Iqbal et al. 2012) and rice (Farooq et al. 2011). The improved root growth and seedling emergence due to seed priming indicated increased availability of nutrients (e.g. Zn) to meet the demand of germinating seed compared with soil nutrient application (Prom-u-thai et al. 2012).

Nutrient seed priming has proved effective in facilitating early root growth; however, the concentration of a priming nutrient solution is critical because seed priming with high nutrient concentration may inhibit the germination and root growth due to toxicity (Farooq et al. 2011; Rehman et al. 2015). Furthermore, nutrient seed priming in combination with plant-growth-promoting rhizobacteria (PGPR) (e.g. *Pseudomonas* sp.) improved root growth and enhanced organic acid anion exudation from roots, aiding nutrient solubilization

Table 8.1 Seed-priming-mediated root growth and crop productivity.

Crops	Priming methods	Growth conditions	Improvements (%) in root growth over controls	Benefits gained	References
Wheat	Seed priming with MWCNT (MW70)	Field, irrigated	↑ RHD, ↑ RL	↑ uptake of essential minerals (e.g. P and K), ↑ cell elongation, ↑ xylem and phloem dilation, ↑ water uptake, ↑ stomatal conductance, ↑ grain yield	Joshi et al. (2018)
	Seed priming with MWCNT (MW80)	Field, irrigated	↑ RHD, ↑ RL		Joshi et al. (2018)
	Seed priming with MWCNT (MW90)	Field, irrigated	↑ RHD, ↑ RL		Joshi et al. (2018)
	Osmopriming ($CaCl_2$) ($\psi s = -1.25$ MPa)	Field, ploughed, wheat after DSR		↑ stand establishment, ↑ grain yield, ↑ water productivity, ↑ economic profitability	Nawaz et al. (2016)
	Seed priming with 50 mg l^{-1} ascorbic acid solution	Farmer's field, salinity stress		↑ stand establishment, ↑ grain yield, ↑ water productivity, ↑ economic profitability, ↑ leaf K content, ↑ TSP, ↑ α-amylase activity	Jafar et al. (2012)
	Osmopriming ($CaCl_2$) ($\psi s = -1.25$ MPa)	Field, irrigated		↑ grain yield, ↑ economic benefits, ↑ allometric traits and WUE	Farooq et al. (2017)
	Hydropriming Osmopriming (KCl) ($\psi s = -1.25$ MPa) Osmopriming ($CaCl_2$) ($\psi s = -1.25$ MPa)	Field, arid (rain-fed), chilling stress		↑ stand establishment, ↑ allometry, ↑ grain yield	Farooq et al. (2008)
Rice	Nutripriming with Zn (2.5 mM)	Low-zinc soil	↑ RL (~36), ↑ LRN (33), ↑ RDW (82)	↑ germination, ↑ seed vigour index, ↑ shoot growth, ↑ biomass, ↓ Zn fertilizer input	Prom-u-thai et al. (2012)

(Continued)

Table 8.1 (Continued)

Crops	Priming methods	Growth conditions	Improvements (%) in root growth over controls	Benefits gained	References
	Nutripriming with Zn (5 mM)	Low-zinc soil	↑ LRN (26), ↑ RDW (23)	↑ shoot growth, ↑ biomass, ↑ Zn uptake, ↓ Zn fertilizer input	Prom-u-thai et al. (2012)
	Biopriming with *Bacillus cereus*	Greenhouse	↑ RL (64), ↑ RDW (60),	↑ plant height, ↑ leaf area, ↑ leaf phosphate, ↑ chitinase, ↑ β-1,3 glucanase, ↑ PAAL, ↑ POX, ↑ IAA content	Chakraborty et al. (2011)
	Hydropriming	Field, flooded	↑ RL (71), ↑ LRN (46)	↑ emergence, ↓ MET, ↑ α-amylase activity, ↑ reducing sugars, ↑ N, P, and K content, ↑ grain yield	Farooq et al. (2007)
	Osomohardening (KCl) (ψs = −1.25 MPa)	Field, flooded	↑ RL (10), ↑ LRN (62)	↑ emergence, ↓ MET, ↑ α-amylase activity, ↑ reducing sugars, ↑ N, P, and K content, ↑ grain yield	Farooq et al. (2007)
	Osomohardening ($CaCl_2$) (ψs = −1.25 MPa)	Field, flooded	↑ RL (73), ↑ LRN (79)	↑ emergence, ↓ MET, ↑ α-amylase activity, ↑ reducing sugars, ↑ N, P, and K content, ↑ grain yield	Farooq et al. (2007)
	Seed priming with 10 mg l^{-1} ascorbic acid solution	Field, flooded	↑ RL (116), ↑ LRN (54)	↑ emergence, ↓ MET, ↑ α-amylase activity, ↑ reducing sugars, ↑ N, P, and K content, ↑ grain yield	Farooq et al. (2007)
	Hardening (tap water)	Field, flooded	↑ RL (8), ↑ LRN (64)	↑ emergence, ↓ MET, ↑ α-amylase activity, ↑ reducing sugars, ↑ N, P, and K content, ↑ grain yield	Farooq et al. (2007)
	Nutripriming (0.5 M Zn solution)	Field, DSR		↑ yield, ↑ grain Zn concentration, ↓ Zn fertilizer input	Farooq et al. (2018)
	Nutripriming (0.1 mM B solution)	Field, DSR		↑ growth, ↑ water relationships, ↑ morphological yield-related traits, ↑ grain yield, ↓ B fertilizer input, ↑ economic benefits	Rehman et al. (2016)

	Nutripriming (0.1 mM B solution)	Field, AWD		↑ growth, ↑ water relationships, ↑ morphological yield-related traits, ↑ grain yield, ↓ B fertilizer input, ↑ economic benefits	Rehman et al. (2016)
	Nutripriming (0.1 mM B solution)	Field, flooded		↑ growth, ↑ water relationships, ↑ morphological yield-related traits, ↑ grain yield, ↓ B fertilizer input, ↑ economic benefits	Rehman et al. (2016)
Maize	Hormonal priming with GA_3 (400 mg l^{-1})	Accelerated aging (40 °C, relative humidity 90–95%, four and seven days	↑ RL	↑ germination, ↑ shoot growth	Siadat et al. (2011)
	Biopriming with *Trichoderma sp.*	Field	↑ RL (82), ↑ RHA (48), ↑ RFW (59)	↑ shoot length, ↑ β-1,3 glucanase, ↑ exochitinase, ↑endochitinase	Harman et al. (2004)
	Nutripriming with B (5 mM B)	Hydroponic culture and field conditions	↑ RDW (~10)	↑ biomass, ↑ grain yield, ↑ B, Mn, and Zn uptake, ↓ B fertilizer input	Imran et al. (2015)
	Nutripriming with Zn + Mn (4 mM Zn + 2.5 mM Mn)	Hydroponic culture and field conditions	↑ RDW (~100)	↑ biomass, ↑ grain yield, ↑ Zn, Mn, and P uptake, ↓ Mn and Zn fertilizer input	Imran et al. (2015)
	Nutripriming with P (0.2 M P)	Hydroponic culture and field conditions	↑ RDW (~90)	↑ biomass, ↑ grain yield, ↑ Zn, Mn, and P uptake, ↓ P fertilizer input	Imran et al. (2015)
	Hydropriming	Low root zone temperature, controlled conditions and field	↑ RL (~10)	↑ grain yield	Imran et al. (2013)
	Nutripriming with Zn + Mn (4 mM Zn + 2.5 mM Mn)	Low root zone temperature, controlled conditions and field	↑ RL (~100), ↑ RD (137), ↑ RFW (42)	↑ Zn, Mn, Fe, and P uptake, ↑ biomass, ↑ grain yield	Imran et al. (2013)

(Continued)

Table 8.1 (Continued)

Crops	Priming methods	Growth conditions	Improvements (%) in root growth over controls	Benefits gained	References
	Nutripriming with Fe (8.5 mM Fe)	Low root zone temperature, controlled conditions and field	↑ RL (~100), ↑ RD (146), ↑ RFW (23)	↑ Zn, Mn, Fe, and P uptake, ↑ biomass, ↑ grain yield	Imran et al. (2013)
	Nutripriming ($0.1\,mg\,l^{-1}$ B + $5\,mg\,l^{-1}$ Zn)	Field, semi-arid conditions		↑ emergence, ↑ grain yield, ↑ protein content	Rasool et al. (2019)
	Nutripriming ($0.1\,mg\,l^{-1}$ B + $1\,mg\,l^{-1}$ Mn)				
	Nutripriming ($0.1\,mg\,l^{-1}$ B + $5\,mg\,l^{-1}$ Zn + $1\,mg\,l^{-1}$ Mn)				
Sunflower (*Helianthus annuus* L.)	Seed priming with tocopherol ($9\,mg\,l^{-1}$)	Accelerated aging (40 °C, humidity 95–100%) for 72 h	↑ RL	↓ abnormal seedlings, ↑ shoot growth	Draganic and Lekic (2012)
	Seed priming with tocopherol ($9\,mg\,l^{-1}$)	Cold stress (10 °C for 7 days)	↑ RL	↓ abnormal seedlings, ↑ shoot growth	Draganic and Lekic (2012)
	Biopriming with *Pseudomonas fluorescens* (UTPf76 and UTPf86)	Laboratory, greenhouse	↑ RL (54), ↑ RDW (131), ↑ LRN (123)	↑ germination, ↑ seed vigour index, ↑ seedling growth	Moeinzadeh et al. (2010)
Canola (*Brassica napus* L.)	Hydropriming	Semi-arid	↑ RL (32-38), ↑ RD (49-88), ↑ RSAD (108), ↑ RDW (326), ↑ RV (169)	↑ grain yield	Nazari et al. (2017)
	Osmopriming with $ZnSO_4$	Semi-arid	↑ RL (41-53), ↑ RDW (296), ↑ RV (237)	↑ grain yield	Nazari et al. (2017)

Sorghum (*Sorghum bicolor* (L.) Moench)	On-farm priming (water)	Field		↑ grain yield, ↑ economic returns	Aune and Ousman (2011)
Millet (*Pennisetum glaucum* L.)	On-farm priming (water)	Field		↑ grain yield, ↑ economic returns	Aune and Ousman (2011)
Vigna radiata (L.) R. Wilczek	Biopriming with *Bacillus cereus*	Greenhouse	↑ RL (141), ↑ RDW (29),	↑ plant height, ↑ leaf area, ↑ leaf length ↑ leaf phosphate, ↑ chitinase, ↑ β-1,3 glucanase, ↑ PAAL, ↑ POX, ↑ IAA content	Chakraborty et al. (2011)
Chickpea *Cicer arietinum* L.	Biopriming with *Bacillus cereus*	Greenhouse	↑ RL (233), ↑ RDW(23)	↑ plant height, ↑ leaf area, ↑ leaf length, ↑ leaf phosphate, ↑ chitinase, ↑ β-1,3 glucanase, ↑ PAAL, ↑ POX, ↑ IAA content	Chakraborty et al. (2011)
Bromegrass	Seed priming with auxin (indole 3-butyric acid) ($100\,mg\,l^{-1}$)	Laboratory, aged seeds	↑ RL (3.6)	↑ chlorophyll, ↑ leaf area, ↑ biomass	Eisvand et al. (2010)

↑, increase; ↓, decrease; DSR, direct seeded rice; AWD, alternate wetting and drying; MWCNT, multi-walled carbon nanotubes; RL, root length; RD, root diameter; RV, root volume; RHA, root hair area; RHD, root hair density; RFW, root fresh weight; RDW, root dry weight; RSAD, root surface area density; LRN, lateral root number; WUE, water use efficiency; MET, mean emergence time; PAAL, phenylalanine ammonia lyase; TSP, total soluble proteins; POX, peroxidase; IAA, indole-3-acetic acid.

and acquisition (Rehman et al. 2018a). Bio-priming with PGPR (*Pseudomonas* sp.) and fungal isolates (*Clonostachys rosea* IK726d11 or *Trichoderma harzianum* T22) increased the soil microbial biomass, root development, and plant growth in carrot *Daucus carota* (L.) subsp. *sativus* (Hoffm.) (Bennett and Whipps 2008).

8.3.3 Seed Priming in Root Branching and Root Hair Growth

Lateral roots production continues throughout the plant life cycle, and root branching pattern determines the root architecture. Exogenous application of phytohormones and nutrient availability influence root branching. For instance, seed priming with IAA reduced the root length and increased the lateral root number (Eisvand et al. 2010). Auxin is needed to initiate lateral root development because the auxin-dependent *PIN-1* gene regulates lateral root initiation (Benková et al. 2003). Seed priming with GA_3 (20 mg l^{-1}) and ascorbic acid (20 mg l^{-1}) reduced the seedling root length and increased the number of seminal roots and root dry weight (Khan et al. 2011).

Root hairs increase root surface area, contribute to resource acquisition, and provide a site for plant-microbial interactions (Wasson et al. 2012). Early germination and deeper roots with dense root hairs result in faster plant growth and higher yields of crop plants (Figure 8.1). Dense root hairs enhance water uptake and nutrient acquisition. For instance, seed priming of wheat with multiwalled carbon nanotubes (MWCNTs) augment uptake of water and nutrients (e.g. P and K), which resulted in a 63% increase in crop yield over nonprimed control plants. Moreover, seed priming with MWCNT increased the cell elongation (80%) and dilated the size of the phloem and xylem vessels by 85% and 83% respectively, which enhanced the capacity of roots to transport more nutrients and water (Joshi et al. 2018; Figure 8.1). Likewise, seed priming with melatonin (300 μM) increased the lateral root formation, root diameter, and root viability in cucumber (Zhang et al. 2013). Seed priming with melatonin plays a key role in initiating the lateral root formation, with melatonin-primed seeds having 121 genes upregulated and 196 genes downregulated (Zhang et al. 2013). Melatonin may influence the appearance of root primordia in the pericycle cells in etiolated hypocotyls of *Lupinus albus* L., thus modulating the distribution pattern of lateral roots, their density, and length (Arnao and Hernandez-Ruiz 2007).

Seed priming with PGPRs or fungi helps in improving the lateral root growth and root hair formation (Table 8.1). For instance, seed priming with *Trichoderma* sp. in maize (Harman et al. 2004) and *Arabidopsis thaliana* (Contreras-Cornejo et al. 2009) modulated root system architecture and enhanced the plant growth because *Trichoderma* sp. synthesized IAA that increased later root formation and root hair growth. Seed priming with PGPR enhanced the lateral root growth because PGPR initiates the postembryonic root system architecture alterations by promoting root hair and lateral root development and by inhibiting the primary root length (Zamioudis et al. 2013). Seed biopriming with PGPRs increased IAA concentration in plant roots, thus improving the numbers of roots, root hairs, and lateral and adventitious roots, and expanding root surface area (Spaepen and Vanderleyden 2011).

Phosphorus is an essential plant nutrient that influences the lateral root morphology, root branching, root:shoot ratio, and root branching (Lopez-Bucio et al. 2002). Biopriming with P-solubilizing microbes (e.g. *Azotobacter, Bacillus, Enterobacter*, and *Pseudomonas*) released organic acid and other organic anions, and enzymes (phosphatases) that increased the solubility and availability of P (Zaidi et al. 2009; Rehman et al. 2018b) and hence improve the lateral root growth because low P supply changed the sensitivity of plants to auxin through the expression of auxin-responsive genes (*TIR1* and *ARF19*) related to the pericycle cell division and lateral root development (Pérez-Torres et al. 2008).

8.4 Seed Priming and Root Development Under Abiotic Stresses

Roots are key plant organs that sense and respond to external stimuli due to their direct interaction with available nutrients and water in soil. The root system perceives stress signals and makes necessary alterations at the genetic level to survive stress (Lynch 1995). Exposure of seed to abiotic stresses (drought, heavy metals, heat, salinity, and low temperature) resulted in lower seed viability, decreased germination, and poor seedling establishment owing to poor root development (Li et al. 2014; Figure 8.3). Drought stress reduced the lateral root initiation and growth (Zhan et al. 2015), limited root activity, and enhanced the oxidative damage by reactive oxygen species (ROSs), resulting in diminished membrane integrity with exacerbated lipid peroxidation (Sun et al. 2020). Low root zone temperature limited root activity, root length, diameter, and root dry weight, thus resulting in reduced nutrient acquisition (Imran et al. 2013). Salt stress inhibited the seminal root growth and biomass production, with increased Na and low K concentrations in barley roots (Shelden et al. 2013). Salinity-induced osmotic stress caused a drop in water potential

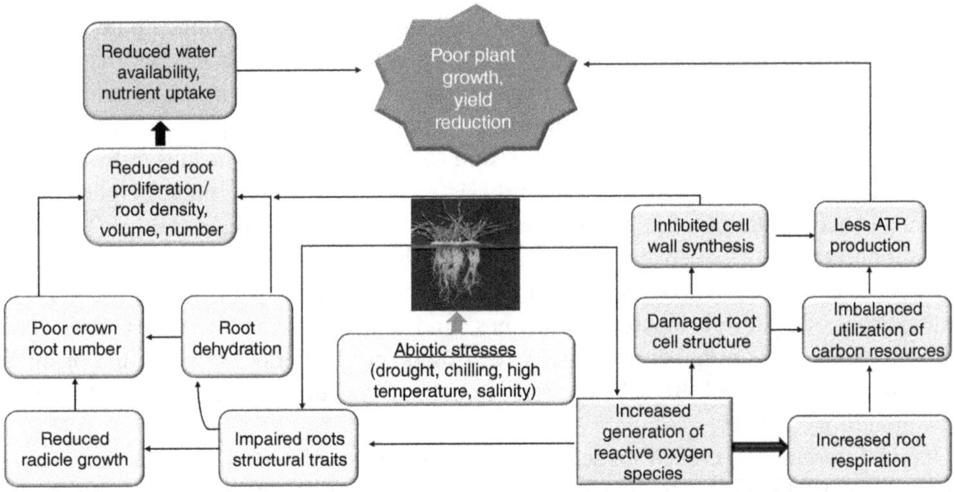

Figure 8.3 Abiotic stresses mediate growth and yield reduction in plants via root growth inhibition. Environmental stresses enhance generation of the reactive oxygen species (ROS), causing oxidative stress that damages the root cell membranes, increases root respiration, and decreases root elongation and branching. The reduced root proliferation is associated with poor water and nutrient uptake and translocation, resulting in diminished photosynthesis, plant growth, and yield.

in the rhizosphere, leading to dehydration, loss of cell turgor, and inhibition of cell expansion and cell division, which ultimately reduced the photosynthesis and crop yield (Munns and Tester 2008).

Pretreatment of seed with promising chemicals enhances the systemic tolerance against multiple abiotic stresses. Moreover, exposure of crop plants that emerge from primed seeds to abiotic stresses has a smaller impact on the root growth and physiology of plants compared to plants from nonprimed seed (Savvides et al. 2016; Figure 8.1). Primed seeds have higher anti-oxidant capacity; in addition, they show rapid improvement in α-amylase and β-amylase activities that leads to starch degradation and sugar accumulation, resulting in increased respiration rate, germination, and seedling establishment due to better root development (Li et al. 2013, 2014; Figure 8.1).

Seed priming with ascorbic acid under aluminium (Al) stress improved root growth of maize by upregulating the lignin-synthesis gene *COMT1* (Xue et al. 2008). Compared to nonprimed seed, the roots of primed seed showed higher expression of *COMT1* and higher tolerance against Al stress. Lignin forms the complex with Al ions and alleviates the harmful effects of Al in roots, and also acts as a sink for reactive oxygen species, leading to stress tolerance (Ezaki et al. 2005).

Priming seed with chemical agents activates the signalling pathway in roots, potentially resulting in systemic accumulation of the dormant tolerance signals (Figure 8.1). Seed priming with $ZnSO_4$ improved root length and root fresh and dry weight of chickpea under water-deficit conditions (Mahmood et al. 2019). In a recent study, Gavassi et al. (2019) found that seed priming with a low concentration of sodium nitroprusside improved drought tolerance in soybean through enhanced proline accumulation, resulting in improved length, diameter, density, surface area, and dry weight of roots.

The maintenance of root elongation and cell division under salt stress is associated with increased synthesis of sugars, organic acids, and amino acids, suggesting a potential role of these metabolic pathways in salinity tolerance and root elongation under salt stress (Shelden et al. 2016). Seed priming with spermine and spermidine improved root growth under salt stress in rice through enhanced expression of genes coding for the anti-oxidant system components (*ANS, CAT, SOD, APX,* and *GR*) in roots, osmolyte biosynthesis genes (*P5CS, PDH,* and *BADH1*), ABA biosynthesis enzyme (*NCED3*) and *LEA* gene, with enhanced accumulation of polyamines compared with the nonprimed control (Paul and Roychoudhury 2017). Seed pretreatment with polyamines (spermine and spermidine) influenced metabolic pathways such as nonenzymatic and enzymatic antioxidants, ion transporters, polyamine metabolic enzymes, and the expression of ABA biosynthetic enzyme genes, thus enhancing abiotic stress tolerance. The effect of spermine priming was more pronounced in salt-sensitive than salt-tolerant rice genotype (Paul and Roychoudhury 2017).

Rapid emergence is crucial for seedlings to avoid/minimize abiotic stresses and ensure strong crop establishment (Gardarin et al. 2016). Srivastava et al. (2010) reported that the application of thiourea in *B. juncea* enhanced root growth and imparted tolerance to salt stress by stabilizing the water homeostasis through coordinated regulation of plasma membrane intrinsic proteins (PIPs). Sowing NaCl-primed seeds improved the osmotic adjustment in plants and increased the concentration of sugars and organic acids in roots, enhancing root growth in the saline environment (Cayuela et al. 1996). Another study reported that priming with calcium chloride (CaCl2) and ascorbate in rice increased the number of roots and their fresh and dry weight because seed priming changed the nitrogen and calcium homeostasis and activity of α-amylase (Farooq et al. 2006a). Moreover, phytohormonal and vitamin seed priming enhanced the meristem activity in root tips (Shakirova et al. 2003), with the incorporation of growth regulators in priming increasing cell divisions in the apical meristem of the seedling roots. Similarly, treating seed with different phytohormones increased the IAA and cytokinin concentrations in the plant tissues, which enhanced the cell divisions resulting in a greater number of roots (Sakhabutdinova et al. 2003) (Figure 8.1).

Nutrient acquisition and particularly root growth are adversely affected at low root-zone temperature. Nutripriming can help in increasing the nutrient concentration in roots and improving root growth at low-temperature stress. Seed priming of maize with Fe and Zn + Mn increased the seed concentration of respective nutrients many-folds, resulting in improved root length, lateral root growth, and root biomass production compared with nonprimed seeds (Imran et al. 2013). Limited micronutrient availability at low root-zone temperature may exacerbate oxidative damage in root cells; however, micronutrient supplementation through seed priming contributes to increased nutrient acquisition by stimulation of root growth (Imran et al. 2013; Rehman et al. 2015).

8.5 Seed Priming and Sustainable Crop Production

Seed priming can help in sustainable crop production under optimal and suboptimal conditions because it improves nutrient-use efficiency through reduced metabolic costs and water used, increased water productivity, and high yield and quality of harvested produce (Carrillo-Reche et al. 2018; Table 8.1). The increased water- and nutrient-use efficiencies are due to the priming-induced improvement in root growth. Roots are key plant organs for nutrient acquisition and water uptake, and their spread and distribution substantially influence plant growth and yield (Doussan et al. 2006). Improved root growth contributes to efficient resource acquisition, thus improving stomatal conductance, photosynthesis, plant growth, biomass production, and grain yield (Rehman et al. 2018a, 2019).

Improved crop performance due to seed priming can have an important knock-on effect for the small-land-holding farmers (Table 8.1). Seed priming and improved yield are often accompanied by an increased yield of straw that can be used in mixed crop-livestock systems. Moreover, increased yield with seed priming in different crops under optimal and suboptimal growth conditions is associated with (i) elevated speed and frequency of seed emergence, (ii) increased root growth that facilitates nutrient and water uptake, (iii) increased leaf area and number of fertile tillers, (iv) enhanced photosynthetic activity, and (v) early flowering (Farooq et al. 2019; Rehman et al. 2018a,b) (Figure 8.1).

Delivery of micronutrients via seed priming is a cost-effective and eco-friendly approach that contributes to enhanced economic benefits and sustainable crop yields. Seed priming with $ZnSO_4$ improved chickpea germination and root growth that resulted in increased water uptake, leading to increased photosynthetic pigment concentration and higher grain yield (Mahmood et al. 2019). In wheat, the external Zn supply via seed priming increased root elongation, volume, surface area, and dry weight, thus improving nutrient and water uptake, photosynthetic activity, chlorophyll content, and biomass production under suboptimal growth conditions (Rehman et al. 2019). Wheat seed priming with Zn in combination with PGPR improved root growth via enhanced exudation of organic acid anions (e.g. citrate, malate, etc.) that increased nutrient solubility and uptake, improved the photosynthetic traits, enhanced grain Zn concentration in all seed parts (embryo, aleurone, and endosperm), lowered grain phytate concentration, and increased grain yield and net profitability (Rehman et al. 2018a,b). The seeds harvested from the Zn+bioprimed treatments had higher Zn concentration that, upon sowing, helped improve root growth, nutrient and water uptake, and grain yield of wheat under suboptimal growth conditions (drought, waterlogging, salinity, or heat) (Faran et al. 2019).

In rice, seed priming with micronutrients increased the rooting depth, root biomass, and soil exploration that enhanced nutrient uptake and translocation to seed (Prom-u-thai et al. 2012), and improved plant water relations, growth, grain yield, and economic benefits in conventional as well as conservation rice-production systems (Rehman et al. 2016). In maize, nutripriming with B, Zn, and Mn improved growth, water relations, grain yield, and grain protein content under semi-arid conditions (Rasool et al. 2019). Similarly, maize seed priming with B, Zn + Mn, Fe, or P improved root growth, biomass production, and accumulation of respective nutrients in roots and shoots, and increased marketable grain yield under low temperature in the root zone (Imran et al. 2013, 2015). As expected, the primed seeds had several-fold higher nutrient concentrations than the nonprimed seeds (Imran et al. 2015).

Seed priming with melatonin improved the cucumber root growth and water uptake under water-deficit conditions. Similarly, Akbari et al. (2019) found improvement in seed yield, oil yield, and harvest index of safflower (*Carthamus tinctorius* L.) under drought stress. The external supply of melatonin improved the root system architecture, resource acquisition, and photosynthesis in a number of plant species (Zhang et al. 2015).

Kamran et al. (2018) demonstrated that seed priming of maize with paclobutrazol regulated root physiology and morphology in the semi-arid environment by increasing root activity, root elongation, diameter, and dry weight density in most of the soil profile (0–70 cm), which contributed to enhanced nutrient and water uptake. They further found that grain yield of maize had a strong positive correlation with root growth.

Seed priming of wheat with ascorbate improved the stand establishment and grain yield, increasing economic benefits on salt-affected soils (Jafar et al. 2012) due to better root growth that was associated with greater water and nutrient uptake. Likewise, osmopriming with $CaCl_2$ improved the allometric traits, water-use efficiency, grain yield, and economic value of wheat not just under optimal growth conditions, but also under early as well as terminal drought stress (Farooq et al. 2017).

In a recent meta-analysis, Carrillo-Reche et al. (2018) reported that primed seeds had a 22% faster emergence rate, 11% higher final emergence, and a 21% increase in grain yield compared with nonprimed seeds. Moreover, on-farm seed priming produced higher yield (22–28%) under suboptimal growth conditions (nutrient deficiency, salinity, and water deficit). Agriculture inputs are often not accessible or affordable to the smallholder and resource-poor farmers (Tonitto and Ricker-Gilbert 2016). Therefore, the cost-effective and sustainable approaches to ameliorate the unfavourable conditions and minimize inputs are highly relevant to ensure food security in semi-arid environments (Branca et al. 2013). One such approach is seed priming that can help the resource-poor farmers working in low-input agriculture systems where crop yield is frequently reduced by various stresses. Seed priming could be adopted by the resource-poor farmers to improve food security on marginal agricultural lands.

8.6 Conclusions and Perspectives

Roots are the first structure emerging from seeds, taking up nutrients and water. Poor emergence results in scanty root growth that affects the growth and yield of crop plants. Seed priming, through the induction of a series of biochemical, physiological, molecular, and subcellular changes, is beneficial in improving seed germination and root growth that leads to better resource acquisition and improved plant growth and grain yield. Seed priming also helps in root development under environmental stresses by triggering the anti-oxidant system to strengthen plant defence mechanisms and maintain membrane integrity, thus protecting the root cells from oxidative damage. Seed priming with micronutrients improves root development and plant growth under nutrient-deficient conditions. Seed priming is a cost-effective and eco-friendly approach that contributes to agricultural sustainability and food security through improved yield and socio-economic conditions under changing climatic conditions.

During the last few years, seed priming has emerged as a highly efficient technology that improves stress tolerance in plants along with enhanced yield benefits. In the modern era, with the adoption of precision agricultural technologies to increase food security, every seed should be healthy and germinate quickly to produce vigorous seedling with high tolerance against abiotic stresses and ensure high crop yield. Seed priming is important for the complex trait of seed quality that influences seed germination and seedling vigour and is positively associated with the plant responses to abiotic stresses. The knowledge of priming-induced modulations of metabolic processes during germination and their effect on root growth is limited. In future studies, the molecular and physiological mechanisms that influence the embryonic and post-embryonic root development need to be characterized. Moreover, physical methods of seed priming still require extensive research for determining their impact on seed germination and root development in the stress environments.

References

Achard, P., Gusti, A., Cheminant, S. et al. (2009). Gibberellin signaling controls cell proliferation rate in *Arabidopsis*. *Current Biology* 19: 1188–1193.

Aichinger, E., Kornet, N., Friedrich, T., and Laux, T. (2012). Plant stem cell niches. *Annual Review of Plant Biology* 63: 615–636.

Akbari, G., Sanavy, S.A., and Yousefzadeh, S. (2007). Effect of auxin and salt stress (NaCl) on seed germination of wheat cultivars (*Triticum aestivum* L.). *Pakistan Journal of Biological Sciences* 10: 2557–2561.

Akbari, G.A., Heshmati, S., Soltani, E. et al. (2019). Influence of seed priming on seed yield, oil content and fatty acid composition of safflower (*Carthamus tinctorius* L.) grown under water deficit. *International Journal of Plant Production* 14: 245–258.

Arnao, M. and Hernandez-Ruiz, J. (2007). Melatonin promotes adventitious- and lateral root regeneration in etiolated hypocotyls of *Lupinus albus* L. *Journal of Pineal Research* 42: 147–152.

Aune, J.B. and Ousman, A. (2011). Effect of seed priming and micro-dosing of fertilizer on sorghum and pearl millet in western Sudan. *Experimental Agriculture* 47: 419–430.

Benková, E., Michniewicz, M., Sauer, M. et al. (2003). Local, efflux-dependent auxin gradients as a common module for plant organ formation. *Cell* 115: 591–602.

Bennett, A.J. and Whipps, J.M. (2008). Dual application of beneficial microorganisms to seed during durum priming. *Applied Soil Ecology* 38: 83–89.

Blunk, S., De Heer, M.I., Malik, A.H. et al. (2019). Seed priming enhances early growth and improves area of soil exploration by roots. *Environmental and Experimental Botany* 158: 1–11.

Bradford, K.J. (1986). Manipulation of seed water relations via osmotic priming to improve germination under stress conditions. *Horticultural Science* 21: 1105–1112.

Branca, G., Lipper, L., McCarthy, N. et al. (2013). Food security, climate change, and sustainable land management – a review. *Agronomy for Sustainable Development* 33: 635–650.

Cakmak, I., Yilmaz, A., Ekiz, H. et al. (1996). Zinc deficiency as a critical nutritional problem in wheat production in Central Anatolia. *Plant and Soil* 180: 165–172.

Cao, Q., Li, G., Cui, Z. et al. (2019). Seed priming with melatonin improves the seed germination of waxy maize under chilling stress via promoting the antioxidant system and starch metabolism. *Scientific Report* 9: 1–12.

Carrillo-Reche, J., Vallejo-Marín, M., and Quilliam, R.S. (2018). Quantifying the potential of 'on-farm' seed priming to increase crop performance in developing countries. A meta-analysis. *Agronomy for Sustainable Development* 38: 64. https://doi.org/10.1007/s13593-018-0536-0.

Cayuela, E., Perez-Alfocea, F., Caro, M. et al. (1996). Priming of seeds with NaCl induces physiological changes on tomato plants grown under salt stress. *Physiologia Plantarum* 96: 231–236.

Chakraborty, A.P., Dey, P., Chakraborty, B. et al. (2011). Plant growth promotion and amelioration of salinity stress in crop plants by a salt-tolerant bacterium. *Recent Research in Science and Technology* 3: 61–70.

Chen, Q., Qi, W.B., Reiter, R.J. et al. (2009). Exogenously applied melatonin stimulates root growth and raises endogenous indoleacetic acid in roots of etiolated seedlings of *Brassica juncea*. *Journal of Plant Physiology* 166: 324–328.

Clowes, F.A.L. (1961). Apical Meristems. Oxford, UK: Blackwell Science.

Contreras-Cornejo, H.A., Macías-Rodríguez, L., Cortés-Penagos, C. et al. (2009). *Trichoderma virens*, a plant beneficial fungus, enhances biomass production and promotes lateral root growth through an auxin-dependent mechanism in *Arabidopsis*. *Plant Physiology* 149: 1579–1592.

De Smet, I., Lau, S., Mayer, U. et al. (2010). Embryogenesis: the humble beginnings of plant life. *The Plant Journal* 61: 959–970.

De Smet, S., Cuypers, A., Vangronsveld, J. et al. (2015). Gene networks involved in hormonal control of root development in *Arabidopsis thaliana*: a framework for studying its disturbance by metal stress. *International Journal of Molecular Sciences* 16: 19195–19224.

Dotto, L. and Silva, V.N. (2017). Beet seed priming with growth regulators. *Semina: Ciências Agrárias* 38: 1785.

Doussan, C., Pierret, A., Garrigues, E. et al. (2006). Water uptake by plant roots: II–modelling of water transfer in the soil root-system with explicit account of flow within the root system–comparison with experiments. *Plant and Soil* 283: 99–117.

Draganic, I. and Lekic, S. (2012). Seed priming with antioxidants improves sunflower seed germination and seedling growth under unfavorable germination conditions. *Turkish Journal of Agriculture and Forestry* 36: 421–428.

Eisvand, H.R., Alizadeh, M.A., and Fekri, A. (2010). How hormonal priming of aged and nonaged seeds of bromegrass affects seedling physiological characters. *Journal of New Seeds* 11: 52–64.

Ezaki, B., Sasaki, K., and Matsumot, H. (2005). Functions of two genes in aluminium (Al) stress resistance: repression of oxidative damage by the AtBCB gene and promotion of efflux of Al ions by the NtGDI1gene. *Journal of Experimental Botany* 56: 2661–2671.

Faran, M., Farooq, M., Rehman, A. et al. (2019). High intrinsic seed Zn concentration improves abiotic stress tolerance in wheat. *Plant and Soil* 437: 195–213.

Farooq, M., Basra, S.M.A., Tabassum, R. et al. (2006a). Evaluation of seed vigour enhancement techniques on physiological and biochemical techniques on physiological basis in coarse rice (*Oryza sativa* L.). *Seed Science and Technology* 34: 741–750.

Farooq, M., Basra, S.M.A., and Hafeez, K. (2006b). Seed invigoration by osmohardening in fine and course rice. *Seed Science and Technology* 34: 181–186.

Farooq, M., Basra, S.M.A., and Khan, M.B. (2007). Seed priming improves growth of nursery seedlings and yield of transplanted rice. *Archives of Agronomy and Soil Science* 53: 315–326.

Farooq, M., Basra, S.M.A., Rehman, H. et al. (2008). Seed priming enhances the performance of late sown wheat (*Triticum aestivum* L.) by improving chilling tolerance. *Journal of Agronomy and Crop Science* 194: 55–60.

Farooq, M., Wahid, A., Kobayashi, N. et al. (2009). Plant drought stress: effects, mechanisms and management. *Agronomy for Sustainable Development* 29: 185–212.

Farooq, M., Rehman, A., and Aziz, T. (2011). Boron nutripriming improves the germination and early seedling growth of rice (*Oryza sativa* L.). *Journal of Plant Nutrition* 34: 1507–1515.

Farooq, M., Wahid, A., and Siddique, K.H.M. (2012). Micronutrient application through seed treatments – a review. *Journal of Soil Science and Plant Nutrition* 12: 125–142.

Farooq, S., Hussain, M., Jabran, K. et al. (2017). Osmopriming with $CaCl_2$ improves wheat (*Triticum aestivum* L.) production under water-limited environments. *Environmental Science and Pollution Research* 24: 13638–13649.

Farooq, M., Ullah, A., Rehman, A. et al. (2018). Application of zinc improves the productivity and biofortification of fine grain aromatic rice grown in dry seeded and puddled transplanted production systems. *Field Crops Research* 216: 53–62.

Farooq, M., Usman, M., Nadeem, F. et al. (2019). Seed priming in field crops: potential benefits, adoption and challenges. *Crop and Pasture Science* 70: 731–771.

Gardarin, A., Coste, F., Wagner, M.H. et al. (2016). How do seed and seedling traits influence germination and emergence parameters in crop species? A comparative analysis. *Seed Science Research* 26: 317–331.

Gavassi, M.A., Gaion, L.A., Monteiro, C.C. et al. (2019). Seed priming with sodium nitroprusside attenuates the effects of water deficit on soybean seedlings. *Communication Science* 10: 176–184.

Guan, Y.J., Hu, J., Wang, X.J. et al. (2009). Seed priming with chitosan improves maize germination and seedling growth in relation to physiological changes under low temperature stress. *Journal of Zheijang University Science B* 10: 427–433.

Harman, G.E., Petzoldt, R., Comis, A. et al. (2004). Interactions between *Trichoderma harzianum* strain T22 and maize inbred line Mo17 and effects of these interactions on diseases caused by *Pythium ultimum* and *Colletotrichum graminicola*. *Phytopathology* 94: 147–153.

Harris, D., Rashid, A., Miraj, G. et al. (2007). 'On-farm'seed priming with zinc sulphate solution—a cost-effective way to increase the maize yields of resource-poor farmers. *Field Crops Research* 102 (2): 119–127.

Imran, M., Mahmood, A., Römheld, V. et al. (2013). Nutrient seed priming improves seedling development of maize exposed to low root zone temperatures during early growth. *European Journal of Agronomy* 49: 141–148.

Imran, M., Maria, K., Romheld, V. et al. (2015). Impact of nutrient seed priming on germination, seedling development, nutritional status and grain yield of maize. *Journal of Plant Nutrition* 38: 1803–1821.

Iqbal, S., Farooq, M., Nawaz, A. et al. (2012). Optimizing boron seed priming treatments for improving the germination and early seedling growth of wheat. *Journal of Agriculture and Social Sciences* 8: 57–61.

Jafar, M.Z., Farooq, M., Cheema, M.A. et al. (2012). Improving the performance of wheat by seed priming under saline conditions. *Journal of Agronomy and Crop Science* 198: 38–45.

Joshi, A., Kaur, S., Dharamvir, K. et al. (2018). Multi-walled carbon nanotubes applied through seed-priming influence early germination, root hair, growth and yield of bread wheat (*Triticum aestivum* L.). *Journal of the Science of Food and Agriculture* 98: 3148–3160.

Kamran, M., Wennan, S., Ahmad, I. et al. (2018). Application of paclobutrazol affect maize grain yield by regulating root morphological and physiological characteristics under a semi-arid region. *Scientific Report* 8: 1–5.

Kathiresan, K., Kalyani, V., and Gnanarethinam, J.L. (1984). Effect of seed treatments on field emergence, early growth and some physiological processes of sunflower (*Helianthus annuus* L.). *Field Crops Research* 9: 215–217.

Khan, M.B., Gurchani, M.A., and Hussain, M. (2011). Wheat seed enhancement by vitamin and hormonal priming. *Pakistan Journal of Botany* 43: 1495–1499.

Li, Z.G., Yang, S.Z., Long, W.B. et al. (2013). Hydrogen sulphide may be a novel downstream signal molecule in nitric oxide-induced heat tolerance of maize (*Zea mays* L.) seedlings. *Plant, Cell and Environment* 36: 1564–1572.

Li, Z., Peng, Y., Zhang, X.Q. et al. (2014). Exogenous spermidine improves seed germination of white clover under water stress via involvement in starch metabolism, antioxidant defenses and relevant gene expression. *Molecules* 19: 18003–18024.

Lopez-Bucio, J., Hernandez-Abreu, E., Sanchez-Calderon, L. et al. (2002). Phosphate availability alters architecture and causes changes in hormone sensitivity in the *Arabidopsis* root system. *Plant Physiology* 129: 244–256.

Lynch, J. (1995). Root architecture and plant productivity. *Plant Physiology* 109: 7–13.

Mahmood, A., Kanwal, H., Kausar, A. et al. (2019). Seed priming with zinc modulate growth, pigments and yield of chickpea (*Cicer arietinum* L.) under water deficit conditions. *Applied Ecology and Environmental Research* 17: 147–160.

Moeinzadeh, A., Sharif-Zadeh, F., Ahmadzadeh, M. et al. (2010). Biopriming of sunflower ('*Helianthus annuus*' L.) seed with *Pseudomonas fluorescens* for improvement of seed invigoration and seedling growth. *Australian Journal of Crop Science* 4: 564.

Munns, R. and Tester, M. (2008). Mechanisms of salinity tolerance. *Annual Review of Plant Biology* 59: 651–681.

Nawaz, A., Farooq, M., Ahmad, R. et al. (2016). Seed priming improves stand establishment and productivity of no till wheat grown after direct-seeded aerobic and transplanted flooded rice. *European Journal of Agronomy* 76: 130–137.

Nazari, S., Aboutalebian, M.A., and Golzardi, F. (2017). Seed priming improves seedling emergence time, root characteristics and yield of canola in the conditions of late sowing. *Agronomy Research* 15: 501–514.

Nieuwland, J., Stamm, P., Wen, B. et al. (2016). Re-induction of the cell cycle in the Arabidopsis post-embryonic root meristem is ABA-insensitive, GA-dependent and repressed by KRP6. *Scientific Reports* 6: 1–10.

Olatunji, D., Geelen, D., and Verstraeten, I. (2017). Control of endogenous auxin levels in plant root development. *International Journal of Molecular Sciences* 18: 2587.

Oracz, K. and Karpinski, S. (2016). Phytohormones signaling pathways and ROS involvement in seed germination. *Frontiers in Plant Science* 7: 864.

Ozturk, L., Yazici, M.A., Yucel, C. et al. (2006). Concentration and localization of zinc during seed development and germination in wheat. *Physiologia Plantarum* 128: 144–152.

Paul, S. and Roychoudhury, A. (2017). Seed priming with spermine and spermidine regulates the expression of diverse groups of abiotic stress–responsive genes during salinity stress in the seedlings of indica rice varieties. *Plant Gene* 11: 124–132.

Péret, B., De Rybel, B., Casimiro, I. et al. (2009). *Arabidopsis* lateral root development: an emerging story. *Trends in Plant Science* 14: 399–408.

Pérez-Torres, C.A., López-Bucio, J., Cruz-Ramírez, A. et al. (2008). Phosphate availability alters lateral root development in Arabidopsis by modulating auxin sensitivity via a mechanism involving the TIR1 auxin receptor. *Plant Cell* 20: 3258–3272.

Prom-u-thai, C., Rerkasem, B., Yazici, A. et al. (2012). Zinc priming promotes seed germination and seedling vigor of rice. *Journal of Plant Nutrition and Soil Science* 175: 482–488.

Rasool, T., Ahmad, R., and Farooq, M. (2019). Seed priming with micronutrients for improving the quality and yield of hybrid maize. *Gesunde Pflanzen* 71: 37–44.

Rehman, A., Farooq, M., Ahmad, R. et al. (2015). Seed priming with zinc improves the germination and early seedling growth of wheat. *Seed Science and Technology* 43: 262–268.

Rehman, A., Farooq, M., Nawaz, A. et al. (2016). Improving the performance of short-duration basmati rice in water-saving production systems by boron nutrition. *Annals of Applied Biology* 168: 19–28.

Rehman, A., Farooq, M., Naveed, M. et al. (2018a). *Pseudomonas*-aided zinc application improves the productivity and biofortification of bread wheat. *Crop and Pasture Science* 69: 659–672.

Rehman, A., Farooq, M., Naveed, M. et al. (2018b). Seed priming of Zn with endophytic bacteria improves the productivity and grain biofortification of bread wheat. *European Journal of Agronomy* 94: 98–107.

Rehman, A., Farooq, M., Asif, M. et al. (2019). Supra-optimal growth temperature exacerbates adverse effects of low Zn supply in wheat. *Journal of Plant Nutrition and Soil Science* 182: 656–666.

Rogers, E.D. and Benfey, P.N. (2015). Regulation of plant root system architecture: implications for crop advancement. *Current Opinion in Biotechnology* 32: 93–98.

Sakhabutdinova, A.R., Fatkhutdinova, D.R., Beazrukova, M.V. et al. (2003). Salicylic acid prevents the damaging action of stress factor of wheat plants. *Bulgarian Journal of Plant Physiology* 21: 314–319.

Savvides, A., Ali, S., Tester, M. et al. (2016). Chemical priming of plants against multiple abiotic stresses: mission possible? *Trends in Plant Science* 21: 329–340.

Seago, J.L. Jr. and Fernando, D.D. (2013). Anatomical aspects of angiosperm root evolution. *Annals of Botany* 112: 223–238.

Shakirova, F.M., Sakhabutdinova, A.R., Bezrokuva, M.V. et al. (2003). Changes in the hormonal status of wheat seedlings induced by salicylic acid and salinity. *Plant Science* 164: 317–322.

Shelden, M.C., Roessner, U., Sharp, R.E. et al. (2013). Genetic variation in the root growth response of barley genotypes to salinity stress. *Functional Plant Biology* 40: 516–530.

Shelden, M.C., Dias, D.A., Jayasinghe, N.S. et al. (2016). Root spatial metabolite profiling of two genotypes of barley (*Hordeum vulgare* L.) reveals differences in response to short-term salt stress. *Journal of Experimental Botany* 67: 3731–3745.

Siadat, S.A., Moosavi, S.A., Zadeh, M.S. et al. (2011). Effects of halo and phytohormone seed priming on germination and seedling growth of maize under different duration of accelerated ageing treatment. *African Journal of Agricultural Research* 6: 6453–6462.

Spaepen, S. and Vanderleyden, J. (2011). Auxin and plant-microbe interactions. *Cold Spring Harbor Perspectives in Biology* 3: a001438.

Srivastava, A.K., Suprasanna, P., Srivastava, S. et al. (2010). Thiourea mediated regulation in the expression profile of aquaporins and its impact on water homeostasis under salinity stress in *Brassica juncea* roots. *Plant Science* 178: 517–522.

Sun, M., Peng, F., Xiao, Y. et al. (2020). Exogenous phosphatidylcholine treatment alleviates drought stress and maintains the integrity of root cell membranes in peach. *Scientia Horticulturae* 259: 108821.

Tonitto, C. and Ricker-Gilbert, J.E. (2016). Nutrient management in African sorghum cropping systems: applying meta-analysis to assess yield and profitability. *Agronomy for Sustainable Development* 36: 1–19.

Ullah, A., Farooq, M., Nadeem, F. et al. (2020). Zinc seed treatments improve productivity, quality and grain biofortification of desi and kabuli chickpea (*Cicer arietinum*). *Crop and Pasture Science* 71 (7): 668–678.

Wasson, A.P., Richards, R.A., Chatrath, R. et al. (2012). Traits and selection strategies to improve root systems and water uptake in water-limited wheat crops. *Journal of Experimental Botany* 63: 3485–3498.

White, P.J. and Brown, P.H. (2010). Plant nutrition for sustainable development and global health. *Annals of Botany* 105: 1073–1080.

Wolny, E., Betekhtin, A., Rojek, M. et al. (2018). Germination and the early stages of seedling development in *Brachypodium distachyon*. *International Journal of Molecular Sciences* 19 (10): 2916.

Xue, Y., Tao, L., and Yang, Z. (2008). Aluminum-induced cell wall peroxidase activity and lignin synthesis are differentially regulated by jasmonate and nitric oxide. *Journal of Agricultural and Food Chemistry* 56: 9676–9684.

Zaidi, A., Khan, M.S., Ahemad, M. et al. (2009). Plant growth promotion by phosphate solubilizing bacteria. *Acta Microbiologica et Immunologica Hungarica* 56: 263–284.

Zamioudis, C., Mastranesti, P., Dhonukshe, P. et al. (2013). Unraveling root developmental programs initiated by beneficial *Pseudomonas* spp. bacteria. *Plant Physiology* 162: 304–318.

Zhan, A., Schneider, H., and Lynch, J.P. (2015). Reduced lateral root branching density improves drought tolerance in maize. *Plant Physiology* 168: 1603–1615.

Zhang, N., Zhao, B., Zhang, H.J. et al. (2013). Melatonin promotes water-stress tolerance, lateral root formation, and seed germination in cucumber (*Cucumis sativus* L.). *Journal of Pineal Research* 54: 15–23.

Zhang, N., Sun, Q., Zhang, H. et al. (2015). Roles of melatonin in abiotic stress resistance in plants. *Journal of Experimental Botany* 66: 647–656.

9

Water Uptake in Drying Soil

Balancing Supply and Demand

Ian C. Dodd[1] *and Guzel Kudoyarova*[2]

[1] *Lancaster Environment Centre, Lancaster University, Lancaster, UK*
[2] *Department of Plant Physiology, Ufa Institute of Biology, Russian Academy of Sciences, Ufa, Russian Federation*

9.1 Putting Soil Drying in Context

Except when crop plants are grown in hydroponics, they are potentially exposed to drying soil on a daily basis. Evaporation from the soil and plant transpiration cause localized drying in the soil that surrounds the root system (the rhizosphere). If soil water availability and hydraulic conductivity are not limiting, any soil water depletion in the rhizosphere may be restricted to periods of increased evaporative demand due to high rates of plant transpiration. Nevertheless, plants can be highly sensitive to small changes in soil water content, even when only part of their root system is exposed to drying soil. Such localized soil drying can trigger adaptive responses such as increased root growth and enhanced root hydraulic conductance, thereby minimizing metabolic disturbance to the shoot.

However, prolonged drought periods afflict crop production in many parts of the world, and a lack of water inflicts the bulk of crop yield losses in the field (Boyer 1982; Tester and Langridge 2010). The duration, intensity, and rapidity of soil drying, along with plant developmental stage when this drying occurs, determine the magnitude of crop yield losses. These stresses are expected to become increasingly common due to the impacts of climate change, expansion of croplands into more marginal (lower rainfall) environments and greater water withdrawals by an expanding human population that may limit water supplies to agriculture. Society expects agricultural scientists to develop solutions to these issues, whether they be delivered by improved crop genotypes that can yield well even in drying soil and/or management solutions (e.g. irrigation techniques) that enhance crop water-use efficiency. Irrespective of the solution(s) developed, understanding how drying soil regulates root water uptake is necessary to appreciate future opportunities to enhance the sustainability and yields of both rain-fed and irrigated agriculture.

The Root Systems in Sustainable Agricultural Intensification, First Edition.
Edited by Zed Rengel and Ivica Djalovic.

9.2 Balancing Water Uptake with Water Demand

Water moves from the soil into the plant and ultimately into the atmosphere according to gradients in water potential (Ψ). The ease of water movement is determined by hydraulic resistances to flow. Measuring soil-plant-atmosphere water potentials is necessary to understand the direction of water flow. Soil water potential (Ψ_{soil}) approaches zero (its maximum theoretical value) when soil is well-watered, but drainage (due to gravity) and solute accumulation (in the soil solution) lowers Ψ_{soil} (to −10 to −30 kPa) depending on soil texture. Root water potential (Ψ_{root}) must be lower (more negative) than the Ψ_{soil} for water to flow from the soil to the root. This is facilitated by the root actively accumulating solutes, which lowers Ψ_{root}. Likewise, water potential of the shoot must be lower (more negative) than Ψ_{root} for water to flow from root to the shoot. And again, the water potential of the atmosphere must be lower (more negative) than the water potential of the shoot for the shoot to transpire (= water flow from shoot to atmosphere). Unless the plant regulates this transpirational water loss by partially closing the stomata, these losses may not be fully replaced, decreasing plant water status.

Because roots can actually lose water into the soil (if $\Psi_{root} > \Psi_{soil}$), soil-root water exchange is limited to regions close to the root tips. In older roots, impermeable barriers (composed of suberin, a hydrophobic compound) are deposited on the root surface to prevent water (and solute) exchange. Although the *Arabidopsis horst* mutant with reduced amounts of suberin and delayed development of Casparian bands (suberized cell wall material in the endodermis) had increased root hydraulic conductance compared to wild-type (WT) plants, the *esb* mutants with enhanced suberin deposition had similar values to the WT (Ranathunge and Schreiber 2011). Enhanced suberization occurred in the distal parts of hydroponically grown barley (*Hordeum vulgare*) seminal roots exposed to an osmotic stress of −0.4 MPa (Kreszies et al. 2019a), although the physiological impacts of these constitutive or adaptive changes in root suberization on shoot water status have attracted little attention.

Even though $\Psi_{root} < \Psi_{soil}$ is typical in a transpiring plant, specific circumstances can reverse this gradient. In very dry soil, Ψ_{soil} may be lower than Ψ_{root}, in which case water would flow from the roots to the soil. When plants are growing in deep, water-retentive soils, often there are substantial vertical soil water gradients, with shallow roots in dry soil and deeper roots in moist soil. Overnight, when the stomata close, the transpirational flow of water to the atmosphere is minimal, and shallow roots may lose water into dry surface soil layers. Meanwhile, roots in the deeper, moister soil layers are at a higher water potential than those in the surface layers, so water can move from the deeper to the shallow roots and into the soil. This phenomenon is known as hydraulic lift. In the morning, when the stomata re-open and the plant starts to intensively transpire again, the water potential of the shallow roots declines below that of the soil that was moistened overnight, so the normal water potential gradients that facilitate root water uptake are re-established. Thus, keeping the roots growing and exploring the soil is critical for sustained plant water uptake.

9.3 Regulation of Root Growth in Drying Soil

Root cell water potential depends on cellular solute content (Ψ_s) and the (turgor) pressure of the cellular contents exerted against the cell wall (Ψ_p). Water in a root cell exerts a pressure, known as turgor, on neighbouring cells and its own cell wall. Root cells enlarge (increase in volume) if the turgor pressure is high enough to overcome resistance of the cell wall. If this enlargement is accompanied by the import of solutes into the cell, the osmotic potential of the cell is lowered, which drives more water into the cell. Thus, root elongation depends upon cellular turgor overcoming resistance of the cell wall, and on the continued uptake of water and solutes. Root growth can be altered by changing cellular turgor and/or cell wall extensibility, both of which change in drying soil.

Generally, root growth is much less sensitive to soil drying than shoot growth, and plant roots can grow at substrate or tissue water potentials that prevent growth of aerial organs such as leaves and stems (Westgate and Boyer 1985). In maize (*Zea mays*), each 0.2 MPa decrease in leaf water potential (Ψ_{leaf}) roughly halved leaf elongation rate, with growth ceasing at Ψ_{leaf} of −1.0 MPa, whereas nodal roots at Ψ_{root} of −1.0 MPa maintained the same elongation rate as at Ψ_{root} of −0.50 MPa, and growth was still measurable at Ψ_{root} of −1.4 MPa (Westgate and Boyer 1985). Maize primary roots continued to grow when Ψ_{soil} was less than −0.8 MPa, whereas shoots had already ceased growing (Sharp et al. 1988). Similarly, soybean (*Glycine max*) primary roots maintained the same growth rates across a range of Ψ_{soil} values (−0.01 to −0.37 MPa) that decreased hypocotyl growth by 75% (Creelman et al. 1990). Such root growth maintenance depends on both rapid osmotic adjustment to allow partial turgor recovery and re-establishing water potential gradients to allow root water uptake, but also on increased cell wall loosening of growing root cells (Hsiao and Xu 2000).

When maize primary roots were transplanted from moistened vermiculite (Ψ_{soil} of −0.03 MPa) to dry vermiculite (Ψ_{soil} of −1.6 MPa), root cell turgor decreased by about 0.4 MPa throughout the root elongation zone (Spollen and Sharp 1991). At the root tip, roots in both moistened and dry vermiculite maintained the same local growth rate despite the turgor difference (Figure 9.1), indicating increased apical cell wall extensibility in roots transplanted to Ψ_{soil} of −1.6 MPa. This increased extensibility was associated with increased activity of xyloglucan endo-transglycosylase (Wu et al. 1994) and increased activity of, and susceptibility to, expansin (Wu et al. 1996). Local growth rate peaked 1 mm closer to the root tip in roots transplanted to Ψ_{soil} of −1.6 MPa (Figure 9.1), but at half the value (Spollen and Sharp 1991). Thereafter, local growth rate declined, presumably because cell walls became less extensible, with roots transplanted to Ψ_{soil} of −1.6 MPa ceasing elongation 3 mm closer to the tip than roots maintained at Ψ_{soil} of −0.03 MPa. A similar shortening of the elongation zone occurred in wheat (*Triticum aestivum*) seminal roots growing at Ψ_{soil} of −1.10 MPa (Figure 9.1), even if maximum local growth rates were halved compared to roots growing in moist vermiculite (Akmal and Hirasawa 2004). Hence, cellular turgor determined absolute root growth rate, whereas its local effects on growth rate were modulated by cell wall extensibility.

Changes in cellular metabolism that increase cellular solute concentrations (and thus lower root osmotic potential) are termed osmotic adjustment. Active solute accumulation resulting from metabolic changes is distinct from any increase in solute concentration occurring due to cellular dehydration. Lowering the osmotic potential maintains water uptake into the cells, thus maintaining root turgor. In maize primary roots growing at Ψ_{soil} of −1.6 MPa, different solutes made different contributions to tissue osmotic potential with distance from the root tip. In the apical 3 mm (where local root growth rates were equivalent in dry and moistened vermiculite), proline accounted for approximately half of the osmotic adjustment (Voetberg and Sharp 1991), whereas soluble carbohydrates (especially hexose) and potassium were more important in the basal 7 mm of the root elongation zone (Sharp

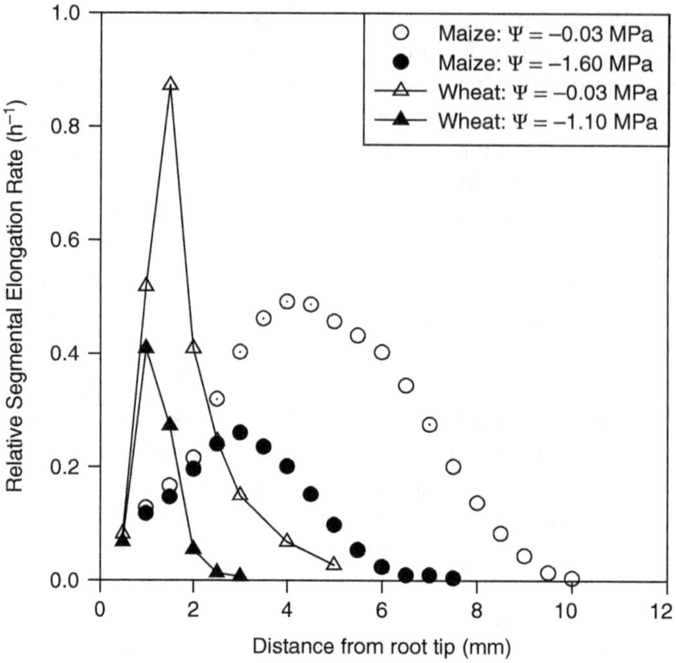

Figure 9.1 Local root growth rate (relative segmental elongation rate) of maize primary (circles) and wheat seminal roots (triangles) grown in moist (hollow symbols) and dry (filled symbols) vermiculite. Moist vermiculite was held at Ψ of −0.03 MPa in both studies, with dry vermiculite being at a Ψ of −1.10 and −1.60 MPa in the wheat (Akmal and Hirasawa 2004) and maize (Sharp et al. 1988) studies, respectively. *Sources:* Akmal and Hirasawa (2004); Sharp et al. (1988).

et al. 1988). Even though differences in water deposition rates accounted for much of the changes in the tissue osmotic potential, spatial differences in the deposition rates of soluble carbohydrates and potassium along the growing root tip (Sharp et al. 1988) highlighted tight regulation of osmotic adjustment. Accumulation of the plant hormone abscisic acid (ABA) in the roots seems critical in regulating such osmotic adjustment because proline accumulation was greatly attenuated (especially at the root tip) in ABA-deficient roots (Ober and Sharp 1994).

ABA's role in regulating root elongation depends on tissue water status because it maintains root growth in dry substrates, but inhibits root growth in moist substrates. In moist vermiculite at Ψ_{soil} of −0.03 MPa, supplying sufficient exogenous ABA (to raise root ABA concentrations to the levels found in roots grown in dry vermiculite) decreased maize primary root elongation (Sharp et al. 1994). Conversely, transplanting ABA-deficient maize mutants or ABA-deficient roots (treated with the carotenoid, and thus ABA, biosynthesis inhibitor fluridone) to dry vermiculite (Ψ_{soil} of −1.60 MPa) decreased root elongation by about 70% after 90 hours, when compared to WT or untreated roots (Saab et al. 1990). Further studies revealed that root ABA accumulation limits synthesis of the growth inhibitory phytohormone ethylene, thereby maintaining growth (Spollen et al. 2000). Treating ABA-deficient roots with ethylene antagonists or biosynthesis inhibitors allowed continued root growth in drying soil.

Despite clear evidence from such experiments with ABA-deficient mutants that ABA is required to maintain primary root growth in drying soil, cultivar variation in root ABA concentration was not associated with differences in absolute root growth rates at low Ψ_{soil} (Leach et al. 2011). Indeed, other plant hormones may also be involved in regulating primary root elongation. In the week following germination, primary roots of maize mutants (*Zmacs6*) with constitutively low ethylene production grew faster than WT plants when grown in vermiculite, a mechanically weak substrate. However, after four weeks of growing in mechanically stronger soil, *Zmacs6* plants had 20% less root biomass than WT plants (Gallie et al. 2009).

Cytokinins (CKs) may also be involved in regulating root growth under water deficit because enhanced root-specific CK degradation in transgenic *Arabidopsis thaliana* and tobacco plants increased primary root elongation, with transgenic plants having a higher survival rate than the WT after severe drought treatment (Werner et al. 2010). Whether changes in root phytohormone profiles can explain the considerable genetic diversity in the capacity of roots from different plant cultivars to elongate in drying soil requires further investigation, especially because these phytohormone profiles have been proposed as selection criteria for shoot responses to drying soil (Wilkinson et al. 2012; Valluru et al. 2016). Theoretically and intuitively, maintaining root growth in drying soil seems a key plant adaptation to allow continued water uptake.

Even though maximum rooting depth can vary substantially (by 0.5 m) between cultivars of the same species (Dardanelli et al. 1997), whether these roots access agronomically significant quantities of soil water has been more difficult to establish. Comprehensive studies of plant water balance demonstrated an extra 11 mm of root water uptake below 1.3 m in the soil profile, which allowed wheat grown in a dryland environment to generate an extra 0.6 t ha^{-1} of yield (Kirkegaard et al. 2007). Despite such empirical examples, there are concerns that many roots at depth are predominantly located in large soil pores generated from previous vegetation, and may be partially hydraulically isolated from the bulk soil (White and Kirkegaard 2010). In view of the technical challenges of different field measurement techniques (core-break versus root washing) to discriminate genotypic differences in root location (Hodgkinson et al. 2017), it may be more useful to measure root water uptake at different depths in the soil profile. Despite some progress in measuring vertical profiles of soil water implying genetic differences in root activity (Whalley et al. 2017), more attention has focused on measuring canopy temperature (reflecting differences in shoot transpiration) as a proxy for genetic differences in root growth (Lopes and Reynolds 2010). Whether genetic variation in canopy temperature reflects differences in soil water availability *per se*, or differences in plant transpirational response to soil water availability (see Section 9.4) is of limited relevance to the plant breeder interested in screening large numbers of divergent genotypes.

Multiple root adaptations (in both architecture and function) can help maintain root water uptake in drying soil. Apart from elongation of primary roots, changes in root morphology (e.g. lateral root proliferation) can also be important in maximizing water extraction from the soil. Selecting genotypes for lateral root proliferation in response to drying soil has attracted little attention, yet constitutive differences in root angle exist in many cereal species (Uga et al. 2013; Richard et al. 2015) that influence the extent of lateral root growth. Genotypes with a wide seminal root angle are expected to have more roots in the upper part of the soil profile, whereas those with a narrow seminal root angle are expected to reach deeper parts of the soil profile (Manschadi et al. 2008). Even though technical improvements in phenotyping can facilitate screening for differences in root angle, demonstrating increased water uptake in genotypes with narrow seminal root angle has been more challenging.

A steep (root angle), cheap (implying minimal metabolic costs), and deep rooting strategy can efficiently acquire soil water at depth by accelerating subsoil exploration (Lynch 2013). In maize, deep rooting was associated with reduced lateral root branching density enabling increased drought resistance (Zhan et al. 2015). Similarly, drought-induced inhibition of lateral root development was identified as an adaptive response in *Arabidopsis* (Xiong et al. 2006). Nevertheless, the relationship between root branching and drought resistance can be variable, and an *Arabidopsis* mutant with improved drought tolerance designated as *enhanced drought tolerance1* had more lateral roots than the WT (Yu et al. 2008). Although longer roots allow water uptake from deeper soil layers, extensive root branching in the upper soil layers may exploit small and/or frequent rainfall events that fail to infiltrate the soil to a great depth (Hodge 2010). Indeed, moderate soil drying and rewatering can greatly promote root extension and initiation, especially of secondary roots (Liang et al. 1996).

Lateral root development is related to local auxin accumulation in tips of developing lateral roots (Du and Scheres 2018) (see also Chapters 2 and 4), as determined by using auxin-sensitive reporter genes and immunolocalization (Benková et al. 2003). Decreased auxin accumulation in the lateral root primordia correlated with the inhibition of lateral root emergence (Krouk et al. 2010). Drought-induced inhibition of lateral root development was partly ABA-mediated (Xiong et al. 2006), with ABA treatment of *Arabidopsis* seedlings decreasing auxin response in the lateral roots (De Smet et al. 2003). Moreover, overexpression of a *Brassica napus NCED* gene (catalysing *de novo* ABA biosynthesis) in *A. thaliana* halved lateral root initiation *in vitro* (Xu and Cai 2017). ABA effects on auxin level in the lateral roots due to ABA-induced activation of indole acetic acid (IAA)-oxidase (Li et al. 2014), inhibition of polar auxin transport (Shkolnik-Inbar and Bar-Zvi 2010) or increased activity of the *GH3* gene involved in conjugating active free auxins (Seo et al. 2009) may play a critical role in root growth under drought.

Considerable turnover of lateral roots when plants are grown in drying soil indicates their ephemeral nature (Anderson et al. 2003), yet their appearance substantially increases the root surface area over which water can be taken up. Furthermore, root hairs (single-celled cylindrical protrusions from the root epidermis of most plant species) can also greatly enhance the absorptive area of the root system. Soil drying stimulates both root hair elongation and density (Li et al. 2017). Nevertheless, WT and a root hairless barley mutant had similar transpiration rates (per unit leaf area) when grown in drying soil at low evaporative demand (<1.5 kPa) (Dodd and Diatloff 2016). Even though the mutant sustained transpiration rate at high evaporative demand (>3 kPa), tissue water potential substantially decreased, indicating the importance of root hairs in maintaining water status at times of peak demand (Carminati et al. 2017). Thus, changes in all types of roots (primary, secondary or lateral roots, and root hairs) can help sustain water uptake in drying soil, but incur energetic costs to the plant (via their production as well as maintenance respiration). Additionally, drying soil can augment root function (specifically hydraulic conductance) to enhance water transport to the shoot.

9.4 Phytohormonal Regulation of Root Hydraulic Conductance

The physiological significance of changes in root hydraulic conductance depends on soil water availability and evaporative demand. When root hydraulic conductance (K_{root}) limits water uptake more than soil hydraulic conductivity, upregulating K_{root} can enhance water transport to the shoot. Conversely, when soil hydraulic conductivity limits water uptake more than K_{root}, downregulating K_{root} can prevent water loss to the soil. Suberization of the root endodermis limits water flows from the root system to dry soil, and root shrinkage (due to turgor loss) forms air gaps between the roots and the soil, further decreasing the hydraulic conductivity at the root–soil interface. Different parts of the root system are exposed to different levels of soil drying and thus show differences in hydraulic conductance, but ultimately it is the whole root system that determines the capacity to supply the shoots with water. Mild soil drying (half the root system was allowed to dry the soil) increased K_{root} of maize, likely due to root ABA accumulation (Zhang et al. 1995), whereas more severe soil drying decreased K_{root} in the same species (Ruiz-Lozano et al. 2009). Moreover, different species can show contrasting responses of K_{root} to the same level of osmotic stress (Kreszies et al. 2019b), which likely is of adaptive significance.

Aquaporins (AQPs) are membrane-based proteins that facilitate water transport across the membrane according to the prevailing water potential gradients. They typically are the products of large, multigene families, the expression of which produces plasma membrane intrinsic (PIP) and tonoplast intrinsic (TIP) proteins in the plasma membrane and tonoplast, respectively. Water deficit often changes transcript abundance of *PIP* genes, which can alter water uptake. Osmotic stress (of −0.82 MPa) stimulated gene expression of a specific AQP in upland (but not lowland) rice (*Oryza sativa*), and further upregulating this AQP under control of a stress-inducible promoter increased K_{root}, Ψ_{leaf} and transpiration under osmotic stress (Lian et al. 2004). Similarly, constitutive overexpression of a ginseng (*Panax ginseng*) AQP (*PgTIP*) in *A. thaliana* enhanced stomatal conductance (g_s), which accelerated soil drying but also increased root elongation rate (Peng et al. 2007). Constitutive overexpression of a tobacco (*Nicotinia tabacum*) AQP (*NtAQP1*) in tomato (*Solanum lycopersicum*) also enhanced g_s, but only enhanced K_{root} when plants were exposed to salt (100 mM NaCl) stress (Sade et al. 2010). In contrast, ectopic expression of a drought-inducible apple (*Malus x domestica*) AQP (*MdPIP1;3*) in tomato slowed the rate of water loss in both detached leaves and plants grown in drying soil, without compromising the endogenous upregulation of native AQPs (*SlPIP1* and *SlPIP2*) as the soil dried (Wang et al. 2017). Thus, it is difficult to generalize about the physiological impacts of transgenically perturbing AQP expression.

Similarly, there is considerable complexity in the responses of individual AQP genes in response to soil drying. For example, in (non-mycorrhizal) maize, soil drying decreased K_{root} by 75%, while *ZmPIP1;1* expression more than doubled, *ZmPIP1;5* expression more than halved, and *ZmPIP1;2* expression remained stable (Ruiz-Lozano et al. 2009). Paradoxically, soil application of ABA increased expression of all of these genes, with accumulated evidence indicating that ABA regulates K_{root} in a dose-dependent manner (Dodd 2013). Moderate exogenous ABA concentrations (0.1–10 μM) applied to hydroponically-grown plants stimulated K_{root}, whereas high ABA concentrations (>10 μM) decreased K_{root}.

Constitutive overexpression of an ABA biosynthesis (*NCED*) gene in tomato approximately doubled root ABA concentrations and K_{root} (Thompson et al. 2007). Similarly, transgenic maize plants with variation (both sense and antisense) in *NCED* gene expression showed sixfold variation in K_{root}, likely due to variation in expression of *PIP* genes (Parent et al. 2009). Constitutive overexpression of an *Arabidopsis* molybdenum cofactor sulfurase (*AtLOS5*) in maize enhanced root ABA concentration and K_{root} when plants were exposed to an osmotic stress (−0.46 MPa), coincident with increased *ZmPIP1* and *ZmPIP2* gene expression (Zhang et al. 2017). Thus, overexpressing ABA biosynthesis genes increased endogenous ABA concentrations sufficiently to enhance K_{root}, whereas excessively high exogenous ABA applications decreased K_{root}.

ABA is also important in regulating root hydraulic changes in response to increased atmospheric evaporative demand, which commonly accompanies soil drying. Higher evaporative demand increased root ABA accumulation, which contributed to increased water flow to the shoots by increasing K_{root}, thereby restoring Ψ_{leaf} and leaf growth (Kudoyarova et al. 2011). However, the ABA-deficient barley mutant *Az34* failed to increase K_{root} and maintained a lower Ψ_{leaf} under these conditions (Veselov et al. 2018), but hydroponic ABA treatment (10 μM) of the mutant increased its hydraulic conductance (Sharipova et al. 2016). Both exogenous ABA (Sharipova et al. 2016) and high evaporative demand (Veselov et al. 2018) increased abundance of the *HvPIP2;2* aquaporin concurrently with increased K_{root}. Thus, ABA-mediated changes in aquaporin expression seem important in balancing root water uptake with shoot water losses.

Other plant hormones can also regulate K_{root} by affecting aquaporin expression and/or activity. For example, methyl jasmonate (MeJA) treatment (10–100 μM) elevated K_{root}, with gene expression of several aquaporins increasing concurrently (Sanchez-Romera et al. 2014). Moreover, mutants defective in jasmonic acid (JA) biosynthesis (*S. lycopersicum def*-1 mutant) had lower K_{root} that MeJA treatment restored (Sanchez-Romera et al. 2014). Most aquaporin genes were repressed by auxin treatment during lateral root emergence (Péret et al. 2012). Modelling predicted that auxin promotes root branching by regulating the spatial and temporal distribution of aquaporin-dependent root tissue water transport (Péret et al. 2012). Although CK regulation of AQPs expression in responses to environmental stress has been mentioned (Kapilan et al. 2018; Pavlů et al. 2018), the exact role of these plant hormones in regulating root hydraulic function has not been studied. As with root elongation, multiple changes in plant hormones have also been implicated in regulating root hydraulic function. Whether these changes affect shoot water status to a large extent depends on the regulation of stomatal conductance.

9.5 Phytohormonal Root-to-Shoot Signalling

Traditionally, soil drying was thought to decrease leaf (and thus guard cell) turgor, thus causing stomatal closure that regulates water uptake. However, field-grown cowpea (*Vigna unguiculata*) plants exposed to slowly developing soil water deficits had Ψ_{leaf} similar to, but stomatal conductance (g_s) lower than, well-watered plants (Bates and Hall 1981). Furthermore, when maize plants were grown with their root systems split between two pots, and water was withheld from one pot, partial stomatal closure occurred without a decrease in Ψ_{leaf} (Blackman and Davies 1985). To account for these observations, it was proposed that the roots sensed drying soil, changed their cellular metabolism and thus loading of phytohormones into the xylem, thereby altering root-to-shoot chemical signalling to induce stomatal closure (reviewed in Dodd 2005). Most attention has focused on the role of the phytohormone ABA, although xylem sap concentrations of CKs (Alvarez et al. 2008), the ethylene precursor aminocyclopropane carboxylic (ACC) (Belimov et al. 2009) and JA (De Ollas et al. 2018) also change with soil drying. Changes in xylem sap hormone concentrations are often interpreted as evidence of root-to-shoot signalling, but measuring authentic *in vivo* concentrations can be technically challenging (Netting et al. 2012). Firstly, evidence that these compounds can act as root-to-shoot signals is reviewed.

Several criteria must be satisfied to regard a compound as a *bona fide* root-to-shoot signal of drying soil (after Jackson 1993). The compound must be:

a) synthesized in the roots
b) transported in the xylem, and
c) capable of inducing a quantitative response in the shoot.

Evidence for and against each proposition is considered in turn, with respect to the most-studied of these putative signals, ABA.

Although roots usually accumulate ABA as the soil dries, root ABA concentrations at any point in time depend on changes in the expression of ABA biosynthesis and catabolism genes, but also ABA export from the roots via the xylem, and ABA import to the roots via the phloem. Thus, more specific experiments are needed than simply measuring root ABA concentrations. Detached roots can accumulate ABA if they are dehydrated (Simonneau et al. 1998), even if they accumulate less ABA than shoot tissues (Zhang et al. 2018). Transplanting just-germinated maize seedlings (before coleoptile emergence) into dry vermiculite (Ψ_{soil} of −1.6 MPa) stimulated root ABA accumulation >10-fold, even though shoot development had not yet occurred (Saab et al. 1990). Moreover, soil drying of plants in the vegetative stage decreased Ψ_{root}, thereby stimulating root ABA accumulation (Zhang and Davies 1989; Puertolas et al. 2013), suggesting that soil drying increases root ABA biosynthesis. Analyses of root gene expression have confirmed substantial upregulation of *NCED* (the rate-limiting step in ABA biosynthesis) gene expression in drying soil (Thompson et al. 2000; Speirs et al. 2013), often to a greater extent than foliar *NCED* gene expression. Thus, roots seem capable of synthesizing ABA, although their contribution to root-to-shoot signalling may be easily dismissed because they have much lower endogenous ABA concentrations than shoots.

Root ABA concentrations are linearly related to the concentrations of ABA detected in xylem sap, suggesting that roots are an important source of xylem ABA (Liang et al. 1997). Root ABA export is most accurately assessed by applying a pneumatic pressure to the roots to induce sap flow rates equivalent to whole-plant transpiration rates, as de-topping a root system can result in variable sap flow rates (or none at all) according to the soil water deficit and root pressure (Else et al. 1994). Xylem ABA concentrations can be diluted or enriched in transit from roots to shoots, although care must be applied in interpreting these differences if the methods of xylem sap collection differ between roots and shoots. Perhaps the best evidence of such gradients comes from applying pneumatic pressure to the roots, and simultaneously collecting xylem sap from multiple leaves in the canopy, by placing a small incision in the midrib of each leaf. This technique demonstrated that xylem ABA concentrations of grapevine more than halved from leaves near the shoot apex to more basal leaves (Soar et al. 2004), suggesting that leaf and stem vasculature might modify the root-to-shoot signal. Imaging fluorescent marker genes indicated vascular gene expression of ABA catabolism (8-OH hydroxylase, Okamoto et al. 2009), which theoretically can modify xylem ABA concentrations, although the significance of vasculature-localized gene expression to xylem ABA concentrations has not been assessed. Thus, the physiological significance of root phytohormone export requires paired measurements with whole-plant gas exchange measurements.

The ABA concentrations detected in leaf xylem sap are sufficient to partially close the stomata of maize (Zhang and Davies 1991) and garden pea (*Pisum sativum*, Rothwell et al. 2015) leaves, as indicated by measuring the transpiration of detached leaves supplied with synthetic ABA via the xylem. The relationship between stomatal conductance and xylem ABA concentration determined in such detached leaf experiments are similar to those occurring *in vivo* when plants are exposed to drying soil (Boyle et al. 2016; Zhang and Davies 1991). However, xylem ABA concentrations *in vivo* cannot explain stomatal closure in other species such as wheat (Munns and King 1988), suggesting that other xylem-borne anti-transpirants exist (Wilkinson and Davies 1997) or that local processes regulate stomatal closure.

Depending on how fast soil water deficits are imposed (determined by evaporative demand and plant size), root ABA accumulation can either precede leaf ABA concentration (Zhang and Davies 1989) or occur simultaneously (Thompson et al. 2007). However, adding osmotica ($\Psi = -1.0\,\text{MPa}$) to the roots of *A. thaliana* caused shoot ABA accumulation prior to any root ABA accumulation (Christmann et al. 2007). Nevertheless, the measurement/sampling frequency in such experiments is often critical to establish whether any tissue-specific differences exist in the timing of ABA accumulation.

Alternatively, reciprocal grafting experiments with WT and ABA-deficient mutants provide more specific evidence of the importance of root ABA accumulation in mediating shoot responses. Soil drying caused stomatal closure of WT tomato scions independent of rootstock genotype, despite leaf xylem ABA concentration of WT/mutant (scion/rootstock) plants being fourfold lower than in WT self-grafts (Holbrook et al. 2002), indicating that leaf-autonomous ABA production is sufficient to elicit stomata closure. Even though some studies indicate that WT/mutant plants can show a higher stomatal conductance than WT self-grafts, the consensus response (compiling data from six independent studies with tomato plants exposed to different edaphic stresses) is that root ABA status has no significant effect on g_s of WT plants, but that a WT rootstock can significantly decrease g_s of mutant scions (Li et al. 2018). Additionally, guard cell-autonomous ABA biosynthesis causes stomatal closure at high evaporative demands (Bauer et al. 2013), although stomatal responses to soil drying were not investigated in this genotype. Taken together, these experiments confirm the importance of ABA in mediating stomatal closure, but ascribe little role to the roots in such regulation of transpiration.

Other reciprocal grafting experiments have demonstrated important roles for other phytohormones in stomatal regulation. Soil drying decreased root (Goicoechea et al. 1997) and xylem (Masia et al. 1994) CK concentrations, and detached leaf transpiration experiments showed that xylem-borne *trans*-zeatin (tZ) could antagonize ABA-induced stomatal closure (Hansen and Dorffling 2003). More convincing evidence of the key role of roots in CK-mediated stomatal responses was obtained by grafting WT plants onto a rootstock overexpressing *IPT* (rate-limiting enzyme in CK biosynthesis). These plants maintained a higher g_s than WT self-grafts when the root zone was salinized (Ghanem et al. 2011). Moreover, plants grafted onto the *IPT*-overexpressing rootstock grew better, highlighting the importance of root-sourced CKs in overcoming effects of below-ground stress. Furthermore, the *Arabidopsis* transporter G14 (member of the ATP-binding cassette superfamily) was root-expressed and played a significant role in regulating CK transport to the shoot (Ko et al. 2014). Abolishing the gradient of protons across the root plasma membrane by applying the protonophore carbonyl cyanide chlorophenylhydrazone (CCCP) inhibited CK uptake from the root medium, but increased root CK export and shoot CK accumulation (Kudoyarova et al. 2014). Thus, changes in CK uptake from the soil solution, its transport within the roots, and root CK biosynthesis are all involved in regulating root CK export to the shoot.

Soil drying can also enhance xylem jasmonate (JA) concentrations, and supplying 1 μM JA to detached leaves via the xylem decreased WT transpiration but increased transpiration in *def-1* (a mutant that fails to accumulate JA as the soil dries) (De Ollas et al. 2018). This unusual effect was attributed to JA decreasing leaf OPDA (cis-(+)-12-oxo-phytodienoic acid) concentrations (an oxylipin with an even greater anti-transpirant effect than JA – Savchenko et al. 2014) in *def-1*. Genotypic differences in the concentrations of various oxylipins made it difficult to interpret stomatal responses to soil drying in reciprocal grafts of WT and *def-1* plants. However, stomata of *def-1* self-grafts were relatively insensitive to soil drying compared to all other graft combinations (De Ollas et al. 2018), suggesting that root-supplied JA was necessary for normal responses to soil drying. Furthermore, compared to all other graft combinations, foliar ABA accumulation of *def-1* self-grafts was attenuated at the same soil water status. Thus, ABA and JA may act together to induce stomatal closure in WT plants (Brossa et al. 2011).

Soil drying downregulates root expression of the strigolactone (SL) biosynthesis (carotenoid cleavage dioxygenase, *CCD7* and *CCD8*) genes (Visentin et al. 2016), which presumably decreases root SL export to the shoot, even though if xylem SL concentrations have been difficult to measure. Constitutive SL deficiency increased g_s of well-watered plants (Liu et al. 2015; Visentin et al. 2016), whereas grafting a WT shoot to a SL-deficient rootstock decreased g_s of well-watered tomato plants (Visentin et al. 2016). Nevertheless, both shoot and root SL status had little impact on the sensitivity of g_s to osmotic stress (Liu et al. 2015) or soil drying (Visentin et al. 2016). Indeed, all graft combinations showed remarkably similar root and shoot ABA accumulation as tissue water status declined in spite of WT scions decreasing ABA levels and *NCED* gene expression in SL-deficient rootstocks (Visentin et al. 2016), consistent with a primary role for ABA in eliciting stomatal closure. Differences in shoot ABA concentration between WT and *SlCCD7*-silenced scions only occurred at a Ψ_{leaf} that was significantly lower than that inducing full stomatal closure. Even though SL status alters plant water use of well-watered plants, stomatal sensitivity to soil drying seems little affected.

9.6 Shoot-to-Root Signalling

Empirical flow models highlight both root-to-shoot and shoot-to-root phytohormone transport (Slovik et al. 1995), with reciprocal grafting and girdling experiments allowing the significance of basipetal phytohormone transport to be investigated when plants are exposed to drying soil. Girdling (removal of stem phloem tissue at the root-shoot junction) eliminates shoot-to-root phytohormone transport and affects root ABA accumulation in response to soil drying (Cornish and Zeevaart 1985; Liang et al. 1997; Manzi et al. 2015; Castro et al. 2019). Variable responses occurred, depending on the species and duration of the girdling treatment (Figure 9.2). Root ABA accumulation in girdled *Xanthium strumarium* and tomato plants was similar to intact plants, with dehydrated roots of stem-girdled plants showing 80% of the root ABA accumulation (averaged across both species) in intact plants (Cornish and Zeevaart 1985). In contrast, girdling almost halved root ABA accumulation of citrus plants (*Citrus sinensis* × *Poncirus trifoliata*) compared to intact plants the first time the soil dried, and eliminated root ABA accumulation altogether the second time the soil dried (Manzi et al. 2015). More recent studies show that the time elapsed since girdling has a pronounced effect on soybean root ABA accumulation in response to drying soil. Girdling decreased root ABA concentration of well-watered plants

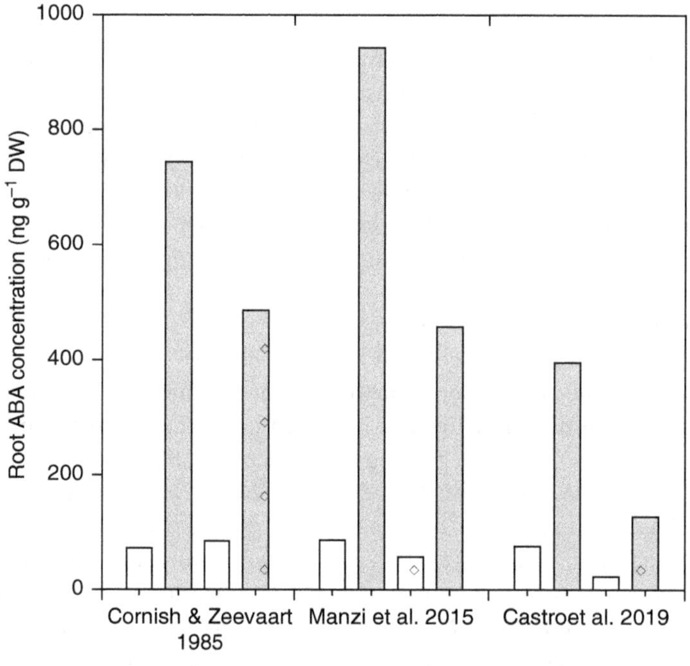

Figure 9.2 Root ABA concentration in well-watered (hollow bars) and stressed (filled bars) in intact (no shading) and girdled (shaded bars) of tomato (Cornish and Zeevaart 1985), citrus (Manzi et al. 2015) and soybean (Castro et al. 2019) plants that had lost 15% of their root fresh weight after being girdled 6 hours ago (Cornish and Zeevaart 1985) or had been exposed to drying soil for 3 days after girdling (Manzi et al. 2015; Castro et al. 2019). Sources: Cornish and Zeevaart (1985); Manzi et al. (2015); Castro et al. (2019).

fourfold within 24 hours (indicating the importance of shoot-sourced ABA in regulating root ABA status), but thereafter root ABA concentration of well-watered, girdled plants increased sixfold after five days (Castro et al. 2019). Thus, after five days, root ABA concentration of girdled plants was independent of soil water status, perhaps regulated by carbohydrate depletion. Two days earlier in the experiment, girdled plants exposed to drying soil had a root ABA concentration that was 5.8-fold higher than girdled, well-watered plants, even if absolute root ABA concentrations were only 34% of intact plants exposed to drying soil (Castro et al. 2019). Even though girdling clearly demonstrates the potential for shoot-to-root signalling to alter root hormone homeostasis, its actual contribution *in vivo* is more difficult to assess because it may induce a stress response in the roots.

Reciprocal grafting experiments also provide an opportunity to assess whether shoot-to-root signalling alters root growth and hormonal responses. Compiling data from six independent studies that reciprocally grafted WT and ABA-deficient tomato plants revealed that shoot ABA status had no significant effect on root ABA concentrations of ABA-deficient rootstocks, whereas mutant/WT (scion/rootstock) plants seemed to have lower root ABA concentrations than WT self-grafts (Li et al. 2018). Nevertheless, an ABA-deficient scion caused pronounced taproot-sourced lateral root proliferation in pea or adventitious rooting in tomato independently of rootstock ABA status, indicating the importance of foliage-derived ABA in suppressing root initiation (McAdam et al. 2016). High root auxin concentrations were implicated in this response, with shoot-derived auxin regulating lateral root development (Reed et al. 1998). This was suggested by experiments showing that local application of the inhibitor of auxin transport, naphthylphthalamic acid, at the shoot base decreased the number and density of lateral roots, auxin concentrations in the root, and transport of labelled exogenous auxin into the root. Shoot excision or dark exposure also inhibited lateral root development in *Arabidopsis*, which was reversible by IAA. Thus, shoot-to-root hormonal signalling also involves multiple phytohormones.

Even though CKs seem predominantly root-synthesized (Puig et al. 2012), some *IPT* genes (controlling *de novo* CK synthesis) are expressed in the leaves (Miyawaki et al. 2004), suggesting that the aerial tissue can be an important source of CKs transported from the shoots to roots. Indeed, shoot-to-root CK translocation is involved in vascular patterning in the root meristem (Bishopp et al. 2011), whereas shoot-derived CKs systemically suppress nodulation in roots (Sasaki et al. 2014). Whether changes in shoot-to-root CK transport are important in regulating plant responses to drying soil is yet to be established.

9.7 Partial Root-zone Drying: a Case Study

To understand how the processes described above might be integrated within the plant, a specific water-saving irrigation technique will be introduced. Partial root-zone drying (PRD) waters only part of the root zone, allowing the remainder to dry, ostensibly to alter phytohormone accumulation within the roots and their export to the shoots. These changes aim to partially close the stomata, thereby improving water-use efficiency, restrict vegetative growth, and modify photoassimilate distribution between source and sink tissues to change fruit and grain quality. Wet and dry parts of the root zone are either maintained throughout the irrigation season (Fixed PRD = PRD-F) or regularly alternated following each drying cycle (Alternate PRD = PRD-A), ostensibly to maintain root viability and root-to-shoot phytohormone signalling. Because plants are commonly exposed to soil water gradients and repeated drying and rewetting events in nature, PRD may provide a useful example of how plants integrate water uptake with demand.

Early studies in grapevine (*Vitis vinifera*) highlighted that PRD substantially increased root development in moist soil layers, independently of whether they occurred in the irrigated or drying parts of the root zone (Dry et al. 2000). Whether this change is of adaptive significance depends on whether increased root mass improves root (hydraulic) function, or results from denser (suberin-rich) tissues being laid down to effectively "waterproof" the root system. By regularly harvesting containerized PRD tomato plants at the end of each drying cycle, Mingo et al. (2004) determined the importance of drying and rewetting cycles to root development. After a single drying cycle, root biomass did not statistically differ between the wet and dry parts of the root system. However, rewatering the previously dry side substantially enhanced root biomass accumulation in this compartment. This "'priming effect" may have resulted from phytohormone accumulation during soil drying enhancing sink strength following rewatering. Prioritizing assimilate partitioning to the roots may be of limited value in crops (such as lettuce) where the shoot is the harvested portion and/or if assimilate partitioning to the reproductive parts of the plant (in grain and fruit crops) is diminished. Nevertheless, greater root development can improve water capture by roots deeper in the soil profile, and may allow irrigation volumes to be decreased.

Many studies have considered the impacts of PRD on root growth, but relatively few have considered its impacts on root function. PRD effects on maize root system hydraulic conductance (K_{root}) increased with experiment duration and when the wet and dry parts of the root zone were fixed. After three complete wetting and drying cycles (each lasting 10 days), PRD-A and PRD-F decreased K_{root} by 12% and 21%, respectively (Hu et al. 2011). These relatively modest changes in whole root system response disguised considerable (and rapid – within five days) changes in the hydraulic conductance of individual parts of the root system exposed to different soil moisture conditions. Root hydraulic conductance of the non-irrigated root zone (K_{dry}) of PRD plants decreased by about 80% compared to well-watered roots, which was almost perfectly compensated by a similar increase in the root hydraulic conductance of the irrigated root zone (K_{wet}) of PRD plants. Even though K_{dry} of PRD-F plants did not change during the experiment, K_{dry} of PRD-A plants increased 3.8-fold (Hu et al. 2011). Whether these changes were due to morphological (e.g. increased suberin deposition, changes in xylem anatomy, root proliferation of more conductive roots in PRD-A plants) or physiological (e.g. changes in aquaporin expression and activity) was not investigated in that study. When a wetland tree species (*Melaleuca argentea*) was exposed to PRD by draining one of the two compartments its roots were placed in, K_{dry} did not change, but K_{wet} increased threefold within 24 hours (McLean et al. 2011). This upregulation of K_{wet} was consistent with increased water uptake by these roots, and a transient increase in the abundance of PIP1 aquaporins. Whether this change in root gene expression was hormonally regulated was not assessed, as root ABA concentration was not measured. Taken together, these two studies that investigated K_{root} responses to PRD indicate discrete, dynamic adjustments in different parts of the root zone to maintain transpiration rate despite localized soil drying.

Differential root hydraulic conductance responses of wet and dry parts of the root zone may be associated with changes in root ABA accumulation, although this has yet to be explicitly assessed. Roots in the dry part of the root zone can accumulate 2- to 10-fold more ABA than roots in the wet part of the root zone (Khalil and Grace 1993; Stoll et al. 2000; Pérez-Pérez et al. 2018), even without Ψ_{leaf} changing. Indeed, a transient decrease (of 0.3 MPa) in Ψ_{leaf} following PRD alternation did not induce additional root ABA accumulation in the non-irrigated root zone of potted citrus (*Citrus macrophylla*) seedlings exposed to PRD (Pérez-Pérez et al. 2018), suggesting that local changes in root ABA biosynthesis determined root ABA accumulation in the dry part of the root zone. In contrast, in the wet part of the root zone, only prolonged soil drying cycles (12 days rather than three or six days) promoted ABA accumulation coincident with decreased Ψ_{leaf} and elevated leaf ABA concentration (Pérez-Pérez et al. 2018). Given that Ψ_{soil} and Ψ_{root} were similar in the wet compartment of all plants irrespective of the duration of soil drying, it seems probable that differential shoot-to-root ABA signalling (with roots in wet soil acting as a sink for ABA) caused this ABA accumulation. Thus, increased root *NCED* gene expression in drying soil and differential shoot-to-root ABA signalling both seem necessary to account for spatial changes in root ABA concentration in plants exposed to PRD.

The dynamics of root water uptake between wet and dry parts of the root system is of considerable importance to plants grown with PRD, because it determines both whole-plant transpiration rate and the movement of root-to-shoot signals. The proportional contributions from both parts of the root system were conveniently assessed by attaching sap flow gauges to "two-roots, one-shoot" grafted sunflower (*Helianthus annuus*) plants, then collecting xylem sap from both the root systems and the leaves of plants exposed to PRD. Root xylem ABA concentration ([X-ABA$_{root}$]) and sap flow from the irrigated root zone was stable throughout the experiment, but continued soil drying of part of the root zone decreased sap flow and increased [X-ABA$_{root}$] (Dodd et al. 2008). Thus, leaf xylem ABA concentration ([X-ABA$_{leaf}$]) of PRD-F plants showed a unique response to soil drying compared to conventionally deficit-irrigated plants (DI) where both parts of the root system were allowed to dry the soil. In DI plants, [X-ABA$_{leaf}$] increased exponentially as the soil dried. In PRD-F plants, initial soil drying increased [X-ABA$_{leaf}$] (presumably because root ABA export from the drying root zone increased with moderate decreases in sap flux), whereas continued soil drying decreased [X-ABA$_{leaf}$] (presumably because root ABA export from the drying root zone decreased as sap flux ceased). Although these responses were supported by a model that explained [X-ABA$_{leaf}$] of PRD-F plants as a function of relative sap flux and [X-ABA$_{root}$] from the different parts of the root zone (Dodd et al. 2008; Pérez-Pérez and Dodd 2015), little empirical evidence of differences in root-to-shoot signalling between PRD-F and DI plants has emerged.

When [X-ABA$_{leaf}$] (Dodd 2007) or [X-ABA$_{root}$] (Wang et al. 2012) of PRD or DI tomato plants were compared on a daily basis (with four to six replicates per treatment), it was often difficult to distinguish treatment differences. Nevertheless, at the end of a second drying cycle, [X-ABA$_{root}$] was 10-fold lower in PRD-A than DI plants, consistent with the much higher Ψ_{root} of PRD plants (Wang et al. 2012), suggesting that root water status regulated [X-ABA$_{root}$] because Ψ_{leaf} of PRD plants did not differ between treatments. When both Ψ_{root} and Ψ_{leaf} of PRD and DI plants were equivalent after three drying and rewetting cycles, [X-ABA$_{root}$] did not differ between treatments (Wang et al. 2012). PRD-F and DI

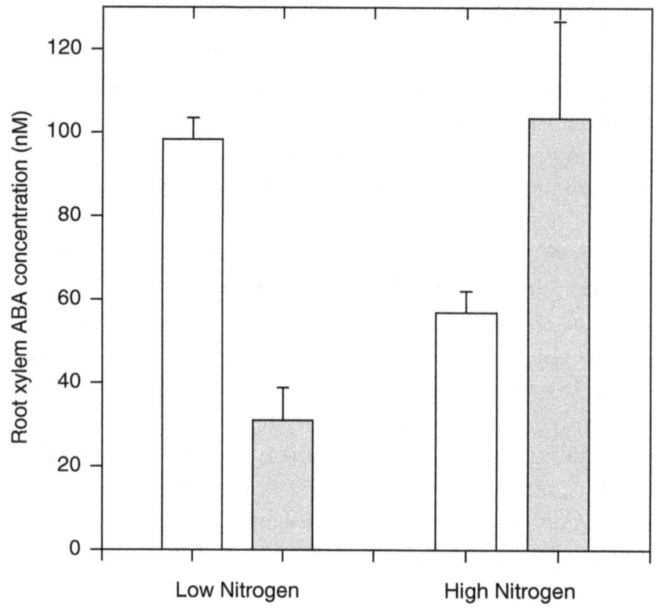

Figure 9.3 Root xylem ABA concentration of containerized, split-root potato plants grown at low or high nitrogen and exposed to either deficit irrigation (irrigating both parts of the root zone – filled bars) or partial root-zone drying (irrigating only one part of the root zone – hollow bars). Data are means ± SE. Source: Yan et al. (2012). © 2012, Elsevier.

plants repeatedly showed different responses of [X-ABA$_{leaf}$] to soil water content, with [X-ABA$_{leaf}$] of PRD-F plants exceeding that of DI plants with moderate soil drying and the reverse occurring with more severe soil drying (Dodd 2007), consistent with the model described above.

Variation in other soil properties may modulate the impact of these irrigation treatments on root-to-shoot ABA signalling. After potato (*Solanum tuberosum*) plants were exposed to three drying cycles, [X-ABA$_{root}$] of PRD-A plants was approximately double that of DI plants if plants were grown at low nitrogen supply (Figure 9.3), but the opposite response was detected when plants were grown at high nitrogen supply (Yan et al. 2012). Effects of plant nitrogen status on root ABA production (Brewitz et al. 1995) clearly interact with the relative contributions of different parts of the root zone to root ABA export in PRD plants, making it difficult to predict whether PRD will enhance or attenuate root-to-shoot ABA signalling.

Although these treatment differences in xylem ABA concentrations were correlated with changes in stomatal conductance (Wang et al. 2012; Yan et al. 2012), other studies show treatment differences in stomatal conductance (g_s) that were independent of leaf ABA concentration. Thus, g_s of PRD tomato plants was 10% higher than in DI plants in one of two cultivars investigated, yet leaf ABA concentrations were comparable (Valerio et al. 2017). Furthermore, when potted grapevines received irrigation equivalent to 50% of full evapotranspiration, g_s of PRD plants was consistently higher than in DI plants, whereas more severe soil water deficits caused similar leaf ABA concentrations in both treatments (Beis and Patakas 2015). Because such treatment differences in stomatal response occurred without changes in ABA concentration, and stomatal closure of PRD plants can occur without ABA increments (Pérez-Pérez et al. 2012), other xylem-borne antitranspirants that modulate stomatal closure have been investigated.

PRD can enhance leaf ethylene evolution (Sobeih et al. 2004) and decrease leaf CK concentration (Kudoyarova et al. 2007), yet whether these responses differ between PRD and DI plants, and whether root-to-shoot ACC or CK signalling is involved, has scarcely been investigated. Nevertheless, increased foliar CK concentrations (of PRD plants irrigated at 50% of full evapotranspiration) may account for their enhanced g_s compared to DI plants (Beis and Patakas 2015) because CKs antagonize ABA-induced stomatal closure (Hansen and Dorffling 2003; Dodd 2005). Indeed, stomatal closure of PRD plants was more sensitive to the ABA/CK ratio than in DI plants (Beis and Patakas 2015). Whether the model of root-to-shoot signalling elaborated above can explain xylem CK concentrations of PRD-F and DI plants should be investigated because there has been some uncertainty as to the relative importance of root-synthesized CKs in regulating plant responses (cf. Faiss et al. 1997; Ghanem et al. 2011).

Even though most of the literature attempting to explain different agronomic responses of PRD-F and DI plants (Dodd 2009) has focused on stomatal regulation and root-to-shoot signalling, it is important to recognize that these techniques also affect vegetative vigour, with the direction of response dependent on absolute soil water availability (Romero et al. 2012). Thus, PRD plants were more vigorous than DI plants at moderate levels of soil water availability, but the opposite response was detected at more severe soil water deficits. Whether these differences can be attributed to root-to-shoot signalling is less certain because much of the literature on shoot growth regulation has emphasized the importance of tissue water relations (Tardieu et al. 2010). Following a midday depression of stem water potential (Ψ_{stem}), PRD almond (*Prunus dulcis*) trees recovered their water status in the afternoon earlier than DI trees exposed to similar soil water availability (Egea et al. 2011). Nevertheless, in dicotyledonous plants most leaf growth occurs at night (Schmundt et al. 1998); thus, midday Ψ_{stem} may be poorly related to treatment differences in vegetative growth. Indeed, leaf area expansion of PRD apple seedlings can be inhibited without Ψ_{leaf} changing (Gowing et al. 1990), again suggesting the importance of root-to-shoot signalling in regulating vegetative growth. Much more work needs to be done on the importance of shoot-to-root signalling in potentially regulating shoot growth of PRD plants.

9.8 Conclusions

Ultimately, both rain-fed crops and plants exposed to PRD (or any other form of deficit irrigation that decreases soil water availability) must adjust water uptake to match their water demands to prevent leaf water deficits. Even though root proliferation and increased hydraulic conductance can enhance root water uptake over variable time-scales, water demands can be regulated more rapidly in response to step changes in environmental conditions by inducing partial stomatal closure. The relative importance of root and shoot systems in regulating these responses likely depends on which tissue is the first to perceive water deficit under the specific environmental conditions imposed. Irrespective of whether roots or shoots are the first to perceive water deficit, long-distance communication between roots and shoots seems to play an important role in balancing water uptake with water demands.

Acknowledgments

ICD thanks the Global Challenge Research Funding (Project RECIRCULATE, ES/P010857/1) and the EU-H2020 program (SHui Project, Number 773903) for continued support of research on partial root-zone drying.

References

Akmal, M. and Hirasawa, T. (2004). Growth responses of seminal roots of wheat seedlings to a reduction in the water potential of vermiculite. *Plant and Soil* 267: 319–328.

Alvarez, S., Marsh, E.L., Schroeder, S.G. et al. (2008). Metabolomic and proteomic changes in the xylem sap of maize under drought. *Plant, Cell and Environment* 31: 325–340.

Anderson, L.J., Comas, L.H., Lakso, A.N. et al. (2003). Multiple risk factors in root survivorship: a 4 year study in Concord grape. *New Phytologist* 158: 489–501.

Bates, L.M. and Hall, A.E. (1981). Stomatal closure associated with soil water depletion not associated with changes in bulk leaf water status. *Oecologia* 50: 62–65.

Bauer, H., Ache, P., Lautner, S. et al. (2013). The stomatal response to reduced relative humidity requires guard cell-autonomous ABA synthesis. *Current Biology* 23: 53–57.

Beis, A. and Patakas, A. (2015). Differential physiological and biochemical responses to drought in grapevines subjected to partial root drying and deficit irrigation. *European Journal of Agronomy* 62: 90–97.

Belimov, A.A., Dodd, I.C., Hontzeas, N. et al. (2009). Rhizosphere bacteria containing ACC deaminase increase yield of plants grown in drying soil via both local and systemic hormone signalling. *New Phytologist* 181: 413–423.

Benková, E., Michniewicz, M., Sauer, M. et al. (2003). Local, efflux-dependent auxin gradients as a common module for plant organ formation. *Cell* 115: 591–602.

Bishopp, A., Lehesranta, S., Vaten, V. et al. (2011). Phloem–transported cytokinin regulates polar auxin transport and maintains vascular pattern in the root meristem. *Current Biology* 21: 927–932.

Blackman, P.G. and Davies, W.J. (1985). Root to shoot communication in maize plants of the effects of soil drying. *Journal of Experimental Botany* 36: 39–48.

Boyer, J.S. (1982). Plant productivity and environment. *Science* 218: 443–448.

Boyle, R.K.A., McAinsh, M.R., and Dodd, I.C. (2016). Stomatal closure of *Pelargonium x hortorum* in response to soil water deficit is associated with decreased leaf water potential only under rapid soil drying. *Physiologia Plantarum* 156: 84–96.

Brewitz, E., Larsson, C.M., and Larsson, M. (1995). Influence of nitrate supply on concentrations and translocation of abscisic acid in barley (*Hordeum vulgare*). *Physiologia Plantarum* 95: 499–506.

Brossa, R., Lopez-Carbonell, M., Jubany-Mari, T. et al. (2011). Interplay between abscisic acid and jasmonic acid and its role in water-oxidative stress in wild-type, ABA-deficient, JA-deficient, and ascorbate-deficient *Arabidopsis* plants. *Journal of Plant Growth Regulation* 30: 322–333.

Carminati, A., Passioura, J.B., Zarebanadkouki, M. et al. (2017). Root hairs enable high transpiration rates in drying soils. *New Phytologist* 216: 771–781.

Castro, P., Puertolas, J., and Dodd, I.C. (2019). Stem girdling uncouples soybean stomatal conductance from leaf water potential by enhancing leaf xylem ABA concentration. *Environmental and Experimental Botany* 159: 149–156.

Christmann, A., Weiler, E.W., Steudle, E. et al. (2007). A hydraulic signal in root-to-shoot signalling of water shortage. *The Plant Journal* 52: 167–174.

Cornish, K. and Zeevaart, J.A.D. (1985). Abscisic acid accumulation by roots of *Xanthium strumarium* L. and *Lycopersicon esculentum* Mill. in relation to water stress. *Plant Physiology* 79: 653–658.

Creelman, R.A., Mason, H.S., Bensen, R.J. et al. (1990). Water deficit and abscisic acid cause differential inhibition of shoot versus root growth in soybean seedlings: analysis of growth, sugar accumulation, and gene expression. *Plant Physiology* 92: 205–214.

Dardanelli, J.L., Bachmeier, O.A., Sereno, R. et al. (1997). Rooting depth and soil water extraction patterns of different crops in a silty loam Haplustoll. *Field Crops Research* 54: 29–38.

De Ollas, C., Arbona, V., Gómez-Cadenas, A. et al. (2018). Attenuated accumulation of jasmonates modifies stomatal responses to water deficit. *Journal of Experimental Botany* 69: 2103–2116.

De Smet, I., Signora, L., Beeckman, T. et al. (2003). An abscisic acid-sensitive checkpoint in lateral root development of *Arabidopsis*. *Plant Journal* 33: 543–555.

Dodd, I.C. (2005). Root-to-shoot signalling: assessing the roles of "up" in the up and down world of long-distance signalling *in planta*. *Plant and Soil* 274: 257–275.

Dodd, I.C. (2007). Soil moisture heterogeneity during deficit irrigation alters root-to-shoot signalling of abscisic acid. *Functional Plant Biology* 34: 439–448.

Dodd, I.C. (2009). Rhizosphere manipulations to maximise "crop per drop" during deficit irrigation. *Journal of Experimental Botany* 60: 2454–2459.

Dodd, I.C. (2013). Abscisic acid and stomatal closure: a hydraulic conductance conundrum? *New Phytologist* 197: 6–8.

Dodd, I.C. and Diatloff, E. (2016). Enhanced root growth of the *brb* (bald root barley) mutant in drying soil allows similar shoot physiological responses to soil water deficit as wild-type plants. *Functional Plant Biology* 43: 199–206.

Dodd, I.C., Egea, G., and Davies, W.J. (2008). Abscisic acid signalling when soil moisture is heterogeneous: decreased photoperiod sap flow from drying roots limits ABA export to the shoots. *Plant, Cell and Environment* 31: 1263–1274.

Dry, P.R., Loveys, B.R., and During, H. (2000). Partial drying of the rootzone of grape. II. Changes in the pattern of root development. *Vitis* 39: 9–12.

Du, Y. and Scheres, B. (2018). Lateral root formation and the multiple roles of auxin. *Journal of Experimental Botany* 69: 155–167.

Egea, G., Dodd, I.C., González-Real, M.M. et al. (2011). Partial rootzone drying improves almond tree leaf-level water use efficiency and afternoon water status compared to regulated deficit irrigation. *Functional Plant Biology* 38: 372–385.

Else, M.A., Davies, W.J., Whitford, P.N. et al. (1994). Concentrations of abscisic-acid and other solutes in xylem sap from root systems of tomato and castor-oil plants are distorted by wounding and variable sap flow-rates. *Journal of Experimental Botany* 45: 317–323.

Faiss, M., Zalubilova, M., Strnad, M. et al. (1997). Conditional transgenic expression of the IPT gene indicates a function for cytokinins in paracrine signaling in whole tobacco plants. *The Plant Journal* 12: 401–415.

Gallie, D.R., Geisler-Lee, J., Chen, J. et al. (2009). Tissue-specific expression of the ethylene biosynthetic machinery regulates root growth in maize. *Plant Molecular Biology* 69: 195–211.

Ghanem, M.E., Albacete, A., Smigocki, A.C. et al. (2011). Root-synthesised cytokinins improve shoot growth and fruit yield in salinised tomato (*Solanum lycopersicum* L.). *Journal of Experimental Botany* 62: 125–140.

Goicoechea, N., Antolin, M.C., and Sanchez-Diaz, M. (1997). Gas exchange is related to hormone balance in mycorrhizal or nitrogen-fixing alfalfa subjected to drought. *Physiologia Plantarum* 100: 989–997.

Gowing, D.J.G., Davies, W.J., and Jones, H.G. (1990). A positive root-sourced signal as an indicator of soil drying in apple, *Malus x domestica* Borkh. *Journal of Experimental Botany* 41: 1535–1540.

Hansen, H. and Dorffling, K. (2003). Root-derived trans-zeatin riboside and abscisic acid in drought-stressed and rewatered sunflower plants: interaction in the control of leaf diffusive resistance? *Functional Plant Biology* 30: 365–375.

Hodge, A. (2010). Roots: the acquisition of water and nutrients from the heterogeneous soil environment. *Progress in Botany* 71: 307–337.

Hodgkinson, L., Dodd, I.C., Binley, A. et al. (2017). Root growth in field-grown winter wheat (*Triticum aestivum* L.): some effects of soil conditions, season and genotype. *European Journal of Agronomy* 91: 74–83.

Holbrook, N.M., Shashidhar, V.R., James, R.A. et al. (2002). Stomatal control in tomato with ABA-deficient roots: response of grafted plants to soil drying. *Journal of Experimental Botany* 53: 1503–1514.

Hsiao, T.C. and Xu, L.K. (2000). Sensitivity of growth of roots versus leaves to water stress: biophysical analysis and relation to water transport. *Journal of Experimental Botany* 51: 1595–1616.

Hu, T., Kang, S.Z., Li, F.S. et al. (2011). Effects of partial root-zone irrigation on hydraulic conductivity in the soil-root system of maize plants. *Journal of Experimental Botany* 62: 4163–4172.

Jackson, M.B. (1993). Are plant hormones involved in root to shoot communication? *Advances in Botanical Research* 19: 103–187.

Kapilan, R., Vaziri, M., and Zwiazek, J.J. (2018). Regulation of aquaporins in plants under stress. *Biological Research* 51: 4.

Khalil, A.A.M. and Grace, J. (1993). Does xylem sap ABA control the stomatal behaviour of water-stressed sycamore (*Acer pseudoplatanus* L.) seedlings? *Journal of Experimental Botany* 44: 1127–1134.

Kirkegaard, J.A., Lilley, J.M., Howe, G.N. et al. (2007). Impact of subsoil water use on wheat yield. *Australian Journal of Agricultural Research* 58: 303–315.

Ko, D., Kang, J., Kiba, T. et al. (2014). *Arabidopsis* ABCG14 is essential for the root-to-shoot translocation of cytokinin. *Proceedings of the National Academy of Sciences of the United States of America* 111: 7150–7155.

Kreszies, T., Shellakkutti, N., Osthoff, A. et al. (2019a). Osmotic stress enhances suberization of apoplastic barriers in barley seminal roots: analysis of chemical, transcriptomic and physiological responses. *New Phytologist* 221: 180–194.

Kreszies, T., Eggels, S., Kreszies, V. et al. (2019b). Seminal roots of wild and cultivated barley differentially respond to osmotic stress in gene expression, suberization, and hydraulic conductivity. *Plant, Cell and Environment* 43: 344–357. https://doi.org/10.1111/pce.13675.

Krouk, G., Lacombe, B., Bielach, A. et al. (2010). Nitrate-regulated auxin transport by NRT1.1 defines a mechanism for nutrient sensing in plants. *Developmental Cell* 18: 927–937.

Kudoyarova, G.R., Vysotskaya, L.B., Cherkozyanova, A. et al. (2007). Effect of partial rootzone drying on the concentration of zeatin-type cytokinins in tomato (*Lycopersicon esculentum* Mill.) xylem sap and leaves. *Journal of Experimental Botany* 58: 161–168.

Kudoyarova, G., Veselova, S., Hartung, W. et al. (2011). Involvement of root ABA and hydraulic conductivity in the control of water relations in wheat plants exposed to increased evaporation demand. *Planta* 233: 87–94.

Kudoyarova, G.R., Korobova, A.V., Akhiyarova, G.R. et al. (2014). Accumulation of cytokinins in roots and their export to the shoots of durum wheat plants treated with the protonophore carbonyl cyanide m-chlorophenylhydrazone (CCCP). *Journal of Experimental Botany* 65: 2287–2294.

Leach, K.A., Hejlek, L.G., Hearne, L.B. et al. (2011). Primary root elongation rate and abscisic acid levels of maize in response to water stress. *Crop Science* 51: 157–172.

Li, S.W., Leng, Y., Feng, L. et al. (2014). Involvement of abscisic acid in regulating antioxidative defense systems and IAA-oxidase activity and improving adventitious rooting in mung bean [*Vigna radiata* (L.) Wilczek] seedlings under cadmium stress. *Environmental Science and Pollution Research* 21: 525–537.

Li, T., Yang, H., Zhang, W. et al. (2017). Comparative transcriptome analysis of root hairs proliferation induced by water deficiency in maize. *Journal of Plant Biology* 60: 26–34.

Li, W., de Ollas, C., and Dodd, I.C. (2018). Long-distance ABA transport can mediate distal tissue responses by affecting local ABA concentrations. *Journal of Integrative Plant Biology* 60: 16–33.

Lian, H.L., Yu, X., Ye, Q. et al. (2004). The role of aquaporin RWC3 in drought avoidance in rice. *Plant and Cell Physiology* 45: 481–489.

Liang, J., Zhang, J., and Wong, M.H. (1996). Effects of air-filled soil porosity and aeration on the initiation and growth of secondary roots of maize (*Zea mays*). *Plant and Soil* 186: 345–254.

Liang, J., Zhang, J., and Wong, M.H. (1997). How do roots control xylem sap ABA concentration in response to soil drying? *Plant and Cell Physiology* 38: 10–16.

Liu, J., He, H., Vitali, M. et al. (2015). Osmotic stress represses strigolactone biosynthesis in *Lotus japonicus* roots: exploring the interaction between strigolactones and ABA under abiotic stress. *Planta* 241: 1435–1451.

Lopes, M.S. and Reynolds, M.P. (2010). Partitioning of assimilates to deeper roots is associated with cooler canopies and increased yield under drought in wheat. *Functional Plant Biology* 37: 147–156.

Lynch, J.P. (2013). Steep, cheap and deep: an ideotype to optimize water and N acquisition by maize root systems. *Annals of Botany* 112: 347–357.

Manschadi, A.M., Hammer, G.L., Christopher, J.L. et al. (2008). Genotypic variation in seedling root architectural traits and implications for drought adaptation in wheat (*Triticum aestivum* L.). *Plant and Soil* 303: 115–129.

Manzi, M., Lado, J., Rodrigo, M.J. et al. (2015). Root ABA accumulation in long-term water-stressed plants is sustained by hormone transport from aerial organs. *Plant and Cell Physiology* 56: 2457–2466.

Masia, A., Pitacco, A., Braggio, L. et al. (1994). Hormonal responses to partial drying of the root system of *Helianthus annuus*. *Journal of Experimental Botany* 45: 69–76.

McAdam, S.A., Brodribb, T.J., and Ross, J.J. (2016). Shoot-derived abscisic acid promotes root growth. *Plant, Cell and Environment* 39: 652–659.

McLean, E.H., Ludwig, M., and Grierson, P.F. (2011). Root hydraulic conductance and aquaporin abundance respond rapidly to partial root-zone drying events in a riparian Melaleuca species. *New Phytologist* 212: 954–963.

Mingo, D.M., Theobald, J.C., Bacon, M.A. et al. (2004). Biomass allocation in tomato (*Lycopersicon esculentum*) plants grown under partial rootzone drying: enhancement of root growth. *Functional Plant Biology* 31: 971–978.

Miyawaki, K., Matsumoto-Kitano, M., and Kakimoto, T. (2004). Expression of cytokinin biosynthetic isopentenyltransferase genes in *Arabidopsis*: tissue specificity and regulation by auxin, cytokinin, and nitrate. *Plant Journal* 37: 128–138.

Munns, R. and King, R.W. (1988). Abscisic acid is not the only stomatal inhibitor in the transpiration stream of wheat plants. *Plant Physiology* 88: 703–708.

Netting, A.G., Theobald, J.C., and Dodd, I.C. (2012). Xylem sap collection and extraction methodologies to determine *in vivo* concentrations of ABA and its bound forms. *Plant Methods* 8: 11.

Ober, E.S. and Sharp, R.E. (1994). Proline accumulation in maize (*Zea mays*) primary roots at low water potentials 1. Requirement for increased levels of abscisic acid. *Plant Physiology* 105: 981–987.

Okamoto, M., Tanaka, Y., Abrams, S.R. et al. (2009). High humidity induces abscisic acid 8-hydroxylase in stomata and vasculature to regulate local and systemic abscisic acid responses in *Arabidopsis*. *Plant Physiology* 149: 825–834.

Parent, B., Hachez, C., Redondo, E. et al. (2009). Drought and abscisic acid effects on aquaporin content translate into changes in hydraulic conductivity and leaf growth rate: a trans-scale approach. *Plant Physiology* 149: 2000–2012.

Pavlů, J., Novák, J., Koukalová, V. et al. (2018). Cytokinin at the crossroads of abiotic stress signalling pathways. *International Journal of Molecular Sciences* 198: 2450.

Peng, Y.H., Lin, W.L., Cai, W.M. et al. (2007). Overexpression of a *Panax ginseng* tonoplast aquaporin alters salt tolerance, drought tolerance and cold acclimation ability in transgenic *Arabidopsis* plants. *Planta* 226: 729–740.

Péret, B., Li, G., Zhao, J. et al. (2012). Auxin regulates aquaporin function to facilitate lateral root emergence. *Nature Cell Biology* 14: 991–998.

Pérez-Pérez, J.G. and Dodd, I.C. (2015). Sap fluxes from different parts of the rootzone modulate xylem ABA concentration during partial rootzone drying and re-wetting. *Journal of Experimental Botany* 66: 2315–2324.

Pérez-Pérez, J.G., Dodd, I.C., and Botía, P. (2012). Partial root-zone drying increases water use efficiency of lemon 'Fino 49' trees independently of root-to-shoot ABA signalling. *Functional Plant Biology* 39: 366–378.

Pérez-Pérez, J.G., Navarro, J.M., Robles, J.M. et al. (2018). Prolonged drying cycles stimulate ABA accumulation in *Citrus macrophylla* seedlings exposed to partial rootzone drying. *Agricultural Water Management* 210: 271–278.

Puertolas, J., Alcobendas, R., Alarcon, J.J. et al. (2013). Long-distance abscisic acid signalling under different vertical soil moisture gradients depends on bulk root water potential and average soil water content in the root zone. *Plant, Cell and Environment* 36: 1465–1475.

Puig, J., Pauluzzi, G., Guiderdoni, E. et al. (2012). Regulation of shoot and root development through mutual signaling. *Molecular Plant* 5: 974–983.

Ranathunge, K. and Schreiber, L. (2011). Water and solute permeabilities of *Arabidopsis* roots in relation to the amount and composition of aliphatic suberin. *Journal of Experimental Botany* 62: 1961–1974.

Reed, R.C., Brady, S.B., and Muday, G.K. (1998). Inhibition of auxin movement from the shoot into the root inhibits lateral root development in *Arabidopsis*. *Plant Physiology* 118: 1369–1378.

Richard, C.A.I., Hickey, L.T., Fletcher, S. et al. (2015). High–throughput phenotyping of seminal root traits in wheat. *Plant Methods* 11: 13.

Romero, P., Dodd, I.C., and Martinez–Cutillas, A. (2012). Contrasting physiological effects of partial root-zone drying in field-grown grapevine (*Vitis vinifera* L. cv. Monastrell) according to total soil water availability. *Journal of Experimental Botany* 63: 4071–4083.

Rothwell, S.A., Elphinstone, E.D., and Dodd, I.C. (2015). Liming can decrease legume crop yield and leaf gas exchange by enhancing root to shoot ABA signalling. *Journal of Experimental Botany* 66: 2335–2345.

Ruiz-Lozano, J.M., del Mar Alguacil, M., Barzana, G. et al. (2009). Exogenous ABA accentuates the differences in root hydraulic properties between mycorrhizal and non mycorrhizal maize plants through regulation of PIP aquaporins. *Plant Molecular Biology* 70: 565–579.

Saab, I.N., Sharp, R.E., Pritchard, J. et al. (1990). Increased endogenous abscisic acid maintains primary root growth and inhibits shoot growth of maize seedlings at low water potentials. *Plant Physiology* 93: 1329–1336.

Sade, N., Gebretsadik, M., Seligmann, R. et al. (2010). The role of tobacco aquaporin1 in improving water use efficiency, hydraulic conductivity, and yield production under salt stress. *Plant Physiology* 152: 245–254.

Sanchez-Romera, B., Ruiz-Lozano, J.M., Li, G. et al. (2014). Enhancement of root hydraulic conductivity by methyl jasmonate and the role of calcium and abscisic acid in this process. *Plant, Cell and Environment* 37: 995–1008.

Sasaki, T., Soyano, T., Kojima, M. et al. (2014). Shoot-derived cytokinins systemically regulate root nodulation. *Nature Communications* 5: 4983.

Savchenko, T., Kolla, V.A., Wang, C.Q. et al. (2014). Functional convergence of oxylipin and abscisic acid pathways controls stomatal closure in response to drought. *Plant Physiology* 164: 1151–1160.

Schmundt, D., Stitt, M., Jahne, B. et al. (1998). Quantitative analysis of the local rates of growth of dicot leaves at a high temporal and spatial resolution, using image sequence analysis. *Plant Journal* 16: 505–514.

Seo, P.J., Xiang, F., Qiao, M. et al. (2009). The MYB96 transcription factor mediates abscisic acid signaling during drought stress response in *Arabidopsis*. *Plant Physiology* 151: 275–289.

Sharipova, G., Veselov, D., Kudoyarova, G.R. et al. (2016). Exogenous application of abscisic acid (ABA) increases root and cell hydraulic conductivity and abundance of some aquaporin isoforms in the ABA deficient barley mutant Az34. *Annals of Botany* 118: 777–785.

Sharp, R.E., Silk, W.K., and Boyer, J.S. (1988). Growth of the maize primary root at low water potentials. I. Spatial distribution of expansive growth. *Plant Physiology* 87: 50–57.

Sharp, R.E., Wu, Y.J., Voetberg, G.S. et al. (1994). Confirmation that abscisic acid accumulation is required for maize primary root elongation at low water potentials. *Journal of Experimental Botany* 45: 1743–1751.

Shkolnik-Inbar, D. and Bar-Zvi, D. (2010). ABI4 mediates abscisic acid and cytokinin inhibition of lateral root formation by reducing polar auxin transport in *Arabidopsis*. *Plant Cell* 22: 3560–3573.

Simonneau, T., Barrieu, P., and Tardieu, F. (1998). Accumulation rate of ABA in detached maize roots correlates with root water potential regardless of age and branching order. *Plant, Cell and Environment* 21: 1113–1122.

Slovik, S., Daeter, W., and Hartung, W. (1995). Compartmental redistribution and long-distance transport of abscisic acid (ABA) in plants as influenced by environmental changes in the rhizosphere – a biomathematical mode. *Journal of Experimental Botany* 46: 881–894.

Soar, C.J., Speirs, J., Maffei, S.M. et al. (2004). Gradients in stomatal conductance, xylem sap ABA and bulk leaf ABA along canes of *Vitis vinifera* cv. Shiraz: molecular and physiological studies investigating their source. *Functional Plant Biology* 31: 659–669.

Sobeih, W.Y., Dodd, I.C., Bacon, M.A. et al. (2004). Long-distance signals regulating stomatal conductance and leaf growth in tomato (*Lycopersicon esculentum*) plants subjected to partial rootzone drying. *Journal of Experimental Botany* 55: 2353–2363.

Speirs, J., Binney, A., Collins, M. et al. (2013). Expression of ABA synthesis and metabolism genes under different irrigation strategies and atmospheric VPDs is associated with stomatal conductance in grapevine (*Vitis vinifera* L. cv Cabernet Sauvignon). *Journal of Experimental Botany* 64: 1907–1916.

Spollen, W.G. and Sharp, R.E. (1991). Spatial distribution of turgor and root growth at low water potentials. *Plant Physiology* 96: 438–443.

Spollen, W.G., LeNoble, M.E., Samuels, T.D. et al. (2000). Abscisic acid accumulation maintains maize primary root elongation at low water potentials by restricting ethylene production. *Plant Physiology* 122: 967–976.

Stoll, M., Loveys, B., and Dry, P. (2000). Hormonal changes induced by partial rootzone drying of irrigated grapevine. *Journal of Experimental Botany* 51: 1627–1634.

Tardieu, F., Parent, B., and Simonneau, T. (2010). Control of leaf growth by abscisic acid: hydraulic or non-hydraulic processes. *Plant, Cell and Environment* 33: 636–647.

Tester, M. and Langridge, P. (2010). Breeding technologies to increase crop production in a changing world. *Science* 327: 818–822.

Thompson, A.J., Jackson, A.C., Parker, R.A. et al. (2000). Abscisic acid biosynthesis in tomato: regulation of zeaxanthin epoxidase and 9-*cis*-epoxycarotenoid dioxygenase mRNAs by light/dark cycles, water stress and abscisic acid. *Plant Molecular Biology* 42: 833–845.

Thompson, A.J., Andrews, J., Mulholland, B.J. et al. (2007). Overproduction of abscisic acid in tomato increases transpiration efficiency and root hydraulic conductivity and influences leaf expansion. *Plant Physiology* 143: 1905–1917.

Uga, Y., Sugimoto, K., Ogawa, S. et al. (2013). Control of root system architecture by DEEPER ROOTING 1 increases rice yield under drought conditions. *Nature Genetics* 145: 1097.

Valerio, M., Lovelli, S., Sofo, A. et al. (2017). Root and leaf abscisic acid concentration impact on gas exchange in tomato (*Lycopersicon esculentum* Mill) plants subjected to partial root-zone drying. *Italian Journal of Agronomy* 12: 25–32.

Valluru, R., Davies, W.J., Reynolds, M. et al. (2016). Foliar abscisic acid-to-ethylene accumulation and response regulate shoot growth sensitivity to mild drought in wheat. *Frontiers in Plant Science* 7: 461.

Veselov, D.S., Sharipova, G.V., Veselov, S. et al. (2018). Rapid changes in root *HvPIP2;2* aquaporins abundance and ABA concentration are required to enhance root hydraulic conductivity and maintain leaf water potential in response to increased evaporative demand. *Functional Plant Biology* 45: 143–149.

Visentin, I., Vitali, M., Ferrero, M. et al. (2016). Low levels of strigolactones in roots as a component of the systemic signal of drought stress in tomato. *New Phytologist* 212: 954–963.

Voetberg, G.S. and Sharp, R.E. (1991). Growth of the maize primary root at low water potentials. Ill. Role of increased proline deposition in osmotic adjustment. *Plant Physiology* 96: 1125–1130.

Wang, Y., Liu, F., and Jensen, C.R. (2012). Comparative effects of deficit irrigation and alternate partial root-zone irrigation on xylem pH, ABA and ionic concentrations in tomatoes. *Journal of Experimental Botany* 63: 1907–1917.

Wang, L., Li, Q.T., Lei, Q. et al. (2017). Ectopically expressing *MdPIP1;3*, an aquaporin gene, increased fruit size and enhanced drought tolerance of transgenic tomatoes. *BMC Plant Biology* 17: 426.

Werner, T., Nehnevajova, E., Köllmer, I. et al. (2010). Root-specific reduction of cytokinin causes enhanced root growth, drought tolerance and leaf mineral enrichment in *Arabidopsis* and tobacco. *Plant Cell* 22: 3905–3920.

Westgate, M.E. and Boyer, J.S. (1985). Osmotic adjustment and the inhibition of leaf, root, stem and silk growth at low water potentials in maize. *Planta* 164: 540–549.

Whalley, W.R., Binley, A., Watts, C.W. et al. (2017). Methods to estimate changes in soil water for phenotyping root activity in the field. *Plant and Soil* 415: 407–422.

White, R.G. and Kirkegaard, J.A. (2010). The distribution and abundance of wheat roots in a dense, structured subsoil – implications for water uptake. *Plant, Cell and Environment* 33: 133–148.

Wilkinson, S. and Davies, W.J. (1997). Xylem sap pH increase: a drought signal received at the apoplastic face of the guard cell that involves the suppression of saturable abscisic acid uptake by the epidermal symplast. *Plant Physiology* 113: 559–573.

Wilkinson, S., Kudoyarova, G.R., Veselov, D.S. et al. (2012). Plant hormone interactions: innovative targets for crop breeding and management. *Journal of Experimental Botany* 63: 3499–3509.

Wu, Y., Spollen, W.G., Sharp, R.E. et al. (1994). Root growth maintenance at low water potentials: increased activity of xyloglucan endotransglycosylase and its possible regulation by abscisic acid. *Plant Physiology* 106: 607–615.

Wu, Y., Sharp, R.E., Durachko, D.M. et al. (1996). Growth maintenance of the maize primary root at low water potentials involves increases in cell-wall extension properties, expansion activity, and wall susceptibility to expansions. *Plant Physiology* 111: 765–772.

Xiong, L., Wang, R.G., Mao, G. et al. (2006). Identification of drought tolerance determinants by genetic analysis of root response to drought stress and abscisic acid. *Plant Physiology* 142: 1065–1074.

Xu, P.P. and Cai, W.M. (2017). Functional characterization of the BnNCED3 gene in *Brassica napus*. *Plant Science* 256: 16–24.

Yan, F., Sun, Y., Song, F. et al. (2012). Differential responses of stomatal morphology to partial root–zone drying and deficit irrigation in potato leaves under varied nitrogen rates. *Scientia Horticulturae* 145: 76–83.

Yu, H., Chen, X., Hong, Y.Y. et al. (2008). Activated expression of an *Arabidopsis* HD-START protein confers drought tolerance with improved root system and reduced stomatal density. *Plant Cell* 20: 1134–1151.

Zhan, A., Schneider, H., and Lynch, J.P. (2015). Reduced lateral root branching density improves drought tolerance in maize. *Plant Physiology* 168: 1603–1615.

Zhang, J. and Davies, W.J. (1989). Sequential response of whole plant water relations to prolonged soil drying and the involvement of xylem sap ABA in the regulation of stomatal behaviour of sunflower plants. *New Phytologist* 113: 167–174.

Zhang, J. and Davies, W.J. (1991). Antitranspirant activity in xylem sap of maize plants. *Journal of Experimental Botany* 42: 317–321.

Zhang, J., Zhang, X., and Liang, J. (1995). Exudation rate and hydraulic conductivity of maize roots are enhanced by soil drying and abscisic acid treatment. *New Phytologist* 131: 329–336.

Zhang, J., Yu, H., Zhang, Y. et al. (2017). Increased abscisic acid levels in transgenic maize overexpressing *AtLOS5* mediated root ion fluxes and leaf water status under salt stress. *Journal of Experimental Botany* 67: 1339–1355.

Zhang, F.P., Sussmilch, F., Nichols, D.S. et al. (2018). Leaves, not roots or floral tissue, are the main site of rapid, external pressure-induced ABA biosynthesis in angiosperms. *Journal of Experimental Botany* 69: 1261–1267.

10

Winter Wheat and Summer Maize Roots in Agro-Ecosystems on the North China Plain

Fadong Li, Zhao Li, Yunfeng Qiao, Nong Zhu, Kun Du, Peifang Leng, and Shanbao Liu

Key Laboratory of Ecosystem Network Observation and Modeling, Institute of Geographic Sciences and Natural Resources Research, Chinese Academy of Sciences, Beijing, China; Shandong Yucheng Agro-ecosystem National Observation and Research Station, Ministry of Science and Technology, Yucheng, China; University of Chinese Academy of Sciences, Beijing, China

10.1 Background

Food security is a considerable challenge (Cole et al. 2018; FAO 2017; FAO, IFAD, and WFP 2015) to realize the Sustainable Development Goals (SDGs) of the United Nations by 2030 (United Nations 2015). One of the most severe threats to food security is insufficient water resources (Grayson 2013), and this threat is being exacerbated continuously. Drought responses of crops, especially responses of roots, improve crop resilience against drought (Grayson 2013; Rey et al. 2017; Gupta et al. 2020).

North China Plain (NCP) supports food security in China (Godfray 2010; Piao et al. 2010; SCIO_PRC 2019) as one of the critical grain-producing areas; however, it is also one of the areas with scarce water resources (Jeong et al. 2014). The average annual precipitation is about 500 mm, but the annual water requirement of wheat and maize is more than 900 mm (Hu et al. 2010; Kendy et al. 2003; Wang et al. 2008). In addition, there is a mismatch between water resources and the distribution of population, which results in two critical water problems: (i) the drying up of the main course of the lower Yellow River reaches and (ii) eco-environmental degradation (Liu and Xia 2004). Due to its unique environment and food requirements, the water security and food security in NCP are particularly important. Therefore, the research on crop roots, soil nutrient utilization, soil structure, and water-use efficiency is essential to the growth and productivity of crops (Deng et al. 2006; Fan et al. 2011; Zhang et al. 2008). Studies have shown that the basis for ensuring high and stable crop yield is large biomass, and the biomass amount is closely related to the development of crop roots. In the crop models, sufficient rooting depth is defined as 1.5 m for winter wheat (Ouyang and Luo 2002) and 1.0 m for maize in NCP (Liu and Luo 2010), with roots located mainly in the upper 0.2–0.3 m (Mo and Liu 2001).

The Root Systems in Sustainable Agricultural Intensification, First Edition.
Edited by Zed Rengel and Ivica Djalovic.

The main objectives of this chapter were to present the crop root research in the NCP, to explore the mechanisms of root water uptake and various relevant nutritional and metabolic processes, and to characterize the improved water absorption and fertilizer utilization capacity of crop roots under the premise of constant (or a small increase) in allocation and consumption of photosynthetic products.

10.2 General Information on the Study Area

10.2.1 North China Plain (NCP)

The NCP is one of the three major plains in China. It is an essential part of the Great Plains in Eastern China (Figure 10.1). It is also known as the Huang-Huai-Hai Plain (meaning that the three main rivers: Huanghe (Yellow) River, Huaihe River, and Haihe River flow through this area), and has an area of 300 000 km^2. It has a flat terrain, convenient transportation, and numerous watercourses and lakes. It has been China's political, economic, and cultural centre since ancient times. Its population and cultivated land each account for about 20% of the whole country.

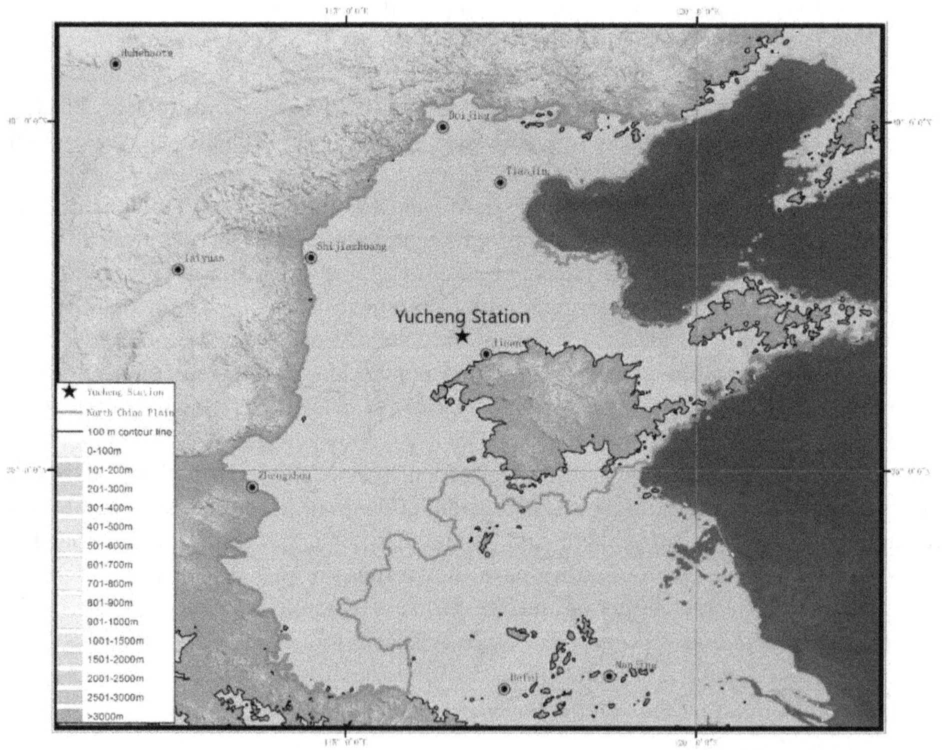

Figure 10.1 Topographic map of the North China Plain. *Source:* Drawn by the authors using the DEM (Digital Elevation Model) data downloaded from: http://srtm.csi.cgiar.org/srtmdata/

Table 10.1 The percentages of wheat, maize, and total grain production and the corresponding planted areas in the North China Plain as proportions of the China's totals.

Year	Wheat grain production (%)	Wheat planting area (%)	Maize grain production (%)	Maize planting area (%)	Total grain production (%)
2009	56	47	39	28	24
2010	56	47	36	28	24
2011	56	47	34	27	23
2012	56	48	32	27	23
2013	55	47	22	23	22
2014	55	47	22	22	21
2015	55	47	21	22	21
2016	54	47	21	22	21
2017	57	50	27	27	24

The fluvo-aquic soil is the most important cultivated soil type in the NCP. It has suitable tillage properties, is rich in mineral nutrients, and has a great potential in agricultural utilization. The NCP has warm, temperate monsoon climate with pronounced seasonal variations. The Huaihe River Basin in the south of NCP is a transition area to the subtropical zone, which has higher temperature and precipitation than areas in the north. The average annual temperature on the plain is 8–15 °C, with cold and dry winters.

The cropping systems are mostly double-crop rotation in one year and three crops in two years only in some parts of the northern area. The main food crops are wheat and maize. In the NCP, the wheat production and sown areas are about 55% and 47%, respectively, of the whole China, and those of maize were about 22.6% and 23.2%, respectively. The total grain production and sown area account for 21.8% and 20.4%, respectively, of the whole of China (Table 10.1).

10.2.2 Shandong Yucheng Agro-Ecosystem National Observation and Research Station (SYA-NORS)

Shandong Yucheng Agro-Ecosystem National Observation and Research Station (SYA-NORS), also called Yucheng Comprehensive Experiment Station (YCES), is one of more than 50 stations supported by the Ministry of Science and Technology, the P.R. of China, and is also one of stations of the Chinese Academy of Sciences; it is located in Dezhou (36°56′N, 116°40′E, 23 m a.s.l.), Shandong Province. It has a warm, temperate semiarid climate. The annual mean air temperature is 13.3 °C, and the annual precipitation is 560 mm according to the long-term observations in SYA-NORS (1980–2018). Precipitation is distributed unevenly, with 70% of the annual precipitation falling between June and September. The annual sunshine is 2640 hours. The soil type is aquents (poorly drained soils formed from human-transported material or on excavated landscapes) with high salinity (Tu and Li 2017).

The research direction of the SYA-NORS is aimed at the rational use of natural resources such as water, soil, and biology, and the sustainable development of the region. It concentrates on the mechanisms and models of energy cycling and conversion in the Earth's surface layer. Experimental research is carried out on development of spatial energy conversion methods; innovative measurement methods and improvement and development of instruments; combining the theories, methods, and means of geography, ecology, and agronomy in studying the structure and function of agro-ecosystems; and optimizing the management models and developing pilot demonstrations of supporting technologies.

10.3 Methods

Given the importance of the root systems and the need to understand them better, it is necessary first to develop the appropriate root system research methods to investigate root morphology and physiology. Ideally, researchers would like to use the simple, destructive, or, even better, non-destructive research methods to determine quickly, accurately, and quantitatively the morphological and physiological characteristics of crop roots. At present, due to a wide variety of research methods on roots and the significant differences in measuring methods, it is not easy to classify the existing root research methods accurately. Based on different research purposes and requirements, the crop root research methods can be divided into the following categories:

i) Direct sampling methods in the field, such as excavation, drilling, coring, sectioning, etc.
ii) Direct observation methods, such as root chambers, glass walls, rhizotron cameras, etc.
iii) Indirect observation methods, such as dyeing techniques, non-radioactive tracers, etc., and
iv) Other methods, such as using *in situ* containers, tube cultivation, micro root canal system observations, etc.

10.4 Root Research on Winter Wheat and Maize in the North China Plain: a Brief Overview

Wheat and maize are the main food crops in the NCP. Given the critical importance of the root systems, many studies have been conducted on wheat and maize roots in the NCP. The growth and development of above-ground biomass depend on the acquisition of soil nutrients and water by roots (Ju et al. 2015; Qi et al. 2012; Wang et al. 2014; Wheeler et al. 1993).

10.4.1 Wheat Root Studies

The characteristics of the crop production in NCP are a lack of water and excessive use of nitrogen fertilizers. Water-use efficiency and nitrogen fertilizer-use efficiency of winter wheat production must urgently be improved in the NCP. Recently, there have been many studies about the relationship between root systems and other factors, such as different irrigation and nitrogen regimes, tillage practices, and rhizosphere microbiome (Liu et al. 2018b).

The results showed that both the irrigation method and schedule influenced root development, the root distribution pattern down the soil profile, and the dynamics of root water uptake (Guo et al. 2018; Jha et al. 2017; Li et al. 2010; Liu et al. 2018a; Lv et al. 2009, 2015; Shao et al. 2009; Xu et al. 2016). Irrigation in the stem elongation stage resulted in high grain yield and water-use efficiency and offered a sound basis for developing the deficit irrigation regimes in north China (Shao et al. 2009).

The effects of different tillage practices (plough tillage, rotary tillage, and no-tillage) on root growth, water consumption characteristics, grain yield, water use, and water-use efficiency were evaluated under rain-fed conditions in a field with 20 years of rotary tillage history (Ali et al. 2018). Zhao et al. (2014) studied the effects of tillage (moldboard plough to a maximum depth of 15 cm, deep moldboard plough to a maximum depth of 30 cm, and chisel plough to a maximum depth of 30 cm) and crop residue management on soil respiration and its mechanisms in the NCP area. Compared with the moldboard plough + crop residue removed treatment, the root dry weight density of the deep moldboard plough + crop residue retained treatment and the chisel plough + crop residue removed treatment increased by 45% and 39%, respectively.

The rhizosphere harbours complex microbial communities, whose dynamic associations are considered critical for plant growth and health (De Vries et al. 2020). Distinct microbial co-occurrence patterns exist in the wheat rhizosphere, which could be associated with various agricultural ecosystem properties (Fan et al. 2018).

10.4.1.1 Winter Wheat Root Distribution down the Soil Profile

Root length distribution changed with the growth stages (Figure 10.2) (Zhang et al. 2004b). Root length distribution in the upper 80 cm declined from 94% in the booting stage to 80% at maturity. The treatments with large amount of irrigation water had more roots than the treatments with less irrigation and the rain-fed treatments (Zhang et al. 2004b).

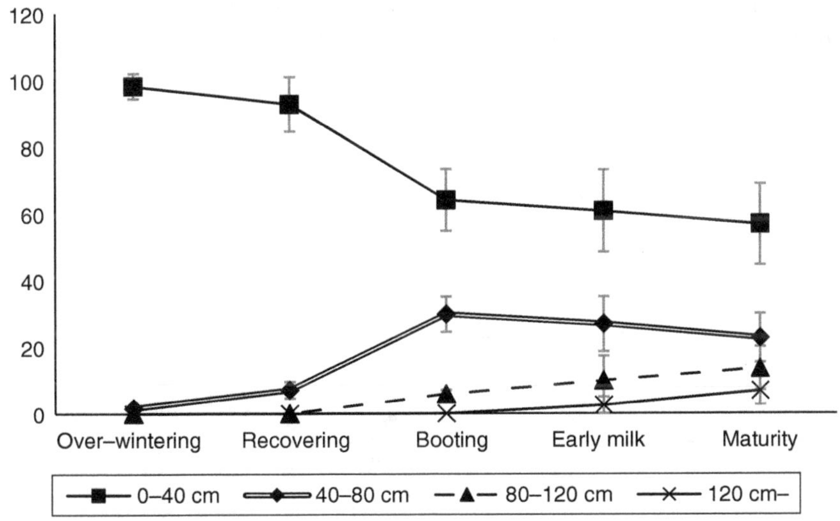

Figure 10.2 Root length distribution (%) of winter wheat along the soil profile (0–40, 40–80, 80–120, and 120 cm and deeper) in the North China Plain. Means ± SE. *Source:* Data from Zhang et al. (2004b).

10.4.1.2 Influence of Cropland Management on Winter Wheat Root Growth

Agronomic management influenced the distribution of winter wheat roots along the soil profile (Benjamin and Nielsen 2006; Zhang et al. 2004b). In particular, the irrigation methods altered root distribution (Lv et al. 2010; Phene et al. 1991; Proffitt et al. 1985). Wang et al. (2014) found that irrigation and nitrogen fertilization influence root growth and grain yield of winter wheat, with grain yield strongly correlated with root weight density (positively at 20–60 cm, but negatively in deep soil layers) in the Huanghuai wheat production area of China.

No significant difference in root length density of winter wheat was found between shallow and deep tillage (Zhang et al. 2004b). However, deep tillage was conducive to root growth in deeper soil layers (40–120 cm). Furthermore, more roots were found under the rows than between the rows (Zhang et al. 2004b) in the 0–20 cm layer. No significant difference was recorded in the other soil layers.

Winter wheat roots reached the 2.0 m depth (Zhou et al. 2008), with the maximum depth achieved at flowering. The ^{15}N isotope data showed that deep-rooted winter wheat could use soil nitrate up to 2.0 m depth. A significant decrease in nitrate content at the 1.4–2.0 m soil depth was observed at winter wheat maturity (Figure 10.3).

The research on nitrogen fertilization is often combined with irrigation. Wang et al. (2014) evaluated the effects of irrigation and nitrogen regimes on wheat root growth under high-yielding conditions in the Huanghuai wheat production area of China by performing field experiments at three locations in Henan Province. Liu et al. (2018b) investigated the effects of irrigation and N regimes on root development and their relationship with soil water and N use in different soil layers.

10.4.2 Maize Root Studies

Another critical food crop in the NCP is maize. There have been many studies on the maize root systems in the recent years, including grain productivity, nitrate leaching, soil water, and soil N content. The effective and modern cultivation model for high maize productivity

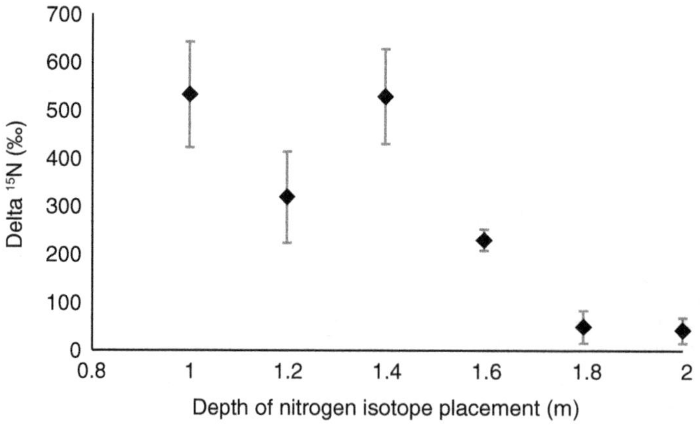

Figure 10.3 ^{15}N in winter wheat grain from the tracer placed at six soil depths. Means ± SE. *Source:* Data from Zhou et al. (2008)

in the semiarid areas is plastic film mulching that enhanced ear and grain properties, and root spatial and temporal distribution down the soil profile, resulting in increased grain yields and reduced productivity risks in dryland farming systems (Jia et al. 2018).

Nitrate that leached below the maize root zone was still available to deep-rooted wheat in the summer maize-winter wheat rotation in the NCP (Zhou et al. 2008). Overuse of inorganic N fertilizers and high net mineralization and nitrification, together with a predominance of rainfall in the summer season and light soil texture, are the main controlling factors responsible for a high nitrate leaching loss in this soil-crop-climate system (Huang et al. 2017).

Root length density of fine roots (diameter <0.2 mm) in the 0–20 cm soil layer was significantly and positively correlated with soil water content and negatively with soil mineral N content. The root length density of fine roots, as well as roots with the diameter >0.4 mm in the 20–60 cm soil layer, was positively correlated with shoot N uptake (Zhang et al. 2018).

10.4.2.1 Summer Maize Root Distribution along the Soil Profile

Figure 10.4 shows the root dry weight density and distribution down the soil profile. Compared with the roots of winter wheat, the summer maize roots were distributed mainly in the shallow layer, with more than 90% growing at 0–40 cm depth (Zhou et al. 2008).

Usually, maize root dry weight reaches a maximum at the tasseling (VT) stage (Figure 10.5) (Qi et al. 2012). For the two most popular maize genotypes, the maximum root dry weight of DH661 was 35.2 and 34.5 g plant^{-1} in 2009 and 2010, respectively, and that of ZD958 was 19.2 and 18.8 g plant^{-1} in 2009 and 2010, respectively (Qi et al. 2012).

10.4.2.2 Influence of Cropland Management on Summer Maize Root Growth

The summer maize roots grew up to 1.2 m soil depth (Zhou et al. 2008), reaching the maximum depth at silking. Maize root length density varied in different growing stages (Figure 10.6) (Ren et al. 2018).

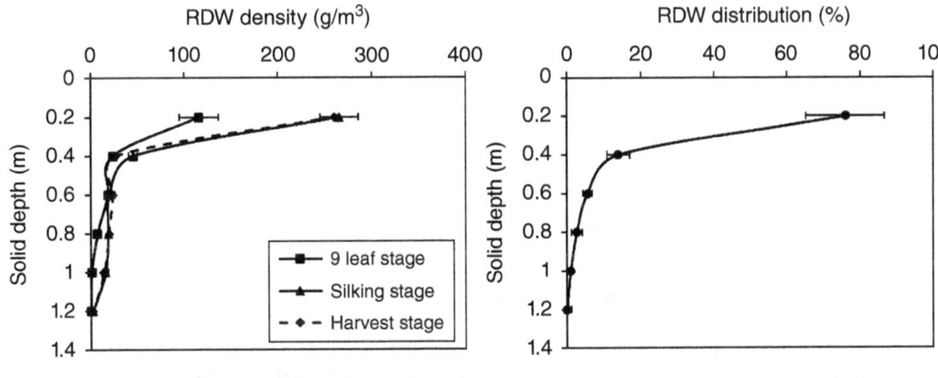

Figure 10.4 Root dry weight (RDW) density at three growth stages and the relative RDW distribution at harvest of summer maize in the North China Plain. Means ± SE. *Source:* Data from Zhou et al. (2008).

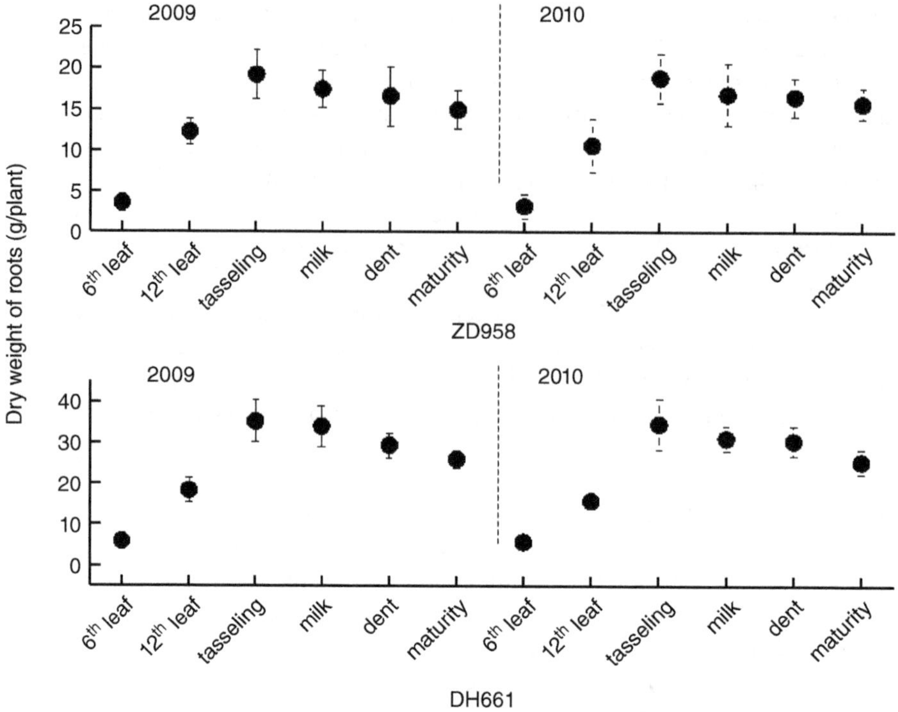

Figure 10.5 Variation in root dry weight of two maize genotypes (ZD958 and DH661) at various growing stages in the North China Plain in 2009 and 2010. Means ± SE. *Source:* Data from Qi et al. (2012).

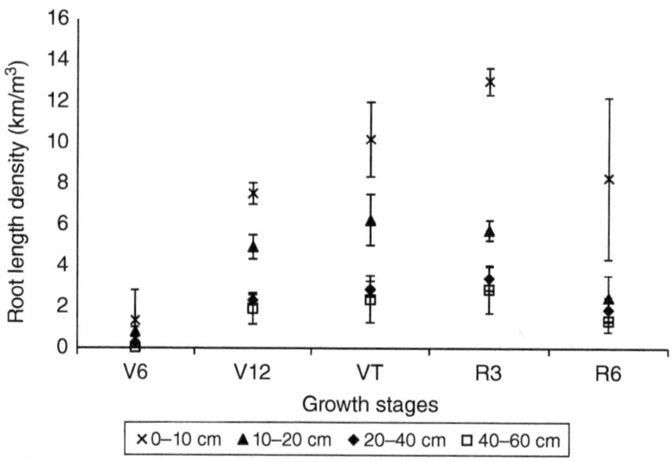

Figure 10.6 Maize root length density at various soil depths and in various growth stages: V6 (6th leaf), V12 (12th leaf), VT (tasseling), R3 (milk), and R6 (physiological maturity). Means ± SE. *Source:* Modified from Ren et al. (2018).

In the NCP, one reason for the wheat-maize rotation system being popular is that the nitrate leached during the winter wheat season remained in the maize root zone, maintaining relatively high nitrate content without N fertilization for the maize season.

10.5 Relationship Between Roots, Water Use, and Crop Yield

10.5.1 Wheat

The current research on crop roots in the NCP is mainly through the combination of field experiments and stable isotope techniques to explore the relationship between plant roots and rhizosphere microorganisms, irrigation methods, water-use efficiency, and nitrogen utilization. After decades of research, it has been proposed to enhance crop growth and yield by promoting the growth of roots.

At the SYA-NORS, located in the NCP, we have accomplished a substantial body of work on wheat and maize root systems. Zhao et al. (2018) evaluated water uptake models (Molz–Remson model, Selim–Iskandar model, and Feddes model) by using the precise field observation data. The hydrogen and oxygen stable isotopes revealed dynamic characteristics of soil water absorption by roots of winter wheat and summer maize during the whole growth period.

We have done long-term observations of the root system in the field trials to collect data on roots, grain yield, meteorological parameters, and soil-related indicators in the 2009–2014 period, monitoring wheat and maize root density three times each year. In the wheat growth period, we chose the stages of stem elongation (March), booting (April), and grain ripening (May and June). In the maize growth period, we chose the stages of seedling growth (July), stem elongation (August), and grain ripening (September). Figure 10.7 shows the root density of wheat in five years (2009–2010 and 2012–2014), with clear differences in both intra- and inter-season variability.

The data on wheat root density and production were shown in Table 10.2. We studied the potential relationship between production and root density during three periods of wheat. The results showed that the maximum root density at grain ripening was 341 $g\,m^{-3}$, and the period of intensive growth of wheat roots was mainly in the stem elongation to booting stages (Table 10.2). The irrigation in this period of intensive root growth was a significant factor in enhancing wheat growth.

Figure 10.8 shows the relationships between wheat root density and grain yield in the 5-year period (see Table 10.2). The rainfall in 2010 was higher than in the other years; hence, the 2010 data were excluded in Figure 10.8a. There was an obvious relationship between wheat root density (at stem elongation and ripening) and grain yield (Figure 10.8a,c). However, there was no correlation between grain production and the root density at booting (Figure 10.8b).

The positive relationship between wheat root density and yield at ripening (Figure 10.8c) was similar to that in the other studies (Ehdaie et al. 2011; Liu et al. 2018a). The results further indicated that the optimization of wheat root density would increase the final yield of wheat. In addition, other relevant studies have also shown that, under experimental

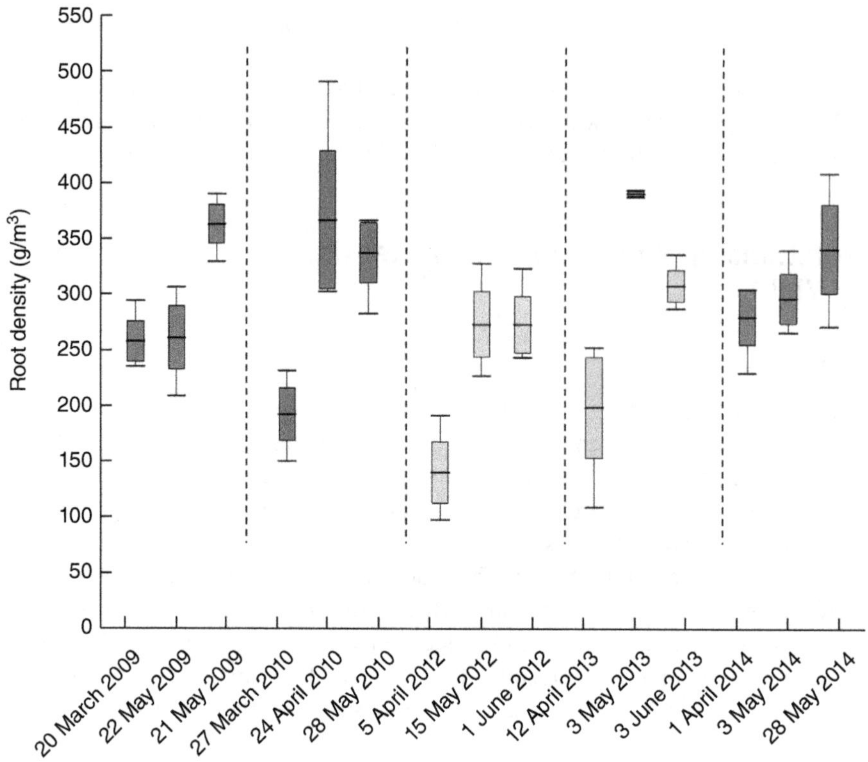

Figure 10.7 The root density of wheat in the top 1 m of soil during five growing seasons on the North China Plain. The central line in each box represents the mean value.

Table 10.2 Wheat root system density in the top 1 m of soil and grain yield in five seasons on the North China Plain.

Sampling time	Root density (g m^{-3})	Grain yield (t ha^{-1})
20 March 2009	258 ± 32	6.8 ± 0.29
22 April 2009	261 ± 49	
21 May 2009	363 ± 31	
27 March 2010	192 ± 18	6.9 ± 0.82
24 April 2010	367 ± 108	
28 May 2010	337 ± 47	
5 April 2012	140 ± 47	6.3 ± 0.58
15 May 2012	274 ± 51	
1 June 2012	273 ± 44	
12 April 2013	199 ± 78	6.4 ± 0.51
3 May 2013	391 ± 3	
3 June 2013	308 ± 25	
1 April 2014	280 ± 43	7.3 ± 0.38
3 May 2014	297 ± 39	
28 May 2014	341 ± 69	

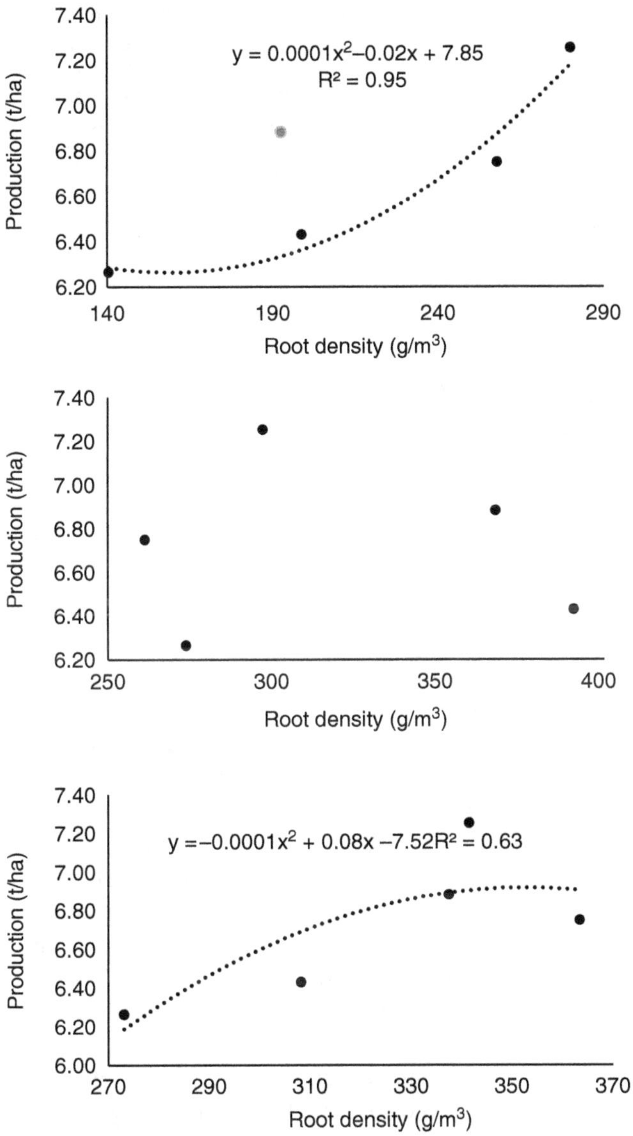

Figure 10.8 Wheat roots density in the top 1 m of soil in various growth stages and grain production on North China Plain over a 5-year period. (a) at stem elongation; (b) at booting; and (c) at grain ripening. The light-grey symbol in (a) was an outlier and was excluded from the analysis.

conditions, rational application of fertilizers and irrigation can improve water-use efficiency and soil N-use efficiency by optimizing root density of wheat (Man et al. 2016; Plett et al. 2020).

Effects of irrigation on wheat root system have been a research focus on the NCP for many years. Wheat in the rain-fed treatment and the irrigated treatment (with a few irrigation events) tended to have relatively greater root density in deeper layers of soil than the highly irrigated treatments. In the topsoil layer, this trend was reversed (Zhang et al. 2004b), with wheat in the non-irrigated treatment and the treatment with a few irrigation events

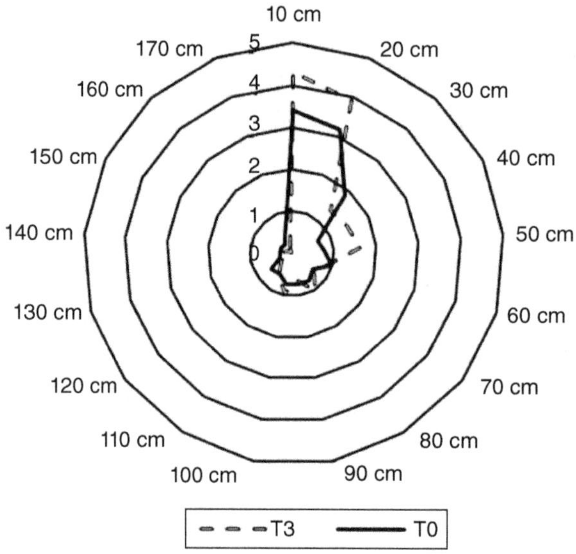

Figure 10.9 Distribution of root length density down the soil profile depth (in cm) at maturity for rain-fed (T0) and irrigated winter wheat (T3; three irrigation events at 60 mm water each). *Source:* Data from Zhang et al. (2004b).

having lower root density than in the more frequently irrigated treatments. At grain ripening, wheat root length density below 140 cm was higher in the rain-fed treatment (T0) than the treatment receiving three irrigations (T3) (Figure 10.9). In contrast, in the topsoil layer, the latter had a higher root length density than the former.

In the field trials at the Yucheng Comprehensive Experimental Station, in order to assess relative root growth, rate of biomass increase, and grain yield, we designed a range of treatments in which soil water contents were kept at 80%, 65%, and 50% of the field capacity by applying irrigation at depths of 0, 30, 50, and 100 cm (Li et al. 2002). The results showed that the 0–30 cm soil layer had a larger wheat root length density when a smaller amount of water was supplied (Figure 10.10a), suggesting the adaptive plant responses to increasing root absorptive capacity. In addition, the distribution of wheat roots was related to the type of irrigation, with surface irrigation associated with roots being mainly in the topsoil and root length density decreasing exponentially with the soil depth (Figure 10.10b).

When deep irrigation was applied, wheat root length density in the top 20 cm of soil was low, and root growth in deeper layers was stimulated compared with the surface irrigation (Figure 10.10b). Hence, manipulating water supply can change root distribution density down the soil profile.

To quantify the effects of shallow water tables on the distribution of root dry weight density of winter wheat at the Yucheng Comprehensive Experimental Station, we set a field trial under rain-fed conditions with lysimeters and water table at different depths (40, 70, 110, or 150 cm) (Liu and Luo 2011). Figure 10.11 shows the distribution of wheat root dry weight density in the soil profile at harvest in the treatments with different groundwater table depths in 2010. The root dry weight within 0–100 cm layer ranged from 55 to 185 $g\,m^{-3}$ and increased with an increase in the groundwater table depth. The groundwater table at 150 cm was associated with the largest root dry matter weight, having 238% more root dry matter compared with the groundwater table at 40 cm depth.

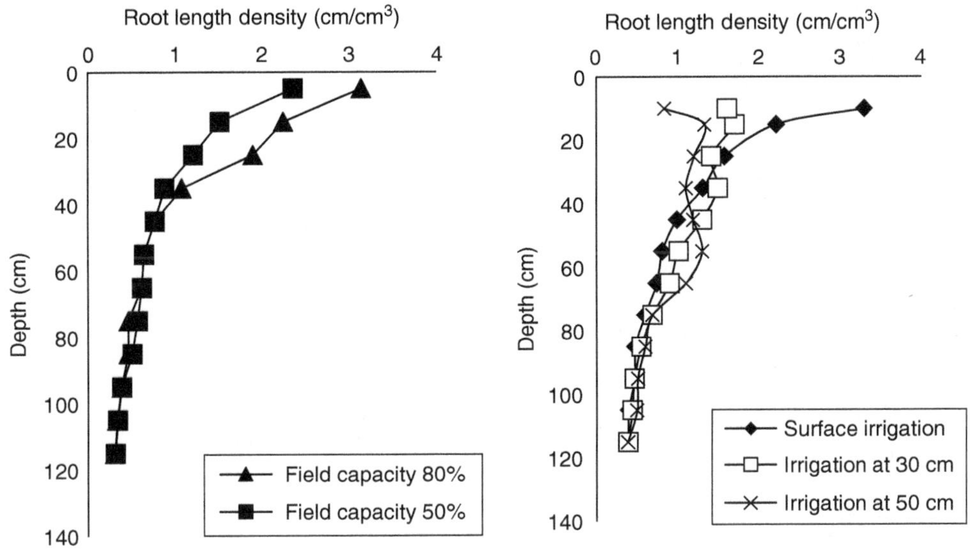

Figure 10.10 Distribution of wheat root length density down the soil profile as influenced by differential water availability and the irrigation type. (a) surface irrigation; (b) water content = 80% of field capacity supplied by three types of irrigation. *Source:* Data from Li et al. (2002).

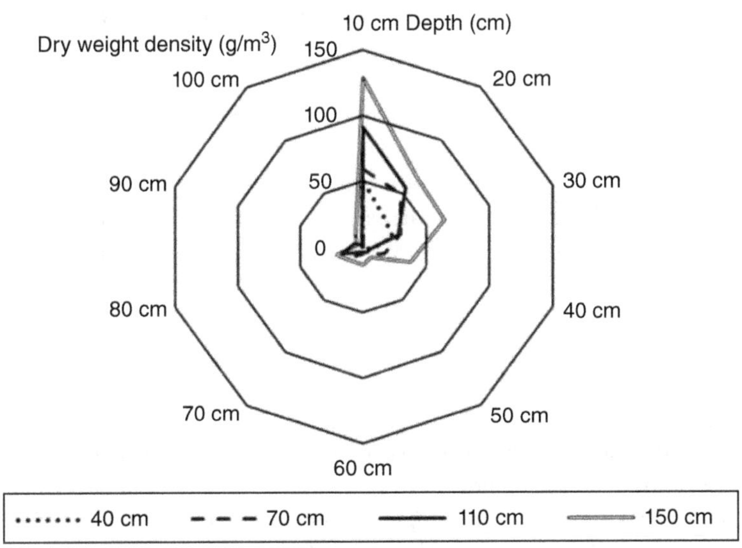

Figure 10.11 Distribution of wheat root dry weight density in the soil profile at harvest as dependent on the water table depths (40, 70, 110, or 150 cm). *Source:* Data from Liu and Luo (2011)

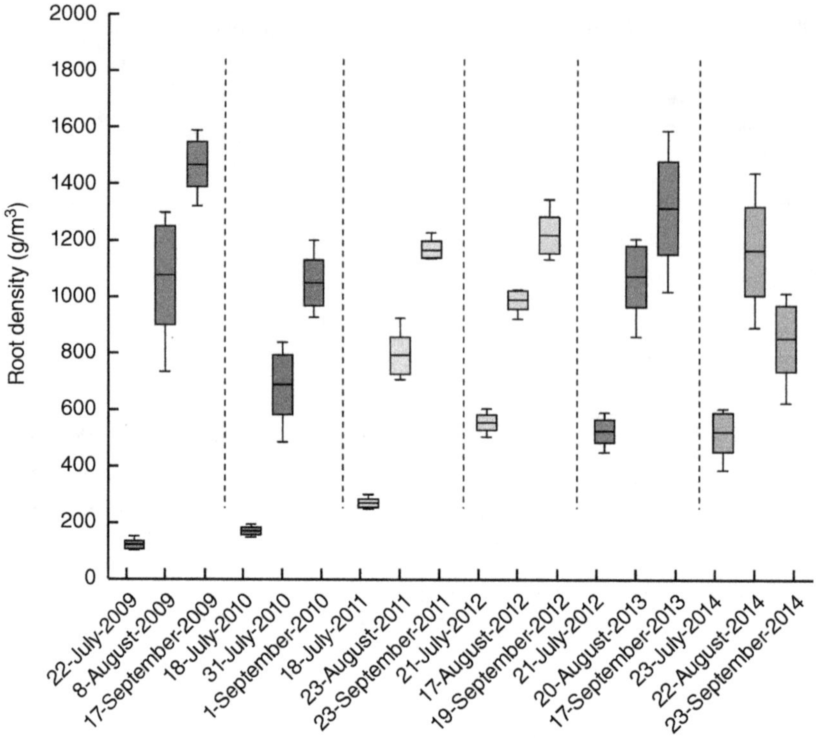

Figure 10.12 The intra- and inter-seasonal variability in maize root density over the 5-year period.

10.5.2 Maize

Across the five field trials, maize roots grew until the grain ripening stage (Figure 10.12), with the maximum root density of 1317 $g\,m^{-3}$. The irrigation was supplied only at sowing.

In the same trials over the five seasons, maize grain yield ranged from 7.4 to 9.7 $t\,ha^{-1}$ (Table 10.3).

10.6 Conclusions

The NCP represents the critical food production area (winter wheat-summer maize cropping) with scarce water resources. Understanding temporal and spatial dynamics of wheat and maize root growth is an essential factor in achieving efficient use of water ad nutrient resources to ensure food security. The main growth period of wheat roots was in the stem elongation and booting stages, whereas the maize roots had a longer growth period extending all the way to grain ripening.

Given that root density was positively correlated with grain yield, it was suggested that increases in root biomass could be linked to an enhanced nutrient and water absorption. The distribution of root density along the soil profile plays a critical role in soil water

Table 10.3 Maize root system density and grain yield over five seasons on the North China Plain.

Sampling time	Root density (g m^{-3})	Grain yield (t ha^{-1})
22 July 2009	1214 ± 263	9.1 ± 0.73
8 August 2009	1468 ± 136	
17 September 2009	1246 ± 300	
18 July 2010	170 ± 24	7.4 ± 0.37
31 July 2010	688 ± 182	
1 September 2010	973 ± 139	
25 September 2010	717 ± 117	
18 July 2011	269 ± 27	–
23 August 2011	792 ± 117	
23 September 2011	1166 ± 51	
21 July 2012	556 ± 49	9.7 ± 0.91
17 August 2012	1218 ± 112	
19 September 2012	989 ± 57	
21 July 2013	525 ± 70	–
20 August 2013	1074 ± 186	
17 September 2013	1317 ± 286	
23 July 2014	522 ± 118	9.7 ± 2.05
22 August 2014	1165 ± 273	
23 September 2014	854 ± 203	

utilization. The findings suggested that optimization of the root system contributed to increased grain yield.

We suggest that the future research should focus on the interactions between crop roots and irrigation, climate, crop genotype, fertilizer application, and root-associated soil microbiome to develop and utilize a water-soil-crop-energy flow model suited for the crops grown on the NCP. This model is likely to contribute to food security, environmental safety, economic development, and the sustainable intensification of agricultural production on the NCP.

Funding: This research was funded by the National Natural Science Foundation of China (No. 41771292, No. U1906219, No. 31170414), National Natural Science Foundation of China, and United Nations Environment Programme (No. 41761144053).

References

Ali, S., Xu, Y., Ahmad, I. et al. (2018). Tillage and deficit irrigation strategies to improve winter wheat production through regulating root development under simulated rainfall conditions. *Agricultural Water Management* 209: 44–54.

Benjamin, J.G. and Nielsen, D.C. (2006). Water deficit effects on root distribution of soybean, field pea and chickpea. *Field Crops Research* 97: 248–253.

Cole, M.B., Augustin, M.A., Robertson, M.J. et al. (2018). The science of food security. *Science of Food* 2: 14.

Deng, X., Lun, S., Zhang, H. et al. (2006). Improving agricultural water use efficiency in arid and semiarid areas of China. *Agricultural Water Management* 80: 23–40.

Ehdaie, B., Layne, A., and Waines, J. (2011). Root system plasticity to drought influences grain yield in bread wheat. *Euphytica* 186: 219–232.

Fan, M., Shen, J., Yuan, L. et al. (2011). Improving crop productivity and resource use efficiency to ensure food security and environmental quality in China. *Journal of Experimental Botany* 63: 13–24.

Fan, K., Weisenhorn, P., Gilbert, J.A. et al. (2018). Wheat rhizosphere harbors a less complex and more stable microbial co-occurrence pattern than bulk soil. *Soil Biology and Biochemistry* 125: 251–260.

FAO (2017). The Future of Food and Agriculture—Trends and Challenges. Rome, Italy: Food and Agriculture Organization of the United Nations.

FAO, IFAD and WFP (2015). The state of food Insecurity in the world 2015. Meeting the 2015 international hunger targets: taking stock of uneven progress. Rome, Italy.

Godfray, H.C.J. (2010). Food security: the challenge of feeding 9 billion people. *Science* 327: 812–818.

Grayson, M. (2013). Agriculture and drought. *Nature* 501: S1.

Guo, X., Sun, X., Ma, J. et al. (2018). Simulation of the water dynamics and root water uptake of winter wheat in irrigation at different soil depths. *Water* 10: 1033.

Gupta, A., Rico Medina, A., and Caño Delgado, A.I. (2020). The physiology of plant responses to drought. *Science* 368: 266.

Hu, Y., Moiwo, J.P., Yang, Y. et al. (2010). Agricultural water-saving and sustainable groundwater management in Shijiazhuang Irrigation District, North China Plain. *Journal of Hydrology* 393: 219–232.

Huang, T., Ju, X., and Yang, H. (2017). Nitrate leaching in a winter wheat-summer maize rotation on a calcareous soil as affected by nitrogen and straw management. *Scientific Reports* 7: 42247–42247.

Jeong, S.-J., Ho, C., Piao, S. et al. (2014). Effects of double cropping on summer climate of the North China Plain and neighbouring regions. *Nature Climate Change* 4: 615–619.

Jha, S.K., Gao, Y., Liu, H. et al. (2017). Root development and water uptake in winter wheat under different irrigation methods and scheduling for North China. *Agricultural Water Management* 182: 139–150.

Jia, Q., Chen, K., Chen, Y. et al. (2018). Mulch covered ridges affect grain yield of maize through regulating root growth and root-bleeding sap under simulated rainfall conditions. *Soil and Tillage Research* 175: 101–111.

Ju, C., Buresh, R.J., Wang, Z. et al. (2015). Root and shoot traits for rice varieties with higher grain yield and higher nitrogen use efficiency at lower nitrogen rates application. *Field Crops Research* 175: 47–55.

Kendy, E., Molden, D.J., Steenhuis, T.S. et al. (2003). Policies Drain the North China Plain: Agricultural Policy and Ground-water Depletion in Luancheng County, 1949-2000. Colombo, Sri Lanka: International Water Management Institute.

Li, Y., Wang, L., Liu, S. et al. (2002). The influence of different amounts of water supplied at different depths in soil-root ointerface on root distribution and yield of winter wheat. *Acta Ecologica Sinica* 22 (10): 1680–1687.

Li, Q., Dong, B., Qiao, Y. et al. (2010). Root growth, available soil water, and water-use efficiency of winter wheat under different irrigation regimes applied at different growth stages in North China. *Agricultural Water Management* 97: 1676–1682.

Liu, Y. and Luo, Y. (2010). A consolidated evaluation of the FAO-56 dual crop coefficient approach using the lysimeter data in the North China Plain. *Agricultural Water Management* 97: 31–40.

Liu, T. and Luo, Y. (2011). Effects of shallow water tables on the water use and yield of winter wheat (*Triticum aestivum* L.) under rain-fed condition. *Australian Journal of Crop Science* 5: 1692–1697.

Liu, C. and Xia, J. (2004). Water problems and hydrological research in the Yellow River and the Huai and Hai River basins of China. *Hydrological Processes* 18: 2197–2210.

Liu, W., Wang, J., Wang, C. et al. (2018a). Root growth, water and nitrogen use efficiencies in winter wheat under different irrigation and nitrogen regimes in North China Plain. *Frontiers in Plant Science* 9: 1798–1798.

Liu, W., Ma, G., Wang, C. et al. (2018b). Irrigation and nitrogen regimes promote the use of soil water and nitrate nitrogen from deep soil layers by regulating root growth in wheat. *Frontiers in Plant Science* 9: 32–32.

Lv, G., Kang, Y., Li, L. et al. (2010). Effect of irrigation methods on root development and profile soil water uptake in winter wheat. *Irrigation Science* 28: 387–398.

Lv, G., Song, J., Bai, W. et al. (2015). Effects of different irrigation methods on micro-environments and root distribution in winter wheat fields. *Journal of Integrative Agriculture* 14: 1658–1672.

Man, J., Shi, Y., Yu, Z. et al. (2016). Root growth, soil water variation, and grain yield response of winter wheat to supplemental irrigation. *Plant Production Science* 19: 1–13.

Mo, X. and Liu, S. (2001). Simulating evapotranspiration and photosynthesis of winter wheat over the growing season. *Agricultural and Forest Meteorology* 109: 203–222.

Ouyang, Z. and Luo, Y. (2002). A conceptual model of predicting soil water storage of winter wheat root zone. *Progress in Geography* 21: 593–599.

Phene, C.J., Davis, K.R., Hutmacher, R.B. et al. (1991). Effect of high frequency surface and subsurface drip irrigation on root distribution of sweet corn. *Irrigation Science* 12: 135–140.

Piao, S., Philippe, C., Yao, H. et al. (2010). The impacts of climate change on water resources and agriculture in China. *Nature* 467: 43–51.

Plett, D.C., Ranathunge, K., Melino, V.J. et al. (2020). The intersection of nitrogen nutrition and water use in plants: new paths toward improved crop productivity. *Journal of Experimental Botany* 71: 4452–4468.

Proffitt, A.P.B., Berliner, P.R., and Oosterhuis, D.M. (1985). A comparative study of root distribution and water extraction efficiency by wheat grown under high- and low-frequency irrigation. *Agronomy Journal* 77: 655–662.

Qi, W., Liu, H., Liu, P. et al. (2012). Morphological and physiological characteristics of corn (Zea mays L.) roots from cultivars with different yield potentials. *European Journal of Agronomy* 38: 54–63.

Ren, B., Li, X., Dong, S. et al. (2018). Soil physical properties and maize root growth under different tillage systems in the North China Plain. *The Crop Journal* 6: 669–676.

Rey, D., Holman, I.P., and Knox, J.W. (2017). Developing drought resilience in irrigated agriculture in the face of increasing water scarcity. *Regional Environmental Change* 17: 1527–1540.

SCIO_PRC (2019). Food security in China. http://www.scio.gov.cn/zfbps/32832/Document/1666228/1666228.htm. (accessed July 2020).

Shao, L., Zhang, X., Chen, S. et al. (2009). Effects of irrigation frequency under limited irrigation on root water uptake, yield and water use efficiency of winter wheat. *Irrigation and Drainage* 58: 393–405.

Tu, C. and Li, F. (2017). Responses of greenhouse gas fluxes to experimental warming in wheat season under conventional tillage and no-tillage fields. *Journal of Environmental Sciences* 54: 314–327.

United Nations (2015) Transforming our world: the 2030 agenda for sustainable development. http://www.un.org/ga/search/view_doc.asp?symbol=A/RES/70/1&Lang=E. (accessed July 2015).

de Vries, F.T., Griffiths, R.I., Knight, C.G. et al. (2020). Harnessing rhizosphere microbiomes for drought-resilient crop production. *Science* 368: 270.

Wang, E., Yu, Q., Wu, D. et al. (2008). Climate, agricultural production and hydrological balance in the North China Plain. *International Journal of Climatology* 28: 1959–1970.

Wang, C., Liu, W., Li, Q. et al. (2014). Effects of different irrigation and nitrogen regimes on root growth and its correlation with above-ground plant parts in high-yielding wheat under field conditions. *Field Crops Research* 165: 138–149.

Wheeler, R.M., Wl, B., Mackowiak, C. et al. (1993). A data base of crop nutrient use, water use, and carbon dioxide exchange in a 2O square meter growth chamber: I. Wheat as a case study. *Journal of Plant Nutrition* 16 (10): 1881–1915.

Xu, C., Tao, H., Tian, B. et al. (2016). Limited-irrigation improves water use efficiency and soil reservoir capacity through regulating root and canopy growth of winter wheat. *Field Crops Research* 196: 268–275.

Zhang, X., Pei, D., and Chen, S. (2004b). Root growth and soil water utilization of winter wheat in the North China Plain. *Hydrological Processes* 18: 2275–2287.

Zhang, X., Chen, S., Sun, H. et al. (2008). Dry matter, harvest index, grain yield and water use efficiency as affected by water supply in winter wheat. *Irrigation Science* 27: 1–10.

Zhang, Y., Li, T., Bei, S. et al. (2018). Growth and distribution of maize roots in response to nitrogen accumulation in soil profiles after long-term fertilization management on a calcareous soil. *Sustainability* 10: 1–16.

Zhao, Y., Xue, Z., Guo, H. et al. (2014). Effects of tillage and crop residue management on soil respiration and its mechanism. *Transactions of the Chinese Society of Agricultural Engineering* 30: 155–165.

Zhao, X., Li, F., Ai, Z. et al. (2018). Stable isotope evidences for identifying crop water uptake in a typical winter wheat–summer maize rotation field in the North China Plain. *Science of The Total Environment* 618: 121–131.

Zhou, S., Wu, Y., Wang, Z. et al. (2008). The nitrate leached below maize root zone is available for deep-rooted wheat in winter wheat–summer maize rotation in the North China Plain. *Environmental Pollution* 152: 723–730.

11

Intercropping to Maximize Root–Root Interactions in Agricultural Plants

Soil–Root Interface Processes

Jun Wasaki[1,2] *and D.M.S.B. Dissanayaka*[2,3]

[1]*Graduate School of Integrated Sciences for Life, Hiroshima University, Higashihiroshima, Japan*
[2]*Graduate School of Biosphere Science, Hiroshima University, Higashihiroshima, Japan*
[3]*Department of Crop Science, Faculty of Agriculture, University of Peradeniya, Kandy, Sri Lanka*

11.1 Introduction

Modern agriculture is dominated by monoculture cropping systems characterized by high inputs, high outputs, low nutrient-use efficiencies, and increasing environmental pressure (Zhang et al. 2010; Li et al. 2013). Extraction of the same nutrients from soil for long periods has caused these cropping systems to be highly dependent on external fertilizer inputs and may result in nutrient imbalances in soils and crops (Zhang et al. 2010). In contrast to monocultures, intercropping/mixed cropping systems consist of two or more crop species or genotypes growing together and co-existing for a period of time on the same parcel of land (Li et al. 2007; Brooker et al. 2015). Different forms of intercropping include mixed intercropping (simultaneously growing two or more crops with no, or limited, distinct arrangement), relay intercropping (planting a second crop before the first crop is mature), and strip intercropping (growing two or more crops simultaneously in strips, allowing crop interactions and independent cultivation) (Brooker et al. 2015) (see also Chapter 12). Intercropping is often practiced by farmers that adopt low-input, low-yield systems on small parcels of land (Ngwira et al. 2012). Under these conditions, intercropping can help increase yields per unit input, minimize crop failures and alleviate market fluctuations, produce a variety of food types, and ultimately increase the farmers' income (Rusinamhodzi et al. 2012).

Intercropping is an age-old practice that has been applied widely around the world, mainly in tropical, small-scale subsistence farming systems (Lithourgidis et al. 2011). In some regions of the world, intercropping is the dominant agricultural system. For example, the area under agroforestry (a mixed cropping system) accounts for more than 1 billion ha (Zomer et al. 2009). Smallholder farmers in Latin America grow 70–90% of beans (*Phaseolus vulgaris*) intercropped with maize (*Zea mays*), potato (*Solanum tuberosum*), and other crops; 60% of the maize-growing areas of the region are under intercropping (Francis 1986).

The Root Systems in Sustainable Agricultural Intensification, First Edition.
Edited by Zed Rengel and Ivica Djalovic.

In Africa, 98% of cowpea (*Vigna unguiculata*) is intercropped, and in Colombia 90% of beans are intercropped (Vandermeer 1989).

Integration of plant species on a same piece of land can have several benefits. Intercropping may be a means to solve key constraints associated with modern farming practices, such as pest and pathogen accumulation, soil degradation, and environmental deterioration (Vandermeer 1989), thereby achieving sustainable crop production (Lithourgidis et al. 2011). Increasing crop quality and yield through efficient utilization of resources is an additional important requirement of intercropping (Brooker et al. 2015). Niche complementarity and interspecific facilitation are the two major ecological principles that are used to explain the underlying combined yield increase (overyielding) of intercropping compared with monocropping systems (Fridley 2001; Hauggaard-Nielsen and Jensen 2005; Li et al. 2007). According to the niche complementarity hypothesis, overyielding in intercropping is attained as a result of niche differentiation between the intercropped species in time, space, and the forms of a given resource, thereby achieving complementary use of resources (Tilman et al. 2001) (see also Chapter 12).

In interspecific facilitation, individuals of one species positively influence the growth and survival of individuals of the co-cultivated species (Callaway 1995). Below-ground interspecific facilitation may result from direct root-mediated facilitation or indirect microbe-mediated facilitation (Sekiya et al. 2011; Li et al. 2014a). Direct root-mediated facilitation is based on mobilization of sparingly soluble/non-available nutrients via root exudates, whereas under microbe-mediated facilitation beneficial soil microbes are stimulated to release nutrients during mineralization of organic matter and transfer nutrients (for example through mycorrhizal networks that may link co-existing crop species) (Bainard et al. 2012; Walder et al. 2012; Ehrmann and Ritz 2014; Sun et al. 2013; Tong et al. 2015).

In this chapter, information required to understand the potential, advantages, and disadvantages of intercropping systems regarding root–root interactions is presented. In addition, important phenomena associated with root–root interactions are discussed.

11.2 Concept of Rhizosphere Sharing

It is important to understand the specific environment around roots, i.e. the rhizosphere. The term 'rhizosphere' has been defined as 'the soil influenced by the root' in the literature related to the roles of soil microorganisms in nutrient dynamics (Hiltner 1904; Hartmann et al. 2008). The concept of the rhizosphere is now well established. Interestingly, the number of publications including the term 'rhizosphere' as a keyword has increased in recent decades from 398 publications in 1997 to 773 publications in 2007 and 1747 publications in 2017 (data from Web of Science). Why is the rhizosphere, a microscale environment, an important topic? The rhizosphere is vital as the location of water and nutrient uptake by plants, and the influence of microorganisms in this location is indispensable for plants (Mommer et al. 2016). Given that plants take up water through roots, the direction of mass-flow in soils is towards roots. Different mineral ions move by mass-flows and/or diffusion and are transferred into root cells via specific transporters at the root surface. In the case of nitrogen (N) nutrition, both ammonium (NH_4^+) and nitrate

(NO_3^-) ions are influenced by soil microbial activities (Coskun et al. 2017). Phosphorus (P), one of the essential macronutrients for plants, is mainly taken up as orthophosphate by plants. The mobility of orthophosphate in soil is extremely limited because it binds with metal ions, such as aluminium (Al), iron (Fe), and/or calcium (Ca) (depending on soil pH), on the surface of soil particles (Maruyama and Wasaki 2017). Potassium (K) ions are relatively immobile in soil (move mainly by diffusion) and have a weak capacity for exchange reactions with soil minerals. The zone of K depletion in the rhizosphere is wider than that of P, but not wider that of soil mobile ions such as N, Ca, and Mg (Marschner 1991; Nord and Lynch 2008; Hinsinger et al. 2011).

It is well known that plant roots release a large amount of organic compounds as 'rhizodeposits' (Kuzyakov 2002). The rhizodeposits can be classified as secretions, exudates, mucilages, mucigels, and cell lysates, and comprise passively released compounds and actively transported substances. High-molecular-weight compounds in the rhizodeposits include polysaccharides and proteins. The roles of polysaccharides released from roots remain unknown, but have been proposed to be important for biofilm formation by rhizobacteria (Somers et al. 2004). Antifungal enzymes, such as chitinase and glucanase, and P-cycling enzymes, i.e. acid phosphatases, play important roles in P dynamics (Wasaki et al. 2003, 2005; Weisskopf et al. 2006). Rhizodeposits are important for interactions between plants and microbes in the rhizosphere, and 40–90% of photosynthates transported to roots are used for rhizodeposition (Kennedy 1998).

The rhizosphere represents a hotspot of soil microorganism density and in many cases contains more than two orders of magnitude higher densities compared with bulk soil samples. The rhizosphere microorganisms have varied influences on plants and may show neutral, beneficial, or deleterious (e.g. pathogenic) effects on plants (Somers et al. 2004). The beneficial microroganisms in the rhizosphere are generally called plant growth-promoting rhizobacteria [PGPR]).

Plants can communicate with neighbouring plants via a 'below-ground highway', i.e. root exudates in which the end result may be either inhibition or facilitation (in terms of nutrient acquisition and growth) of their neighbours (Callaway 2002). Root–root interactions (or root intermingling) between two plants with contrasting nutrient acquisition strategies play an important role in mixed cultures (Dissanayaka et al. 2015; Li et al. 2016). Both direct (root-mediated) and indirect (microbe-mediated) facilitation are important in sharing rhizosphere functions between two intercropped species (Figure 11.1).

The concept of 'rhizosphere sharing' is considered to be fundamentally important to understand rhizosphere functions and how nutrient dynamics is influenced by interactions among organisms in the rhizosphere.

11.3 Dynamics of Mineral Nutrients in the Rhizosphere

Crop production in many agricultural systems is limited by N, P, and/or water availability. Moreover, crop production on 70% of the world's agricultural land may be restricted by the phytoavailability of Fe, Zn, and copper (Cu) on alkaline and calcareous soils, or by toxicity of Al or manganese (Mn) on acidic soils (White and Greenwood 2013).

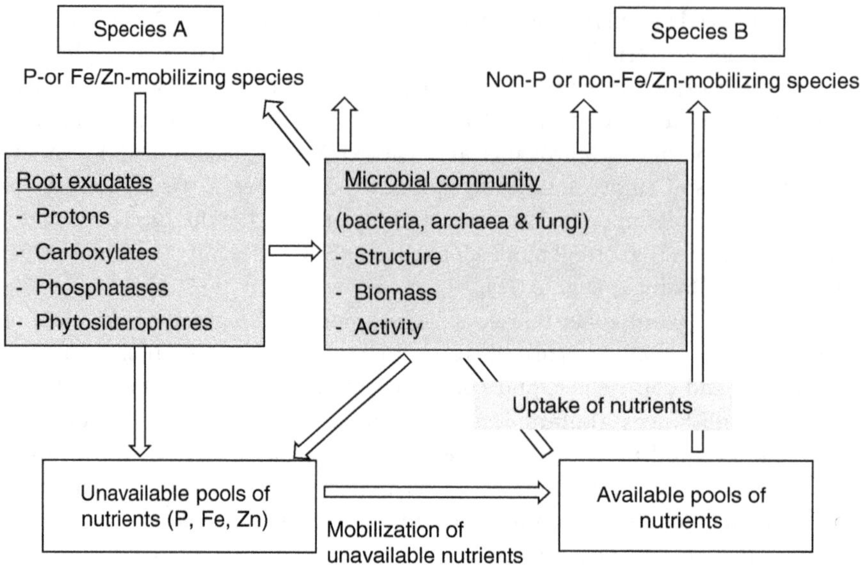

Figure 11.1 Direct root-mediated facilitation and indirect microbe-mediated facilitation of shared rhizosphere functions in intercropping systems. Phosphorus (P) or iron (Fe)/zinc (Zn) mobilizing species (species A) mobilizes P that is initially non-available to species B through root exudation (direct root-mediated facilitation). Root exudation by species A may cause changes in the microbial community structure, biomass, and/or activity, thereby resulting in alteration of available and non-available nutrient pools. Increases in the size of the available nutrient pool at the expense of the non-available pool indicate indirect microbe-mediated facilitation in the intercropping system. Strong root intermingling between the two intercropped species would strengthen the sharing of mobilized resources.

Intercropping of plant species capable of mobilizing non-available nutrients with plant species that lack this capacity can help increase phytoavailability and acquisition of limiting nutrients by the latter species (Dissanayaka et al. 2017). For example, in P-limited soil, P benefit can be achieved by integration into intercropping systems of plant species that use sparingly available phosphate and organic P (Dissanayaka et al. 2015). Under such conditions, P-mobilizing plant species enhance the growth and P acquisition of companion cereals (Simpson et al. 2011). Enhanced P acquisition and growth have been demonstrated previously for different cereals in mixed culture systems with legumes, e.g. wheat (*Triticum aestivum*)–chickpea (*Cicer arietinum*) (Li et al. 2003), wheat–white lupin (*Lupinus albus*) (Cu et al. 2005; Kamh et al. 1999), wheat–faba bean (*Vicia faba*) (Song et al. 2007), maize–chickpea (Li et al. 2004), maize–white lupin (Dissanayaka et al. 2015), barley (*Hordeum vulgare*)–chickpea (Gunes et al. 2007), and sorghum (*Sorghum bicolor*)–pigeon pea (*Cajanus cajan*) (Ae et al. 1990) mixed culture systems. Improved management of root-rhizosphere interactions can further enhance resource-use efficiency by crops (Zhang et al. 2010; Shen et al. 2013; White et al. 2013; Ehrmann and Ritz 2014; Li et al. 2014b) in an intercropping system compared with the corresponding monocultures.

11.3.1 Direct Effects of Organic Acid Anions on Phosphorus Dynamics

Root exudates are the most important factor in understanding the effect of rhizosphere sharing on P dynamics. Sugars, amino acids, and organic acid anions are abundant as low-molecular-weight root exudates (Jones 1998). In particular, the role of organic acid anions on nutrient mobilization in the rhizosphere soil has been well investigated. The roles of organic acid anions are dependent on their chelating capacity. Root secretes mostly di- or tri-carboxylates (Table 11.1). Carboxyl groups in a carboxylate can form coordinate bonds with multivalent cations such as Al^{3+}, Fe^{3+}, Fe^{2+}, and Ca^{2+}. As a result, the minerals are solubilized. Given that these multivalent cations frequently immobilize phosphate in the soil, solubilization of the minerals by carboxylates also contributes to P mobilization and uptake by plants. Plants that form cluster roots, such as white lupin and members of the family Proteaceae, release from roots relatively large amounts of a variety of organic acid anions (Neumann and Martinoia 2002; Lambers et al. 2011).

A variety of carboxylates, such as citrate, malate, and oxalate, released from roots (Table 11.1) into the rhizosphere play an important role also in Al detoxification to chelate toxic Al^{3+} ions (Ma et al. 2001; Ryan et al. 2001). Sasaki et al. (2004) isolated a key malate transporter, TaALMT1, from an Al-tolerant cultivar of wheat. The key transporter for malate exudation under low-P conditions has recently been isolated from Arabidopsis (AtALMT3; Maruyama et al. 2018). Candidates for the activated malate transporter (ALMT) and multidrug and toxin extrusion (MATE; for citrate transporters) family transporters were found in transcriptomic analyses of white lupin (O'Rourke et al. 2013; Secco et al. 2014; Wang et al. 2014). The mechanisms of carboxylate exudation under P-deficient conditions are yet to be completely elucidated.

Table 11.1 Organic acid anions identified in root exudates and the rhizosphere soil under P-deficient conditions.

Sample type	Plant species	Detected organic acid anions	References
Root exudates	*Lupinus albus*	Citrate, malate	Neumann and Römheld (1999)
	Vigna unguiculata	Citrate, malonate, malate, oxalate	Neumann and Römheld (1999)
	Triticum aestivum	Malate, oxalate	Neumann and Römheld (1999) and Sasaki et al. (2004)
	Solanum lycopersicum	Citrate, malate, fumarate, shikimate	Neumann and Römheld (1999)
	Cajanus cajan	Piscidate	Ae et al. (1990)
Rhizosphere soils	*Lupinus sp.*	Citrate, fumarate, malate	Dinkelaker et al. (1989), Wasaki et al. (2005), and Wang et al. (2013)
	Trifolium repens	Acetate, malate	Bolan (1994)
	Glycine max	Citrate	Nwoke et al. (2008)
	Vigna unguiculata	Citrate	Veneklaas et al. (2003) and Nwoke et al. (2008)

11.3.2 Effect of Organic Acid Anions on Dynamics of Metal Nutrients

Some plant species can also be integrated in intercropping systems specifically to improve Fe and Zn availability in the rhizosphere and to share these mobilized ions between two co-cultivated plants via root intermingling. This strategy is extremely important because in most soils Fe exists predominantly in the form of Fe^{3+}, which is poorly soluble and insufficiently available to many plants, especially dicotyledonous and non-graminaceous monocotyledonous species (Li et al. 2014b). Mechanisms for Fe mobilization graminaceous monocotyledonous species involve the release of phytosiderophores, which are chelators with high affinity for Fe^{3+}. This facilitates diffusion of Fe^{3+} towards the root surface, where the entire Fe-phytosiderophore complex is absorbed (Ma 2005). Phytosiderophore exudation increases not only Fe availability but also that of Zn, as demonstrated in barley under Zn deficiency (Zhang et al. 1991). When a phytosiderophore-exuding species and a species that lacks this capacity are grown together, it is likely that the phytosiderophore-exuding species enhances the mobility of Fe^{3+} in soil, and thus enhances availability of Fe to the neighbouring plants (Li et al. 2014b). This response has been clearly shown in a peanut (*Arachis hypogaea*)–maize intercropping system (Inal et al. 2007). Maize roots release phytosiderophores (e.g. deoxymugineic acid) and mobilize Fe^{3+}, which may then diffuse to the root surface of peanut. The Fe^{3+} is reduced at the root surface of peanut and taken up (Inal et al. 2007; Li et al. 2014b). In addition, the complex of Fe^{3+}-deoxymugineic acid may be absorbed directly by the roots of intercropped peanut plants (Xiong et al. 2013).

Specific Mn accumulation by cluster-root-forming Proteaceae species has been well demonstrated (Lambers et al. 2014). It is thought that high Mn accumulation is caused by the solubilization of Mn-containing minerals by carboxylates secreted by cluster roots. In a recent study, we demonstrated accumulation of high concentrations of Mn and Al in mature leaves of *Helicia cochinchinensis*, a member of the Proteaceae family (unpublished data). We also demonstrated that Mn concentration in white lupin tended to be higher in P-deficient conditions. These findings imply that the intercropping of crops with cluster-root-forming plants may enhance root–root interactions not only regarding P but also other minerals, especially Mn and Zn. Hence, intercropping may alter mineral accumulation in the edible parts, which needs to be taken into consideration.

11.4 Microbial Interactions in the Rhizosphere

Microorganisms are present in soils in spatially and temporally varied densities and diversities. Roots influence microbial density in the rhizosphere via rhizodeposition. Here, we discuss (i) the effects of mycorrhizal fungi on P uptake by plants, and (ii) the relationship with other microorganisms influencing P and N dynamics in the rhizosphere.

11.4.1 Arbuscular Mycorrhizal Fungi

In addition to root exudate-driven P mobilization and acquisition, mycorrhizal symbiosis, especially of arbuscular mycorrhizal fungi (AMF), contributes substantially to P acquisition in many crop plants (Hu et al. 2010; Yoneyama et al. 2007). This is because the soil volume exploited can be increased by AMF colonization of roots, forming an extensive external network of hyphae (Smith and Read 2008). AMF can colonize more than 80% of plant species (Berruti et al. 2016). AMF provide minerals (especially P) to plants and receive photosynthates from plants. Increased plant growth and P uptake following AMF inoculation in the field has been reported for diverse crops, e.g. pepper (*Capsicum annuum*), tomato, eggplant (*Solanum melongena*), melon (*Cucumis melo*), watermelon (*Citrullus lanatus*), cucumber (*Cucumis sativus*), maize, cotton (*Gossypium hirsutum*), faba bean, chickpea, soybean (*Glycine max*), and garlic (*Allium sativum*) (Ortas 2012), as well as wheat (Hu et al. 2010), Egyptian clover (*Trifolium alexandrinum*), and maize (Pellegrino et al. 2011), and Japanese leek (*Allium fistulosum*) (Tawaraya et al. 2012).

Plant roots release strigolactones to enhance the symbiosis with AMF (Akiyama et al. 2005). These compounds were first isolated as a 'branching factor' released from roots of *Lotus japonicus*. Once strigolactones are produced in plants as phytohormones, AMF penetrate into root cells of the host plants and produce arbuscules, which are important for the transport of P and carbon between plant cells and AMF.

Network of AMF hyphae has been shown to play a dominant role in shaping plant–plant interactions. The cultivation of AMF-symbiotic crops in the preceding season promoted mycorrhizal formation in maize and increased P uptake and grain yield (Arihara and Karasawa 2000). Similar results have been reported for cover crops (García-González et al. 2018). Given that the host specificity of AMF is not as strict as that observed in rhizobium-legume symbiosis, AMF can form symbiosis with a number of plant species and almost any neighbouring plants may be connected via AMF hyphae. Thus, AMF colonization may contribute strongly to plant-plant interactions.

Some P-mobilizing legumes are non-mycorrhizal; therefore, root intermingling between mycorrhizal and non-mycorrhizal plant species is possible in an intercropping system. However, the influence of a mycorrhizal plant species intercropped with a P-mobilizing non-mycorrhizal plant is poorly understood. Vierheilig et al. (2003) demonstrated that the addition of root exudates of non-mycorrhizal lupin did not affect the AMF colonization of mycorrhizal cucumber plants. Additional studies on this aspect are warranted to identify compatible crop combinations to improve P acquisition in intercropping systems. An additional line of evidences has been advanced by the recent plant-microorganism interaction studies that highlight the potential of mycorrhizal fungi to improve intercropping systems (Ehrmann and Ritz 2014) via transport of allelochemicals to neighbouring plants (Barto et al. 2012).

The 'mycorrhizosphere' concept has been proposed (Johansson et al. 2004) because secretions and nutrients release from mycorrhizal fungi increase microbial populations and their activities in the soil around mycorrhizal fungi without direct effects from plant roots. In mycorrhizosphere, there can be mycorrhiza-helper bacteria that support the functioning of AMF (Frey-Klett et al. 2007). Apart from helping plants to acquire P from soil patches where roots cannot penetrate, AMF-derived phosphatase mobilizes organic P mobilization (Sato et al. 2015. Therefore, the complex mechanisms in the rhizosphere of AMF-colonized plants should be considered in the root–root interactions in intercropping systems.

11.4.2 Interactions Between Plants and Microorganisms Influence P Dynamics

There are many diverse microorganisms in the rhizosphere, including AMF and PGPRs. In addition to direct root-mediated facilitation in the rhizosphere, indirect microbe-mediated facilitation can also occur as a result of changes in the microbial community structure, abundance, and/or activity (Hinsinger et al. 2011). There are increasing evidences suggesting that soil chemical properties, such as pH, have a major influence on microbial diversity in the rhizosphere (Philippot et al. 2009). Roots may exude diverse secondary metabolites that change soil pH and some signalling molecules that could shape microbial communities and influence root–microbe and root–root communication (Walker et al. 2003). Moreover, differences in the type and concentration of root exudates could determine the associated microbial community (Hartmann et al. 2009; Dennis et al. 2010). Differences in the fungal community structure in the rhizosphere of white lupin were associated with citrate exudation, whereas differences in the bacterial community structure were linked to exudation of *cis*-aconitate, citrate, and malate (Marschner et al. 2002). Hence, strong possibility exists for an organic acid anion-exuding plant species to influence microbial communities of co-occurring (intercropped) species with intermingling root systems.

Phytate is generally the most abundant form of organic P in soils (Turner 2006). Adams and Pate (1992) showed that the availability of phytate-P to white lupin reflected solubility in the soil, and acid phosphatase secreted by white lupin roots had little effect on phytate (Li and Tadano 1996; Miller et al. 2001). Therefore, it is reasonable to consider that phytate is solubilized by phytase secreted by microorganisms. Unno et al. (2005) isolated many phytate-utilizing strains of *Burkholderia* spp. from the rhizosheath and rhizoplane of white lupin. Interestingly, these authors suggested that phytate-utilizing bacteria isolated from the rhizosheath showed a higher rate of phosphate release compared with isolates from the rhizoplane. This finding raises the possibility of using intercropping with white lupin to release P from phytate by phytate-utilizing bacteria in the rhizosphere.

Root-secreted carboxylates may be readily consumed by soil microorganisms (Bürgmann et al. 2005; Oburger et al. 2009). To maintain the long-term function of carboxylates, continuous release that would exceed consumption, or protection of carboxylates from consumption by microorganisms, is required. White lupin has complex strategies to protect secreted carboxylates in the rhizosphere of cluster roots (Weisskopf et al. 2006). Acidification by subsequent proton release accompanying carboxylate exudation limits the growth of bacteria. Increased chitinase and glucanase activities limit the growth of fungi. Isoflavonoid exudation causes sporulation of fungi (Weisskopf et al. 2006). This trait is promising for utilization of P in the shared rhizosphere of an intercropping system. However, the effects of root exudates on interactions between crop plants and microorganisms are worth further attention.

11.4.3 Interactions Between Plants and Microorganisms Influence N Dynamics

Nitrogen forms available to plants are NH_4^+ and NO_3^-. If available N is low in the environment, N_2-fixing bacteria play an important role in increasing the pools of biologically-available N. The presence of various free-living N_2-fixing bacteria in soil is expected to increase N pools in the rhizosphere by assimilation of carbon from rhizodeposits. However, addition of artificial root exudates to soil did not result in an increase in N_2-fixing activity of free-living diazotrophs (Bürgmann et al. 2005).

Direct root-mediated facilitation can be achieved by integration of plants that increase plant availability of N, P, or micronutrients (Fe, Zn, and Cu) (Zhang et al. 2010; Shen et al. 2013; White et al. 2013; Li et al. 2014b). Nitrogen benefit is achieved by combining N_2-fixing species with species that lack this capacity (Zhang et al. 2010). The nodulation and N_2 fixation in leguminous monoculture systems are reduced with N application (Fang et al. 2011). An indirect increase in the soil N pool is expected in legume-grass intercropping. However, when legumes are intercropped with cereals, rhizosphere N availability decreases (Fang et al. 2013), and this reduction presumably increased nodulation and N_2 fixation in legumes (Fan et al. 2006; Li et al. 2009). However, the effect of cereals on nodulation is inconsistent and depends on the crop combination used in the cropping system (Fan et al. 2006). Li et al. (2016) found that a faba bean–maize intercropping system enhanced productivity, nodulation, and N_2 fixation of faba bean via interspecific root interactions. In this intercropping system, maize promoted N_2 fixation of faba bean, whereby root exudates from maize increased root hair deformation and nodulation in faba bean, doubled exudation of flavonoids (signalling compounds for rhizobia), and up-regulated the expression of (i) a chalcone–flavanone isomerase gene involved in flavonoid synthesis and (ii) genes that mediate nodulation (Li et al. 2016).

11.5 Examples of Intercropping Experiments Aimed at Improving Nutrient Accumulation in Crops

Plant species show different functional traits, including differences in P mobilization (Lambers et al. 2008), Fe and Zn mobilization (Kraemer et al. 2006), and N_2 fixation (Fornara and Tilman 2008). 'Mobilizing' plant species are defined as those that can convert a particular non-available soil nutrient pool into an available form that can be taken up readily (Li et al. 2014b). These ecological principles apply in the shared rhizosphere to enhance nutrient acquisition by non-mobilizing plant species under nutrient-limited conditions. Occurrence of these phenomena is well-illustrated in different cropping systems (Table 11.2). Interspecific facilitation operates through enhanced P acquisition in intercropping systems, whereby P-mobilizing plants increase P availability in the rhizosphere and thus P acquisition and growth of non-P-mobilizing species (Table 11.2). Other possibilities to improve tolerance to abiotic and biotic stresses are shown based on direct and indirect root–root interactions.

11.5.1 Enhanced P Acquisition in Intercropping Systems

The majority of cultivated plants grow poorly on low-P soils. Some of these low-P soils contain a considerable amount of P that is unavailable to most species. Organic P represents 30–70% of the total P pool in agricultural soils (Dalal 1978; Macklon et al. 1994; Turner 2006). Plants are unable to acquire organic P directly from soil; first, the organic P fraction needs

Table 11.2 Evidence for facilitation of nutrient uptake (nitrogen [N], phosphorus [P], iron [Fe], zinc [Zn], and manganese [Mn]) in mixed cultures of plant species that show different functional traits.

Plant combination	Evidence	References
White lupin and wheat	Improved P uptake by intercropped wheat	Gardner and Boundy (1983)
White lupin and wheat	Improved growth and P uptake by intercropped wheat	Horst and Waschkies (1987)
Pigeon pea and sorghum	Enhanced P uptake by intercropped sorghum	Ae et al. (1990)
White lupin and wheat	Improved growth and P uptake by intercropped wheat	Kamh et al. (1999)
Maize and guava	Improved uptake of Fe and Zn by intercropped guava	Kamal et al. (2000)
Peanut and maize	Improved Fe and Zn nutrition of intercropped peanut	Zuo et al. (2000), Inal et al. (2007) and Xiong et al. (2013)
Peanut and maize	Improved P uptake by intercropped maize	El Dessougi et al. (2003)
Chickpea and maize	Improved P uptake by intercropped maize	Li et al. (2003, 2004)
White lupin and wheat	Improved growth and P uptake by intercropped wheat	Cu et al. (2005)
Faba bean and maize	Improved P uptake by intercropped maize	Li et al. (2007)
White lupin and maize	No facilitation	Li et al. (2010)
White lupin and wheat	No facilitation	Wang et al. (2012)
White lupin and maize	Improved growth and P uptake by intercropped maize	Lelei and Onwonga (2014)
Banksia attenuata and *Scholtzia involucrata*	Improved growth and Mn uptake by neighbouring *Scholtzia involucrata*	Muler et al. (2014)
White lupin and maize	Improved growth and P uptake by intercropped maize	Dissanayaka et al. (2015)
Faba bean and maize	Enhanced nodulation and symbiotic N_2 fixation by intercropped faba bean	Li et al. (2016)

to be hydrolysed by phytase released by microorganisms. After that, P-mobilizing species can access a portion of the organic P and insoluble inorganic P pools by releasing carboxylates, protons, and enzymes by roots. Hence, P acquisition by non-P-mobilizing species may be improved when they are grown together with P-mobilizing species. This phenomenon was evident in several intercropping systems, namely wheat–white lupin (Horst and Waschkies 1987), sorghum–pigeon pea (Ae et al. 1990), wheat–chickpea (Li et al. 2003), peanut–maize (El Dessougi et al. 2003), maize–chickpea (Li et al. 2004), faba bean–maize (Li et al. 2007), and white lupin–maize (Dissanayaka et al. 2015).

The mechanisms responsible for interspecific facilitation of P acquisition in intercropping systems are increased P acquisition from the organic P sources (Li et al. 2003, 2004,

Dissanayaka et al. 2015), inorganic P hydroxide complexes or other non-available P forms (Horst and Waschkies 1987; Ae et al. 1990; Li et al. 2007; Dissanayaka et al. 2015) due to P-mobilizing plant species releasing phosphatases, carboxylates, and/or protons into the rhizosphere. As a result, the available P pool in the soil is increased, and P can be taken up by both P-mobilizing and non-P-mobilizing species in the intercropping system. Co-occurring plants would acquire a portion of the mobilized nutrients if root intermingling is strong. This response has been demonstrated in maize–white lupin intercropping (Dissanayaka et al. 2015). Given that root systems of the two species intermingle in the intercropping system, maize plants can take up relatively large quantities of P made available by white lupin. In this system, strong root intermingling was evident from pH reduction in the rhizosphere of intercropped maize (Dissanayaka et al. 2015). However, in a P-impoverished soil (Andosol), companion maize plants could not acquire P mobilized by white lupin because of scant root intermingling due to poor root growth of both species (Dissanayaka et al. 2015). These results highlight the importance of overlap in the rhizosphere of the two co-cultivated species for the benefits of intercropping to be shared, especially in low-P soils.

11.5.2 Improvement in Tolerance to Abiotic and Biotic Stresses in Intercropping Systems

In addition to facilitation of nutrient uptake, intercropping systems may also increase water availability and enhance water-use efficiency (Morris and Garrity 1993; Xu et al. 2008). This outcome is mainly attributed to enhanced water uptake through complementary root distribution in the soil profile (Shackel and Hall 1984; Mao et al. 2012), hydraulic lift (or hydraulic redistribution) of water by deep-rooted crops or mycorrhizal networks (Caldwell et al. 1998; Prieto et al. 2012), and reduced surface runoff (Van Duivenbooden et al. 2000). Thus, selection of crop species with complementary root architectures for the most effective use of water stored in the soil profile is important for improved water acquisition in intercropping systems (Zegada-Lizarazu et al. 2006; De Barros et al. 2007). A potential therefore exists for selection of specific root traits, or mycorrhizal associations, to enhance capture and movement of water to benefit shallow-rooted or non-mycorrhizal plants, especially in arid environments (Burgess 2011).

Gao et al. (2014) suggested that underground root interactions in intercropping systems may be an efficient ecological strategy to control soil-borne plant diseases because maize–soybean intercropping suppressed the occurrence of soybean red crown rot, a severe soil-borne disease caused by *Cylindrocladium parasiticum*. The inhibitory effects decreased with increasing distance between two intercropped plants under both low and high P supply, which highlights the importance of root intermingling for the suppressive effect on soybean red crown rot (Gao et al. 2014).

Strigolactone biosynthesis appears to be a general phenomenon in plants that induce not only branching of AMF hyphae but also influence germination of parasitic plant genera *Orobanche* and *Striga* (Cardoso et al. 2011). Therefore, an increase in strigolactone release may be a common signal for branching of AMF or roots approaching seeds of a parasitic species. Even though white lupin releases only small amounts of strigolactones, in maize–white lupin intercropping the extent of damage caused by *Striga hermonthica* parasitism was decreased (Weisskopf et al. 2009). This clearly showed that the parasitic invasion could be reduced in some intercropping systems with the selection of compatible crop combinations.

11.6 Conclusions and Perspectives

This chapter covers the background and principles of root–root interactions in intercropping. In particular, intercropping systems enhance P acquisition and show potential for decreased usage of inorganic fertilizers for sustainable agricultural production. To develop intercropping systems further, additional studies on soil properties, interactions between crops, and plant-microbe interactions are required. Interactions between nutrients and soil particles and/or organic matter, effects of soil pH, and mineral interactions including targeted (e.g. P or N) and non-targeted minerals (e.g. Mn or Zn) remain unclear. The effects of phytochemicals on interactions between crops may be important. Regarding biotic interactions, the effects of soil fauna (e.g. earthworms and insects) in intercropping remain to be clarified.

Acknowledgments

This work was supported partly by KAKENHI Grant Numbers JP17H03783 and JP17KK0156 from the Japanese Society for the Promotion of Science. We thank Robert McKenzie, PhD, from Edanz Group (www.edanzediting.com/ac) for editing a draft of this manuscript.

References

Adams, M.A. and Pate, J.S. (1992). Availability of organic and inorganic forms of phosphorus to lupins (*Lupinus* spp.). *Plant and Soil* 145: 107–113.

Ae, N., Arihara, J., Okada, K. et al. (1990). Phosphorus uptake by pigeon pea and its role in cropping systems of the Indian subcontinent. *Science* 248: 477–480.

Akiyama, K., Matsuzaki, K., and Hayashi, H. (2005). Plant sesquiterpenes induce hyphal branching in arbuscular mycorrhizal fungi. *Nature* 435: 824–827.

Arihara, J. and Karasawa, T. (2000). Effect of previous crops on arbuscular mycorrhizal formation and growth of succeeding maize. *Soil Science and Plant Nutrition* 46: 43–51.

Bainard, L.D., Koch, A.M., Gordon, A.M. et al. (2012). Temporal and compositional differences of arbuscular mycorrhizal fungal communities in conventional monocropping and tree-based intercropping systems. *Soil Biology and Biochemistry* 45: 172–180.

Barto, E.K., Weidenhamer, J.D., Cipollini, D. et al. (2012). Fungal superhighways: do common mycorrhizal networks enhance below ground communication? *Trends in Plant Science* 17: 633–637.

Berruti, A., Lumini, E., Balestrini, R. et al. (2016). Arbuscular mycorrhizal fungi as natural biofertilizers: let's benefit from past successes. *Frontiers in Microbiology* 6: 1559.

Bolan, N.S., Naidu, R., and Mahimairaja, S. (1994). Influence of low-molecular-weight organic-acids on the solubilization of phosphates. *Biology and Fertility of Soils* 18: 311–319.

Brooker, R.W., Bennett, A.E., Cong, W.F. et al. (2015). Improving intercropping: a synthesis of research in agronomy, plant physiology and ecology. *New Phytologist* 206: 107–117.

Burgess, S.S.O. (2011). Can hydraulic redistribution put bread on our table? *Plant and Soil* 341: 25–29.

Bürgmann, H., Meier, S., and Bunge, M. (2005). Effects of model root exudates on structure and activity of a soil diazotroph community. *Environmental Microbiology* 7: 1711–1724.

Caldwell, M.M., Dawson, T.E., and Richards, J.H. (1998). Hydraulic lift: consequences of water efflux from the roots of plants. *Oecologia* 113: 151–161.

Callaway, R.M. (1995). Positive interactions among plants. *Botanical Review* 61: 306–349.

Callaway, R.M. (2002). The detection of neighbors by plants. *Trends in Ecology and Evolution* 17: 104–105.

Cardoso, C., Ruyter-Spira, C., and Bouwmeester, H.J. (2011). Strigolactones and root infestation by plant-parasitic *Striga*, *Orobanche* and *Phelipance* spp. *Plant Science* 180: 414–420.

Coskun, D., Britto, D.T., Shi, W. et al. (2017). How plant root exudates shape the nitrogen cycle. *Trends in Plant Science* 22: 661–673.

Cu, S., Hutson, J., and Schuller, K.A. (2005). Mixed culture of wheat (*Triticum aestivum* L.) with white lupin (*Lupinus albus* L.) improves the growth and phosphorus nutrition of the wheat. *Plant and Soil* 272: 143–151.

Dalal, R.C. (1978). Organic phosphorus. *Advances in Agronomy* 29: 83–117.

De Barros, I., Gaiser, T., Lange, F.M. et al. (2007). Mineral nutrition and water use patterns of a maize/cowpea intercrop on a highly acidic soil of the tropic semiarid. *Field Crops Research* 101: 26–36.

Dennis, P.G., Miller, A.J., and Hirsch, P.R. (2010). Are root exudates more important than other sources of rhizodeposits in structuring rhizosphere bacterial communities? *FEMS Microbiology Ecology* 72: 313–327.

Dinkelaker, B., Römheld, V., and Marschner, H. (1989). Citric acid excretion and precipitation of calcium citrate in the rhizosphere of white lupin (*Lupinus albus* L.). *Plant, Cell & Environment* 12: 285–292.

Dissanayaka, D.M.S.B., Maruyama, H., Masuda, G. et al. (2015). Interspecific facilitation of P acquisition in intercropping of maize with white lupin in two contrasting soils as influenced by different rates and forms of P supply. *Plant and Soil* 390: 223–236.

Dissanayaka, D.M.S.B., Wickramasinghe, W.M.K.R., and Marambe, B. (2017). Phosphorus-mobilization strategy based on carboxylate exudation in lupin (*Lupinus*, Fabaceae): a mechanism facilitating the growth and phosphorus acquisition of neighboring plants under phosphorus-limited conditions. *Experimental Agriculture* 53: 308–319.

Ehrmann, J. and Ritz, K. (2014). Plant: soil interactions in temperate multi-cropping production systems. *Plant and Soil* 376: 1–29.

El Dessougi, H., Dreele, A.Z., and Claassen, N. (2003). Growth and phosphorus uptake of maize cultivated alone, in mixed culture with other crops or after incorporation of their residues. *Journal of Plant Nutrition and Soil Science* 166: 254–261.

Fan, F., Zhang, F., Song, Y. et al. (2006). Nitrogen fixation of faba bean (*Vicia faba* L.) interacting with a non-legume in two contrasting intercropping systems. *Plant and Soil* 283: 275–286.

Fang, S.Q., Gao, X., Deng, Y. et al. (2011). Crop root behavior coordinates phosphorus status and neighbors: from field studies to three dimensional in situ reconstruction of root system architecture. *Plant Physiology* 155: 1277–1285.

Fang, S., Clark, R.T., Zheng, Y. et al. (2013). Genotypic recognition and spatial responses by rice roots. *Proceedings of the National Academy of Sciences of the United States of America* 110: 2670–2675.

Fornara, D.A. and Tilman, D. (2008). Plant functional composition influences rates of soil carbon and nitrogen accumulation. *Journal of Ecology* 96: 314–322.

Francis, C.A. (1986). Introduction: distribution and importance of multiple cropping. In: Multiple Cropping Systems (ed. C.A. Francis), 1–19. New York, NY: MacMillan Publishing Co.

Frey-Klett, P., Garbaye, J., and Tarkka, M. (2007). The mycorrhiza helper bacteria revisited. *New Phytologist* 176: 22–36.

Fridley, J.D. (2001). The influence of species diversity on ecosystem productivity: how, where, and why? *Oikos* 93: 514–526.

Gao, X., Wu, M., and Xu, R. (2014). Root interactions in a maize/soybean intercropping system control soybean soil-borne disease, red crown rot. *PLoS ONE* 9: e95031.

García-González, I., Quemada, M., and Gabriel, J.L. (2018). Legacy of eight-year cover cropping on mycorrhizae, soil and plants. *Journal of Plant Nutrition and Soil Science* 181: 818–826.

Gardner, W.K. and Boundy, K.A. (1983). The acquisition of phosphorus by *Lupinus albus* L. IV. The effect of interplanting wheat and white lupin on the growth and mineral composition of the two species. *Plant and Soil* 70: 391–402.

Gunes, A., Bagci, E.G., and Inal, A. (2007). Interspecific facilitative root interactions and rhizosphere effects on phosphorus and iron nutrition between mixed grown chickpea and barley. *Journal of Plant Nutrition* 30: 1455–1469.

Hartmann, A., Rothballer, M., and Schmid, M. (2008). Lorenz Hiltner, a pioneer in rhizosphere microbial ecology and soil bacteriology research. *Plant and Soil* 312: 7–14.

Hartmann, A., Schmid, M., van Tuinen, D. et al. (2009). Plant-driven selection of microbes. *Plant and Soil* 321: 235–257.

Hauggaard-Nielsen, H. and Jensen, E.S. (2005). Facilitative root interactions in intercrops. *Plant and Soil* 274: 237–250.

Hiltner, L. (1904). Über neuere Erfahrungen und Probleme auf dem Gebiete der Bodenbakteriologie unter besonderer BerÜcksichtigung der GrÜndÜngung und Brache. *Arbeiten der Deutschen Landwirtschaftlichen Gesellschaft* 98: 59–78. (in German).

Hinsinger, P., Betencourt, E., and Bernard, L. (2011). P for two, sharing a scarce resource: soil phosphorus acquisition in the rhizosphere of intercropped species. *Plant Physiology* 156: 1078–1086.

Horst, W.J. and Waschkies, C. (1987). Phosphorus-nutrition of spring wheat (*Triticum aestivum* L.) in mixed culture with white lupin (*Lupinus albus* L.). *Journal of Plant Nutrition and Soil Science* 150: 1–8.

Hu, J., Lin, X., Wang, J. et al. (2010). Arbuscular mycorrhizal fungus enhances P acquisition of wheat (*Triticum aestivum* L.) in a sandy loam soil with long-term inorganic fertilization regime. *Applied Microbiology and Biotechnology* 88: 781–787.

Inal, A., Gunes, A., and Zhang, F. (2007). Peanut/maize intercropping induced changes in rhizosphere and nutrient concentrations in shoots. *Plant Physiology and Biochemistry* 45: 350–356.

Johansson, J.F., Paul, L.R., and Finlay, R.D. (2004). Microbial interactions in the mycorrhizosphere and their significance for sustainable agriculture. *FEMS Microbiology Ecology* 48: 1–13.

Jones, D.L. (1998). Organic acids in the rhizosphere – a critical review. *Plant and Soil* 205: 25–404.

Kamal, K., Hagagg, L., and Awad, F. (2000). Improved Fe and Zn acquisition by guava seedlings grown in calcareous soils intercropped with graminaceous species. *Journal of Plant Nutrition* 23: 2071–2080.

Kamh, M., Horst, W.J., and Amer, F. (1999). Mobilization of soil and fertilizer phosphate by cover crops. *Plant and Soil* 211: 19–27.

Kennedy, A.C. (1998). The rhizosphere and spermosphere. In: Principles and Applications of Soil Microbiology (eds. D.M. Sylvia, J.J. Fuhrmann, P.G. Hartel and D.A. Zuberer), 389–407. New York, NY: Prentice Hall.

Kraemer, S.M., Crowley, D.C., and Kretzschmar, R. (2006). Geochemical aspects of phytosiderophore-promoted iron acquisition by plant. *Advances in Agronomy* 91: 1–46.

Kuzyakov, Y. (2002). Review: Factors affecting rhizosphere priming effects. *Journal of Plant Nutrition and Soil Science* 165: 382–396.

Lambers, H., Chapin, F.S. III, and Pons, T.L. (eds.) (2008). Plant Physiological Ecology, 2nd e. New York, NY: Springer.

Lambers, H., Finnegan, P.M., Laliberté, E. et al. (2011). Phosphorus nutrition of Proteaceae in severely phosphorus-impoverished soils: are there lessons to be learned for future crops? *Plant Physiology* 156: 1058–1066.

Lambers, H., Hayes, P.E., and Laliberté, E. (2014). Leaf manganese accumulation and phosphorus-acquisition efficiency. *Trends in Plant Science* 20: 83–90.

Lelei, J.J. and Onwonga, R.N. (2014). White lupin (*Lupinus albus* L. cv. Amiga) increases solubility of Minjingu phosphate rock, phosphorus balances and maize yields in Njoro Kenya. *Sustainable Agriculture Research* 3: 37–49.

Li, M. and Tadano, T. (1996). Comparison of characteristics of acid phosphatases secreted from roots of lupin and tomato. *Soil Science and Plant Nutrition* 42: 753–763.

Li, L., Tang, C., Rengel, Z. et al. (2003). Chickpea facilitates phosphorus uptake by intercropped wheat from an organic phosphorus source. *Plant and Soil* 248: 297–303.

Li, S.M., Li, L., Zhang, F.S. et al. (2004). Acid phosphatase role in chickpea/maize intercropping. *Annals of Botany* 94: 297–303.

Li, L., Li, S.M., Sun, J.H. et al. (2007). Diversity enhances agricultural productivity via rhizosphere phosphorus facilitation on phosphorus-deficient soils. *Proceedings of the National Academy of Sciences of the United States of America* 104: 11192–11196.

Li, Y.Y., Yu, C.B., Cheng, X. et al. (2009). Intercropping alleviates the inhibitory effect of N fertilization on nodulation and symbiotic N_2 fixation of faba bean. *Plant and Soil* 323: 295–308.

Li, H., Shen, J., Zhang, F. et al. (2010). Phosphorus uptake and rhizosphere properties of intercropped and monocropped maize, faba bean, and white lupin in acidic soil. *Biology and Fertility of Soils* 46: 79–91.

Li, L., Zhang, L.Z., and Zhang, F.Z. (2013). Crop mixtures and the mechanisms of overyielding. In: Encyclopedia of Biodiversity, 2nd ed, vol. 2 (ed. S.A. Levin), 382–395. Waltham, MA: Academic Press.

Li, H., Zhang, F., Rengel, Z. et al. (2014a). Rhizosphere properties in monocropping and intercropping systems between faba bean (*Vicia faba* L.) and maize (*Zea mays* L.) grown in a calcareous soil. *Crop and Pasture Science* 64: 976–984.

Li, L., Tilman, D., Lambers, H. et al. (2014b). Plant diversity and overyielding: insights from belowground facilitation of intercropping in agriculture. *New Phytologist* 203: 63–69.

Li, B., Li, Y.Y., and Wu, H.M. (2016). Root exudates drive interspecific facilitation by enhancing nodulation and N_2 fixation. *Proceedings of the National Academy of Sciences of the United States of America* 113: 6496–6501.

Lithourgidis, A.S., Dordas, C.A., Damalas, C.A. et al. (2011). Annual intercrops: an alternative pathway for sustainable agriculture. *Australian Journal of Crop Science* 5: 396–410.

Ma, J.F. (2005). Plant root responses to three abundant soil minerals: silicon, aluminum and iron. *Critical Reviews in Plant Sciences* 24: 267–281.

Ma, J.F., Ryan, P.R., and Delhaize, E. (2001). Aluminium tolerance in plants and the complexing role of organic acids. *Trends in Plant Science* 6: 273–278.

Macklon, A.E.S., Mackie-Dawson, L.A., Sim, A. et al. (1994). Soil P resources, plant growth and rooting characteristics in nutrient poor upland grassland. *Plant and Soil* 163: 257–266.

Mao, L.L., Zhang, L.Z., Li, W.Q. et al. (2012). Yield advantage and water saving in maize/pea intercrop. *Field Crops Research* 138: 11–20.

Marschner, H. (1991). Mechanisms of adaptation of plants to acid soils. *Plant and Soil* 134: 1–20.

Marschner, P., Neumann, G., Kania, A. et al. (2002). Spatial and temporal dynamics of the microbial community structure in the rhizosphere of cluster roots of white lupin (*Lupinus albus* L.). *Plant and Soil* 246: 167–174.

Maruyama, H. and Wasaki, J. (2017). Transgenic approaches for improving phosphorus use efficiency in plants. In: Plant Macro-Nutrient Use Efficiency: Molecular and Genomic Perspectives (eds. M.A. Hossain, T. Kamiya, D.J. Burrit, et al.), 323–338. London: Academic Press.

Maruyama, H., Sasaki, T., Yamamoto, Y. et al. (2019). AtALMT3 is involved in malate efflux induced by phosphorus deficiency in *Arabidopsis thaliana* root hairs. *Plant and Cell Physiology* 60: 107–115.

Miller, S.S., Liu, J., Allan, D.L. et al. (2001). Molecular control of acid phosphatase secretion into the rhizosphere of proteoid roots from phosphorus-stressed white lupin. *Plant Physiology* 127: 594–606.

Mommer, L., Kirkegaard, J., and van Ruijven, J. (2016). Root–root interactions: towards a rhizosphere framework. *Trends in Plant Science* 21: 209–217.

Morris, R.A. and Garrity, D.P. (1993). Resource capture and utilization in intercropping – water. *Field Crops Research* 34: 303–317.

Muler, A.L., Oliveira, R., and Lambers, H. (2014). Does cluster-root activity benefit nutrient uptake and growth of co-existing species? *Oecologia* 174: 23–31.

Neumann, G. and Martinoia, E. (2002). Cluster roots – an underground adaptation for survival in extreme environments. *Trends in Plant Science* 7: 162–167.

Neumann, G. and Römheld, V. (1999). Root excretion of carboxylic acids and protons in phosphorus-deficient plants. *Plant and Soil* 211: 121–130.

Ngwira, A.R., Aune, J.B., and Mkwinda, S. (2012). On-farm evaluation of yield and economic benefit of short term maize legume intercropping systems under conservation agriculture in Malawi. *Field Crops Research* 132: 149–157.

Nord, E.A. and Lynch, J.P. (2008). Delayed reproduction in *Arabidopsis thaliana* improves fitness in soil with suboptimal phosphorus availability. *Plant, Cell & Environment* 31: 1432–1441.

Nwoke, O.C., Diels, J., and Abaidoo, R. (2008). Organic acids in the rhizosphere and root characteristics of soybean (*Glycine max*) and cowpea (*Vigna unguiculata*) in relation to phosphorus uptake in poor savannah soils. *African Journal of Biotechnology* 7: 3620–3627.

O'Rourke, J.A., Yang, S.S., Miller, S.S. et al. (2013). An RNA-seq transcriptome analysis of orthophosphate-deficient white lupin reveals novel insights into phosphorus acclimation in plants. *Plant Physiology* 161: 705–724.

Oburger, E., Kirk, G.J.D., and Wenzel, W.W. (2009). Interactive effects of organic acids in the rhizosphere. *Soil Biology and Biochemistry* 41: 449–457.

Ortas, I. (2012). The effect of mycorrhizal fungal inoculation on plant yield, nutrient uptake and inoculation effectiveness under long–term field conditions. *Field Crops Research* 125: 35–48.

Pellegrino, E., Bedini, S., and Avio, L. (2011). Field inoculation effectiveness of native and exotic arbuscular mycorrhizal fungi in a Mediterranean agricultural soil. *Soil Biology and Biochemistry* 43: 367–376.

Philippot, L., Cuhel, J., Saby, N.P.A. et al. (2009). Mapping field-scale spatial patterns of size and activity of the denitrifier community. *Environmental Microbiology* 11: 1518–1526.

Prieto, I., Armas, C., and Pugnaire, F.I. (2012). Water release through plant roots: new insights into its consequences at the plant and ecosystem level. *New Phytologist* 193: 830–841.

Rusinamhodzi, L., Corbeels, M., Nyamangara, J. et al. (2012). Maize–grain legume intercropping is an attractive option for ecological intensification that reduces climatic risk for smallholder farmers in central Mozambique. *Field Crops Research* 136: 12–22.

Ryan, P.R., Delhaize, E., and Jones, D.L. (2001). Function and mechanism of organic anion exudation from plant roots. *Annual Review of Plant Physiology and Plant Molecular Biology* 52: 527–560.

Sasaki, T., Yamamoto, Y., Ezaki, B. et al. (2004). A wheat gene encoding an aluminum-activated malate transporter. *Plant Journal* 37: 645–653.

Sato, T., Ezawa, T., Cheng, W. et al. (2015). Release of acid phosphatase from extraradical hyphae of arbuscular mycorrhizal fungus *Rhizophagus clarus*. *Soil Science and Plant Nutrition* 61: 269–274.

Secco, D., Shou, H., Whelan, J. et al. (2014). RNA-seq analysis identifies an intricate regulatory network controlling cluster root development in white lupin. *BMC Genomics* 15: 230.

Sekiya, N., Araki, H., and Yano, K. (2011). Applying hydraulic lift in an agroecosystem: forage plants with shoots removed supply water to neighboring vegetable crops. *Plant and Soil* 341: 39–50.

Shackel, K.A. and Hall, A.E. (1984). Effect of intercropping on the water relations of sorghum and cowpea. *Field Crops Research* 8: 381–387.

Shen, J.B., Li, C.J., Mi, G.H. et al. (2013). Maximizing root/rhizosphere efficiency to improve crop productivity and nutrient use efficiency in intensive agriculture of China. *Journal of Experimental Botany* 64: 1181–1192.

Simpson, R.J., Oberson, A., Culvenor, R.A. et al. (2011). Strategies and agronomic interventions to improve the phosphorus-use efficiency of farming systems. *Plant and Soil* 349: 89–120.

Smith, S.E. and Read, D.J. (2008). Mycorrhizal Symbiosis. London: Academic Press.

Somers, E., Vanderleyden, J., and Srinivasan, M. (2004). Rhizosphere bacterial signaling: a love parade beneath our feet. *Critical Reviews in Microbiology* 30: 205–240.

Song, Y.N., Zhang, F.S., Marschner, P. et al. (2007). Effect of intercropping on crop yield and chemical and microbiological properties in rhizosphere of wheat (*Triticum aestivum* L.), maize (*Zea mays* L.), and faba bean (*Vicia faba* L.). *Biology and Fertility of Soils* 43: 565–574.

Sun, Y., Zhang, N., Wang, E.T. et al. (2013). Influence of intercropping and intercropping plus rhizobial inoculation on microbial activity and community composition in rhizosphere of alfalfa (*Medicago sativa* L.) and Siberian wild rye (*Elymus sibiricus* L.). *FEMS Microbiology Ecology* 70: 218–226.

Tawaraya, K., Hirose, R., and Wagatsuma, T. (2012). Inoculation of arbuscular mycorrhizal fungi can substantially reduce phosphate fertilizer application to *Allium fistulosum* L. and achieve marketable yield under field condition. *Biology and Fertility of Soils* 48: 839–843.

Tilman, D., Reich, P.B., Knops, J. et al. (2001). Diversity and productivity in a long-term grassland experiment. *Science* 294: 843–845.

Tong, Y., Gabriel-Neumann, E., Krumbein, A. et al. (2015). Interactive effects of arbuscular mycorrhizal fungi and intercropping with sesame (*Sesamum indicum*) on the glucosinolate profile in broccoli (*Brassica oleracea* var. Italica). *Environmental and Experimental Botany* 109: 288–296.

Turner, B.L. (2006). Inositol phosphates in soil: amounts, forms and significance of the phosphorylated inositol stereoisomers. In: Inositol Phosphates: Linking Agriculture and the Environment (eds. B.L. Turner, A.E. Richardson and E.J. Mullaney), 186–206. Wallingford: CABI Publishing.

Unno, Y., Okubo, K., Wasaki, J. et al. (2005). Plant growth promotion abilities and microscale bacterial dynamics in the rhizosphere of lupin analysed by phytate utilization ability. *Environmental Microbiology* 7: 396–404.

Van Duivenbooden, N., Pala, N., and Studer, C. (2000). Cropping systems and crop complementarity in dryland agriculture to increase soil water use efficiency: a review. *NJAS – Wageningen Journal of Life Sciences* 48: 213–236.

Vandermeer, J.H. (1989). The ecology of intercropping. Cambridge: Cambridge University Press.

Veneklaas, E.J., Stevens, J., Cawthray, G.R. et al. (2003). Chickpea and white lupin rhizosphere carboxylates vary with soil properties and enhance phosphorus uptake. *Plant and Soil* 248: 187–197.

Vierheilig, H., Lerat, S., and Picho, Y. (2003). Systemic inhibition of arbuscular mycorrhiza development by root exudates of cucumber plants colonized by *Glomus mosseae*. *Mycorrhiza* 13: 167–170.

Walder, F., Niemann, H., Natarajan, M. et al. (2012). Mycorrhizal networks: common goods of plants shared under unequal terms of trade. *Plant Physiology* 159: 789–797.

Walker, T.S., Bais, H.P., Grotewold, E. et al. (2003). Root exudation and rhizosphere biology. *Plant Physiology* 132: 44–51.

Wang, Y., Marschner, P., and Zhang, F. (2012). Phosphorus pools and other soil properties in the rhizosphere of wheat and legumes growing in three soils in monoculture or as mixture of wheat and legume. *Plant and Soil* 354: 283–298.

Wang, X., Pearse, S.J., and Lambers, H. (2013). Cluster-root formation and carboxylate release in three Lupinus species as dependent on phosphorus supply, internal phosphorus concentration and relative growth rate. *Annals of Botany* 112: 1449–1459.

Wang, Z., Straub, D., Yang, H. et al. (2014). The regulatory network of cluster-root function and development in phosphate-deficient white lupin (*Lupinus albus*) identified by transcriptome sequencing. *Physiologia Plantarum* 151: 323–338.

Wasaki, J., Yamamura, T., Shinano, T. et al. (2003). Secreted acid phosphatase is expressed in cluster roots of lupin in response to phosphorus deficiency. *Plant and Soil* 248: 129–136.

Wasaki, J., Rothe, A., Kania, A. et al. (2005). Root exudation, P acquisition and microbial diversity in the rhizosphere of *Lupinus albus* as affected by P supply and atmospheric CO_2 concentration. *Journal of Environmental Quality* 34: 2157–2166.

Weisskopf, L., Abou-Mansour, E., Fromin, N. et al. (2006). White lupin has developed a complex strategy to limit microbial degradation of secreted citrate required for phosphate acquisition. *Plant, Cell & Environment* 29: 919–927.

Weisskopf, L., Akello, P., Milleret, R. et al. (2009). White lupin leads to increased maize yield through a soil fertility-independent mechanism: a new candidate for fighting *Striga hermonthica* infestation? *Plant and Soil* 319: 101–114.

White, P.J. and Greenwood, D.J. (2013). Properties and management of cationic elements for crop growth. In: Soil Conditions and Plant Growth (eds. P.J. Gregory and S. Nortcliff), 160–194. Oxford: Blackwell Publishing.

White, P.J., George, T.S., Gregory, P.J. et al. (2013). Matching roots to their environment. *Annals of Botany* 112: 207–222.

Xiong, H., Kakei, Y., Kobayashi, T. et al. (2013). Molecular evidence for phytosiderophore-induced improvement of iron nutrition of peanut intercropped with maize in calcareous soil. *Plant, Cell & Environment* 36: 1888–1902.

Xu, B.C., Li, F.M., and Sham, L. (2008). Switchgrass and milkvetch intercropping under 2:1 row-replacement in semiarid region, northwest China: aboveground biomass and water use efficiency. *European Journal of Agronomy* 28: 485–492.

Yoneyama, K., Xie, X., Kusumoto, D. et al. (2007). Nitrogen deficiency as well as phosphorus deficiency in sorghum promotes the production and exudation of 5-deoxystrigol, the host recognition signal for arbuscular mycorrhizal fungi and root parasites. *Planta* 227: 125–132.

Zegada-Lizarazu, W., Izumi, Y., and Iijima, M. (2006). Water competition of intercropped pearl millet with cowpea under drought and soil compaction stresses. *Plant Production Science* 9: 123–132.

Zhang, F.S., Treeby, M., Römheld, V. et al. (1991). Mobilization of iron by phytosiderophores as affected by other microelements. *Plant and Soil* 130: 173–178.

Zhang, F., Shen, J., Zhang, J. et al. (2010). Rhizosphere processes and management for improving nutrient use efficiency and crop productivity: implications for China. *Advances in Agronomy* 107: 1–32.

Zomer, R.J., Trabucco, A., Coe, R. et al. (2009). Trees on farm: analysis of global extent and geographical patterns of agroforestry. ICRAF Working Paper no. 89. Nairobi, Kenya: World Agroforestry Centre.

Zuo, Y.M., Zhang, F.S., Li, X.L. et al. (2000). Studies on the improvement in iron nutrition of peanut by intercropping with maize on a calcareous soil. *Plant and Soil* 220: 13–25.

12

Intercropping to Maximize Root–Root Interactions in Agricultural Plants

Agronomic Aspects

Long Li[1], *Yi-Xiang Liu*[2], *and Xiao-Fei Li*[3]

[1] *China Agricultural University, Beijing, China*
[2] *Yunnan Agricultural University, Kunming, Yunnan, China*
[3] *Institute of Cotton Research, Chinese Academy of Agricultural Sciences, Anyang, Henan, China*

12.1 Intercropping in the World

Intercropping means growing two or more crops simultaneously in the same field. According to the row arrangement and co-growth time, intercropping can be divided into three types: mixed, relay, or strip intercropping (Li et al. 2013b). Mixed intercropping refers to growing two or more crops simultaneously with no distinct row arrangement. Relay intercropping denotes growing two or more crops simultaneously during part of their life cycles. A second crop is planted after the first crop has reached its reproductive stage of growth, but before it is ready for harvest. Strip intercropping is to grow two or more crops simultaneously in different strips that are wide enough to permit independent cultivation, but narrow enough for the crops to interact agronomically (Li et al. 2013b).

Intercropping systems are widely practiced in many parts of the world, including tropical and temperate regions (Francis 1986; Vandermeer 1992). Intercropping is becoming common in the Americas, Asia, Africa, and Europe (Zomer et al. 2009). It is found to play a role in securing the grain supply and raising the income of farmers in developing countries (Yin et al. 2018).

12.1.1 Intercropping in China

Intercropping has been practiced in Chinese agriculture for centuries and still plays an important role in modern agriculture. In China, estimated 2.8×10^7 ha of annually sown area is under intercropping, representing 20–25% of arable land (Li et al. 2007). Species commonly used in intercropping in China include maize, soybean, peanut, potato, wheat, millet, faba bean, tobacco, cotton, sorghum, sesame, garlic, cassava, and vegetables, which

The Root Systems in Sustainable Agricultural Intensification, First Edition.
Edited by Zed Rengel and Ivica Djalovic.

significantly increases the aggregate yields per unit input, insures against crop failure and market fluctuations, and increases farmer income (Li et al. 2001a,b).

Nowadays, the importance of intercropping in China is because of not only tradition and the extent of it but also due to representing ecological and sustainable agriculture compared with monoculture. In the last decade, the studies of intercropping in China have expanded, leading to a better understanding of intercropping systems. Compared with India or Africa, there is little work in China focusing on special breeding of varieties suitable for intercropping. In rural areas, Chinese famers are finding it hard to increase the size of their farm because of paucity of arable land; hence, the only way to maximize yield is to optimize crop management strategies such as using intercropping and rotation, which lead to efficient utilization of natural resources over space and time.

Strip and relay intercropping are the main intercropping systems in China. Strip intercropping means that species in intercropping are cultivated simultaneously in different strips that permit crop independent growth, and at the same time, species interact with each other (Vandermeer 1992). In the relay intercropping system, which could increase total plant density, the preceding crop is planted at the same time as if monoculture, and the follow-up crop is cultivated between the preceding crop while it is maturing (Federer 2012; Keating and Carberry 1993). Strip and relay intercropping are practiced mainly in the north, such as Hetao in Inner Mongolia and the North China Plain, in the northwest such as Xinjiang and the corridor in the Gansu, and along the Yangtze and Yellow rivers.

In China, the multiple crop index (a ratio of intercropping area to total area of cultivated land) increased 30% from 1949 to 1995 (Li 2001). The intercropping area in China is the largest in the word and mainly includes wheat–maize, wheat–cotton, wheat–faba bean, wheat–soybean, maize–soybean, maize–faba bean, and maize–peanut. For example, the area of wheat–maize intercropping in Ningxia was 75 100 ha, producing 43% of total grain yield (Li et al. 2001a). In the Heihe River Basin, 40% of agricultural land is wheat–maize intercropping (Yamazaki et al. 2005). Also, most of peanut in northern China is produced in intercropping (Zuo et al. 2004). The wheat–soybean and maize–faba bean intercropping systems are grown large-scale in northwest of China (Zhang and Li 2003). In the southwest, the area of maize–soybean and maize–potato intercropping has increased over time to become the common agroecosystem model (Wang et al. 2015). In general, the development of intercropping in the north, northeast, and northwest of China showed more potential than in the southwest because the climate conditions in the southwest allow more flexible rotations that replace intercropping systems.

12.1.2 Intercropping in other Countries in Asia

In parts of East Asia and India, the rainfall primarily comes in summer (Kingwell-Banham et al. 2015; Weisskopf et al. 2015). Crop species adapted to this region are thus suited to the summer rainfall, and farming activities are also focused in this period of the year. Studies aimed at unravelling the complexities of cropping strategies in South Asia found that mixed-cropping, double-cropping, and multi-cropping are the main ways to grow multiple crops in the suitable season(s) (Fuller and Madella 2002; Weber 2003; Wright 2010).

The rice–wheat rotation is one of the major multiple-cropping systems in Asia, occupying about 18 million ha, out of which 13.5 million ha are in the Indo-Gangetic Plains of India, Bangladesh, Nepal, and Pakistan (Bhushan et al. 2007). The semi-arid tropics (SAT) in

India extend over 43% of the country's total geographic area and 70% of net cultivated area across diverse agroecological regions. Farmers in this region of India have been involved in intercropping legumes, such as finger millet mixed with black gram or other legumes (Chandra et al. 2013). Intercropping legumes with finger millet (non-legume) often results in higher resource-use efficiency compared to sole cropping (Maikhuri et al. 1996, 1997; Chandra 2007). In central India, cotton (*Gossypium hirsutum* L.) intercropped with pigeonpea (*Cajanus cajan* (L.) Millsp.) is a traditional cropping system practiced, occupying nearly 1.5 million ha (Blaise et al. 2005).

In north-western India, sugarcane is an important crop even though grown on a slightly smaller scale than the dominant rice–wheat system because of long economic returns cycle. Intercropping of short duration *Brassica* spp. and maize (for immature cobs) (Singh et al. 2008) with sugarcane in autumn improved productivity and profitability compared to sole sugarcane. The suitability of intercropping of cabbage, potato, onion, peas, and radish in autumn sugarcane has also been reported (Saini et al. 2003; Al Azad and Alam 2004).

In southwest Bangladesh, the High Ganges River floodplain agro-ecological region has a tropical monsoon climate with the lowest and least reliable rainfall (Manalo 1976). The farmers have tried to minimize the risks of total crop failure by mixed-cropping. Intercropping of mustard, lentil, and chickpea in Jessore region showed higher yield than their monocultures (Hassan et al. 1991). In Bangladesh, maize–sweet potato intercropping increased productivity and ensured food security for resource-poor hill farmers (Islam et al. 2014). In northwest Bangladesh, banana intercropped with potato had higher yield and the land equivalent ratio (LER) compared with monoculture (Rahman et al. 2006).

The intercropping patterns in Nepal and Pakistan are similar to those in India and Bangladesh. Legumes intercropped with nonlegumes are also the main intercropping system (Khan and Khan 2005; Khatri et al. 2014; Khadka et al. 2013). The difference, however, is that maize and finger millet (the two predominant staple crops in the rainfed area of Nepal) are also grown intercropped to increase land productivity and potential economic returns (Chapagain et al. 2018).

12.1.3 Intercropping in North America

The development and use of any land practices is expected to be different in temperate North America than in tropical or less-developed countries given the existing social, environmental, and economic situation (Williams and Gordon 1992). In contrast to Asia and Africa that face serious food security problems, the problems in North America are associated with resource management policies, farming technology, politics, labour costs, and real estate values (Raintree 1991). In the United States, many strip-intercropping systems such as maize–soybean and oat–soybean were reported to increase the productivity and sustainability (Ghaffarzadeh et al. 1994). The legume intercropping with nonlegumes, including canavalia (*mucuna*) –maize and kidney bean–maize, is an important intercropping system in Mexico (Wortmann 1993; Flores-Sanchez et al. 2013).

Agroforestry is the main intercropping system in North America and shows unique advantages in water-quality enhancement, carbon sequestration, and biodiversity conservation compared with the conventional land-use systems (Thevathasan and Gordon 2004). There are more than 140 million ha of marginal or degraded land in North America available for agroforestry establishment (Dixon et al. 1994). In the history of North

America, many types of agroforestry systems, including windbreaks, shelterbelts, silvipastoral systems, integrated riparian forest systems, forest farming systems, and tree–crop intercropping, have been practiced and proved to have a vast potential for economic and environmental benefits (Garrett et al. 2009; Gordon et al. 1997).

In many areas of North America, such as southern Ontario and southwestern Quebec in Canada, and southeastern United States, agroforestry has been beneficial for water-quality enhancement, carbon sequestration, and biodiversity conservation (Thevathasan and Gordon 2004). The crops can also benefit tree production; for example, Rivest et al. (2010) reported soybean intercrop resulted in greater biomass of hybrid poplar. The positive interactions between the woody (tree) and nonwoody (agricultural or annual crop) species play key roles in the success of tree-based intercropping systems. At present, many tree-based intercropping, including maple–maize (or soybean) (Reynolds et al. 2007), hybrid poplar (or Norway spruce)–barley (Peichl et al. 2006), Norway spruce (or red oak, black walnut, or white cedar)–soybean (Wotherspoon et al. 2014), hybrid poplar–soybean (Rivest et al. 2009) in Canada, were reported to have environmental benefits and no negative interactions. Susaeta et al. (2012) studied the economics of alley cropping of loblolly pine and switchgrass in the southern United States and found that landowners would be financially better off adopting intercropping when switchgrass prices were more than $\$30\,t^{-1}$. Hence, tree–crop intercropping is an agroforestry system that shows great potential for this region. These land-management options can be more favourable than conventional agriculture in terms of long-term productivity and sustainability.

12.1.4 Intercropping in Europe

The characteristics of modern agriculture in Western Europe, with most countries being well developed, are high input of fertilizers and pesticides and highly mechanized. The arable crop sector maximizes the potential of agriculture in Western Europe (Licker et al. 2010). However, many environmental problems related to water and soil quality also arise from such intensive cropping systems (Licker et al. 2010). The interest in intercropping in Europe has been revived over the past two decades because of the potential to solve important ecological (especially biodiversity) problems and produce additional market benefits. Importantly, the interaction between species in intercropping could reduce the pesticide and fertilizer usage, which might be one of the reasons behind this renewed interest (Anil et al. 1998). Intercropping system might be an alternative to the current arable agricultural sector in Europe to achieve the natural resource preservation goals. However, inability to use agricultural mechanization in intercropping systems is the key factor restricting intercropping development in Western Europe and throughout the developed world (Vandermeer 1992).

Cereal–legume intercrop combinations have been practiced in northern Europe for fodder production, with a goal to increase protein concentration and reduce the cost of feeding protein concentrates to livestock (Anil et al. 1998). Spring barley–spring pea intercropping was the most frequent out of 27 combinations between pulses and cereals in five European countries, which showed better yield stability, effective weed suppression, and good quality of feed (von Fragstein und Niemsdorff et al. 2008). In field experiments in Europe in the 2002/03, 2003/04, and 2004/05 growing seasons, intercropping wheat with faba bean (in Denmark, Germany, Italy, and UK) or pea (in France) regularly increased the nitrogen and sulfur concentration in cereal grains, hence increasing the wheat quality for bread-making (Gooding et al. 2007). A hypothesis of Manevski et al. (2015) that maize intercropping using fertilizer N rates applicable to monocrop maize decreased N leaching without a significant yield loss was not rejected in intercropping systems such as maize–soybean, maize–groundnut, maize–ryegrass, and maize–lucerne (Nie et al. 2012). Oat (or wheat)–white clover intercropping was reported to maintain yield using lower fertilizer inputs compared with monoculture (Thorsted et al. 2002, 2006). Cereals in these cereal–clover mixtures also had higher crude protein concentrations than monoculture cereals harvested for fodder (Andrzejewska et al. 2014).

12.1.5 Intercropping in Africa and South America

In developing countries in Africa, agricultural development is considered the key to economic development. In most of these countries, agriculture accounts for 35% of gross domestic product, employs about 62% of the population, and produces about 60% of export earnings, thus addressing food security and raw materials for domestic industries (Taylor et al. 1996). However, farmers face numerous constraints, ranging from a lack of appropriate technologies, inputs, and infrastructure to the natural hazards (Jalloh 2001).

As the traditional agricultural system in Africa, mixed cropping accounts for more than 80% of rain-fed uplands. A large number of interspecies interactions occur in these traditional mixed-cropping systems, bringing substantial complementarity, including efficient use of labour, suppression of weeds and insects, control of soil erosion, and reduced risk of crop losses (Ikeorgu et al. 1989).

Promoting technologies that increase farmers' productivity is a major strategy for achieving food security in Africa. Many scientists and extension workers have tried to introduce crop monoculture techniques in Africa as an alternative to traditional mixed farming. However, crop monoculture techniques show no advantage over the traditional practice of mixed cropping (Webster and Wilson 1966). The mixed cropping systems evolved in different areas are so deeply established among farmers that they could hardly accept new cultivation systems (Nnko and Doto 1982).

Over the years, the mixed cropping systems in Africa comprised two to four major crops. In the western part of West Africa, rice is the staple crop that is intercropped mostly with cassava, maize, or millet. In the rest of West Africa and central Africa, cassava and yam predominate and are mostly intercropped with maize. In Eastern and Southern Africa, wheat and beans are the main intercropping species, but lately cassava is gaining importance among the farmers in this region (Jalloh 2001).

12.2 Importance of Root–Root Interactions in Intercropping

Although intercropping can be done with field, pasture, vegetables, and even tree crops, intercropping legumes with cereals is by far the most common (Li et al. 2013b). The root–root interactions between intercropped species inevitably occur when multiple plants coexist in a given soil volume (Vandermeer 1992). A large body of literature has investigated below-ground interactions between intercropped roots in the widespread cereal–legume intercropping systems (Li et al. 1999, 2001b, 2003b, 2007; Hauggaard-Nielsen and Jensen 2005). In this chapter, we focus mainly on how the below-ground processes driven by root–root interactions can be used to sustainably intensify agricultural production by optimizing intercropping systems both agronomically and ecologically (see also Chapter 11).

12.2.1 Root–Root Interactions Play a Vital Role in Yield Advantage of Intercropping

One of the most obvious advantages emphasized when trying to convince farmers to adopt intercropping strategies is to improve yield compared with the respective sole crops (Li et al. 2007, 2013b, 2014; Brooker et al. 2016; Isbell et al. 2017). A generally accepted method for the evaluation of yield advantage of intercropping is LER (Willey and Rao 1980). The LER is the relative land area that is required to produce the same yield in sole crops as in the intercropping. If the LER value is greater than 1, the intercropping is advantageous (Willey 1979). The yield improvement potential has been repeatedly demonstrated in many intercropping systems (Li et al. 2014; Ren et al. 2014; Bedoussac et al. 2015), including cereals–cereals (i.e. wheat–maize, barley–maize, and spring maize–summer maize), cereals–legumes (i.e. maize–soybean, maize–peanut, maize–pea, maize–faba bean, wheat–faba bean, and wheat-soybean), and some cereals–vegetables intercropping (Li et al. 1999, 2001b, 2013b).

Interspecific root-root interactions play an important role in overyielding in the intercropping systems. More specifically, a significant positive yield effect on maize was found when the root systems intermingled freely (no barrier) or partly (400 mesh nylon barrier) compared with no root interaction (plastic sheet barrier) in microplot experiments (Li et al. 1999, 2007; Zhang and Li 2003). When the roots of two species intermingled, LER values based on total yields and grain yields were 1.21 and 1.34, respectively, but when the roots of the two species were separated completely, the intercropping advantage was greatly diminished, and the LER values were reduced to 1.06 and 1.12 (Li et al. 1999). In wheat–maize and wheat–soybean intercropping, almost a half of grain yield increases in wheat were derived from root–root interactions between intercropped species under field conditions (Li et al. 2001b).

12.2.2 Complementarity and Facilitation of Nutrient Acquisition between Intercropped Species Induced by Root–Root Interactions

Complementarity suggests that crops differ in the way they acquire and use resources, thereby overcoming interspecific competition. Complementarity can be categorized as temporal, spatial, or chemical partitioning (Duchene et al. 2017). Experiments in wheat–maize and barley–maize intercropping provided examples of how root–root interactions induce significant temporal niche differentiation in nutrient uptake rate between intercropped species (Zhang et al. 2016b). Spatial complementarity means that the nutrient uptake process occurs in different locations depending on root interactions and the traits of the intercropped plants. Pea–barley intercropping provided evidence for such spatial distribution of roots, with barley having deeper and finer roots than pea, thus being capable of capturing resources in deeper soil layers (Hauggaard-Nielsen et al. 2001). A well-known example of chemical complementarity is the capacity of legumes to symbiotically fix nitrogen from the atmosphere that is not available to colocated nonlegume species. For instance, interspecific interactions increased the percentage of N derived from air (%Ndfa) in the wheat–faba bean system, but not in the maize–faba bean system when no N fertilizer was applied (Fan et al. 2006). Jensen (1996) also demonstrated that N_2 fixation by pea intercropped with barley was 17.7 g N m^{-2} at maturity, whereas it was only 5.1 g N m^{-2} when pea was grown alone.

Interspecific facilitation is an additional outcome of root–root interactions between neighbouring plant species. However, it is difficult to distinguish facilitation from niche differentiation in practice (Loreau and Hector 2001). Facilitation often provides additional agro-ecosystem services by improving growth conditions and resources availability, especially of soil P (Duchene et al. 2017). Hence, positive plant–plant interactions such as root–root facilitation have been shown to play a significant role in P acquisition (Li et al. 2007, 2014; Hinsinger et al. 2011; Zhang et al. 2016a). For instance, total P acquisition by plants was 68% greater with root intermingling and 37% greater with nylon mesh separation than with the solid root barrier (Li et al. 2003a). Similar results were obtained in faba bean–maize (Li et al. 2003b, 2007; Zhang and Li 2003; Mei et al. 2012), white lupin–maize (Dissanayaka et al. 2015), wheat–white lupin (Cu et al. 2005), and spring pea–barley intercropping (Hauggaard-Nielsen et al. 2009).

12.2.3 Root Exudates and Soil Microorganisms as Main Drivers of Root–Root Interactions

In addition to resource competition, both root exudates and the rhizobiome have been recognized as the main drivers of root–root interactions (Mommer et al. 2016). There are two main pathways: via modification of resource availability to the plants (e.g. carboxylates or arbuscular mycorrhizal fungi [AMF]), and by directly affecting root–root interactions (e.g. phytotoxins and fungal pathogens).

Several components in root exudates are known to influence nutrient availability and, thus, have a potential to mediate resource competition (Bais et al. 2006; Mommer et al. 2016). For example, the plant-available concentration of P and Zn in soil increases with exudation of acid phosphatases, protons, and/or carboxylates (Duffnerm et al. 2012; Li et al. 2014). However, plants may also benefit from the root exudates of their neighbours. In intercropping of maize and faba bean, the acidification of rhizosphere due to the release of protons by faba bean enhanced P nutrition and growth of maize (Li et al. 2007, 2013a; Zhang et al. 2016). The role of root exudates is not constrained to increasing nutrient availability to plants. Exudates also directly influence the outcome of root–root interactions in the rhizosphere through allelopathy, whereby some plant species release phytotoxins that reduce the establishment, growth, or survival of their neighbours, thus reducing competition and increasing a share of resources (Bais et al. 2003).

The effects of root exudates on nutrient availability can hardly be evaluated without explicitly considering the role of soil biota because root exudates are an important substrate for microbial communities (Hinsinger et al. 2009). The rhizobiome can change the amount and composition of root exudates (Meier et al. 2013), creating the possibility for feedbacks among plants and soil microorganisms. Vice versa, plants influence their rhizobiomes via exudates (Broeckling et al. 2008; Badri et al. 2013) because these can also contain secondary metabolites, such as antimicrobials that reduce pathogen pressure. Perhaps the clearest example of root–microbe interactions affecting nutrient availability and competition between plants is the symbiosis with AMF. For instance, the presence of AMF enhanced the biomass and nutrient uptake by both component crops in faba bean–maize (Qiao et al. 2016) and chili pepper (*Capsicum annuum*)–sweet maize intercropping (Hu et al. 2018).

12.3 Root–Root Interactions and Interspecific Facilitation Effects on Nutrient Availability

Numerous studies have reported how yield advantages in intercropping systems compared to monocultures are underpinned by below-ground processes (Li et al. 2014). The aim here is to discuss the effects of arable multiple-crop systems on below-ground processes, considering root distribution, biological nitrogen fixation, and phosphorus and micronutrient acquisition.

12.3.1 Root Distribution and Interactions Between Intercropped Species

The spatial distribution and density of root networks in the soil determine the capacity of a crop to acquire nutrients and water. In faba bean–maize intercropping, symmetric interspecific facilitation was observed during co-growth (Li et al. 1999). One of the mechanisms was the compatible deployment of roots of the two crop species, where faba bean had a relatively shallow root distribution, and the roots of intercropped maize spread underneath the roots of faba bean and intermingled with them in the upper layers of the soil profile (Li et al. 2006). In wheat–maize intercropping, asymmetric interspecific facilitation occurred during co-growth (Li et al. 2001b). There was an increase in growth and nutrient uptake by intercropped wheat, but a decrease in maize (Li et al. 2001b). Intercropped wheat not only had a greater root length density (RLD) than sole wheat but also occupied more below-ground space, including that directly under the maize plants. The RLD of maize was increased by intercropping, but the roots of maize could not spread under the wheat roots (Li et al. 2006). However, intercropping with wheat led to greater root weight density and larger below-ground space occupied by intercropped maize in the late growth stages (Li et al. 2011). In addition, the roots of intercropped maize had a longer life span than the roots of sole maize in the late growth stages (Li et al. 2011). These findings may advance understanding of the growth recovery of later-maturing crops in intercropping in the late growth stage (Li et al. 2001a).

12.3.2 Enhanced Nodulation and N_2 Fixation by Root–Root Interactions Driven by Root Exudates

Intercropping of faba bean and maize resulted in overyielding and enhanced nodulation by faba bean. By using permeable and impermeable root barriers in a 2-year field experiment, we showed that root–root interactions between faba bean and maize significantly increased both nodulation and symbiotic N_2 fixation (by 32% and 36%, respectively) in intercropped compared to monoculture faba bean (Li et al. 2016). Nitrogen fertilization decreased dry weight of nodules per plant by 27%, and a proportion of N derived from fixation decreased by 22%. However, root interactions between faba bean and maize enhanced N_2 fixation, even under high N fertilization rate (Li et al. 2009).

Traditionally, a decrease in N availability by cereals that are more competitive for soil inorganic N than legumes was assumed to drive greater nodulation and N_2 fixation of legumes (Fan et al. 2006; Li et al. 2009). However, the effect of cereals on nodulation is not consistent. Nodulation and N_2 fixation were not enhanced when faba bean was intercropped with wheat or barley (Fan et al. 2006). This inconsistency suggests that low soil N availability is not the only driver of increased nodulation of legumes in mixtures. Intercropped maize also caused a twofold increase in exudation of flavonoids (signalling compounds for rhizobia). Roots of faba bean treated with the maize root exudates exhibited an immediate 11-fold increase in the expression of the chalcone–flavanone isomerase (involved in flavonoid synthesis) gene together with significantly increased expression of genes mediating nodulation and auxin responses. After 35 days, faba bean treated with maize root exudate continued to show up-regulation of key nodulation genes, such as early nodulin 93 (*ENOD93*) and increased nitrogen fixation. These findings suggested a mechanism of how intercropped maize promoted nitrogen fixation of faba bean, whereby maize root exudates promoted flavonoid synthesis in faba bean, thus increasing nodulation and stimulating nitrogen fixation (Li et al. 2016).

12.3.3 Interspecific Facilitation of Phosphorus Uptake by Root–Root Interactions

Soil P is poorly available due to adsorption and formation of insoluble P compounds in (Bieleski 1973). A number of plant species have developed mechanisms by which soil P can be mobilized into forms that can be taken up by them and by plants growing in proximity, confirming the root–root interactions (Zhang et al. 2004). Mechanisms by which P cycling and plant uptake is influenced by multiple crop arrangements include complementary P use and P facilitation (Li et al. 2007; Hinsinger et al. 2011). In this section, we focus on changes in the rhizosphere.

Interspecific facilitation of P uptake by intercropped species was examined under inorganic or organic P supplies. Some plant species had a high capacity to acquire insoluble inorganic P (i.e. white lupin and faba bean) and organic P (i.e. chickpea). Mechanisms of P facilitation in intercropping included the following: (i) acidification of the rhizosphere by P-efficient species resulting in increased availability of insoluble inorganic P, such as Ca_{10}-P, $FePO_4$, and $AlPO_4$, which benefited intercropped P-inefficient species with poor capacity to mobilize insoluble soil P; (ii) carboxylates from root exudates of one species chelated Ca, Fe, and Al, consequently mobilizing insoluble soil P, which benefited that species and other intercropped species; and (iii) greater phosphatase activity in the rhizosphere transformed soil organic P into inorganic forms that can be used by both intercropped species in the wheat–chickpea and maize–chickpea systems (Li et al. 2003a,b, 2007; Zhou et al. 2009).

12.3.4 Improved Micronutrient Uptake by Intercropped Species due to Root–Root Interactions

Intercropping legumes with cereals improved micronutrient nutrition of legumes grown in pot experiments and under field conditions. For example, intercropping peanut with barley, oat, wheat, or maize increased the chlorophyll and HCl-extractable Fe and total Fe concentrations in young leaves of peanut, thus alleviating Fe deficiency chlorosis on calcareous soils (Zuo et al. 2000; Inal et al. 2007; Zuo and Zhang 2008; Xue et al. 2016). In addition, Zn concentration in the whole peanut shoots was increased by intercropping of peanut with barley, oat, wheat, or maize in calcareous sandy soils in a glasshouse study (Zuo and Zhang 2008). Similar evidence was reported for other micronutrients such as Fe and Zn (Li et al. 2004; Xia et al. 2013a; Wiche et al. 2016).

12.4 Root–Root Interactions and the Competition-Recovery Principles

12.4.1 The Competition-Recovery Principles in Intercropping Systems

Interspecific competition is inevitable when two crops are grown together (Vandermeer 1992). Such competition usually decreases survival, growth, or reproduction of at least one species (Crawly 1997). Li et al. (2001a,b) reported the "competitive-recovery production principle" in intercropping, whereby interspecific competition enhances the yield of the dominating species (wheat) and impairs the growth of the other species (maize or soybean) during the co-growth period (70–80 days). After the harvest of the dominating species (wheat), the other species (maize or soybean) had about 60 days of growth alone and showed recovery growth. The growth recovery offsets negative border-row effects on the intercropped maize, and this maintains (or sometimes even increases) grain yield.

Similarly, compared with the sole rice, the biomass of rice intercropped with mung bean was significantly higher and showed compensation after mung bean harvest (Aggarwal et al. 1992). Nitrogen uptake by pigeon pea was decreased when intercropped with sorghum but recovered by harvest (Madhavan and Shanmugasundaram 1990). In cassava–pigeon pea intercropping, total production of all cassava cultivars was severely reduced at pigeon pea harvest. The subsequent recovery was slow, and final tuber yield in all cassava cultivars was <25% of the corresponding yield in the monocrop (Cenpukdee and Fukai 1992a,b). The "competitive-recovery production" process has also been observed in other intercropping systems such as maize–pigeon pea (Dalal 1974), cassava–cowpea (Mason et al. 1986), wheat–mung bean (Aggarwal et al. 1992), wheat–soybean (Li et al. 2001a), and maize–turnip (Xia et al. 2013b).

12.4.2 Root Competition and Plasticity

In intercropping, root competition shows a larger effect on crop growth and resource capture than shoot competition (Wilson 1988; Zhang et al. 2001). During the competition period in wheat–maize intercropping, the yield and nutrient uptake advantages of wheat result from its greater competitive capacity to acquire nutrients compared with maize

(Li et al. 2001b). Moreover, Li et al. (2006) found that the productivity of wheat/maize intercropping was partly dependent on root interactions. During the co-growth stage, intercropped wheat had greater RLD and occupied a larger soil volume, including the soil space under the maize, whereas the roots of intercropped maize were limited laterally (Li et al. 2006). However, the root architecture of plant roots shows plasticity depending on the changes in soil environment such as nutrient and water availability.

Phenotypic plasticity is the property of a given genotype that produces different physiological or morphological phenotypes in response to different environmental conditions (Pigliucci 2001). Root behaviour in response to neighbours was significant in a simulated forestry ecosystem with two tree species, sweet gum (*Liquidambar styraciflua* L.) and loblolly pine (*Pinus taeda* L.), indicating the species exhibited plasticity in response to the presence of neighbouring plants (Fort et al. 2014; Mou et al. 1995). Based on meta-analysis, Davidson et al. (2011) concluded that species with invasive capacity generally had greater phenotypic plasticity than co-occurring noninvasive species. Jackson and Caldwell (1996) also found that species with the plasticity of root proliferation could help co-occurring species absorb more N and P in nutrient-rich patches. Liu et al. (2015) reported that the RLD and lateral root distribution of intercropped wheat were more sensitive to the soil N status than those of intercropped maize, indicating higher phenotypic plasticity of roots of intercropped wheat and attesting to a greater competitive capacity of wheat compared with maize.

Physiological plasticity responses that are more carbon efficient than morphological ones can contribute substantially to the capacity of species to grow in diverse and variable environments (Mou et al. 2013). However, relatively few studies have focused on physiological plasticity in resource acquisition. Physiological plasticity in stomatal aperture, leaf angle, and photosynthetic rate allows plants to adapt to variable environmental factors such as light intensity and transpiration demand (Sultan 2000). Cui and Caldwell (1998) found that root P influx rates per unit root length or mass of desert perennials and shrubs was much higher in the P-rich soil patches than predicted based on the P-poor soil patches. Intercropped wheat had greater nitrogen uptake per root length than sole wheat, whereas maize showed the opposite trend, which may be the mechanism underlying the greater competitive capacity of wheat in wheat–maize intercropping (Liu et al. 2015). In addition, physiological plasticity of roots is generally more important for increasing uptake of NO_3^- than phosphate (Robinson et al. 1994).

In summary, the root morphology and physiology of a given species in intercropping are not constant. Root competition between species varies in various environmental conditions, which is determined by the plasticity of root responses to both soil resources and neighbour species. The species with greater root plasticity usually show a greater root competitive capacity in intercropping systems.

12.4.3 Recovery and Below-Ground Interspecific Interactions

Based on the "competitive-recovery production" principle of intercropping systems, growth and nutrient uptake by the late-maturing species recover after the harvest of the early-maturing species. Aggarwal et al. (1992) suggested that an increase in soil volume accessible for N extraction and in the aerial space available led to rice yield compensation after mung bean had been harvested. In pigeon pea–sorghum intercropping, the recovery of total N acquisition by pigeon pea at harvest was a result of a longer growing season (Madhavan and Shanmugasundaram 1990). Furthermore, Li et al. (2011) found that maize intercropped with wheat occupied a greater soil volume in the late than early growth stages, and maize intercropped with either wheat or faba bean had longer root lifetime compared to the maize grown alone, supporting the recovery of late-maturing species generally found in intercropping systems between two crop species differing in growth stages.

The phenotypic plasticity of roots modulates root distribution to enhance growth in heterogeneous and changing environments (Mommer et al. 2012; Mou et al. 2013). In wheat–maize intercropping, in the co-growth stage, the RLD of intercropped maize under the wheat area showed little response to soil N and displayed a significant negative correlation with soil N after wheat harvest (Liu et al. 2019), which played an important role in recovery growth of intercropped maize.

Plants can enhance their nutrient-uptake capacity not only via modulating root architecture but also by increasing the uptake per unit root length or root mass (NURI) (Aziz et al. 2017; Cahill et al. 2010). The difference in NURI between intercropping and monoculture also plays an important role in recovery growth of the late-maturing species. In wheat–maize intercropping, the NURI of intercropped maize after wheat harvest was significantly greater than that of sole-cropped maize, likely due to intercropping maize roots absorbing additional N from the wheat-growing area after wheat harvest (Liu et al. 2019). In contrast, in the co-growth stage, the intercropped maize had a lower NURI than sole-cropped maize because of strong competition for N by wheat (Liu et al. 2015). This reversal of NURI may be one of the mechanisms underlying the growth recovery of intercropped maize.

12.5 Conclusions

Compared with monoculture, intercropping has a number of advantages and therefore is practiced by farmers worldwide. Root–root interactions between species play an important role in determining increased productivity of intercropping, including complementary patterns of resource use and facilitative interactions. Root exudates and soil microbes are considered the main drivers of root–root interactions in intercropping. Also, root–root interactions enhance nitrogen, phosphorus, and micronutrient uptake by intercropped species via the facilitative effects. In addition to complementarity and interspecific facilitation, root morphological and physiological plasticity may be the key mechanisms for overyielding and competition-recovery production in intercropping.

References

Aggarwal, P.K., Garrity, D.P., Liboon, S.P. et al. (1992). Resource use and plant interactions in a rice-mungbean intercrop. *Agronomy Journal* 84: 71–78.

Al Azad, M.A.K. and Alam, M.J. (2004). Popularizing of sugarcane based intercropping systems in Non Millzone. *Journal of Agronomy* 3: 159–161.

Andrzejewska, J., Albrecht, K.A., and Harasimowiczhermann, G. (2014). Intercropping winter cereals with caucasian clover for forage in northern Europe. *Acta Agriculturae Scandinavica* 64: 734–740.

Anil, L., Park, R.H., Phipps, R.H. et al. (1998). Temperate intercropping of cereals for forage: a review of the potential for growth and utilization with particular reference to the UK. *Grass and Forage Science* 53: 301–317.

Aziz, M.M., Palta, J.A., Siddique, K.H.M. et al. (2017). Five decades of selection for yield reduced root length density and increased nitrogen uptake per unit root length in australian wheat varieties. *Plant and Soil* 413: 181–192.

Badri, D.V., Chaparro, J.M., Zhang, R. et al. (2013). Application of natural blends of phytochemicals derived from the root exudates of arabidopsis to the soil reveal that phenolic-related compounds predominantly modulate the soil microbiome. *Journal of Biological Chemistry* 288: 4502–4512.

Bais, H.P., Vepachedu, R., Gilroy, S. et al. (2003). Allelopathy and exotic plant invasion: from molecules and genes to species interactions. *Science* 301: 1377–1380.

Bais, H.P., Weir, T.L., Perry, L.G. et al. (2006). The role of root exudates in rhizosphere interations with plants and other organisms. *Annual Review of Plant Biology* 57: 233–266.

Bedoussac, L., Journet, E.P., Hauggaard-Nielsen, H. et al. (2015). Ecological principles underlying the increase of productivity achieved by cereal-grain legume intercrops in organic farming. a review. *Agronomy for Sustainable Development* 35: 911–935.

Bhushan, L., Ladha, J.K., Gupta, R.K. et al. (2007). Saving of water and labor in a rice–wheat system with no-tillage and direct seeding technologies. *Agronomy Journal* 99: 1288–1296.

Bieleski, R.L. (1973). Phosphate pools, phosphate transport, and phosphate availability. *Annual Review of Plant Physiology* 24: 225–252.

Blaise, D., Bonde, A.N., and Chaudhary, R.S. (2005). Nutrient uptake and balance of cotton+pigeonpea strip intercropping on rainfed Vertisols of central India. *Nutrient Cycling in Agroecosystems* 73: 135–145.

Broeckling, C.D., Broz, A.K., Bergelson, J. et al. (2008). Root exudates regulate soil fungal community composition and diversity. *Applied and Environmental Microbiology* 74: 738–744.

Brooker, R.W., Karley, A.J., Newton, A.C. et al. (2016). Facilitation and sustainable agriculture: a mechanistic approach to reconciling crop production and conservation. *Functional Ecology* 30: 98–107.

Cahill, J.F., McNickle, G.G., Haag, J.J. et al. (2010). Plants integrate information about nutrients and neighbors. *Science* 328: 1657–1657.

Cenpukdee, U. and Fukai, S. (1992a). Cassava/legume intercropping with contrasting cassava cultivars. 1. Competition between component crops under three intercropping conditions. *Field Crops Research* 29: 113–133.

Cenpukdee, U. and Fukai, S. (1992b). Cassava/legume intercropping with contrasting cassava cultivars. 2. Selection criteria for cassava genotypes in intercropping with two contrasting legume crops. *Field Crops Research* 29: 135–149.

Chandra, A. (2007). Traditional agrodiversity management in central Himalayan village ecosystem (Doctoral dissertation, Ph. D. Thesis, University of Delhi, Delhi, India).

Chandra, A., Kandari, L.S., Negi, V.S. et al. (2013). Role of intercropping on production and land use efficiency in the central Himalaya, India. *Environment & We* 8: 105–113.

Chapagain, T., Pudasaini, R., Ghimire, B. et al. (2018). Intercropping of maize, millet, mustard, wheat and ginger increased land productivity and potential economic returns for smallholder terrace farmers in Nepal. *Field Crops Research* 227: 91–101.

Crawly, M.J. (1997). Plant Ecology. Cambridge: Blackwell.

Cu, S.T.T., Hutson, J., and Schuller, K.A. (2005). Mixed culture of wheat (*Triticum aestivum* L.) with white lupin (*Lupinus albus* L.) improves the growth and phosphorus nutrition of the wheat. *Plant and Soil* 272: 143–151.

Cui, M. and Caldwell, M.M. (1998). Nitrate and phosphate uptake by *Agropyron desertorum* and *Artemisia tridentata* from soil patches with balanced and unbalanced nitrate and phosphate supply. *New Phytologist* 139: 267–272.

Dalal, R.C. (1974). Effects of intercropping maize with pigeon peas on grain yield and nutrient uptake. *Experimental Agriculture* 10: 219–224.

Davidson, A.M., Jennions, M., and Nicotra, A.B. (2011). Do invasive species show higher phenotypic plasticity than native species and, if so, is it adaptive? A meta-analysis. *Ecology Letters* 14: 419–431.

Dissanayaka, D.M.S.B., Maruyama, H., Masuda, G. et al. (2015). Interspecific facilitation of P acquisition in intercropping of maize with white lupin in two contrasting soils as influenced by different rates and forms of P supply. *Plant and Soil* 390: 223–236.

Dixon, R.K., Winjum, J.K., Andrasko, K.J. et al. (1994). Integrated land-use systems: Assessment of promising agroforest and alternative land-use practices to enhance carbon conservation and sequestration. *Climatic Change* 27: 71–92.

Duchene, O., Vian, J.F., and Celette, F. (2017). Intercropping with legume for agroecological cropping systems: complementarity and facilitation processes and the importance of soil microorganisms. A review. *Agriculture, Ecosystems & Environment* 240: 148–161.

Duffnerm, A., Hoffland, E., and Temminghoff, E.J.M. (2012). Bioavailability of zinc and phosphorus in calcareous soils as affected by citrate exudation. *Plant and Soil* 361: 165–175.

Fan, F.L., Zhang, F.S., Song, Y.N. et al. (2006). Nitrogen fixation of faba bean (*Vicia faba* L.) interacting with a non-legume in two contrasting intercropping systems. *Plant and Soil* 283: 275–286.

Federer, W.T. (2012). Statistical Design and Analysis for Intercropping Experiments: Volume 1: Two Crops. New York: Springe.

Flores-Sanchez, D., Pastor, A., Lantinga, E.A. et al. (2013). Exploring maize-legume intercropping systems in Southwest Mexico. *Agroecology and Sustainable Food Systems* 37: 739–761.

Fort, F., Cruz, P., and Jouany, C. (2014). Hierarchy of root functional trait values and plasticity drive early-stage competition for water and phosphorus among grasses. *Functional Ecology* 28: 1030–1040.

von Fragstein und Niemsdorff, P Knudsen, M.T., Gooding, M.J. et al. (2008). Experiences with intercropping design–a survey about pulse cereal-combinations in Europe. Paper presented Jun 18–20 at ISOFAR 2008 in Modena, Italy.

Francis, C.A. (1986). Multiple Cropping Systems. New York: Macmillan Publishing Company.

Fuller, D.Q. and Madella, M. (2002). Issues in Harappan archaeobotany: retrospect and prospect. In: In: Indian Archaeology in Retrospect II: Protohistory (eds. S. Settar and R. Korisettar), 317–390. Manohar: New Delhi.

Garrett, H.E., Rietveld, W.J., Fisher, R.F. et al. (2009). North American Agroforestry: An Integrated Science and Practice. Madison, WIS: American Society of Agronomy.

Ghaffarzadeh, M., Préchac, F.G., and Cruse, R.M. (1994). Grain yield response of corn, soybean, and oat grown in a strip intercropping system. *American Journal of Alternative Agriculture* 9: 171–177.

Gooding, M.J., Kasyanova, E., Ruske, R. et al. (2007). Intercropping with pulses to concentrate nitrogen and sulphur in wheat. *The Journal of Agricultural Sciences* 145: 469–479.

Gordon, A.M., Newman, S.M., and Williams, P.A. (1997). Temperate agroforestry: an overview. In: Temperate Agroforestry Systems (eds. A.M. Gordon and S.M. Newman), 1–8. Wallingford: CAB International.

Hassan, M.K., Aktar, M.S., Hussain, M.A. et al. (1991). Intercropping: a sustainable land use system in the rainfed agriculture of south west Bangladesh. *Bangladesh Journal of Agricultural Research* 14: 99–107.

Hauggaard-Nielsen, H. and Jensen, E.S. (2005). Facilitative root interactions in intercrops. *Plant and Soil* 274: 237–250.

Hauggaard-Nielsen, H., Ambus, P., and Jensen, E.S. (2001). Temporal and spatial distribution of roots and competition for nitrogen in pea-barley intercrops – a field study employing ^{32}P technique. *Plant and Soil* 236: 63–74.

Hauggaard-Nielsen, H., Gooding, M., Ambus, P. et al. (2009). Pea–barley intercropping for efficient symbiotic N_2-fixation, soil N acquisition and use of other nutrients in European organic cropping systems. *Field Crops Research* 113: 64–71.

Hinsinger, P., Bengough, A.G., Vetterlein, D. et al. (2009). Rhizosphere: biophysics, biogeochemistry and ecological relevance. *Plant and Soil* 321: 117–152.

Hinsinger, P., Betencourt, E., Bernard, L. et al. (2011). P for two, sharing a scarce resource: soil phosphorus acquisition in the rhizosphere of intercropped species. *Plant Physiology* 156: 1078–1108.

Hu, J., Li, M., Liu, H. et al. (2018). Intercropping with sweet corn (*Zea mays* L. *var. rugosa* Bonaf.) expands P acquisition channels of chili pepper (*Capsicum annuum* L.) via arbuscular mycorrhizal hyphal networks. *Journal of Soils and Sediments* 19: 1632–1639.

Ikeorgu, J.E.G., Unamma, R.P.A., Ezumah, H.C. (1989). Effects of fertilizer and time of introduction of cassava or maize on the productivity of yam and cassava-based intercrops in a humid ultisol. *Second Annual Meeting of the Collaborative Group in Cassava-Based Cropping Systems Research*, Ibadan, Oyo State (Nigeria) (7–10 November 1988). IITA.

Inal, A., Gunes, A., Zhang, F. et al. (2007). Peanut/maize intercropping induced changes in rhizosphere and nutrient concentrations in shoots. *Plant Physiology and Biochemistry* 45: 350–356.

Isbell, F., Adler, P.R., Eisenhauer, N. et al. (2017). Benefits of increasing plant diversity in sustainable agroecosystems. *Journal of Ecology* 105: 871–879.

Islam, M.N., Akhteruzzaman, M., Alom, M.S. et al. (2014). Hybrid maize and sweet potato intercropping: a technology to increase productivity and profitability for poor hill farmers in Bangladesh. *SAARC Journal of Agriculture* 12: 101–111.

Jackson, R. and Caldwell, M. (1996). Integrating resource heterogeneity and plant plasticity: modelling nitrate and phosphate uptake in a patchy soil environment. *Journal of Ecology* 84: 891–903.

Jalloh, A. (2001). Promoting Appropriate Intercropping Technologies for Sustainable Agricultural Production in Africa: a Farmer Centered Approach. Freetown, Siera Leone: Institute of Agricultural Research.

Jensen, E.S. (1996). Grain yield, symbiotic N_2 fixation and interspecific competition for inorganic N in pea-barley intercrops. *Plant and Soil* 182: 25–38.

Keating, B.A. and Carberry, P.S. (1993). Resource capture and use in intercropping: solar radiation. *Field Crops Research* 34: 273–301.

Khadka, R., Joshi, S., and Chaudhary, R.B. (2013). Intercropping of chick pea and mustard on control of botrytis grey mold in western Terai, Nepal. *Agronomy Journal of Nepal* 3: 89–96.

Khan, M. and Khan, R.U. (2005). Mutual effect of legume and cereal intercropping under Rodkohi rainfed conditions of DI Khan [Pakistan]. *Sarhad Journal of Agriculture* 21: 629–632.

Khatri, N., Dahal, K.R., Amgainm, L.P. et al. (2014). Productivity and economic assessment of maize and soybean intercropping under various tillage and residue levels in Chitwan, Nepal. *World Journal of Agricultural Research* 2: 6–12.

Kingwell-Banham, E., Petrie, C.A., and Fuller, D.Q. (2015). Early agriculture in South Asia. In: Cambridge World History (eds. G. Barker and C. Goucher), 261–288. Cambridge: Cambridge University Press.

Li, W.H. (2001). Agro-Ecological Farming Systems in China, Man and the Biosphere Series (ed. J.N.R. Jeffers), 26. Paris: The Parthenon Publishing Group Ltd.

Li, L., Yang, S.C., Li, X.L. et al. (1999). Interspecific complementary and competitive interactions between intercropped maize and faba bean. *Plant and Soil* 212: 105–114.

Li, L., Sun, J.H., Zhang, F.S. et al. (2001a). Wheat/maize or wheat/soybean strip intercropping. II. Recovery or compensation of maize and soybean after wheat harvesting. *Field Crops Research* 71: 173–181.

Li, L., Sun, J.H., Zhang, F.S. et al. (2001b). Wheat/maize or wheat/soybean strip intercropping. I. Yield advantage and interspecific interactions on nutrients. *Field Crops Research* 71: 123–137.

Li, L., Tang, C.X., Rengel, Z. et al. (2003a). Chickpea facilitates phosphorus uptake by intercropped wheat from an organic phosphorus source. *Plant and Soil* 248: 297–303.

Li, L., Zhang, F.S., Li, X.L. et al. (2003b). Interspecific facilitation of nutrient uptake by intercropped maize and faba bean. *Nutrient Cycling in Agroecosystems* 65: 61–71.

Li, L., Tang, C., Rengel, Z. et al. (2004). Calcium, magnesium and microelement uptake as affected by phosphorus sources and interspecific root interactions between wheat and chickpea. *Plant and Soil* 261: 29–37.

Li, L., Sun, J.H., Zhang, F.S. et al. (2006). Root distribution and interactions between intercropped species. *Oecologia* 147: 280–290.

Li, L., Li, S.M., Sun, J.H. et al. (2007). Diversity enhances agricultural productivity via rhizosphere phosphorus facilitation on phosphorus-deficient soils. *Proceedings of the National Academy of Sciences of the United States of America* 104: 11192–11196.

Li, Y.Y., Yu, C.B., Cheng, X. et al. (2009). Intercropping alleviates the inhibitory effect of N fertilization on nodulation and symbiotic N_2 fixation of faba bean. *Plant and Soil* 323: 295–308.

Li, L., Sun, J., and Zhang, F. (2011). Intercropping with wheat leads to greater root weight density and larger below-ground space of irrigated maize at late growth stages. *Soil Science and Plant Nutrition* 57: 61–67.

Li, H.G., Zhang, F.S., Rengel, Z. et al. (2013a). Rhizosphere properties in monocropping and intercropping systems between faba bean (*Vicia faba* L.) and maize (*Zea mays* L.) grown in a calcareous soil. *Crop & Pasture Science* 64: 976–984.

Li, L., Zhang, L., and Zhang, F. (2013b). Crop mixtures and the mechanisms of overyielding. In: Encyclopedia of Biodiversity, 2nde (ed. S.A. Levin). Waltham: Academic Press.

Li, L., Tilman, D., Lambers, H. et al. (2014). Plant diversity and overyielding: insights from belowground facilitation of intercropping in agriculture. *New Phytologist* 203: 63–69.

Li, B., Li, Y.Y., Wu, H.M. et al. (2016). Root exudates drive interspecific facilitation by enhancing nodulation and N_2 fixation. *Proceedings of the National Academy of Sciences of the United States of America* 113: 6496–6501.

Licker, R., Johnston, M., Foley, J.A. et al. (2010). Mind the gap: how do climate and agricultural management explain the "yield gap" of croplands around the world? *Global Ecology and Biogeography* 19: 769–782.

Liu, Y.X., Zhang, W.P., Sun, J.H. et al. (2015). High morphological and physiological plasticity of wheat roots is conducive to higher competitive ability of wheat than maize in intercropping system. *Plant and Soil* 397: 387–399.

Liu, Y.X., Sun, J.H., Zhang, F.F. et al. (2019). The plasticity of root distribution and nitrogen uptake contributes to recovery of maize growth at late growth stages in wheat/maize intercropping. *Plant and Soil* 447: 39–53. https://doi.org/10.3389/fpls.2018.01328.

Loreau, M. and Hector, A. (2001). Partitioning selection and complementarity in biodiversity experiments. *Nature* 412: 72–76.

Madhavan, M. and Shanmugasundaram, V.S. (1990). Effect of population on nutrient uptake of pigeonpea genotypes in sole and intercropped situation with sorghum CO_2. *Acta Agronomica Hungarica* 39: 389–392.

Maikhuri, R.K., Rao, K.S., and Saxena, K.G. (1996). Traditional crop diversity for sustainable development of Central Himalayan agroecosystems. *International Journal of Sustainable Development & World Ecology* 3: 8–31.

Maikhuri, R.K., Semwal, R.L., Rao, K.S. et al. (1997). Eroding traditional crop diversity imperils the sustainability of agricultural systems in Central Himalaya. *Current Science* 73: 777–782.

Manalo, E.B. (1976). Agro-climatic survey of Bangladesh. Los Banos, Philippines: International Rice Research Institute.

Manevski, K., Børgesen, C.D., Andersen, M.N. et al. (2015). Reduced nitrogen leaching by intercropping maize with red fescue on sandy soils in North Europe: a combined field and modeling study. *Plant and Soil* 388: 67–85.

Mason, S.C., Leihner, D.E., and Vorst, J.J. (1986). Cassava-cowpea and cassava-peanut intercropping. I. Yield and land use efficiency. *Agronomy Journal* 78: 43–46.

Mei, P.P., Gui, L.G., Wang, P. et al. (2012). Maize/faba bean intercropping with rhizobia inoculation enhances productivity and recovery of fertilizer P in a reclaimed desert soil. *Field Crops Research* 130: 19–27.

Meier, I.C., Avis, P.G., and Phillips, R.P. (2013). Fungal communities influence root exudation rates in pine seedlings. *FEMS Microbiology Ecology* 83: 585–595.

Mommer, L., van Ruijven, J., Jansen, C. et al. (2012). Interactive effects of nutrient heterogeneity and competition: implications for root foraging theory? *Functional Ecology* 26: 66–73.

Mommer, L., Kirkegaard, J., and van Ruijven, J. (2016). Root-root interactions: towards a rhizosphere framework. *Trends in Plant Science* 21: 209–217.

Mou, P., Jones, R., Mitchell, R. et al. (1995). Spatial distribution of roots in sweetgum and loblolly pine monocultures and relations with above-ground biomass and soil nutrients. *Functional Ecology* 9: 689–699.

Mou, P., Jones, R.H., Tan, Z. et al. (2013). Morphological and physiological plasticity of plant roots when nutrients are both spatially and temporally heterogeneous. *Plant and Soil* 364: 373–384.

Nie, S.W., Eneji, A.E., Chen, Y.Q. et al. (2012). Nitrate leaching from maize intercropping systems with N fertilizer over-dose. *Journal of Integrative Agriculture* 11: 1555–1565.

Nnko, E.N. and Doto, A.L. (1982). Intercropping maize or millet with soybean, with particular reference to planting schedule. In Intercropping. Proceedings of the Second Symposium on Intercropping in Semi-Arid Areas, held at Morogoro, Tanzania, (4–7 August 1980). IDRC, Ottawa, ON, Canada.

Peichl, M., Thevathasan, N.V., Gordon, A.M. et al. (2006). Carbon sequestration potentials in temperate tree-based intercropping systems, southern Ontario, Canada. *Agroforestry Systems* 66: 243–257.

Pigliucci, M. (2001). Phenotypic Plasticity: Beyond Nature and Nurture. Baltimore, MD: JHU Press.

Qiao, X., Bei, S.K., Li, H.G. et al. (2016). Arbuscular mycorrhizal fungi contribute to overyielding by enhancing crop biomass while suppressing weed biomass in intercropping systems. *Plant and Soil* 406: 173–185.

Rahman, M.Z., Rahman, M.H., Haque, M.E. et al. (2006). Banana based intercropping system in North–West part of Bangladesh. *Journal of Agronomy* 5: 228–231.

Raintree, J.B. (1991). Agroforestry in North America: the next generation, keynote address. Proceedings of the First Conference on Agroforestry in North America (ed. P.A. Williams) (13–16 August 1989) Guelph, Ontario, Canada, 1–14.

Ren, W., Hu, L., Zhang, J. et al. (2014). Can positive interactions between cultivated species help to sustain modern agriculture? *Frontiers in Ecology and the Environment* 12: 507–514.

Reynolds, P.E., Simpson, J.A., Thevathasan, N.V. et al. (2007). Effects of tree competition on corn and soybean photosynthesis, growth, and yield in a temperate tree-based agroforestry intercropping system in southern Ontario, Canada. *Ecological Engineering* 29: 362–371.

Rivest, D., Cogliastro, A., Vanasse, A. et al. (2009). Production of soybean associated with different hybrid poplar clones in a tree-based intercropping system in southwestern Québec, Canada. *Agriculture, Ecosystems & Environment* 131: 51–56.

Rivest, D., Cogliastro, A., Bradley, R.L. et al. (2010). Intercropping hybrid poplar with soybean increases soil microbial biomass, mineral N supply and tree growth. *Agroforestry Systems* 80: 33–40.

Robinson, D., Linehan, D.J., and Gordon, D.C. (1994). Capture of nitrate from soil by wheat in relation to root length, nitrogen inflow and availability. *New Phytologist* 128: 297–305.

Saini, L.K., Singh, M., and Kapur, M.L. (2003). Relative profitability of intercropping vegetable crops in autumn planted sugarcane. *Sugar Tech* 5: 95–97.

Singh, A.K., Lal, M., and Suman, A. (2008). Effect of intercropping in sugarcane (Saccharum complex hybrid) on productivity of plant cane-ratoon system. *Indian Journal of Agronomy* 53: 140–144.

Sultan, S.E. (2000). Phenotypic plasticity for plant development, function and life history. *Trends in Plant Science* 5: 537–542.

Susaeta, A., Lal, P., Alavalapati, J. et al. (2012). Economics of intercropping loblolly pine and switchgrass for bioenergy markets in the southeastern United States. *Agroforestry Systems* 86: 287–298.

Taylor, A., Boukambou, G., Dahniya, M. et al. (1996). Strengthening National Agricultural Research Systems in the Humid and Sub-Humid Zones of West and Central Africa. World Bank.

Thevathasan, N.V. and Gordon, A.M. (2004). Ecology of tree intercropping systems in the North temperate region: Experiences from southern Ontario, Canada. *Agroforestry Systems* 61: 257–268.

Thorsted, M.D., Olesen, J.E., and Koefoed, N. (2002). Effects of white clover cultivars on biomass and yield in oat/clover intercrops. *The Journal of Agricultural Science* 138: 261–267.

Thorsted, M.D., Olesen, J.E., and Weiner, J. (2006). Mechanical control of clover improves nitrogen supply and growth of wheat in winter wheat/white clover intercropping. *European Journal of Agronomy* 24: 149–155.

Vandermeer, J.H. (1992). The Ecology of Intercropping. Cambridge: Cambridge University Press.

Wang, X.C., Yang, W.Y., Deng, X.Y. et al. (2015). Difference in maize photosynthetic characteristics and notrogen regulation effects in maize/soybean and maize/soybean and maize/sweet potatp relay strip intercropping. *Chinese Journal of Eco-Agriculture* 23: 141–149.

Weber, S.A. (2003). Archaeobotany at Harappa: Indications for change. In: Indus Ethnobiology: New Perspectives from the Field (eds. S.A. Weber and W.R. Belcher), 175–198. Lanham: Lexington Books.

Webster, C.C. and Wilson, P.N. (1966). Agriculture in the Tropics. New York: Longmans.

Weisskopf, A., Qin, L., Ding, J. et al. (2015). Phytoliths and rice: from wet to dry and back again in the Neolithic Lower Yangtze. *Antiquity* 89: 1051–1063.

Wiche, O., Kummer, N.A., and Heilmeier, H. (2016). Interspecific root interactions between white lupin and barley enhance the uptake of rare earth elements (REEs) and nutrients in shoots of barley. *Plant and Soil* 402: 235–245.

Willey, R.W. (1979). Intercropping – its importance and research needs. Part 1. Competition and yield advantages. *Field Crops Abstracts* 32: 1–10.

Willey, R.W. and Rao, M.R. (1980). A competitive ratio for quantifying competition between intercrops. *Experimental Agriculture* 16: 117–125.

Williams, P.A. and Gordon, A.M. (1992). The potential of intercropping as an alternative land use system in temperate North America. *Agroforestry Systems* 19: 253–263.

Wilson, J.B. (1988). Shoot competition and root competition. *Journal of Applied Ecology* 25: 279–296.

Wortmann, C.S. (1993). Research methods for cereal/legume intercropping. In: Proceedings of a Workshop for Cereal/Legume Intercropping in Eastern and Southern Africa (eds. S.R. Waddington, A.F.E. Palmer and O.T. Edje). Mexico DF, Mexico: CIMMYT.

Wotherspoon, A., Thevathasan, N.V., Gordon, A.M. et al. (2014). Carbon sequestration potential of five tree species in a 25-year-old temperate tree-based intercropping system in southern Ontario, Canada. *Agroforestry Systems* 88: 631–643.

Wright, R.P. (2010). The Ancient Indus: Urbanism, Economy, and Society: Case Studies in Early Societies. Cambridge: Cambridge University Press.

Xia, H.Y., Zhao, J.H., Sun, J.H. et al. (2013a). Dynamics of root length and distribution and shoot biomass of maize as affected by intercropping with different companion crops and phosphorus application rates. *Field Crops Research* 150: 52–62.

Xia, H.Y., Zhao, J.H., Sun, J.J. et al. (2013b). Maize grain concentrations and above-ground shoot acquisition of micronutrients as affected by intercropping with turnip, faba bean, chickpea, and soybean. *Science China Life Sciences* 56: 823–834.

Xue, Y.F., Xia, H.Y., Christie, P. et al. (2016). Crop acquisition of phosphorus, iron and zinc from soil in cereal/legume intercropping systems: a critical review. *Annals of Botany* 117: 363–377.

Yamazaki, Y., Kubota, J., Nakawo, M. et al. (2005). Differences in water budget among wheat, maize and their intercropping field in the heihe river basin of northwest China. AOGS Paper, Asia Oceania Geoscience Society (AOGS) 2nd Annual Meeting, (20–24 June 2005) Singapore.

Yin, W., Guo, Y., Hu, F. et al. (2018). Wheat–maize intercropping with reduced tillage and straw retention: a step towards enhancing economic and environmental benefits in arid areas. *Frontiers in Plant Science* 9: 1328. https://doi.org/10.3389/fpls.2018.01328.

Zhang, F.S. and Li, L. (2003). Using competitive and facilitative interactions in intercropping systems enhances crop productivity and nutrient-use efficiency. *Plant and Soil* 248: 305–312.

Zhang, F.S., Li, L., and Sun, J.H. (2001). Contribution of above- and below-ground interactions to intercropping. In: Plant Nutrition – Food Security and Sustainability of Agroecosystems (eds. W.J. Horst, M.K. Schenk, A. Buerkert, et al.), 978–979. Dordrecht: Kluwer Academic Publishers.

Zhang, F., Shen, J., Li, L. et al. (2004). An overview of rhizosphere processes related with plant nutrition in major cropping systems in China. *Plant and Soil* 260: 89–99.

Zhang, D.S., Zhang, C.C., Tang, X.Y. et al. (2016). Increased soil phosphorus availability induced by faba bean root exudation stimulates root growth and phosphorus uptake in neighbouring maize. *New Phytologist* 209: 823–831.

Zhang, W.P., Liu, G.C., Sun, J.H. et al. (2016b). Temporal dynamics of nutrient uptake by neighboring plant species: evidence from intercropping. *Functional Ecology* 31: 469–479.

Zhou, L.L., Cao, J., Zhang, F.S. et al. (2009). Rhizosphere acidification of faba bean, soybean and maize. *Science of the Total Environment* 407: 4356–4362.

Zomer, R.J., Trabucco, A., Coe, R. et al. (2009). Trees on Farm: Analysis of Global Extent and Geographical Patterns of Agroforestry. ICRAF Working Paper no. 89. Nairobi Kenya, World Agroforestry Centre.

Zuo, Y.M. and Zhang, F.S. (2008). Effect of peanut mixed cropping with gramineous species on micronutrient concentrations and iron chlorosis of peanut plants grown in a calcareous soil. *Plant and Soil* 306: 23–36.

Zuo, Y.M., Zhang, F.S., Li, X.L. et al. (2000). Studies on the improvement in iron nutrition of peanut by intercropping with maize on a calcareous soil. *Plant and Soil* 220: 13–25.

Zuo, Y.M., Liu, Y.X., Zhang, F.S. et al. (2004). A study on the improvement of iron nutrition of peanut intercropping with maize on nitrogen fixation at early stages of growth of peanut on a calcareous soil. *Soil Science and Plant Nutrition* 50: 1071–1078.

13

The 3-D Imaging of Roots Growing in Soil

Sebastian R.G.A. Blaser[1]*, Nicolai Koebernick*[2]*, Steffen Schlüter*[1]*, and Doris Vetterlein*[1,2]

[1] *Department of Soil System Science, Helmholtz-Centre for Environmental Research GmbH – UFZ, Halle (Saale), Germany*
[2] *Institute of Agricultural and Nutritional Sciences, Martin-Luther-University Halle-Wittenberg, Halle (Saale), Germany*

13.1 Why Do We Need 3D Imaging of Roots Growing in Soils?

13.1.1 Roots are the Hidden Half

In 1996, a book with the title *Roots – the Hidden Half* was published by Waisel et al. (1996). This title wonderfully reflects the fact that roots are much more difficult to access than the above-ground organs of the plant; hence, more preparation and advanced techniques are needed to explore and study roots in opaque soil.

Not all roots are hidden, as some plants are able to form above-ground roots, such as aerial roots in epiphytes, mangroves, and some trees such as figs. Nevertheless, the vast majority of the roots are located in soil. This natural habitat is opaque (nontransparent) and, hence, more ingenuity is needed to observe the hidden half, which is the focus of this chapter.

13.1.2 Long-term Interest in Root Growth and Definition of Root System Architecture

The interest of humans in their environment and especially in plants is probably as old as humankind itself – not least because plants have always been part of the food chain. With the beginning of agriculture, soil management and plant nutrition have clearly gained in importance and are currently experiencing a renaissance through the efforts to ensure the nutrition of a growing world population with less fertilization and a better and sustainable soil management.

In addition to anchoring plants in soil, the central task of the root system is to take up water and nutrients in order to maintain the plant vital functions and enable healthy and

The Root Systems in Sustainable Agricultural Intensification, First Edition.
Edited by Zed Rengel and Ivica Djalovic.

lush growth. In order to fulfill these tasks, roots explore the soil in a specific spatial arrangement, which has been termed the root system architecture (RSA). The RSA is defined as the three-dimensional (3D) structure of the root system, including the primary root, branch (lateral) roots, and root hairs (Osmont et al. 2007). The term "branch roots" can be misleading and is less common in the recent literature, although the term "branch" reflects the morphologic appearance of the root system very well. Instead, the term lateral root is preferred.

The primary root is formed during embryogenesis, and lateral roots are formed postembryonically from existing roots (see Chapters 2 and 4). Mainly originating from the pericycle, the process of lateral root formation can take place at roots of different orders and thus create the next level of root order. Apart from these root types, there are adventitious roots (originating from the shoot), aerial roots, crown, storage, or cluster roots, leading to a wide variety of possible root network morphologies and RSAs.

13.1.3 Root Growth Does Not Depend on the Genetic Program Only

The development of root growth and the spatial characteristics of the root architecture depend on many environmental and ontogenetic factors, i.e. the individual origin and development of an organism within its lifetime (Gould 1977). However, the genetic programme of the plant can be described as having the overarching influence.

The morphological diversity of root systems is remarkable across plant species, and hence, a simple classification scheme is not applicable (Fitter 1987). However, in most cases distinction between two major root system morphologies can be made for angiosperms: with taproot (also termed allorhizic) and without (homorhizic systems). Two characteristic representatives of these major types are *Vicia faba* and *Hordeum vulgare*, respectively (Figure 13.1).

Allorhizic systems are typical for dicotyledonous plants. The allorhizic system is dominated by the primary root (termed taproot), from which the first order lateral roots emerge (Esau 1965). These laterals can then also form lateral roots (considered the second-order laterals). If they branch further, the resulting roots are called the third-order lateral roots and so on.

In contrast, the root system of monocotyledonous plants such as *H. vulgare* (barley) or *Triticum aestivum* (wheat) is characterized by the growth of adventitious roots in addition to the primary roots (often termed seminal roots). Some monocots such as *Zea mays* (maize) or *Oryza sativa* (rice) can form shoot-borne roots, termed brace or crown roots (Hochholdinger et al. 2004). As lateral roots can be formed from all of these root types, the morphologic appearance is referred to as "bushy" (Lima et al. 2010; Mounier et al. 2014; Osmont et al. 2007). This can result in a particularly intensive exploration of the soil volume, typically more intensive than by the taproot systems.

13.1.4 Other Biotic and Abiotic Stimuli Influencing Root Growth

The contrast in RSA of dicots and monocots emphasizes the genetic control of the spatial arrangement of the root system. In addition, various biotic and abiotic exogenous factors result in a remarkable developmental plasticity of RSA (Morris et al. 2017; Osmont et al. 2007). The related terms are "morphological plasticity" (Bingham and Bengough 2003; Fransen et al. 1998; Johnson and Biondini 2001; De Kroons and Hutchings 1995) or "the plastic plant" (Hodge 2004, 2006).

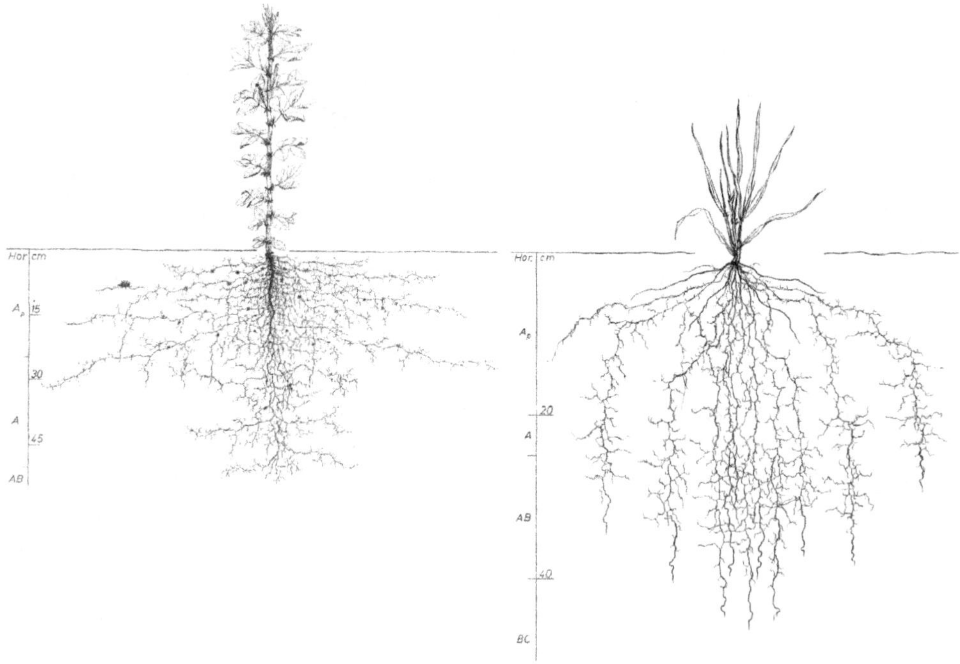

Figure 13.1 Root system architecture drawings of faba bean (*Vicia faba*) on the left and barley (*Hordeum vulgare*) on the right. Please note the different depth axis scales. *Source:* Kutschera et al. (2009) with permission of DLG-Verlag GmbH.

The directional growth response to certain stimuli is called "tropism" (Esmon et al. 2004; Gilroy 2008; Gregory 2008). There are many tropisms; among the most important are gravitropism to gravity, thigmotropism to touch, phototropism to light, and hydrotropism to water. Most of those are known for a long time and trace back to Darwin (1897), but more detailed studies about these tropisms have been conducted recently (Boonsirichai et al. 2002; Correll and Kiss 2002; Eapen et al. 2005; Fitter 1987; Kaspar and Bland 1992; Onderdonk and Ketcheson 1973).

An explicitly defined "tropism" towards nutrients does not exist, but the idea that root growth responds to the availability and concentration of nutrients is generally recognized. This reaction can also be understood as being part of "chemotropism" (Newcombe and Rhodes 1904), and this has been used and applied, e.g. in Leitner et al. (2010) (a simulation study of phosphate uptake by maize roots). The publications of Drew and colleagues in the 1970s (Drew 1975; Drew and Saker 1975, 1978; Drew et al. 1973) are particularly well-known, investigating plant responses to a localized supply of nutrients. Earlier studies date back to Wiersum (1958) or Hackett (1972).

13.1.5 Root Visualization and Analysis

To analyse any type of tropism or root growth response in general, an appropriate setup with controlled experimental conditions is needed, as well as visualization and subsequent analysis. It has always been a challenge to capture root growth, especially when growing in soil due to the opaqueness of this natural growth habitat.

In 1960, Lore Kutschera published the first version of the well-known "Wurzelatlas" (root atlas) with detailed drawings of excavated root systems (Kutschera 1960). Until today – the seventh volume of this atlas series was released in 2009 – a lot of well-deserved attention has been paid to these drawings, especially as root plasticity is illustrated superbly with many root systems of the same plant that are shown for different soils or climatic conditions (Kutschera et al. 2009). While intriguing, these drawings largely lack quantitative information about root length of different root types and orders and especially regarding the 3D arrangement of the RSA in the soil volume.

A knowledge gap regarding quantification of root length can be closed partially via separating the roots from soil by washing them out and measuring their length. Initially, this was done simply by a tape measure or a ruler, before the line intersect method was developed by Tennant (1975). The next large step was the introduction of high-resolution document scanners combined with image analysis software such as WinRHIZO (Regent Instruments Inc., Québec, Canada) in the 1990s (Arsenault et al. 1995). This method made it possible to scan roots washed out of soil and examine their length and diameter automatically with a high-degree of detail on a computer. This software is still widely used today. The advantages are the easy handling and the robust, comparable results. However, the 3D analysis of the undisturbed arrangement of roots in soil is not possible with this method. Some open-source alternatives to WinRHIZO are now available, such as RootNav (Pound et al. 2013), Root System Analyser (Leitner et al. 2014), or DART (Le Bot et al. 2010).

Other methods were applied in the field or in the laboratory, e.g. soil profile wall methods, resin embedding techniques, in-growth bags, transparent artificial growth media, rhizoboxes, or other inclined boxes with the underlying idea that the roots grow along a transparent surface of a box so that they can be detected and analysed (Downie et al. 2012; Iyer-Pascuzzi et al. 2010; Neumann et al. 2009; Smit et al. 2013; Trachsel et al. 2011). Although many of these methods have the advantage of being cost-effective, all of them also have grave disadvantages. Only for some methods recording over time is possible, in some a large proportion of roots is not detectable, so that a conclusion about the 3D characteristics of the root system is hardly possible, and in others, the experimental environments are very different from natural conditions, resulting in artefacts.

13.2 Overview of Noninvasive Imaging Techniques – NT, MRI, and X-Ray CT

To overcome the problems mentioned above, noninvasive imaging methods are now increasingly used, most of which have initially been developed for medical applications. The three techniques mainly used for the 3D imaging of roots are neutron tomography (NT), magnetic resonance imaging (MRI), and particularly industrial X-ray computed tomography (CT). These techniques are summarized in Table 13.1.

All of these methods have the fundamental disadvantage that they cannot be used directly in the field – at least not in the near future. Therefore, it is necessary to prepare appropriate samples in order to carry out the measurements in the laboratory.

Table 13.1 Comparison among the NT, MRI, and industrial X-ray CT methods for the root system architecture characterization.

Method	Neutron tomography (NT)	Magnetic resonance imaging (MRI)	Industrial X-ray computed tomography (CT)
Pioneering work	Willatt et al. (1978)	Brown et al. (1986)	Watanabe et al. (1992)
Type of radiation used	Neutrons ((indirect) ionizing radiation)	Electromagnetic waves (not ionizing)	X-rays (ionizing radiation)
Measurement principles	Strongest neutron attenuation by hydrogen, but most other materials are easily penetrated	Nuclear magnetic resonance effect (NMR). Atomic nuclei are excited and their excitation response is monitored in terms of the emitted radiation in the time domain.	Attenuation and scattering of the X-ray beam are dependent on electron densities as a function of atomic number and bulk density
Accessibility	*	**	***
Advantages	Sensitive to hydrogen and hence good detection of water	Good contrast between soil and roots; water content can be measured as well	Quick, only few restrictions on the root growth media used, good accessibility (apart from synchrotron-CT)
Disadvantages	Evokes radioactivity in the sample and therefore less scope for studies over time, low accessibility and decreasing penetration of the samples with increasing water content	Sensitive to ferromagnetic particles, longer duration, artefacts, mediocre accessibility	Potential influence of radiation on root growth, artefacts, poor contrast between roots and water-filled pores
Image resolution	**	*	***
Sample sizes possible	*	***	**
Typical application	Water distribution in soil and roots	Root growth over time	Root growth over time, simultaneous information on soil structure
Recommended reviews	• Oswald et al. (2008) • Strobl et al. (2009)	• Borisjuk et al. (2012) • Stingaciu et al. (2013) • van Dusschoten et al. (2016)	• Taina et al. (2008) • Mooney et al. (2012) • Helliwell et al. (2013) • Wildenschild and Sheppard (2013)
Comparative Reviews	Li et al. (2014), Downie et al. (2015), Pohlmeier et al. (2018)		

For the ranked attributes: * = low, ** = medium and *** = high.

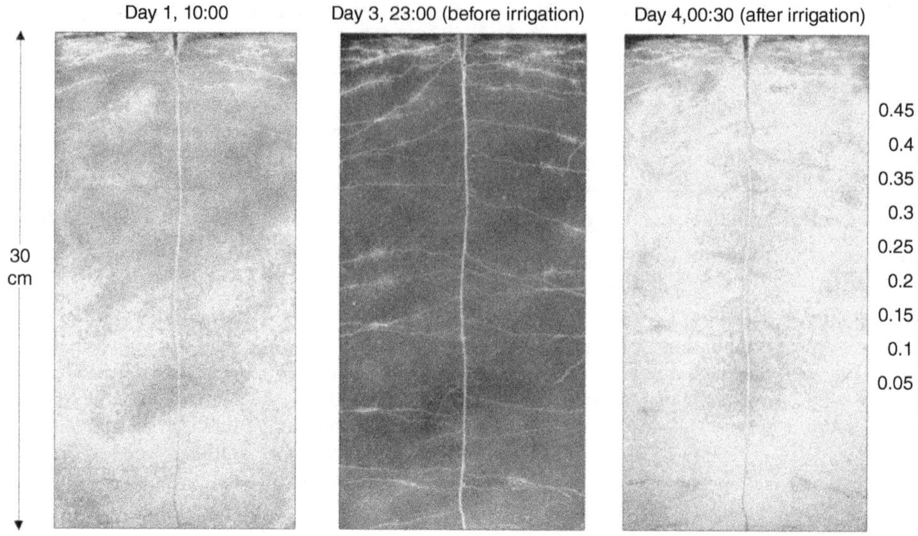

Figure 13.2 Neutron radiography during soil drying and after irrigation. The colour map is proportional to the water content (θ). Note the higher water content near the upper lupin (*Lupinus albus*) roots at day 1. After irrigation (day 4), the rhizosphere of the upper roots, of the tap root, and that of the proximal parts of the lower roots remained markedly drier compared to the rest of the sample. In contrast, a region with high water content is visible around the root tips in the lower parts of the sample. *Source:* Carminati (2013). An open-access article under the terms of Creative Commons CC BY license.

NT relies on the physical principle of neutron absorbance by the sample material. Because hydrogen is attenuating exceptionally strongly, water can be detected and visualized particularly well with NT. A main disadvantage is the low availability of the technology, and usually, only thin slabs can be used for 2D radiography studies (e.g. 1-cm thickness in Zarebanadkouki et al. 2016) and relatively small columns for tomography studies (e.g. 27 mm diameter in Moradi et al. 2011). The initial root research using NT goes back to the late 1970s (Couchat et al. 1980; Willatt and Struss 1979; Willatt et al. 1978). The recent studies used NT to, e.g., capture the root systems and water distributions (Rudolph-Mohr et al. 2017) and analyse the 3D water flow influenced by roots (Totzke et al. 2017). An example of neutron radiography application is shown in Figure 13.2.

MRI is based on the nuclear magnetic resonance effect. Within an external magnetic field, the spin system of the H atom in water is excited by radio frequency. The response signal, coming from the relaxation after the excitation, is recorded and contains the information needed for reconstruction and subsequent analysis. Similarly to NT, MRI is particularly suitable for analysis of the water distribution in the sample. The accessibility of MRI is better than that of NT, but there are limitations with MRI regarding the soil material used (Hall et al. 1997). The basic work in the context of root research with MRI goes back to the late 1980s (Brown et al. 1986; Macfall et al. 1990, 1991). The recent studies have used MRI to study changes in soil water content due to plant water uptake (Pohlmeier et al. 2008, 2010) and to conduct plant root system analysis (Schulz et al. 2012, 2013; van Dusschoten et al. 2016).

The polychromatic X-ray CT (using industrial or medical scanners) has emerged in recent years as the technique most commonly used for *in situ* root analysis in soil due to relatively easy access to scanners, the high spatial resolution, and the relatively few restrictions in the choice of soil for growing plants. Apart from industrial or medical CT scanners, synchrotron radiation is now increasingly used for root–soil imaging. The high brilliance of the third-generation synchrotron sources enables extremely high-resolution images to be taken with a short exposure time (Wildenschild and Sheppard 2013). Because the beam in this case can be monochromatic, the special reconstruction techniques (such as phase-contrast imaging) can be employed to increase the contrast in biological tissues (Chen et al. 2016; Kastner et al. 2017). However, due to the necessary technology (especially a particle accelerator), the accessibility of such devices is considerably lower than with normal industrial X-ray CT and is roughly comparable to the accessibility of NT.

A clear disadvantage of X-ray CT compared to other imaging methods is the (relatively) poor contrast in the images. In particular, the distinction of a root from a water-filled pore is usually not possible by the grey-scale value alone.

In comparing MRI and X-ray CT, Metzner et al. (2015) concluded that CT was advantageous because of the higher spatial resolution, but MRI would be better suited for larger pots due to the high contrast between roots and soil. Figure 13.3 gives the comparison between MRI and CT.

The trade-off between sample size and spatial resolution is an intrinsic problem of most imaging techniques. The larger the sample, the lower the spatial resolution of the images. Therefore, the images of larger samples are often stitched from multiple scans. This is particularly useful for samples in columns because several images can be taken on top of each other and then assembled for analysis (Blaser et al. 2018; Flavel et al. 2012; Koebernick et al. 2014). However, even then, fine structures, especially higher-order lateral roots, cannot always be captured due to their small size. Therefore, either samples with a small diameter are selected (Flavel et al. 2012; Haling et al. 2013; Tracy et al. 2012b), or the samples are not recorded in their entirety, but only in parts (Carminati et al. 2009; Koebernick et al. 2018). The latter is also called the region-of-interest imaging.

The selection of the sample size and the type of image acquisition is therefore linked to the question of what is to be detected – is it more about details, focussing on fine roots, or is it important to image the root system as a whole?

13.3 Image Analysis

After measurement, it is necessary for all methods to process the 3D images in order to extract meaningful information. First, the images are usually filtered to reduce the image noise while preserving the features of interest. Depending on the signal-to-noise ratio and the target quantity to be analysed, different filters are used.

In the next step, roots are distinguished from soil. This process is also referred to as "segmentation," depicted in Figure 13.4. Segmentation is difficult due to overlapping grey-scale values of soil and roots. Segmentation results can be improved by shape information, i.e. the fact that roots form a continuous network (Blaser et al. 2018; Flavel et al. 2012; Koebernick et al. 2014; Tracy et al. 2012b) and that they have an elongated tubular shape

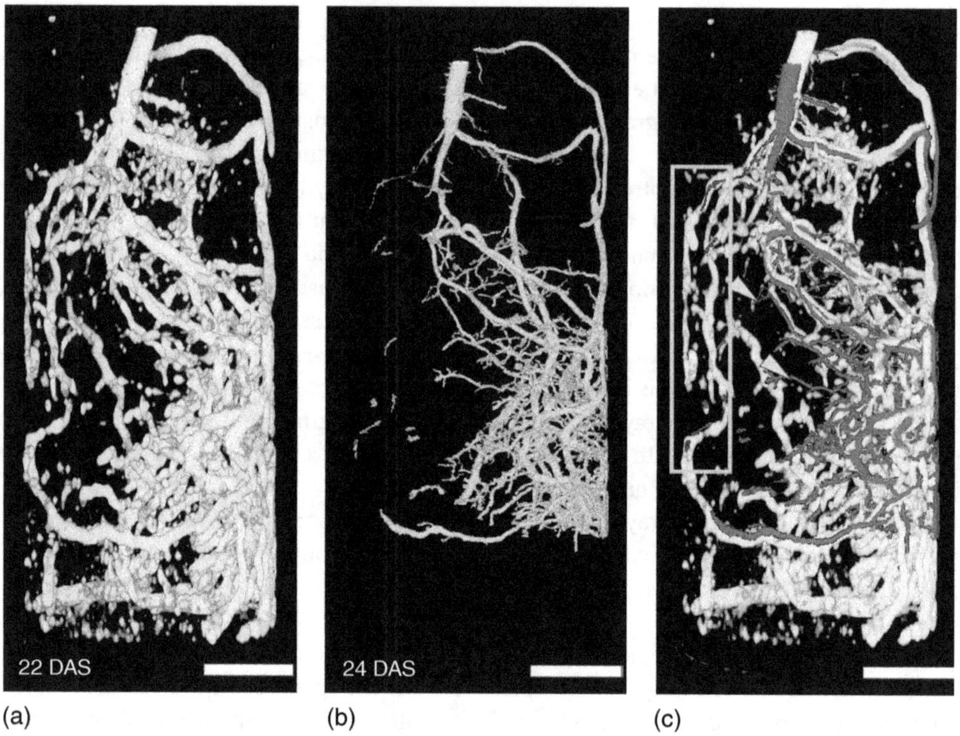

Figure 13.3 MRI and CT images of a bean (*Phaseolus vulgaris*) root system in a soil-filled pot. The same plant grown in a soil-filled pot (inner diameter 56 mm and height 200 mm) was imaged sequentially by MRI and CT. (a) MRI image measured with a voxel size of $375 \times 375 \times 1000\,\mu m$. (b) CT image of the root system measured with a voxel size of $68 \times 68 \times 68\,\mu m$. Roots in the lowest part of the pot could not be segmented and are therefore not shown. (c) CT-MRI coregistration: the CT image is in dark grey (red) and the MRI image in light grey. The arrowheads (yellow) highlight roots visible in CT but not in MRI. Please note that the CT image was taken two days after the MRI image so that it is possible that some of the roots have grown during this period. Box highlights the area where few roots are visible in CT. Scale bar: 10 mm. *Source:* Metzner et al. (2015). An open-access article under the terms of Creative Commons CC BY license.

(Schulz et al. 2013). The current publications with focus on root segmentation are by Gao et al. (2019b), Flavel et al. (2017), and Mairhofer et al. (2012). A more general overview of processing images from X-ray microtomographies is given by Schlüter et al. (2014).

13.4 Typical Applications of 3D Imaging of Roots Growing in Soil

Apart from very general or qualitative studies (Garbout et al. 2012; Nwankwo et al. 2018; Oswald et al. 2015; Totzke et al. 2017), typical applications and main topics of the 3D and 4D information acquired by imaging techniques are root length and architecture, root age, root–soil contact, root hairs, soil–root distances, soil compaction by roots, and creation of data bases for modelling.

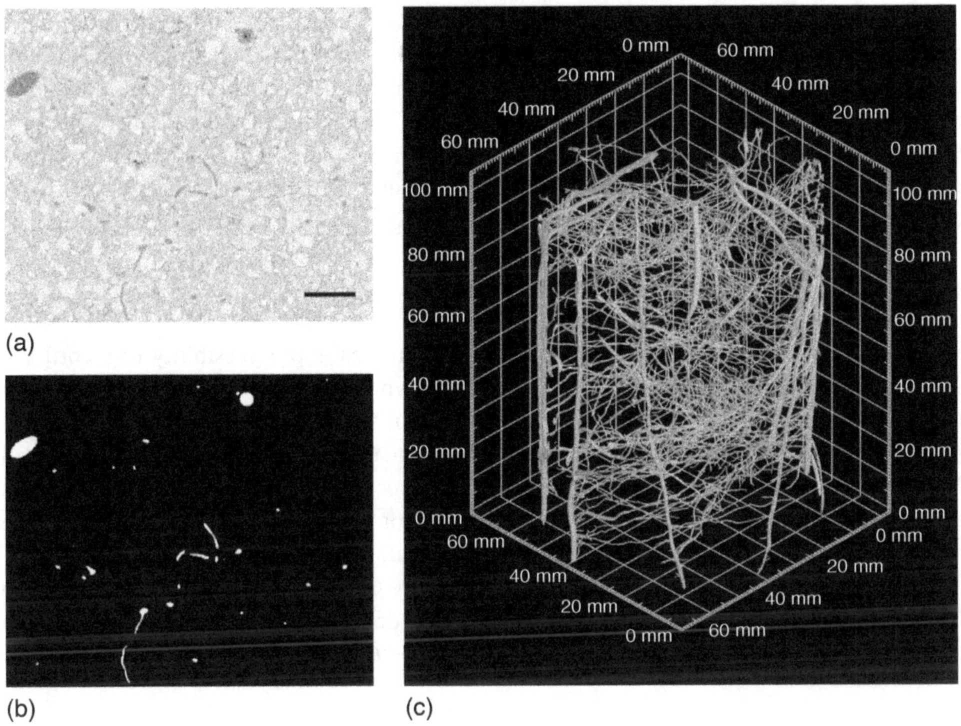

Figure 13.4 Root segmentation of maize (*Zea mays*) with the algorithm Rootine 1.0. (a) 2D projection of the original raw CT image, (b) segmented roots from (a); (c) 3D root network rendered in VG Studio Max 2.1. The scale bar in (a) represents a length of 5 mm. *Source:* Gao et al. (2019b).

Root length or root length density distribution in space are a basic measure, which in part can also be obtained from destructive sampling. In addition to the root length, more spatial information and/or the differentiation according to root orders is included in the analysis of the RSA, as well as information on root demography and transport distances. The central topic is the spatial exploration of the soil volume by the growing, developing root system. Plant species, genotypes, or soils can be compared. Often, however, a much more complex question is in the background, such as the reaction of root growth to locally-placed fertilizers (especially P and N), soil structure, water distribution, physical differences (such as the availability of biopores, or areas in the soil with higher or lower bulk density), and the growth response to plant diseases such as e.g. apple replant disease.

13.4.1 Root Length and Root System Architecture

A multitude of methods to derive quantitative information from root images have been developed over the years. Beginning with relatively simple approaches that have shown the suitability of CT to root research (Tollner et al. 1994, Watanabe et al. 1992), the analysis of root length and RSA has continued to evolve. The further milestone publications on root length analysis using CT were Gregory et al. (2003), Moran et al. (2000), Pierret et al. (1999), and Heeraman et al. (1997), with an increased level of detail and a higher proportion of detected roots. From there on, the technical development was rapid, and recent studies are covering a wide range of topics such as root–soil contact (Koebernick et al. 2018; Schmidt et al. 2012), root shrinkage under drought stress (Carminati 2013; Carminati et al. 2009), root decay (Haling et al. 2013), response to local fertilizer (Ahmed et al. 2016b; Flavel et al. 2014; Gao et al. 2019a), influence of soil structure (Moran et al. 2000), soil compaction (Tracy et al. 2012a,b), biopores (Daly et al. 2015; Pagenkemper et al. 2013), or pathogens (Sturrock et al. 2015), showing root length and RSA in increasing details, despite technical limitations.

13.4.2 Root Age

An essential strength of noninvasive imaging techniques is the possibility to record the same sample several times during growth and thus dynamically record the development of the roots (Figure 13.5). From this, the growth rate can be derived, a parameter that is very labour-intensive and difficult to determine without the use of noninvasive imaging. CT and MRI are better suited to repeated use due to the indirect ionizing radiation when neutrons are used in *e.g.* NT. Nevertheless, caution is advised for the other methods as well, because repeated measurements result in a cumulative radiation dose. There is currently no evidence that this would be an issue for MRI. For X-ray CT, there is mixed evidence for the influence of accumulated radiation on root growth, depending on different factors, especially the plant species (Blaser et al. 2018; Johnson 1936; Zappala et al. 2013).

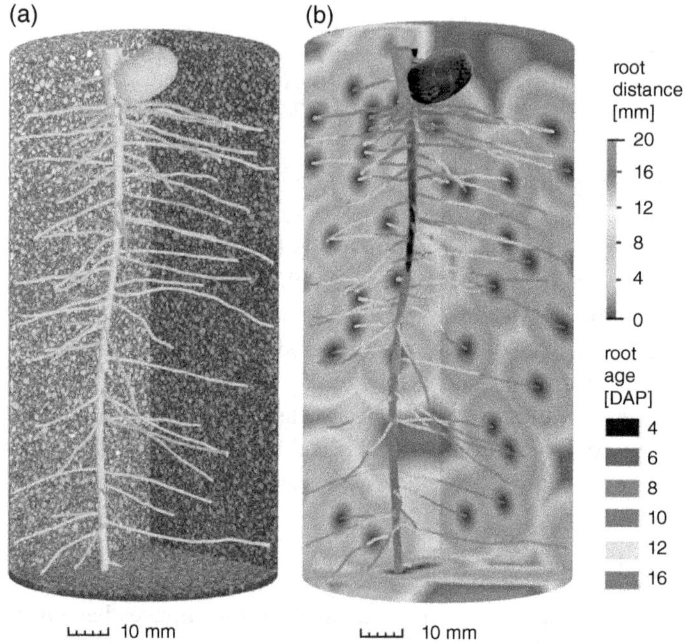

Figure 13.5 Root system of faba bean (*Vica faba*) segmented from the surrounding soil, acquired by repeated application of X-ray CT. (a) root system of faba bean 16 days after planting (DAP), embedded in the soil matrix, virtually cropped for visualization; (b) combined colour-coded representation of root age and soil–root distances 16 days after planting for the sample in (a). *Source:* Schlüter et al. (2018). An open-access article under the terms of Creative Commons CC BY license.

Repeated measurements reveal which parts of the root system developed when (Blaser et al. 2018; Carminati 2013; Koebernick et al. 2014). This allows basic statements to be made on the development of the root system and the roots of different order, but also in such a way that different properties (and potentially functions) can be assigned to the respective root segments. This applies, for example, to the question where roots take up water (Koebernick et al. 2015; Zarebanadkouki et al. 2013) and influence of mucilage on the rhizosphere properties and root water uptake (Ahmed et al. 2014, 2016a). Some of these topics were also discussed in Vetterlein and Doussan (2016).

13.4.3 Root–soil Contact

An optimal contact between root and soil is essential for plants to ensure both anchorage in the soil and sufficient supply of water and nutrients. A complete root–soil contact may reduce gas diffusion from the atmosphere into the soil; in contrast, a low root–soil contact can limit uptake of water and solutes by confining the hydraulic pathway (Veen et al. 1992). When the soil dries or when roots grow into a macropore with a larger diameter than that of the root (Tinker 1976), the contact may be poor. Shrinkage of the root tissue itself may also be involved in the gap formation (Cole and Alston 1974; Huck et al. 1970; North and Nobel 1997a,b).

The phenomenon of air gap formation was observed decades ago, and theoretical considerations were formulated in the 1950s (Bernstein et al. 1959; Philip 1957). Visualization and quantification was practically impossible without noninvasive imaging techniques. With the help of these techniques, it was possible to visualize and measure the gaps (Carminati et al. 2009) as well as to evaluate the root–soil contact by quantifying the contact area (Schmidt et al. 2012; Koebernick et al. 2018). In Figure 13.6, an illustrative example is given of the development of such an air gap and the effect of the water content. By simultaneously measuring stomatal conductance and transpiration, the interactions between the root–soil contact and plant water stress could be analysed. It was concluded that root shrinkage was not the initial cause of water stress, but rather the consequence (Carminati et al. 2013; Koebernick et al. 2018). Moreover, it was shown that plant species differed in the extent of shrinkage of different root orders (Carminati et al. 2013). A recent study was carried out to quantify the seed–soil contact of sugar beet using X-ray CT, showing about double the seed–soil contact for pelleted and coated seeds in comparison to naked (untreated) seeds, which is likely to enhance germination (Blunk et al. 2017).

13.4.4 Root Hairs

Root hairs are tubular outgrowths from the root epidermal cells. They occur in the zone of maturation, are lateral extensions of a single cell, and are rather small (a diameter between 5 and 20 μm and a length of 0.1–1.5 mm) (Gregory 2008). Their role or function has been under discussion for a long time. There are indications they are involved in water and nutrient (especially phosphorus) uptake due to an increase in the effective root surface area (Bates and Lynch 1996; Gahoonia et al. 1997; Carminati et al. 2017). Moreover, they stimulate synergies with microorganisms, help anchor plants in soil and are involved in rhizosheath formation (Haling et al. 2014).

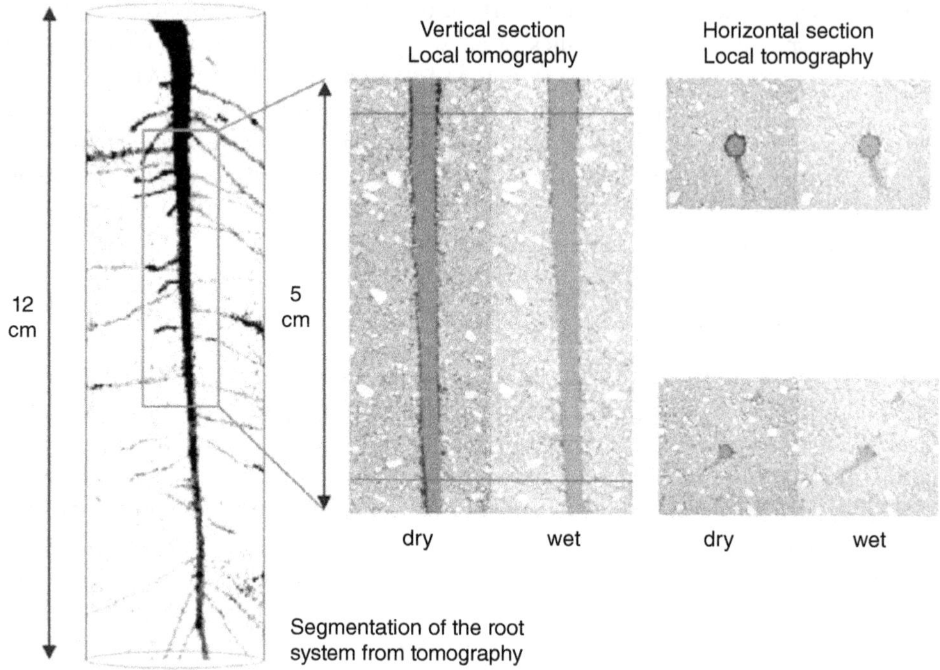

Figure 13.6 X-ray tomography of roots in soil. Left: three-dimensional root architecture of white lupin (*Lupinus albus*) growing in sandy soil. Middle: a two-dimensional vertical section of the local tomography under dry (left) and wet (right) conditions. Right: horizontal sections of the local tomography at two different depths (lines in the centre image correspond to the depths), showing air gaps between roots and soil that are visible under dry conditions (black), but are partially closed under wet conditions. *Source:* Carminati *et al.* (2009), with permission of Soil Science Society of America.

The small size of root hairs, their relatively short lifetime of about two to three weeks, and the penetration into small pores and interaggregate spaces make them difficult to investigate. In the recent studies, it was possible to visualize root hairs of *T. aestivum* and *H. vulgare* using synchrotron radiation to study their role in P uptake (Keyes et al. 2013, 2017a); and the rhizosphere structure formation (Koebernick et al. 2017). A very small sample size in combination with a powerful technique and image processing are needed to achieve the level of detail needed for analysis of root hairs (Figure 13.7). Given the comparable scale, it is also conceivable that mycorrhizal associations with plant roots can be studied by visualizing extraradical hyphae.

13.4.5 Soil–root Distances as a New Perspective of Soil Exploration by Roots

From the root perspective, RSA is typically quantified in terms of root length density and branching patterns. In contrast, from the soil perspective, root distances are a concept to consider regarding RSA (see Figure 13.5). The Euclidean distance from any location in the soil to the nearest root can be expressed as either distance distribution or distance map

Figure 13.7 Digitally rendered 3D volume from a high-resolution synchrotron radiation X-ray tomographic microscopy (SRXTM) at a resolution of approximately 1 µm. The soil phases are partially cut away to reveal a section of a seminal root of *Triticum aestivum*, including lateral roots and root hairs. *Source:* Keyes et al. (2013) with permission of John Wiley & Sons.

(Koebernick et al. 2014). This is a more direct estimate than root length of how efficiently a plant explores the soil, as these distances are a proxy for diffusion lengths of nutrients and water. The root distance distribution of *V. faba* was shown (Figure 13.5) to develop in predictable ways, which could be reproduced with a simple, empirical model (Schlüter et al. 2018). This can be the basis for the sophisticated, dynamic models of root water and nutrient uptake.

13.4.6 Soil Compaction by Roots

As roots grow through the soil, they often use already existing cavities and biopores (Passioura 1991) or in general they preferentially follow a path of low penetration resistance (Sands et al. 1979). Root mucilage serves as a gel-like lubricant that reduces penetration resistance of soil temporarily and in a small region around the root tip (Czarnes et al. 2000; McCully 1999). In the further course of root growth, there is also a shift and re-positioning of soil particles associated with changes in the spatial extension of pores and the pore network (Jones et al. 2004).

Using optical microscopy and an artificial system, Vollsnes et al. (2010) were able to visualize particle displacement in 2D due to root growth of maize in a thin layer of sand between two glass microscope slides. By using X-ray CT, Keyes et al. (2016, 2017b) managed to track root growth and associated rigid-body movement of soil particles in 3D (Figure 13.8).

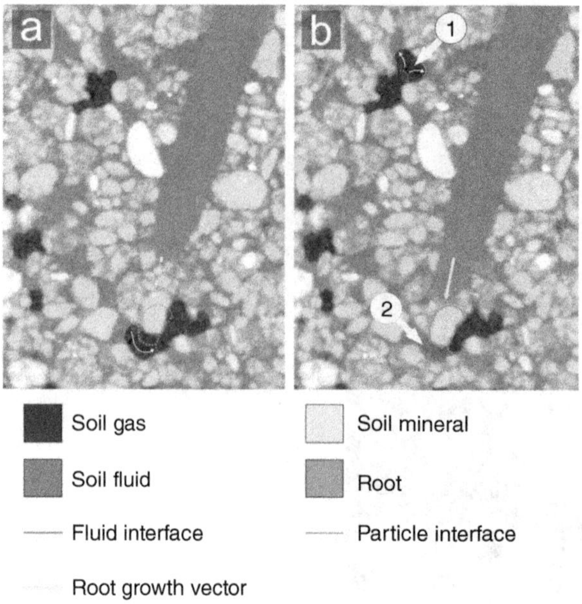

Figure 13.8 Deformation behaviour is shown qualitatively for a 2D section of the X-ray CT data over a single growth step. During the 1-hour interval between (a) and (b), the root tip extended along vector approximated by the arrow (yellow), and the fluid/gas interfaces at locations (1) and (2) moved. At location (1), this interface movement (i.e. Haines jump) occurred without obvious soil matrix deformation in the pore region. At location (2), the interface movement coincided with obvious rigid-body movement of a soil grain driven by the extension of the root tip, resulting in a macropore volume change. *Source:* Keyes et al. (2016) with permission of Elsevier.

Aravena et al. (2011) showed that root-induced particle movement can result in a compression of large interaggregate pores. By numerical modelling based on the actual CT images, they showed that the root-induced deformation of the aggregated soil around a root increased unsaturated water flow in direction of the root.

Recent evidence suggests that soil compaction in the rhizosphere might not be a general trend; instead, roots may also increase the porosity near the root surface by enhancing the aggregation of soil particles (Helliwell et al. 2017; Koebernick et al. 2017, 2018).

13.4.7 Modelling

The 3D modelling of root uptake of water or nutrients from soil requires information on root architecture. Mostly, this information is generated using RSA models because direct measurements have not been available (Dunbabin et al. 2013). Koebernick et al. (2015) were the first to replace modelled root architecture by time-resolved measured root architecture derived from the X-ray CT data. In addition, they used measured soil matric potential dynamics as a reference for the different scenarios addressed by modelling.

For increased precision and a realistic representation of the natural conditions and the resulting consequences, detailed input variables are essential in modelling. This is exactly what imaging with the current methods can provide. The combination of imaging and modelling is therefore taken up in several studies on phosphate uptake (Fang et al. 2009; Keyes et al. 2013), water transfer (Doussan et al. 2006; Javaux et al. 2008), inverse modelling of root hydraulic conductivity (Zarebanadkouki et al. 2016), or influence of root-induced compaction on hydraulic properties in the rhizosphere (Aravena et al. 2011).

Image-based modelling enables the direct application of numerical methods (e.g. finite element method) on the images by converting the segmented structures (e.g. soil matrix, pore space, and pore network) into a textural mesh. This allows the characterization of the physical properties of the pore space (Daly et al. 2018a,b; Koebernick et al. 2017).

Imaging enables the associated modelling of the root system, e.g. with the hydraulic tree approach (Doussan et al. 1998; Landsberg and Fowkes 1978). This allows dynamic modelling of growing root systems that can, for instance, change their hydraulic properties as the roots mature (Koebernick et al. 2015), which can be parameterized by image analysis.

13.5 Future Developments

RSA is measured indirectly because classical imaging normally is dependent on the radiation of the sample and the reception of the resulting beam by a detector. For this reason, imaging usually takes place on the laboratory scale. The application on the field scale as a standard procedure is not yet in sight. However, for MRI, there are initial approaches for use outside the laboratory (Pohlmeier et al. 2018).

The technological progress of the methods described in this chapter has been rapid in the last decades and will likely continue. New instruments with better detectors and faster measurement methods are under development. A shortening of the measurement time is an essential aspect for the future. This is associated with a lower radiation exposure of the samples and a possibility of increasing the frequency of measurements or the number of samples measured in order to make more precise statements about the values of dynamic variables based on improved statistical capability.

Fast phenotyping of RSA is of great interest for many scientific and breeding questions. This is slowly becoming more realistic, but still hindered by time-consuming postprocessing and analyses. Recent advances in developing more automated workflow to run the analyses in a batch mode are promising (Gao et al. 2019b). The research community needs to define standards for root scanning and method comparison concerning the image analyses. The training data sets for algorithm development are now available. The huge amounts of data generated during acquisition, and the complex steps of image processing and analysis are not only very time-consuming, but also require appropriate computing power and equipment to handle the input data. Therefore, image analysis is now increasingly recognized as "the new bottleneck in plant phenotyping" (Minervini et al. 2015), and the faster and more standardized segmentation protocols are needed (Pfeifer et al. 2015). Major hurdles that need to be overcome are manual user interaction, combinations of different software packages in the image-processing workflow, as well as the computational costs of large 3D data sets (Flavel et al. 2017; Mairhofer et al. 2012, 2013).

The machine learning approaches have a huge potential to revolutionize image analysis in the future. However, this has been used seldom so far, with a paucity of case studies such as the "trainable WEKA segmentation" (Keyes et al. 2017a; Koebernick et al. 2017).

The scan time can be described as a further bottleneck because a single scan usually takes a relatively long time, multiplied by a large number of scans required. The scan time can be reduced at the expense of image quality (signal-to-noise ratio). The acquisition time for a typical X-ray CT scan is in the range of 10–120 minutes (Ahmed et al. 2016b; Pfeifer et al. 2015; Tracy et al. 2012b; Koebernick et al. 2018). With a reduction to 8.5 minutes, it was still possible to detect successfully the rather thick *V. faba* root network at an image resolution of 40 μm, whereas the detection of *H. vulgare* roots failed under these conditions (Blaser et al. 2018). Hence, the optimal trade-off between scan time and image quality depends on the plant species, as well as on pot-size dependent image resolution, X-ray CT hardware, and reconstruction software. There are also approaches for faster data acquisition for other methods, for example ultrafast NT with a total acquisition time not exceeding 60 seconds (Totzke et al. 2017).

The methods and algorithms from the other specialized areas such as materials research and of course the medical sector could be useful here, e.g. spiral CT (Kalender et al. 1997) or helical differential X-ray phase-contrast CT (Fu et al. 2014; Marschner et al. 2016). With these methods, an object often no longer has to be irradiated from all 360 angles in order to generate the 3D data set and, therefore, a lot of time can be saved during data acquisition. Furthermore, there are new iterative reconstruction algorithms that require a smaller number of X-ray radiographs than the conventional filtered-back projection algorithms to achieve the same image quality in the 3D scans (Beister et al. 2012; Kazantsev et al. 2016).

Fast-throughput CT scanning is available in the industrial sector with the in-line inspection systems that are fully automatic and even equipped with automatic defect recognition (Brunke et al. 2012). Hence, objects can be scanned in a few seconds, but much higher energies in the range of kilowatt are used, which would not be applicable in root research because such doses would be harmful to plants (Blaser et al. 2018; Johnson 1936). Approaches with lower doses are more likely to come from the clinical and biological CT applications.

References

Ahmed, M.A., Kroener, E., Holz, M. et al. (2014). Mucilage exudation facilitates root water uptake in dry soils. *Functional Plant Biology* 41: 1129–1137.

Ahmed, M.A., Kroener, E., Benard, P. et al. (2016a). Drying of mucilage causes water repellency in the rhizosphere of maize: measurements and modelling. *Plant and Soil* 407: 161–171.

Ahmed, S., Klassen, T.N., Keyes, S. et al. (2016b). Imaging the interaction of roots and phosphate fertiliser granules using 4D X-ray tomography. *Plant and Soil* 401: 125–134.

Aravena, J.E., Berli, M., Ghezzehei, T.A. et al. (2011). Effects of root-induced compaction on rhizosphere hydraulic properties – X-ray microtomography imaging and numerical simulations. *Environmental Science and Technology* 45: 425–431.

Arsenault, J.L., Poulcur, S., Messier, C. et al. (1995). WinRHIZO™, a root-measuring system with a unique overlap correction method. *HortScience* 30: 906–906.

Bates, T.R. and Lynch, J.P. (1996). Stimulation of root hair elongation in *Arabidopsis thaliana* by low phosphorus availability. *Plant, Cell and Environment* 19: 529–538.

Beister, M., Kolditz, D., and Kalender, W.A. (2012). Iterative reconstruction methods in X-ray CT. *Physica Medica* 28: 94–108.

Bernstein, L., Gardner, W., and Richards, L. (1959). Is there a vapor gap around plant roots? *Science* 129: 1750–1753.

Bingham, I.J. and Bengough, A.G. (2003). Morphological plasticity of wheat and barley roots in response to spatial variation in soil strength. *Plant and Soil* 250: 273–282.

Blaser, S., Schluter, S., and Vetterlein, D. (2018). How much is too much?-Influence of X-ray dose on root growth of faba bean (*Vicia faba*) and barley (*Hordeum vulgare*). *PLoS One* 13: e0193669.

Blunk, S., Malik, A.H., De Heer, M.I. et al. (2017). Quantification of differences in germination behaviour of pelleted and coated sugar beet seeds using X-ray Computed Tomography (X-ray CT). *Biomedical Physics and Engineering Express* 3: 044001.

Boonsirichai, K., Guan, C., Chen, R. et al. (2002). Root gravitropism: an experimental tool to investigate basic cellular and molecular processes underlying mechanosensing and signal transmission in plants. *Annual Review of Plant Biology* 53: 421–447.

Borisjuk, L., Rolletschek, H., and Neuberger, T. (2012). Surveying the plant's world by magnetic resonance imaging. *The Plant Journal* 70: 129–146.

Brown, J.M., Johnson, G.A., and Kramer, P.J. (1986). *in vivo* magnetic resonance microscopy of changing water content in *Pelargonium hortorum* roots. *Plant Physiology* 82: 1158–1160.

Brunke, O., Hansen, F., Stuke, I. et al. (2012). A new concept for high–speed flatline and inlinect for up to 100% mass production process control allowing both, 3D metrology and failure analysis. Proceedings of the International Conference on Industrial Computed Tomography, iCT.

Carminati, A. (2013). Rhizosphere wettability decreases with root age: a problem or a strategy to increase water uptake of young roots? *Frontiers in Plant Science* 4: 298.

Carminati, A., Vetterlein, D., Weller, U. et al. (2009). When roots lose contact. *Vadose Zone Journal* 8: 805–809.

Carminati, A., Vetterlein, D., Koebernick, N. et al. (2013). Do roots mind the gap? *Plant and Soil* 367: 651–661.

Carminati, A., Passioura, J.B., Zarebanadkouki, M. et al. (2017). Root hairs enable high transpiration rates in drying soils. *New Phytologist* 216: 771–781.

Chen, R., Xie, H., Deng, B. et al. (2016). X-ray microtomography at Shanghai Synchrotron radiation facility. Developments in X-ray Tomography X. International Society for Optics and Photonics, 99671B.

Cole, P.J. and Alston, A.M. (1974). Effect of transient dehydration on absorption of chloride by wheat roots. *Plant and Soil* 40: 243–247.

Correll, M.J. and Kiss, J.Z. (2002). Interactions between gravitropism and phototropism in plants. *Journal of Plant Growth Regulation* 21: 89–101.

Couchat, P., Moutonnet, P., Houelle, M. et al. (1980). *In situ* study of corn seedling root and shoot growth by neutron radiography. *Agronomy Journal* 72: 321–324.

Czarnes, S., Hallett, P.D., Bengough, A.G. et al. (2000). Root- and microbial-derived mucilages affect soil structure and water transport. *European Journal of Soil Science* 51: 435–443.

Daly, K.R., Mooney, S.J., Bennett, M.J. et al. (2015). Assessing the influence of the rhizosphere on soil hydraulic properties using X-ray computed tomography and numerical modelling. *Journal of Experimental Botany* 66: 2305–2314.

Daly, K.R., Keyes, S.D., and Roose, T. (2018a). Determination of macro-scale soil properties from pore scale structures: image-based modelling of poroelastic structures. *Proceedings of The Royal Society A. Mathematical Physical and Engineering Sciences* 474: 20170745. https://doi.org/10.1098/rspa.2017.0141.

Daly, K.R., Tracy, S.R., Crout, N.M. et al. (2018b). Quantification of root water uptake in soil using X-ray computed tomography and image-based modelling. *Plant, Cell and Environment* 41: 121–133.

Darwin, C. (1897). The Power of Movement in Plants. Appleton.

De Kroons, H. and Hutchings, M.J. (1995). Morphological plasticity in clonal plants: the foraging concept reconsidered. *The Journal of Ecology* 83: 143.

Doussan, C., Pagès, L., and Vercambre, G. (1998). Modelling of the hydraulic architecture of root systems: an integrated approach to water absorption – model description. *Annals of Botany* 81: 213–223.

Doussan, C., Pierret, A., Garrigues, E. et al. (2006). Water uptake by plant roots: II. Modelling of water transfer in the soil root-system with explicit account of flow within the root system – comparison with experiments. *Plant and Soil* 283: 99–117.

Downie, H., Holden, N., Otten, W. et al. (2012). Transparent soil for imaging the rhizosphere. *PLoS One* 7: e44276.

Downie, H.F., Adu, M.O., and Schmidt, S. (2015). Challenges and opportunities for quantifying roots and rhizosphere interactions through imaging and image analysis. *Plant, Cell and Environment* 38: 1213–1232.

Drew, M. (1975). Comparison of the effects of a localised supply of phosphate, nitrate, ammonium and potassium on the growth of the seminal root system, and the shoot, in barley. *New Phytologist* 75: 479–490.

Drew, M. and Saker, L. (1975). Nutrient supply and the growth of the seminal root system in barley. II. Localized, compensatory increases in lateral root growth and rates of nitrate uptake when nitrate supply is restricted to only part of the root system. *Journal of Experimental Botany* 26: 79–90.

Drew, M. and Saker, L. (1978). Nutrient supply and the growth of the seminal root system in barley. III. Compensatory increases in growth of lateral roots, and in rates of phosphate uptake, in response to a localized supply of phosphate. *Journal of Experimental Botany* 29: 435–451.

Drew, M., Saker, L., and Ashley, T. (1973). Nutrient supply and the growth of the seminal root system in barley I. The effect of nitrate concentration on the growth of axes and laterals. *Journal of Experimental Botany* 24: 1189–1202.

Dunbabin, V.M., Postma, J.A., Schnepf, A. et al. (2013). Modelling root–soil interactions using three–dimensional models of root growth, architecture and function. *Plant and Soil* 372: 93–124. https://doi.org/10.1007/s11104-013-1769-y.

Eapen, D., Barroso, M.L., and Ponce, G. (2005). Hydrotropism: root growth responses to water. *Trends in Plant Science* 10: 44–50.

Esau, K. (1965). Plant Anatomy. New York: Wiley.

Esmon, C.A., Pedmale, U.V., and Liscum, E. (2004). Plant tropisms: providing the power of movement to a sessile organism. *International Journal of Developmental Biology* 49: 665–674.

Fang, S., Yan, X., and Liao, H. (2009). 3D reconstruction and dynamic modeling of root architecture in situ and its application to crop phosphorus research. *The Plant Journal* 60: 1096–1108.

Fitter, A.H. (1987). An architectural approach to the comparative ecology of plant-root systems. *New Phytologist* 106: 61–77.

Flavel, R.J., Guppy, C.N., Tighe, M. et al. (2012). Non-destructive quantification of cereal roots in soil using high-resolution X-ray tomography. *Journal of Experimental Botany* 63: 2503–2511.

Flavel, R.J., Guppy, C.N., Tighe, M.K. et al. (2014). Quantifying the response of wheat (*Triticum aestivum* L.) root system architecture to phosphorus in an Oxisol. *Plant and Soil* 385: 303–310.

Flavel, R.J., Guppy, C.N., and Rabbi, S.M.R. (2017). An image processing and analysis tool for identifying and analysing complex plant root systems in 3D soil using non-destructive analysis: root. *PLoS One* 12: e0176433.

Fransen, B., De Kroon, H., and Berendse, F. (1998). Root morphological plasticity and nutrient acquisition of perennial grass species from habitats of different nutrient availability. *Oecologia* 115: 351–358.

Fu, J., Willner, M., Chen, L. et al. (2014). Helical differential X-ray phase-contrast computed tomography. *Physica Medica* 30: 374–379.

Gahoonia, T.S., Care, D., and Nielsen, N.E. (1997). Root hairs and phosphorus acquisition of wheat and barley cultivars. *Plant and Soil* 191: 181–188.

Gao, W., Blaser, S.R.G.A., and Schlüter, S. (2019a). Effect of localised phosphorus application on root growth and soil nutrient dynamics in situ – comparison of maize (*Zea mays* L.) and faba bean (*Vicia faba*) at the seedling stage. *Plant and Soil* 441: 469–483.

Gao, W., Schlüter, S., and Blaser, S.R.G.A. (2019b). A shape-based method for automatic and rapid segmentation of roots in soil from X-ray computed tomography images: rootine. *Plant and Soil* 441: 643–655.

Garbout, A., Munkholm, L.J., Hansen, S.B. et al. (2012). The use of PET/CT scanning technique for 3D visualization and quantification of real-time soil/plant interactions. *Plant and Soil* 352: 113–127.

Gilroy, S. (2008). Plant tropisms. *Current Biology* 18: R275–R277.

Gould, S.J. (1977). Ontogeny and Phylogeny. Cambridge, MA: Harvard University Press.

Gregory, P.J. (2008). Plant Roots: Growth, Activity and Interactions with the Soil. New York, NY: Wiley.

Gregory, P.J., Hutchison, D.J., Read, D.B. et al. (2003). Non-invasive imaging of roots with high resolution X-ray micro-tomography. *Plant and Soil* 255: 351–359.

Hackett, C. (1972). A method of applying nutrients locally to roots under controlled conditions, and some morphological effects of locally applied nitrate on the branching of wheat roots. *Australian Journal of Biological Sciences* 25: 1169–1180.

Haling, R.E., Tighe, M.K., Flavel, R.J. et al. (2013). Application of X-ray computed tomography to quantify fresh root decomposition *in situ*. *Plant and Soil* 372: 619–627.

Haling, R.E., Brown, L.K., Bengough, A.G. et al. (2014). Root hair length and rhizosheath mass depend on soil porosity, strength and water content in barley genotypes. *Planta* 239: 643–651.

Hall, L.D., Amin, M.H.G., and Dougherty, E. (1997). MR properties of water in saturated soils and resulting loss of MRI signal in water content detection at 2 Tesla. *Geoderma* 80: 431–448.

Heeraman, D.A., Hopmans, J.W., and Clausnitzer, V. (1997). Three dimensional imaging of plant roots in situ with X-ray computed tomography. *Plant and Soil* 189: 167–179.

Helliwell, J., Sturrock, C., Grayling, K. et al. (2013). Applications of X-ray computed tomography for examining biophysical interactions and structural development in soil systems: a review. *European Journal of Soil Science* 64: 279–297.

Helliwell, J.R., Sturrock, C.J., Mairhofer, S. et al. (2017). The emergent rhizosphere: imaging the development of the porous architecture at the root-soil interface. *Scientific Reports* 7: 1–10. https://doi.org/10.1038/s41598-017-14904-w.

Hochholdinger, F., Park, W.J., Sauer, M. et al. (2004). From weeds to crops: genetic analysis of root development in cereals. *Trends in Plant Science* 9: 42–48.

Hodge, A. (2004). The plastic plant: root responses to heterogeneous supplies of nutrients. *New Phytologist* 162: 9–24.

Hodge, A. (2006). Plastic plants and patchy soils. *Journal of Experimental Botany* 57: 401–411.

Huck, M.G., Klepper, B., and Taylor, H.M. (1970). Diurnal variations in root diameter. *Plant Physiology* 45: 529–530.

Iyer-Pascuzzi, A.S., Symonova, O., and Mileyko, Y. (2010). Imaging and analysis platform for automatic phenotyping and trait ranking of plant root systems. *Plant Physiology* 152: 1148–1157.

Javaux, M., Schroder, T., Vanderborght, J. et al. (2008). Use of a three-dimensional detailed modeling approach for predicting root water uptake. *Vadose Zone Journal* 7: 1079–1088.

Johnson, E.L. (1936). Susceptibility of seventy species of flowering plants to X-radiation. *Plant Physiology* 11: 319–342.

Johnson, H.A. and Biondini, M.E. (2001). Root morphological plasticity and nitrogen uptake of 59 plant species from the Great Plains grasslands, U.S.A. *Basic and Applied Ecology* 2: 127–143.

Jones, D.L., Hodge, A., and Kuzyakov, Y. (2004). Plant and mycorrhizal regulation of rhizodeposition. *New Phytologist* 163: 459–480.

Kalender, W.A., Engelke, K., and Schaller, S. (1997). Spiral CT: medical use and potential industrial applications. Developments in X-Ray Tomography. International Society for Optics and Photonics, 188–203.

Kaspar, T.C. and Bland, W.L. (1992). Soil temperature and root growth. *Soil Science* 154: 290–299.

Kastner, J., Heinzl, C., Plank, B. et al. (2017). New X-ray computed tomography methods for research and industry. 7th Conference on Industrial Computed Tomography (iCT2017).

Kazantsev, D., Guo, E., Kaestner, A. et al. (2016). Temporal sparsity exploiting nonlocal regularization for 4D computed tomography reconstruction. *Journal of X-Ray Science and Technology* 24: 207–219.

Keyes, S.D., Daly, K.R., Gostling, N.J. et al. (2013). High resolution synchrotron imaging of wheat root hairs growing in soil and image based modelling of phosphate uptake. *New Phytologist* 198: 1023–1029.

Keyes, S.D., Gillard, F., Soper, N. et al. (2016). Mapping soil deformation around plant roots using *in vivo* 4D X-ray computed tomography and digital volume correlation. *Journal of Biomechanics* 49: 1802–1811.

Keyes, S., Zygalakis, K., and Roose, T. (2017a). Explicit structural modelling of root-hair and soil interactions at the micron-scale parameterized by synchrotron x-ray computed tomography. *Bulletin of Mathematical Biology* 79: 2785–2813. https://doi.org/10.1007/s11538-017-0350-x.

Keyes, S.D., Cooper, L., and Duncan, S. (2017b). Measurement of micro-scale soil deformation around roots using four-dimensional synchrotron tomography and image correlation. *Journal of The Royal Society Interface* 14: 20170560. https://doi.org/10.1098/rsif.2017.0560.

Koebernick, N., Weller, U., Huber, K. et al. (2014). *In situ* visualization and quantification of three-dimensional root system architecture and growth using x-ray computed tomography. *Vadose Zone Journal* 13: 1–10. https://doi.org/10.2136/vzj2014.03.0024.

Koebernick, N., Huber, K., Kerkhofs, E. et al. (2015). Unraveling the hydrodynamics of split root water uptake experiments using CT scanned root architectures and three dimensional flow simulations. *Frontiers in Plant Science* 6: 370.

Koebernick, N., Daly, K.R., and Keyes, S.D. (2017). High-resolution synchrotron imaging shows that root hairs influence rhizosphere soil structure formation. *New Phytologist* 216: 124–135.

Koebernick, N., Schluter, S., Blaser, S.R.G.A. et al. (2018). Root-soil contact dynamics of *Vicia faba* in sand. *Plant and Soil* 431: 417–431.

Kutschera, L. (1960). Wurzelatlas mitteleuropaischer Ackerunkrauter und Kulturpflanzen. Frankfurt, Germany: DLG–Verlag.

Kutschera, L., Lichtenegger, E., and Sobotik, M. (2009). Wurzelatlas der Kulturpflanzen gemäßigter Gebiete: mit Arten des Feldgemüsebaues. Frankfurt, Germany: DLG–Verlag.

Landsberg, J. and Fowkes, N. (1978). Water movement through plant roots. *Annals of Botany* 42: 493508.

Le Bot, J., Serra, V., Fabre, J. et al. (2010). DART: a software to analyse root system architecture and development from captured images. *Plant and Soil* 326: 261–273.

Leitner, D., Klepsch, S., Bodner, G. et al. (2010). A dynamic root system growth model based on L-systems. *Plant and Soil* 332: 177–192.

Leitner, D., Felderer, B., Vontobel, P. et al. (2014). Recovering root system traits using image analysis exemplified by two-dimensional neutron radiography images of lupine. *Plant Physiology* 164: 24–35.

Li, L., Zhang, Q., and Huang, D. (2014). A review of imaging techniques for plant phenotyping. *Sensors* 14: 20078–20111.

Lima, J.E., Kojima, S., Takahashi, H. et al. (2010). Ammonium triggers lateral root branching in *Arabidopsis* in an AMMONIUM TRANSPORTER1;3-dependent manner. *Plant Cell* 22: 3621–3633.

Macfall, J.S., Johnson, G.A., and Kramer, P.J. (1990). Observation of a water-depletion region surrounding loblolly pine roots by magnetic resonance imaging. *Proceedings of the National Academy of Sciences of the United States of America* 87: 1203–1207.

Macfall, J.S., Johnson, G.A., and Kramer, P.J. (1991). Comparative water-uptake by roots of different ages in seedlings of loblolly-pine (*Pinus taeda* L). *New Phytologist* 119: 551–560.

Mairhofer, S., Zappala, S., Tracy, S.R. et al. (2012). RooTrak: automated recovery of three-dimensional plant root architecture in soil from x-ray microcomputed tomography images using visual tracking. *Plant Physiology* 158: 561–569.

Mairhofer, S., Zappala, S., Tracy, S. et al. (2013). Recovering complete plant root system architectures from soil via X-ray μ-computed tomography. *Plant Methods* 9: 8.

Marschner, M., Willner, M., Potdevin, G. et al. (2016). Helical X-ray phase-contrast computed tomography without phase stepping. *Scientific Reports* 6: 23953.

Mccully, M.E. (1999). Roots in soil: unearthing the complexities of roots and their rhizospheres. *Annual Review of Plant Physiology and Plant Molecular Biology* 50: 695–718.

Metzner, R., Eggert, A., Van Dusschoten, D. et al. (2015). Direct comparison of MRI and X-ray CT technologies for 3D imaging of root systems in soil: potential and challenges for root trait quantification. *Plant Methods* 11: 17.

Minervini, M., Scharr, H., and Tsaftaris, S.A. (2015). Image analysis: the new bottleneck in plant phenotyping [applications corner]. *IEEE Signal Processing Magazine* 32: 126–131.

Mooney, S.J., Pridmore, T.P., Helliwell, J. et al. (2012). Developing X-ray computed tomography to non-invasively image 3-D root systems architecture in soil. *Plant and Soil* 352: 1–22.

Moradi, A.B., Carminati, A., and Vetterlein, D. (2011). Three-dimensional visualization and quantification of water content in the rhizosphere. *New Phytologist* 192: 653–663.

Moran, C.J., Pierret, A., and Stevenson, A.W. (2000). X-ray absorption and phase contrast imaging to study the interplay between plant roots and soil structure. *Plant and Soil* 223: 101–117.

Morris, E.C., Griffiths, M., and Golebiowska, A. (2017). Shaping 3D root system architecture. *Current Biology* 27: R919–R930.

Mounier, E., Pervent, M., and Ljung, K. (2014). Auxin-mediated nitrate signalling by NRT1.1 participates in the adaptive response of *Arabidopsis* root architecture to the spatial heterogeneity of nitrate availability. *Plant, Cell and Environment* 37: 162–174.

Neumann, G., George, T.S., and Plassard, C. (2009). Strategies and methods for studying the rhizosphere – the plant science toolbox. *Plant and Soil* 321: 431–456.

Newcombe, F.C. and Rhodes, A.L. (1904). Chemotropism of roots. *Botanical Gazette* 37: 22–34.

North, G.B. and Nobel, P.S. (1997a). Drought-induced changes in soil contact and hydraulic conductivity for roots of *Opuntia ficus-indica* with and without rhizosheaths. *Plant and Soil* 191: 249–258.

North, G.B. and Nobel, P.S. (1997b). Root-soil contact for the desert succulent *Agave deserti* in wet and drying soil. *New Phytologist* 135: 21–29.

Nwankwo, C.I., Blaser, S.R., Vetterlein, D. et al. (2018). Seedball-induced changes of root growth and physico-chemical properties – a case study with pearl millet. *Journal of Plant Nutrition and Soil Science* 181: 768–776.

Onderdonk, J.J. and Ketcheson, J.W. (1973). Effect of soil temperature on direction of corn root growth. *Plant and Soil* 39: 177–186.

Osmont, K.S., Sibout, R., and Hardtke, C.S. (2007). Hidden branches: developments in root system architecture. *Annual Review of Plant Biology* 58: 93–113.

Oswald, S.E., Menon, M., and Carminati, A. (2008). Quantitative imaging of infiltration, root growth, and root water uptake via neutron radiography. *Vadose Zone Journal* 7: 1035–1047.

Oswald, S.E., Tötzke, C., Haber-Pohlmeier, S. et al. (2015). Combining neutron and magnetic resonance imaging to study the interaction of plant roots and soil. *Physics Procedia* 69: 237–243.

Pagenkemper, S.K., Peth, S., Puschmann, D.U. et al. (2013). Effects of root-induced biopores on pore space architecture investigated with industrial X-ray computed tomography.

In: Soil–Water–Root Processes: Advances in Tomography and Imaging (eds. S.H. Anderson and J.W. Hopmans), 69–96. Madison, Wis: Soil Science Society of America https://doi.org/10.2136/sssaspecpub61.c4.

Passioura, J.B. (1991). Soil structure and plant-growth. *Australian Journal of Soil Research* 29: 717–728.

Pfeifer, J., Kirchgessner, N., Colombi, T. et al. (2015). Rapid phenotyping of crop root systems in undisturbed field soils using X-ray computed tomography. *Plant Methods* 11: 41.

Philip, J. (1957). The physical principles of soil water movement during the irrigation cycle. 3rd Congress International Commission on Irrigation and Drainage Quest, 8.125–8.154

Pierret, A., Capowiez, Y., Moran, C.J. et al. (1999). X-ray computed tomography to quantify tree rooting spatial distributions. *Geoderma* 90: 307–326.

Pohlmeier, A., Oros-Peusquens, A., Javaux, M. et al. (2008). Changes in soil water content resulting from Ricinus root uptake monitored by magnetic resonance imaging. *Vadose Zone Journal* 7: 1010–1017.

Pohlmeier, A., Vergeldt, F., and Gerkema, E. (2010). MRI in soils: determination of water concent changes due to root water uptake by means of a multi-slice-multi-echo sequence (MSME). *The Open Magnetic Resonance Journal* 3: 69–74.

Pohlmeier, A., Garre, S., and Roose, T. (2018). Noninvasive imaging of processes in natural porous media: from pore to field scale. *Vadose Zone Journal* 17: 1–3. https://doi.org/10.2136/vzj2018.03.0044.

Pound, M.P., French, A.P., and Atkinson, J.A. (2013). RootNav: navigating images of complex root architectures. *Plant Physiology* 162: 1802–1814.

Rudolph-Mohr, N., Totzke, C., Kardjilov, N. et al. (2017). Mapping water, oxygen, and pH dynamics in the rhizosphere of young maize roots. *Journal of Plant Nutrition and Soil Science* 180: 336–346.

Sands, R., Greacen, E., and Gerard, C. (1979). Compaction of sandy soils in radiata pine forests. I. A penetrometer study. *Soil Research* 17: 101–113.

Schlüter, S., Sheppard, A., Brown, K. et al. (2014). Image processing of multiphase images obtained via X- ray microtomography: a review. *Water Resources Research* 50: 3615–3639.

Schlüter, S., Blaser, S.R.G.A., Weber, M. et al. (2018). Quantification of root growth patterns from the soil perspective via root distance models. *Frontiers in Plant Science* 9: 1084.

Schmidt, S., Bengough, A.G., Gregory, P.J. et al. (2012). Estimating root-soil contact from 3D X-ray microtomographs. *European Journal of Soil Science* 63: 776–786.

Schulz, H., Postma, J.A., Van Dusschoten, D. et al. (2012). 3D reconstruction of plant roots from MRI images. *Proceedings of the International Conference on Computer Vision, Theory and Applications VISAPP*, Rome, Italy, 24–33.

Schulz, H., Postma, J.A., Van Dusschoten, D. et al. (2013). Plant root system analysis from MRI images. *Communications in Computer and Information Science* 359: 411–425.

Smit, A.L., Bengough, A.G., Engels, C. et al. (2013). Root Methods: A Handbook. New York, NY: Springer Science and Business Media.

Stingaciu, L., Schulz, H., Pohlmeier, A. et al. (2013). *In situ* root system architecture extraction from magnetic resonance imaging for water uptake modeling. *Vadose Zone Journal* 12 (1): vzj2012.0019.

Strobl, M., Manke, I., Kardjilov, N. et al. (2009). Advances in neutron radiography and tomography. *Journal of Physics D–Applied Physics* 42: 243001.

Sturrock, C.J., Woodhall, J., Brown, M. et al. (2015). Effects of damping-off caused by *Rhizoctonia solani* anastomosis group 2-1 on roots of wheat and oil seed rape quantified using X-ray computed tomography and real–time PCR. *Frontiers in Plant Science* 6: 461.

Taina, I.A., Heck, R.J., and Elliot, T.R. (2008). Application of X-ray computed tomography to soil science: a literature review. *Canadian Journal of Soil Science* 88: 1–20.

Tennant, D. (1975). Test of a modified line intersect method of estimating root length. *Journal of Ecology* 63: 995–1001.

Tinker, P.B. (1976). Roots and water – transport of water to plant roots in soil. *Philosophical Transactions of the Royal Society of London Series B-Biological Sciences* 273: 445–461.

Tollner, E., Ramseur, E., and Murphy, C. (1994). Techniques and approaches for documenting plant root development with X–ray computed tomography. In: Tomography of Soil–Water–Root Processes (eds. S.H. Anderson and J.W. Hopmans), 115–133. Madison, Wis: Soil Science Society of America https://doi.org/10.2136/sssaspecpub36.c10.

Totzke, C., Kardjilov, N., Manke, I. et al. (2017). Capturing 3D water flow in rooted soil by ultra-fast neutron tomography. *Scientific Reports* 7: 6192.

Trachsel, S., Kaeppler, S.M., Brown, K.M. et al. (2011). Shovelomics: high throughput phenotyping of maize (*Zea mays* L.) root architecture in the field. *Plant and Soil* 341: 75–87.

Tracy, S.R., Black, C.R., Roberts, J.A. et al. (2012a). Quantifying the effect of soil compaction on three varieties of wheat (*Triticum aestivum* L.) using X-ray micro computed tomography (CT). *Plant and Soil* 353: 195–208.

Tracy, S.R., Black, C.R., Roberts, J.A. et al. (2012b). Quantifying the impact of soil compaction on root system architecture in tomato (*Solanum lycopersicum*) by X-ray micro-computed tomography. *Annals of Botany* 110: 511–519.

Van Dusschoten, D., Metzner, R., Kochs, J. et al. (2016). Quantitative 3D analysis of plant roots growing in soil using magnetic resonance imaging. *Plant Physiology* 170: 1176–1188.

Veen, B., Van Noordwijk, M., De Willigen, P. et al. (1992). Root-soil contact of maize, as measured by a thin-section technique. *Plant and Soil* 139: 131–138.

Vetterlein, D. and Doussan, C. (2016). Root age distribution: how does it matter in plant processes? A focus on water uptake. *Plant and Soil* 407: 145–160.

Vollsnes, A., Futsaether, C., and Bengough, A. (2010). Quantifying rhizosphere particle movement around mutant maize roots using time-lapse imaging and particle image velocimetry. *European Journal of Soil Science* 61: 926–939.

Waisel, Y., Eshel, A., and Kafkafi, U. (1996). Plant Roots: The Hidden Half. New York, NY: Marcel Dekker, Inc.

Watanabe, K., Mandang, T., Tojo, S. et al. (1992). Nondestructive root-zone analysis with X-ray CT scanner. Paper-American Society of Agricultural Engineers (USA). No. 923018.

Wiersum, L. (1958). Density of root branching as affected by substrate and separate ions. *Acta Botanica Neerlandica* 7: 174–190.

Wildenschild, D. and Sheppard, A.P. (2013). X-ray imaging and analysis techniques for quantifying pore-scale structure and processes in subsurface porous medium systems. *Advances in Water Resources* 51: 217–246.

Willatt, S.T. and Struss, R.G. (1979). Germination and early growth of plants studied using neutron radiography. *Annals of Botany* 43: 415–422.

Willatt, S., Struss, R., and Taylor, H. (1978). *In situ* root studies using neutron radiography. *Agronomy Journal* 70: 581–586.

Zappala, S., Helliwell, J.R., and Tracy, S.R. (2013). Effects of X-ray dose on rhizosphere studies using X-ray computed tomography. *PLoS One* 8: e67250.

Zarebanadkouki, M., Kim, Y.X., and Carminati, A. (2013). Where do roots take up water? Neutron radiography of water flow into the roots of transpiring plants growing in soil. *New Phytologist* 199: 1034–1044.

Zarebanadkouki, M., Meunier, F., and Couvreur, V. (2016). Estimation of the hydraulic conductivities of lupine roots by inverse modelling of high-resolution measurements of root water uptake. *Annals of Botany* 118: 853–864.

14

Simulating the Diversity and Plasticity of Root Systems Using 3D Models of the Root System Architecture

Loïc Pagès

INRA UR PSH 1115 Centre PACA, Avignon cedex 9, France

14.1 Introduction

Several functions performed by the root systems are closely linked to their architecture, that is, to their structure and their spatial distribution (Fitter et al. 1991; Berntson 1994; de Dorlodot et al. 2007; Hinsinger et al. 2011; Dunbabin et al. 2013). This is particularly obvious for soil–plant exchange functions (water and nutrient uptake and exudation of organic compounds) because the nutrients involved have variable and/or limited mobility (Hinsinger et al. 2011) and heterogeneous distribution in soil. Moreover, exchange capacities vary substantially among and along the roots. For example, water uptake fluxes depend on both radial and axial conductances whose values are considerably different in young distal parts (near the apex) and older more proximal parts (Doussan et al. 1998; Couvreur et al. 2012). Root systems are highly dynamic; they continuously develop and adapt their architecture to the plant growth stage and environmental conditions. These dynamics, at various time scales, are vital to understanding the integrated functioning of the system. For this purpose, dynamic simulation models of the root system architecture are the key tools to both understanding and managing the soil–plant systems.

Botanists and ecologists have described root system architecture since at least the beginning of the twentieth century, and several authors (e.g. Cannon 1949; Weaver 1958; Kutschera 1960; Krasilnikov 1968) have produced "models" in the form of precise drawings and textual descriptions of root systems at various stages and in different environmental situations. Computer models appeared later, including the mono-dimensional approach of Lungley (1973) and the three-dimensional models proposed by Diggle (1988) and Pagès and Aries (1988). A large number of sophisticated architectural models have been developed since, combining plant developmental rules and soil characteristics in different ways (Fitter et al. 1991; Pagès et al. 1992; Berntson 1994; Clausnitzer and Hopmans 1994; Jourdan et al. 1995; Lynch et al. 1997; Somma et al. 1998; Thaler and Pagès 1998; Bidel et al. 2000; Dunbabin et al. 2002a, 2002b; Vercambre et al. 2003; Pagès et al. 2004, 2014; Collet et al. 2005; Wu et al. 2007; Leitner et al. 2010; Bingham and Wu 2011; Henke

The Root Systems in Sustainable Agricultural Intensification, First Edition.
Edited by Zed Rengel and Ivica Djalovic.

et al. 2014; Lobet et al. 2014; Postma et al. 2017; Barczi et al. 2018; Gérard et al. 2017; Schnepf et al. 2018). It is worth noting that most models were initially dedicated to a particular species (e.g. wheat: Diggle 1988; maize: Pagès et al. 1989; palm tree: Jourdan et al. 1995) and generic models appeared later (e.g. Somma et al. 1998; Dunbabin et al. 2003; Pagès et al. 2004; Postma et al. 2017). Meanwhile, an increasing number of root–soil interactions have been integrated in these modelling approaches, thereby extending their potential use towards diverse applications (e.g. Somma et al. 1998; Doussan et al. 1998, 2006; Javaux et al. 2008; Draye et al. 2010; Schnepf et al. 2012; Dunbabin et al. 2013; Ndour et al. 2017; Gérard et al. 2017).

In contrast to root-dedicated models, the "crop" models often represented root systems as simple mono-dimensional profiles of root length density versus depth (e.g. STICS model, Brisson et al. 1998). These representations were assumed to be sufficient in homogeneous mono-crop plots with high levels of fertilization and irrigation. Such representation of root systems resulted in a relatively low computational demand, so crop modellers were not keen on including complexities of root system architecture. However, average mono-dimensional profiles are hardly relevant to representing mixed and complex plant populations and dynamic and multi-faceted soil–plant interactions. Moreover, the recent trends in agricultural practices (e.g. use of intercropped species and minimal inputs) require models with realistic and detailed representations of the root systems (Hinsinger et al. 2011). In this context, plant-centred approaches and the architectural models of root systems become essential.

In the present chapter, we emphasize the main principles and specificities of 3D architectural models focused on root system development. We discuss the concepts shared by most models of this type and illustrate some examples provided by these tools. This presentation is complementary to the comprehensive recent review of Dunbabin et al. (2013) that described the functionalities of several 3D models of root architecture in detail. We concentrate on how the current models attempt to consider the combination of various plant species and environmental conditions and simulate plastic root systems whose characteristics integrate both the genetic diversity of plants species and dynamics of variable growing conditions. This will be illustrated by the simulation of simple scenarios with limitations imposed on the root system development due to soil constraints such as temperature and water availability.

14.2 Main Principles

14.2.1 The Root System as a Developing Branching System

Unlike the root density models (Page and Gerwitz 1974; Acock and Pachepsky 1996; De Willigen et al. 2002; Heinen et al. 2003; Dupuy et al. 2005; Bonneu et al. 2012) where the output is an abstract root spatial distribution described as an extensive variable defined in the soil volume (e.g. root length or root biomass per unit of soil volume), architectural models simulate virtual root systems in which root components are explicitly represented. More precisely, they simulate root systems as sets of connected root segments (or nodes) with several morphological attributes: spatial coordinates, age, diameter, etc. Even though fractal or other statistical models also describe root systems as branching patterns (Deans and Ford 1983; Henderson et al. 1983; Spek and Van Noordwijk 1994; Spek 1997;

Ozier-Lafontaine et al. 1999), we focus exclusively on the developmental models of root architecture. The time evolution of root systems is aimed at reproducing the effect of concurrent developmental processes acting on individual root entities, namely root elongation, acropetal branching, adventitious root emergence, radial growth, and root decay.

This modelling strategy, which represents root system architecture as the result of realistic developmental processes, is promising for several reasons:

i) a number of global characteristics of root systems emerge from their developmental construction,
ii) functional properties of root segments are tightly linked to their developmental stage and their position in the developmental processes, and
iii) a developmental vision facilitates straightforward dialogue and collaborations with other scientists working on plants (e.g. morphogenesis, carbon fluxes, nutrient uptake, and hydraulic architecture) and soils (especially biological and geochemical processes in the rhizosphere and bulk soil).

It is important to note that the architectural plant models that also incorporate various functional aspects are often called functional and structural plant models (FSPM). Such functional aspects are important as either inputs to the developmental processes (e.g. carbon fluxes, see Bidel et al. 2000) or outputs predicted by the interactions between the root system architecture and the environment (e.g. nutrient uptake, see Gérard et al. 2017).

14.2.2 Combining Developmental Processes

The dynamics of root system architecture hinges on the simultaneous action of several developmental processes. Part of the apparent complexity of root systems comes from this dynamic combination, in which multiple developmental processes operate simultaneously on a large number of distinct root entities (meristems and root segments) at different positions and in various stages.

The following developmental processes were represented in all architectural models initially: (i) the organized emergence of adventitious roots (since the RootMap model, see Diggle 1988, or the maize model, see Pagès et al. 1989); (ii) the root elongation, and (iii) root acropetal branching. Later on, Vercambre et al. (2003) and Collet et al. (2005) proposed to include additional processes, namely radial growth, root decay, and self-pruning, in order to increase the capacity of architectural models to simulate the diversity of plant species and functions and to represent later developmental stages, especially in trees (for a review see Tobin et al. 2007). These processes (radial growth, root decay, and self-pruning) also have an important impact on the whole root system architecture and functioning. For example, the abscission of dead roots or root parts is an important source of carbon to soil, whereas radial growth of proximal roots is essential for increasing their water-conducting capacity and their performance as plant anchors (Jenik 1978; Danjon et al. 1999; Nicoll et al. 2006).

14.2.3 Representing Diversity of Roots in a Single Root System

One of the major advantages of architectural models is the representation of root diversity that is inherent in every root system (see also Chapter 13). Such diversity (or heterorhizy, *sensu* Sutton and Tinus 1983) was recognized many years ago. It concerns several aspects

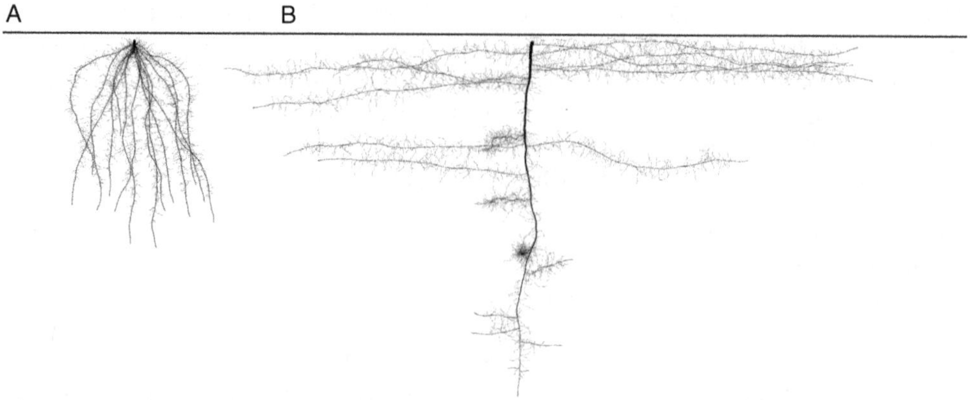

Figure 14.1 Examples of root systems illustrating the diversity of roots in single root systems. (a) Fasciculated root system with strong differentiation between long and thick adventitious roots connected to the stem, and much finer and shorter-branched roots along the adventitious ones; (b) taproot system, featuring the taproot, long and roughly horizontal lateral roots, and short lateral roots growing in all directions.

of the developmental capacity of roots and their subsequent contribution to various functions (e.g. Passot et al. 2016). This point is particularly obvious in many monocotyledonous plant species (Figure 14.1a) that exhibit strong differentiation between (i) the main roots originating from the stem base, with high elongation and branching capacities, and (ii) most of their lateral roots that elongate slowly and over a short term only. Thus, root differentiation is mainly accounted for by the position of a specific root within the branching pattern. Another important example is the root system of young trees (Kahn 1977; Jenik 1978; Danjon et al. 1999; Collet et al. 2005; see also Figure 14.1b), with a contrasting behaviour between a taproot that exhibits strong positive gravitropism (vertical downward direction of growth) and some of the lateral roots exhibiting plagiotropism (horizontal growth). In order to formalize this diversity of roots, the root population is usually divided into a number of categories corresponding to branching orders or to predefined types (see more detailed explanations below). An exception to this general utilization of root categories is the ArchiSimple model (Pagès et al. 2014), in which developmental characteristics of individual roots are determined by the root apical diameter, a continuous variable attributed to each root at emergence, according to the root position in the whole architecture. In this model, the variable apical diameter is taken to represent root diversity.

14.2.4 Representing the Topological Relationships between Root Types

Some topological properties of root systems are obvious, such as the existence of between-root connections, or the fact that lateral branch roots are usually finer and less vigorous than the mother roots. These general (and obvious) properties are captured and made explicit in architectural models, and they are important because of their huge impact on the whole structure of root systems. For example, the fact that roots are connected induces a spatial aggregation (Grabarnik et al. 1998) and important overlaps of the rhizospheres of neighbouring roots (Ge et al. 2000; Pagès 2011).

The topological relationships in the root system were often modelled using developmental branching orders (Diggle 1988; Clausnitzer and Hopmans 1994; Lynch et al. 1997), whereby roots of order i give rise to roots of order $i+1$ by lateral branching, and the total number of branching orders is usually given by a predefined value. In this class of models, one root category is associated with one branching order and is given a set of input parameters to quantify common developmental characteristics. In other models (Jourdan et al. 1995; Pagès et al. 2004), root categories (or types) are defined independently of branching orders. In this case, the relationships associated with branching are defined explicitly: each root type can give rise to a number of other (or identical) root types. Such a formalization is quite general. It enhances the flexibility of the topological relationships that can be modelled, but it may also increase the number of model parameters if the number of root types is large. In contrast to the above models, the ArchiSimple model (Pagès et al. 2014) does not use the concept of root categories; instead, the diversity of roots is linked to the variations in apical diameter.

14.2.5 Various Roots Encounter Different Endogenous and Exogenous Conditions

Another important aspect that increases the complexity and plasticity of real root systems is that local and transient environmental conditions modulate the developmental processes in each root. Most of the recent 3D architectural models make it possible to represent the dynamic responses of root entities (root segments and/or root meristems) to these conditions during root development. The environment can be defined through endogenous variables (e.g. photoassimilate availability) and through time-fluctuating and spatially variable exogenous soil conditions (e.g. soil water potential, nutrient concentration, and temperature) at the location of the root entities. The recordings, in computer memory, of the root entities as objects with a number of individual attributes is a decisive advantage to allow the integration of sub-models of these responses, providing that appropriate information about environmental variables is available. Various models were thus constructed to take into account (i) the effects of temporal variation in photoassimilate availability (e.g. Clausnitzer and Hopmans 1994; Thaler and Pagès 1998; Bidel et al. 2000; Dunbabin et al. 2003) on root elongation, and (ii) the local and potentially transient influences of the soil (Somma et al. 1998; Dunbabin et al. 2002b; Henke et al. 2014; Schnepf et al. 2018). We illustrate this major point below by simulating the scenarios applied to two different species.

14.2.6 Stochastic Representation of Processes

In majority of the existing models of root systems, several processes are represented as stochastic, so that the simulated root systems have a number of random characteristics (Diggle 1988; Pagès et al. 1992; Leitner et al. 2010). For example, root trajectories have stochastic components (e.g. initial angles or deflexions during growth). In some models (Pagès et al. 1992), the intensity of root elongation is also modelled with stochastic components. Introducing stochastic components in the architectural models was an efficient way to produce realistic simulations without considering the fine-scale determinism of the observed variability. For example, the fine-scale geometry of soil pores contributes to the precise root trajectories, but the root-soil model cannot operate at this scale for practical reasons. From a methodological point of view, it remains difficult to map the root environment at the root-tip scale, that is a scale ranging between 0.1 and 1 mm. Regarding root elongation rate, it has been shown in several species that even neighbouring roots emerging acropetally on the same mother root exhibit differences in elongation rate (MacLeod 1990; Forde 2009). According to Forde (2009) or Pagès (2011), these variations among successive laterals are worth considering because they can have a significant influence on the foraging performance. Moreover, the magnitude of this variability might be genetically controlled (Forde 2009).

14.3 Applications

14.3.1 Simulating Various Developmental Strategies

In their classification of root systems, a number of authors (e.g. Cannon 1949; Krasilnikov 1968; Kahn 1977) pointed out different dynamic "strategies" of plants in developing their roots. Many plant species develop the entire root system from a single seminal root (the radicle) that becomes the central taproot (Figure 14.2a). This type of root system is called the primary root system (Cannon 1949). For other species, the seminal root system is a juvenile organ that reaches a limited level of development. It has a transient role and is complemented by vigorous shoot-borne (adventitious) roots that quickly become the essential components of the root system (Cannon 1949). This type of root system (called secondary root system) can give rise to a caespitose system when the adventitious roots appear close to each other, as in many cereals (Figure 14.2b), or it can be dispersed when roots appear along a creeping shoot or an elongating rhizome (Figure 14.2c). Many plant

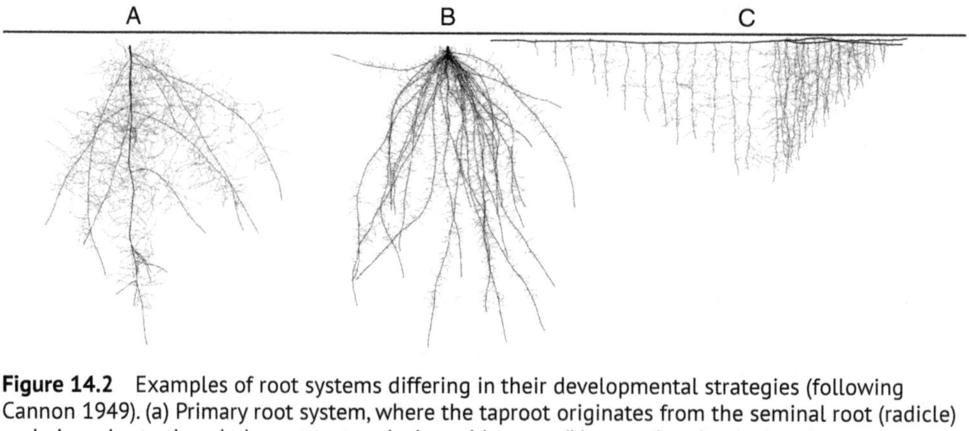

Figure 14.2 Examples of root systems differing in their developmental strategies (following Cannon 1949). (a) Primary root system, where the taproot originates from the seminal root (radicle) and gives rise to the whole root system by branching out; (b) secondary fasciculated root system, in which the seminal root system is complemented by the emergence of strong adventitious roots near the plant collar coordinated with shoot development; (c) secondary dispersed root system, in which the seminal root system is complemented by the emergence of adventitious roots along creeping shoots or rhizomes.

species combine the primary and secondary root systems (mixed root systems), whereby the balance between the two types may vary according to plant species and environmental conditions (Kahn 1977).

Pagès et al. (2004) illustrated how these various root development strategies impact the whole root system structure and could be represented by a single generic model (such as RootTyp). Several species exemplifying the different cases were simulated and compared to the real root systems drawn by Kutschera (1960). Another consequence is shown in Figure 14.3, which presents the distribution of the age of root segments in the top layer of soil (0–20 cm) for the three types of root systems simulated in Figure 14.2. We have chosen to present the distribution of the age of root segments because of the functional importance of root age regarding water and nutrient uptake and exudation of organic and inorganic compounds. In topsoil, the primary root system (Figure 14.3a) has mostly mid-age segments, whereas the secondary dispersed root system (Figure 14.3c) has a majority of young and very young root parts because of the continuous emergence of new main roots from the rhizome. The secondary centralized root system (Figure 14.3b) shows an intermediate age pattern of root segments, dependent on the kinetics of adventitious root emergence. Thus, we can expect differences in the functioning of these three types of root systems not only because of the differences regarding the spatial distribution of their roots but also because the age distribution of root segments is different.

Figures 14.2 and 14.3 illustrate the interest in simulating the dynamic root structures based on the developmental strategies underpinned by the botanical knowledge. These simulated root systems have emergent spatial and structural properties that are important in functional ecology. Hence, this is an example of how a dialogue between botanists and ecologists can be formalized by the modelling approach.

14.3.2 Simulating Root Developmental Responses to the Shoot and Soil Environments Using the ArchiSimple Model

Although structural properties of root systems are determined strongly by the developmental processes, it is also well documented that root systems are plastic to a range of environmental factors that vary in space and time. The magnitude of such variations is usually high in the soil environments, providing evolutionary, and developmental forces that contributed to the general complexity of root systems, and making models absolutely necessary for building synthetic representations of these complex root systems. In this section, we illustrate the capacity of the root models to simulate the plausible effects of two variables that have a demonstrated influence on root system dynamics: soil temperature and soil water content. Our aim is not to accurately represent particular situations but to show the potential benefit of simulation approaches in studying real and complex systems.

14.3.2.1 Simulation of Different Plant Species

Using ArchiSimple model, we simulated two different plant species growing in a homogeneous and favourable soil for 90 days. The potential growth of roots was limited (10% reduction) only by the provision of shoot resources. In this architectural model, potential elongation is calculated at each time step for each individual root from its apical diameter and can be reduced by the limited photoassimilate availability at the whole root system

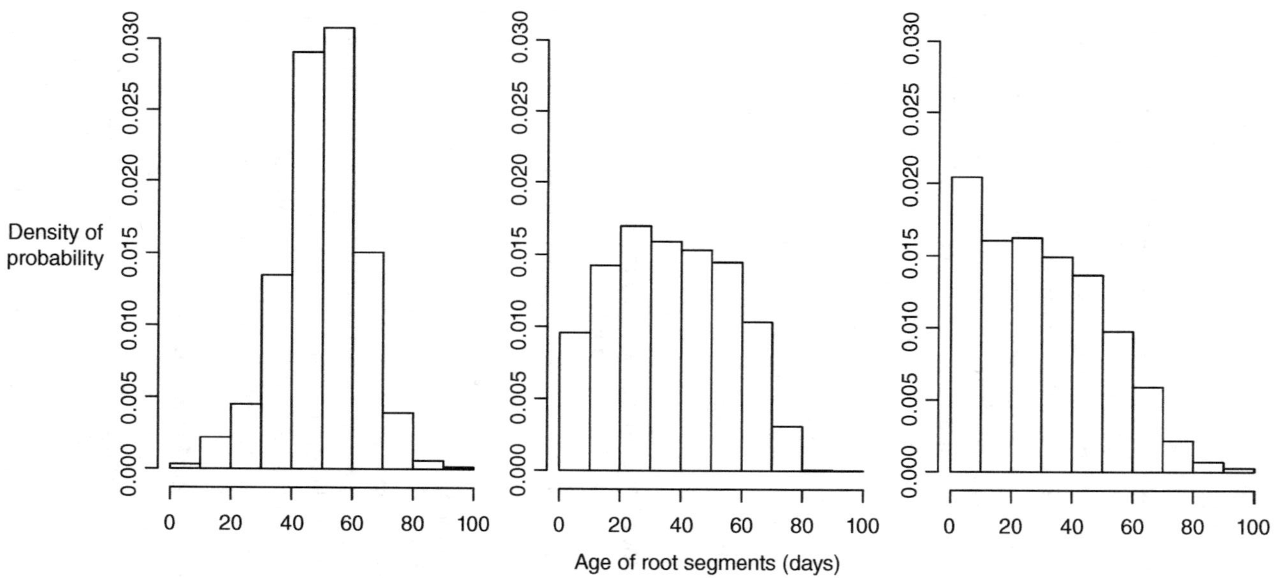

Figure 14.3 Distribution of the ages of root segments in the topsoil layer (between 0 and 20 cm) for the three types of root systems. (a) The primary root system (see Figure 14.2a); (b) the secondary fasciculated root system (see Figure 14.2b); (c) the secondary dispersed root system (see Figure 14.2c).

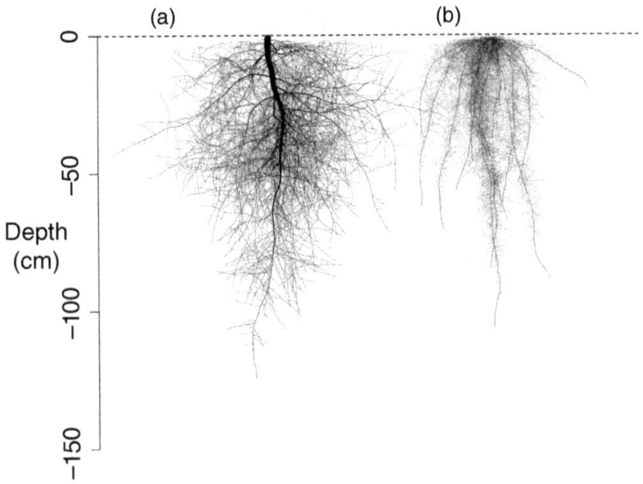

Figure 14.4 Two types of simulated root systems grown in a homogeneous and favourable soil for 90 days. (a) Typical dicotyledonous plant root system (e.g. pea, family Fabaceae); (b) typical monocotyledonous plant root system (e.g. wheat, family Poaceae). Very fine roots are not shown to increase the visibility of the overall root structures.

level. Figure 14.4 presents the orthogonal projections of the 3D structures. The first plant (Figure 14.4a) could represent a dicotyledonous plant species (e.g. from the Fabaceae family), whereas the second (Figure 14.4b) depicts a monocotyledonous species (e.g. from the Poaceae family). The precise values of parameters used in the simulations are not important in this discussion, but the main differences between the two are worth noting in order to interpret the simulations. The root system in Figure 14.4a is a primary system developed from the radicle, whereas the root system in Figure 14.4b is a secondary root system, with the initial seminal system being complemented during plant growth by the emergence of several adventitious roots. Radial growth was activated for the root system (a) but not for the system (b). Radial growth has a strong impact on biomass allocation throughout the root system. Moreover, in the root system (a), the branch and the parent roots have a comparable level of development. In the root system (b), a large differentiation was simulated between the branch and the parent roots, with most branch lateral roots having a determinate growth pattern and remaining short. Lastly, branching density along the roots was higher in the root system (b) than (a).

The root length distribution along the soil profile (see Figure 14.5a) is easy to calculate from the 3D simulations described in the preceding paragraph. The global shapes of these virtual root length profiles are comparable to those obtained experimentally (Pagès et al. 2012). A majority of roots are found in the upper soil layers, especially around 10 cm for the monocotyledonous species (e.g. wheat) simulation, and between 10 and 40 cm for the dicotyledonous species (e.g. pea) simulation. Lower in the soil, a decrease in length density is much steeper for pea than wheat, especially between 40 and 60 cm. However, the maximal rooting depth is higher for pea (about 120 cm) than wheat.

The presented simulations illustrate how root architectural diversity can be analyzed and depicted by considering (i) variations in several developmental processes and

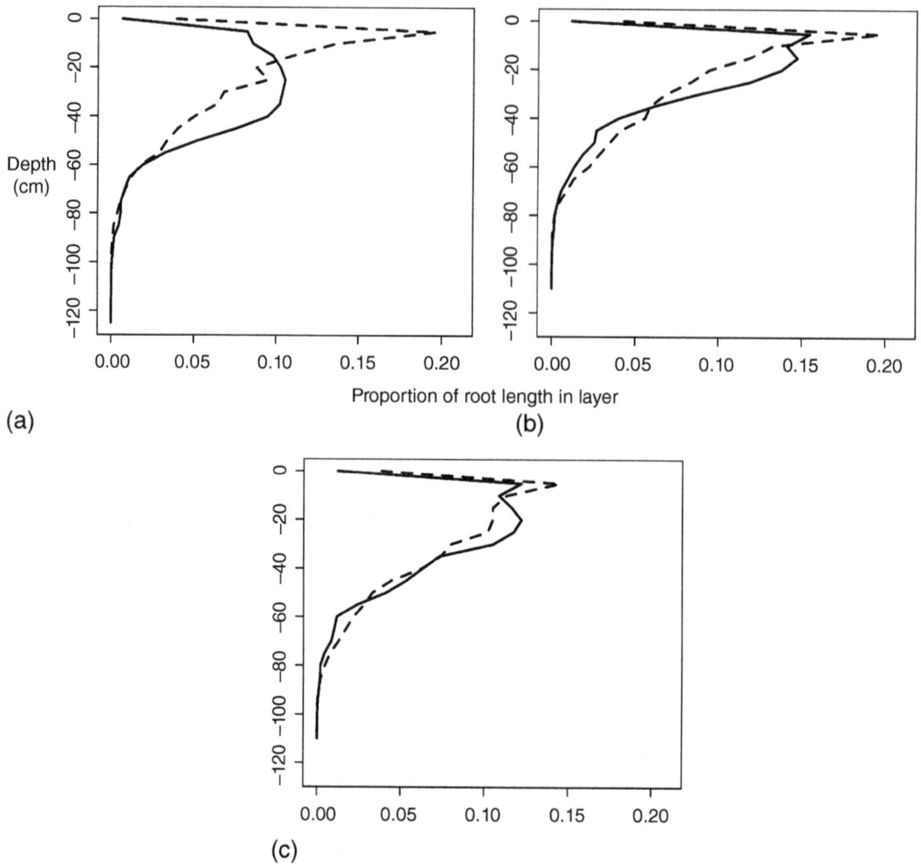

Figure 14.5 Profiles of root length versus depth for the two simulated plant species and the three scenarios. The proportion of length is calculated in 5-cm-thick layers. The solid lines represent a dicotyledonous species (see Figure 14.4a), and the dashed lines represent a monocotyledonous species (see Figure 14.4b). (a) Homogeneous and favourable soil; (b) soil with a gradient of temperature; (c) soil with a gradient of temperature and gradual drying.

(ii) quantitative variation in input parameters. These combinations can give an infinite number of simulated root system architectures. The important developmental processes included in the model are obtained by the dynamic observation of root systems. Realistic sets of input parameters must be defined using sensitivity analyses and calibration in relevant conditions.

14.3.2.2 Simulation of the Temperature Effect

Under temperate climates, the effect of temperature on root growth can be important, given that (i) significant gradients of soil temperature with depth are usually observed during the growth period in spring and summer and (ii) root developmental rates are sensitive to temperature. Thus, the soil temperature gradient is an example of a relevant

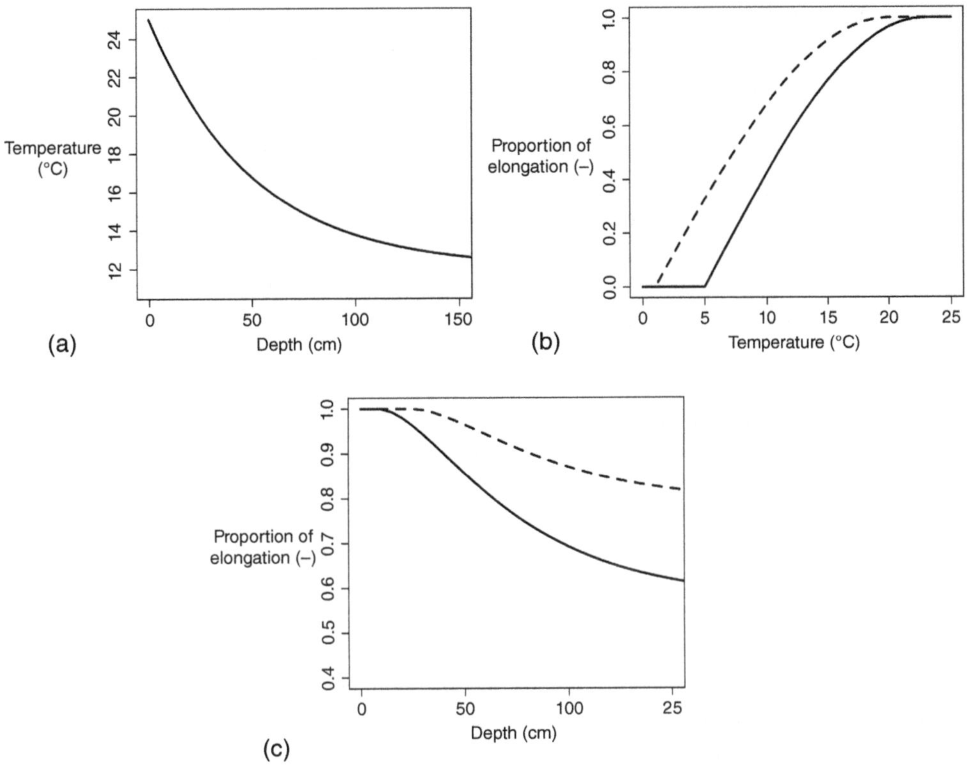

Figure 14.6 Effect of soil temperature. (a) Temperature gradient in soil, typical of a late-spring gradient in central France; (b) response of the elongation rate of individual roots to temperature as a proportion of the maximal elongation rate at optimal temperature; (c) subsequent relationship between proportion of individual root elongation rate and depth. The solid lines represent a dicotyledonous species (see Figure 14.4a), and the dashed lines represent a monocotyledonous species (see Figure 14.4b). Results of the simulation (including the effect of temperature) are given in Figure 14.5b.

environmental variable worth using as an input to simulate the dynamics of root systems in real field situations (Porter et al. 1986). Figure 14.6a presents a typical temperature gradient in the soil in late spring in central France. The putative response of root growth to soil temperature is represented in Figure 14.6b for the two species using a common model response and slightly different parameters: we assume in this case that a monocotyledonous species is better adapted to cooler climates than a dicotyledonous species. It is well known that cultivated species, and even varieties of the same species, exhibit different temperature adaptations. Figure 14.6c presents a resulting impact of the temperature gradient down the soil profile on a potential elongation rate of each individual root.

New simulations with the model using this temperature input and the response model provided new results presented in Figure 14.5b. The total root length of simulated root systems was decreased by the temperature gradient: 10% in a monocotyledonous species (e.g. from family Poaceae) and 25% in a dicotyledonous species (e.g. from family Fabaceae).

When comparing Figure 14.5a, b, we also see the overall impact on root length distribution: rooting depth was more strongly decreased in a Fabaceae species, and the root length distribution was shifted towards upper soil horizons. The effect was less obvious in a Poaceae species.

Such simulations can be used, for example, to compare root distributions of the same species (or genotypes) at different latitudes in order to specifically quantify the effect of temperature on integrated root distribution. It should be borne in mind that the calibration of architectural models is usually done using pot-grown plants that are not subjected to the same temperature variations as in field soils. Thus, it is necessary to take into account this temperature effect to predict the magnitude of the specific variables in field situations.

14.3.2.3 Simulation of Soil Drying

Another common phenomenon affecting most fields under temperate climates is gradual soil drying during the growing season from spring to summer. It is usually due to the combination of a decrease in rainfall and an increase in water extraction from the soil by plant roots. Various models have been proposed to represent the effect of soil water content on root growth (e.g. Jones et al. 1991), but they cannot represent all the relevant situations because this effect depends on soil type, plant species (or genotype) and on the general status of the plant considered. Therefore, we propose to make a speculative simulation considering that soil drying occurred gradually, from the beginning to the end of the simulation, and was linearly related to soil depth between 0 and 80 cm, with more intense drying in the surface layers (more roots and more evaporation). Figure 14.7a presents a decrease in individual root elongation due to soil drying, from the beginning of the simulation (no effect at day 0) to the end (a 50% decrease in root elongation in the surface layer at day 90). We assumed a linear time evolution and a linear effect of depth, as shown in Figure 14.7a. This effect was assumed to be common to both plant species. In addition, we assumed that the combined reduction due to drying and to temperature could be calculated by a simple multiplication of the reduction coefficients (Figure 14.7b).

The results of simulations are presented in Figure 14.5b, c. The maximal rooting depth was not affected by soil drying according to the second scenario (temperature gradient and soil drying, Figure 14.5c) in comparison with the scenario with the temperature gradient alone (Figure 14.5b). Conversely, root distributions were shifted relatively deeper (Figure 14.5b, c), as the consequence of the dynamic response of the root system to its drying environment. From a qualitative point of view, this result could be expected because the elongation of deeper roots was favoured. However, the quantification of the effect at this integrated scale is not straightforward because it is sensitive to (i) the precise dynamics of soil drying and (ii) the root developmental properties.

In the three examples presented above, we can see that modelling and simulation of the root system architecture is an interesting tool to link the knowledge and the data that can be obtained at the root level to the integrated effects at plant and population levels during the whole growing season. This is a major point, especially for root systems, because data are very costly to obtain in field conditions. Sensitivity analyses using simulation models are a classical way for establishing systematic relationships between input parameters, environmental variables, and various outputs.

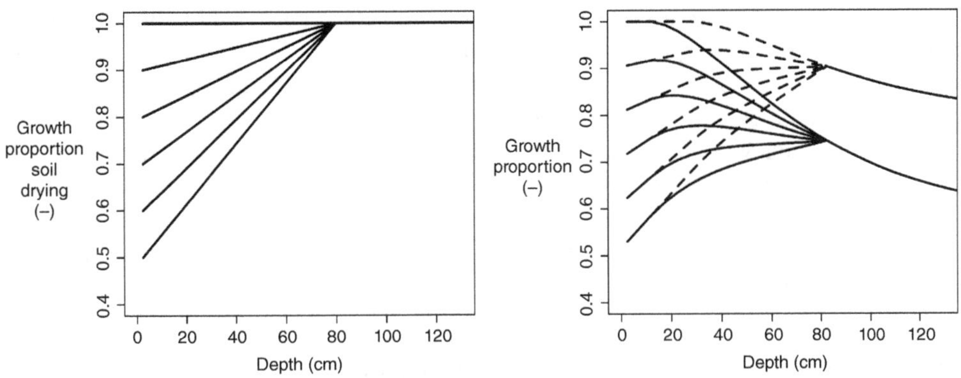

Figure 14.7 Effect of temperature gradient and gradual soil drying. (a) Putative effects of gradual soil drying on the proportion of potential root elongation rate at six different dates spaced uniformly along the growing season (day 0: top line; day 90: bottom line). The response to soil drying is supposed to be the same for a dicotyledonous and a monocotyledonous species. (b) Combination of the effect of soil drying and temperature gradients (similar to that in Figure 14.5) on the proportion of potential elongation rate at six different dates spaced uniformly along the growing season (day 0: top curve; day 90: bottom curve). The solid lines/curves represent a dicotyledonous species (see Figure 14.4a), and the dashed lines/curves represent a monocotyledonous species (see Figure 14.4b). Results of this simulation (including the combined effects of temperature and soil drying) are given in Figure 14.5c.

14.4 Perspectives

14.4.1 Contribution to Plant Breeding in the Frame of New Agro-ecosystems

There is a considerable potential for breeding plants with enhanced root foraging performance based on exploiting root system diversity and plasticity. Indeed, the genotype selection and cultivar breeding during the recent decades was done in the context of high input agriculture without considering the plant hidden half, selecting against plant genotypes that invest a relatively large proportion of their resources in the root system (de Dorlodot et al. 2007). To reverse this trend (and recognize an increasingly important role of roots in providing resources for plant growth in less-than-ideal conditions), the simulation approaches, that synthesize virtual root systems by integrating the developmental (genetically controlled) characteristics and the environmental attributes and cues that modulate root development, have a central role to play. By combining the validated 3D root models with the use of intensive simulation tools and optimization methods, it will be possible to create efficient ideotypes for the target situations, that is, combinations of environments and agricultural practices.

To be effective in this context, root architectural models should fulfil a number of necessary conditions: (i) they must correctly simulate the relevant attributes that determine the foraging performance of root systems and respond to the environmental variables that modulate this performance; (ii) they must be simple enough to allow the estimation of parameters in a large number of potential genotype/edaphic/climatic conditions and allow usage of *ad hoc* calculation techniques; (iii) the input parameters that characterize the simulated species or genotypes must be really representative of the genotype or species and must have low sensitivity to environmental variations in the target range; and (iv) the input parameters should be easy to identify independently of each other.

The challenging work remains to be done to increase the performance of models regarding all these criteria. Moreover, coupling architectural models with other models, (e.g. water uptake) to predict the target functions (e.g. grain yield) may be necessary in order to evaluate the performance of the root systems of interest more accurately (Draye et al. 2010; Meunier et al. 2017).

14.4.2 Methodological Contribution to Phenotyping

In order to conduct more intensive work on root systems, it is obviously important to improve our capacity to characterize them (Chen et al. 2011) and to be able to do so in the relevant growth stages and under relevant growing conditions, that is when they have reached an advanced stage of development and expressed their potential (Pagès and Pellerin 1994).

Several reports (Tsegaye et al. 1995; Pagès and Pellerin 1996; Pellerin and Pagès 1996; Grabarnik et al. 1998; Chen et al. 2011; Garré et al. 2012; Vansteenkiste et al. 2013) have shown the usefulness of models of the root system architecture in facilitating experimental work (i) as a guiding link to appropriate measurements, (ii) to decide on sampling designs, and (iii) to link and interpret the data collected. Reversely, experimental work benefited root architectural models as well, providing the data for model calibration and validation for a variety of purposes, plant species and genotypes, and growing conditions.

When possible, conducting a virtual experiment before a real one allows optimization of treatments, designs, and data collection and analysis. To achieve this, the realistic nature of three-dimensional root system simulations is paramount, emphasizing the quality of the geometrical attributes (root length, root direction, and interbranch distances) of the virtual root systems. Moreover, the use of images has become a central point in root phenotyping. Using the root system images that can be synthesized from the three-dimensional root system simulations, it is possible to compare real and virtual images and to improve the automatic estimation of architectural variables in real images (Lobet et al. 2017) (see also Chapter 13).

A bridge established by these models between the local (easy-to-measure) traits of *roots*, and global traits of *the root system* (generally more difficult and costly to obtain), is particularly helpful. If one wants to use the bridge in both directions, it is again indispensable to use a simple model, with a parsimonious parameterization. The traits that are measured at the root level must be carefully chosen and standardized to promote their biological significance and to test the strength of their genetic determinism (Pagès and Kervella 2018).

14.4.3 Contribution to the Creation of New Models of the Agro-ecosystem Functioning

Usually, it is not easy to integrate detailed models of root system architecture into larger models dedicated to the functioning of plant populations interacting with their environment, without risking a drastic increase in computation time. Moreover, such models of plant populations must be homogeneous regarding the level of detail for different compartments.

This difficulty can be circumvented using two strategies: (i) the integration of a simplified architectural model into a global model as a particular module; or (ii) developing a new generation of simple plant-centred root system model, containing the global and essential attributes of the root system. For the latter strategy, the detailed architectural models can play a role in the design and calibration of individual modules, creating meta-models (Pagès et al. 2012).

14.5 Conclusions

The three-dimensional models of root system architecture have been in existence for about 30 years. They were considered rather weird initially, but nowadays, they attract considerable interest from scientists in many cognate disciplines. The current models can provide detailed and dynamic representation of diverse root systems. Increasingly, sophisticated models that integrate a number of developmental and functional aspects are being produced, simulating the diversity and plasticity of roots subjected to spatially and temporally heterogeneous environments, and extending the range of applications and uses. However, as in other modelling exercises, the simplicity and efficiency in all the incorporated processes remain pivotal to creating the valuable tools. Therefore, from "weird," these models are becoming the "current" or even "indispensable" tools in increasing our understanding of root development and functioning.

References

Acock, B. and Pachepsky, Y.A. (1996). Convective-diffusive model of two-dimensional root growth and proliferation. *Plant and Soil* 180: 231–240.

Barczi, J.F., Rey, H., Griffon, S. et al. (2018). DigR: a generic model and its open source simulation software to mimic three-dimensional root-system architecture diversity. *Annals of Botany* 121: 1089–1104.

Berntson, G.M. (1994). Modelling root architecture: are there tradeoffs between efficiency and potential of resource acquisition? *New Phytologist* 127: 483–493.

Bidel, L.P.R., Pagès, L., Rivière, L.M. et al. (2000). MassFlowDyn I: a carbon transport and partitioning model for root system architecture. *Annals of Botany* 85: 869–886.

Bingham, I.J. and Wu, L. (2011). Simulation of wheat growth using the 3D root architecture model SPACSYS: validation and sensitivity analysis. *European Journal of Agronomy* 34: 181–189.

Bonneu, A., Dumont, Y., and Rey, H. (2012). A minimal continuous model for simulating growth and development of plant root systems. *Plant and Soil* 354: 211–227.

Brisson, N., Mary, B., Ripoche, D. et al. (1998). STICS: a generic model for the simulation of crops and their water and nitrogen balances. I. Theory and parameterization applied to wheat and corn. *Agronomie* 18: 311–346.

Cannon, W.A. (1949). A tentative classification of root systems. *Ecology* 30: 542–548.

Chen, Y.L., Dunbabin, V.M., Postma, J.A. et al. (2011). Phenotypic variability and modelling of root structure of wild *Lupinus angustifolius* genotypes. *Plant and Soil* 348: 345–364.

Clausnitzer, V. and Hopmans, J.W. (1994). Simultaneous modeling of transient three-dimensional root growth and soil water flow. *Plant and Soil* 164: 299–314.

Collet, C., Löf, M., and Pagès, L. (2005). Root system development of oak seedlings analysed using an architectural model. Effects of competition with grass. *Plant and Soil* 258: 1–17.

Couvreur, V., Vanderborght, J., and Javaux, M. (2012). A simple three-dimensional macroscopic root water uptake model based on the hydraulic architecture approach. *Hydrology and Earth System Sciences* 16: 2957–2971.

Danjon, F., Sinoquet, H., Godin, C. et al. (1999). Characterization of structural tree root architecture using 3D digitizing and AMAPmod software. *Plant and Soil* 211: 241–258.

Deans, J.D. and Ford, E.D. (1983). Modelling root structure and stability. *Plant and Soil* 71: 189–195.

Diggle, A.J. (1988). ROOTMAP – a model in three-dimensional coordinates of the growth and structure of fibrous root systems. *Plant and Soil* 105: 169–178.

de Dorlodot, S., Forster, B., Pagès, L. et al. (2007). Root system architecture: opportunities and constraints for genetic improvement of crops. *Trends in Plant Science* 12: 474–482.

Doussan, C., Vercambre, G., and Pagès, L. (1998). Modelling the hydraulic architecture of root systems: an integrated approach of water absorption. II. Distribution of conductances and consequences. *Annals of Botany* 81: 225–232.

Doussan, C., Pierret, A., Garrigues, E. et al. (2006). Water uptake by plant roots: II – Modelling of water transfer in the soil root-system with explicit account of flow within the root system – Comparison with experiments. *Plant and Soil* 283: 99–117.

Draye, X., Kim, Y., Lobet, G. et al. (2010). Model–assisted integration of physiological and environmental constraints affecting the dynamic and spatial patterns of root water uptake from soils. *Journal of Experimental Botany* 61: 2145–2155.

Dunbabin, V., Diggle, A., Rengel, Z. et al. (2002a). Modelling the interactions between water and nutrient uptake and root growth. *Plant and Soil* 239: 19–38.

Dunbabin, V., Diggle, A., and Rengel, Z. (2002b). Simulation of field data by a three-dimensional model of interactive root growth. *Plant and Soil* 239: 39–54.

Dunbabin, V., Diggle, A., and Rengel, Z. (2003). Is there an optimal root architecture for nitrate capture in leaching environments? *Plant Cell Environment* 26: 835–844.

Dunbabin, V.M., Postma, J., Schnepf, A. et al. (2013). Modelling root-soil interactions using three-dimensional models of root growth, architecture and function. *Plant and Soil* 372: 93–124.

Dupuy, L., Fourcaud, T., Stokes, A. et al. (2005). A density-based approach for the modelling of root architecture: application to maritime pine (*Pinus pinaster* Ait.) root systems. *Journal of Theoretical Biology* 236: 323–334.

Fitter, A.H., Stickland, T.R., Harvey, M.L. et al. (1991). Architectural analysis of plant root systems. architectural correlates of exploitation efficiency. *New Phytologist* 118: 375–382.

Forde, B.G. (2009). Is it good noise? The role of developmental instability in the shaping of a root system. *Journal of Experimental Botany* 60: 3989–4002.

Garré, S., Pagès, L., Laloy, E. et al. (2012). Parameterizing the root system development of spring barley using minirhizotron data. *Vadose Zone Journal* 11 https://doi.org/10.2136/vzj2011.0179.

Ge, Z.Y., Rubio, G., and Lynch, J.P. (2000). The importance of root gravitropism for inter-root competition and phosphorus acquisition efficiency: results from a geometric simulation model. *Plant and Soil* 218: 159–171.

Gérard, F., Blitz-Frayret, C., Hinsinger, P. et al. (2017). Modelling the interactions between root system architecture, root functions and reactive transport processes in soil. *Plant and Soil* 413: 161–180.

Grabarnik, P., Pagès, L., and Bengough, A.G. (1998). Geometrical properties of simulated maize root systems: consequences for length density and intersection density. *Plant and Soil* 200: 157–167.

Heinen, M., Mollier, A., and De Willigen, P. (2003). Growth of a root system described as diffusion. II. Numerical model and application. *Plant and Soil* 252: 251–265.

Henderson, R., Ford, E.D., and Renshaw, E. (1983). Morphology of the structural root system of Sitka spruce. Computer simulation of rooting pattern. *Forestry* 56: 137–153.

Henke, M., Sarlikioti, V., Buck-Sorlin, G. et al. (2014). Exploring root developmental plasticity to nitrogen with a three-dimensional architectural model. *Plant and Soil* 385: 49–62.

Hinsinger, P., Brauman, A., Devau, N. et al. (2011). Acquisition of phosphorus and other poorly mobile nutrients by roots. Where do plant nutrition models fail? *Plant and Soil* 348: 29–61.

Javaux, M., Schröder, T., Vanderborght, J. et al. (2008). Use of a three-dimensional detailed modeling approach for predicting root water uptake. *Vadose Zone Journal* 7: 1079–1088.

Jenik, J. (1978). Roots and root systems in tropical trees: morphologic and ecologic aspects. Fourth Cabot symposium. Harvard forest, Petersham, Massachusetts, 1976/04/26-30. In: Tropical Trees as Living Systems, vol. 14 (eds. P.B. Tomlinson and M.H. Zimmermann), 323–349. Cambridge, UK: Cambridge University Press.

Jones, C.A., Bland, W.L., Ritchie, J.T. et al. (1991). Simulation of root growth. In: Modeling Plant and Soil Systems-Agronomy Monograph No. 31, 91–123. Madison, WI: ASA-CSSA-SSSA.

Jourdan, C., Rey, H., and Guédon, Y. (1995). Architectural analysis and modelling of the branching process of the young oil-palm root system. *Plant and Soil* 177: 63–72.

Kahn, F. (1977). Analyse structurale des systèmes racinaires des plantes ligneuses de la forêt tropicale dense humide. *Candollea* 32: 321–358.

Krasilnikov, P.K. (1968). On the classification of the root system of 261 trees and shrubs. In: Methods of Productivity Studies in Root Systems and Rhizosphere Organisms (eds. M.S. Ghilanov, V.A. Kovda, L.N. Novichkova-Ivanova, et al.), 106–114. Leningrad: Nauka (URSS Academy of Sciences).

Kutschera, L. (1960). Wurzelatlas Mitteleuropäischer Ackerunkräuter und Kulturpflanzen. Frankfurt, Germany: DLG Verlag.

Leitner, D., Klepsch, S., Bodner, G. et al. (2010). A dynamic root system growth model based on L-systems. *Plant and Soil* 332: 177–192.

Lobet, G., Pagès, L., and Draye, X. (2014). A modeling approach to determine the importance of dynamic regulation of plant hydraulic conductivities on the water uptake dynamics in the soil-plant-atmosphere system. *Ecological Modelling* 290: 65–75.

Lobet, G., Koevoets, I.T., Nol, M. et al. (2017). Using a structural root system model to evaluate and improve the accuracy of root image analysis pipelines. *Frontiers in Plant Science* 8: 447.

Lungley, D.R. (1973). The growth of root systems – a numerical computer simulation model. *Plant and Soil* 38: 145–159.

Lynch, J.P., Nielsen, K.L., Davis, R.D. et al. (1997). SimRoot: modelling and visualization of root systems. *Plant and Soil* 188: 139–151.

MacLeod, R.D. (1990). Lateral root primordium inception in *Zea mays* L. *Environmental and Experimental Botany* 30: 225–234.

Meunier, F., Couvreur, V., Draye, X. et al. (2017). Towards quantitative root hydraulic phenotyping: novel mathematical functions to calculate plant-scale hydraulic parameters from root system functional and structural traits. *Journal of Mathematical Biology* 75: 1133–1170.

Ndour, A., Vadez, V., Pradal, C. et al. (2017). Virtual plants need water too: functional-structural root system models in the context of drought tolerance breeding. *Frontiers in Plant Science* 8: 1577.

Nicoll, B.C., Berthier, S., Achim, A. et al. (2006). The architecture of *Picea sitchensis* structural root systems on horizontal and sloping terrain. *Trees Structure Function* 20: 701–712.

Ozier-Lafontaine, H., Lecompte, F., and Sillon, J.F. (1999). Fractal analysis of the root architecture of *Gliricidia sepium* for the spatial prediction of root branching, size, and mass. Model development and evaluation in agroforestry. *Plant and Soil* 209: 167–180.

Page, E.R. and Gerwitz, A. (1974). Mathematical models, based on diffusion equations, to describe root systems of isolated plants, row crops, and swards. *Plant and Soil* 41: 243–254.

Pagès, L. and Aries, F. (1988). SARAH: modèle de simulation de la croissance, du développement, et de l'architecture des systèmes racinaires. *Agronomie* 8: 889–896.

Pagès, L., Jordan, M.O., and Picard, D. (1989). Simulation of the three-dimensional architecture of the maize root system. *Plant and Soil* 119: 147–154.

Pagès, L., Chadoeuf, J., and Kervella, J. (1992). Modélisation stochastique de la croissance et du développement racinaire de jeunes pêchers (*Prunus persica* Batsch). 1. Estimation et validation du modèle. *Agronomie* 12: 447–458.

Pagès, L. and Pellerin, S. (1994). Evaluation of parameters describing the root system architecture of field grown maize plants (*Zea mays* L.). II. Density, length, and branching of first-order lateral roots. *Plant and Soil* 164: 169–176.

Pagès, L. and Pellerin, S. (1996). Study of differences between vertical root maps observed in a maize crop and simulated maps obtained using a three-dimensional model of the root system architecture. *Plant and Soil* 182: 329–337.

Pagès, L., Vercambre, G., Drouet, J.L. et al. (2004). RootTyp: a generic model to depict and analyse the root system architecture. *Plant and Soil* 258: 103–119.

Pagès, L. (2011). Links between root developmental traits and foraging performance. *Plant, Cell & Environment* 34: 1749–1760.

Pagès, L., Bruchou, C., and Garré, S. (2012). Links between root length density profiles and models of the root system architecture. *Vadose Zone Journal* 11 https://doi.org/10.2136/vzj2011.0152.

Pagès, L., Bécel, C., Boukcim, H. et al. (2014). Calibration and evaluation of ArchiSimple, a parsimonious model of the root system architecture. *Ecological Modelling* 290: 76–84.

Pagès, L. and Kervella, J. (2018). Seeking stable traits to characterize the root system architecture. Study on 60 species located at two sites in natura. *Annals of Botany* 122: 107–115.

Passot, S., Gnacko, F., Moukouanga, D. et al. (2016). Characterization of pearl millet root architecture and anatomy reveals three types of lateral roots. *Frontiers in Plant Science* 7: 829.

Pellerin, S. and Pagès, L. (1996). Evaluation in field conditions of a three-dimensional architectural model of the maize root system: comparison of simulated and observed horizontal root maps. *Plant and Soil* 178: 101–112.

Porter, J.R., Klepper, B., and Belford, R.K. (1986). A model (WHTROOT) which synchronizes root growth and development with shoot development for winter wheat. *Plant and Soil* 92: 133–145.

Postma, J.A., Kuppe, C., Owen, M.R. et al. (2017). OpenSimRoot: widening the scope and application of root architectural models. *New Phytologist* 215: 1274–1286.

Schnepf, A., Leitner, D., and Klepsch, S. (2012). Modeling phosphorus uptake by a growing and exuding root system. *Vadose Zone Journal* 11: 318–327.

Schnepf, A., Leitner, D., Landl, M. et al. (2018). CRootBox: a structural-functional modelling framework for root systems. *Annals of Botany* 121: 1033–1053.

Somma, F., Hopmans, J.W., and Clausnitzer, V. (1998). Transient three-dimensional modeling of soil water and solute transport with simultaneous root growth, root water and nutrient uptake. *Plant and Soil* 202: 281–293.

Sutton, R.F. and Tinus, R.W. (1983). Root and root system terminology. Forest Science 29. Monograph 24. Washington, DC: Society of American Foresters.

Spek, L.Y. and Van Noordwijk, M. (1994). Proximal root diameter as predictor of total root size for fractal branching models. 2. Numerical model. *Plant and Soil* 164: 119–127.

Spek, L.Y. (1997). Generation and visualization of root-like structures in a three-dimensional space. *Plant and Soil* 197: 9–18.

Thaler, P. and Pagès, L. (1998). Modelling the influence of assimilate availability on root growth and architecture. *Plant and Soil* 201: 307–320.

Tobin, B., Čermák, J., Chiatante, D. et al. (2007). Towards modelling of tree root systems. *Plant Biosystems* 141: 481–501.

Tsegaye, T., Mullins, C.E., and Diggle, A.J. (1995). Modelling pea (*Pisum sativum*) root growth in drying soil. A comparison between observations and model predictions. *New Phytologist* 131: 179–189.

Vansteenkiste, J., Van Loon, J., Garré, S. et al. (2013). Estimating the parameters of a 3-D root distribution function from root observations with the trench profile method: case study with simulated and field-observed root data. *Plant and Soil* 375: 75–88.

Vercambre, G., Pagès, L., Doussan, C. et al. (2003). Architectural analysis and synthesis of the plum tree root system in an orchard using a quantitative modelling approach. *Plant and Soil* 51: 1–11.

Weaver, J.E. (1958). Classification of root systems of forbs of grassland and a consideration of their significance. *Ecology* 39: 393–401.

De Willigen, P., Heinen, M., Mollier, A. et al. (2002). Two-dimensional growth of a root system modelled as a diffusion process. I. Analytical solutions. *Plant and Soil* 240: 225–234.

Wu, L., McGechan, M.B., McRoberts, N. et al. (2007). SPACSYS: integration of a 3D root architecture component to carbon, nitrogen and water cycling – model description. *Ecological Modelling* 200: 343–359.

15

Phenotyping-Modelling Interfaces to Advance Breeding for Optimized Crop Root Systems

Gernot Bodner[1], Daniel Leitner[2], Boris Rewald[3], Jiangsan Zhao[4], Alireza Nakhforoosh[5], and Andrea Schnepf[6]

[1] *Department of Crop Sciences, Division of Agronomy, University of Natural Resources and Life Sciences BOKU Vienna, Tulln an der Donau, Austria*
[2] *Simulationswerkstatt – Services in Computational Sciences, Linz, Austria*
[3] *Department of Forest and Soil Sciences, University of Natural Resources and Life Sciences, Vienna, Austria*
[4] *Division of Biotechnology and Plant Health, Norwegian Institute of Bioeconomy Research (NIBIO), Ås, Norway*
[5] *Global Institute for Food Security, University of Saskatchewan, Saskatoon, Canada*
[6] *Institute of Bio- and Geosciences IBG-3, Forschungszentrum, Jülich, Germany*

15.1 Introduction

Plant roots have historically been neglected in biological and agronomic research, with only some important exceptions (e.g. Weaver and Bruner 1926; Kutschera et al. 2009). Therefore, the root system is still an under-investigated organ. This is in contradiction with the evolutionary importance of root systems: successful plant colonization of a wide range of ecological zones has been related to adapted root systems. For example, Schenk and Jackson (2002) demonstrated the correlation of rooting depth with hydrological regimes of different vegetation zones. Ma et al. (2018) revealed that species from different ecosystems differ in specific root lengths to optimize adaptation to temporal variability in below-ground factors influencing growth. Being sessile organisms, developmental plasticity is a general rule for evolutionary success of plants (Sultan 2000). This is evidenced by the high plasticity of root systems to cope with the high spatio-temporal heterogeneity of soil resources in terrestrial ecosystems (Hodge 2004). Consequently, roots present a challenge for representative sampling, quantification, and comparative analysis: multiple environmental influences and stimuli modify their constitutive genetic programmes of growth and development (e.g. nutrients: Forde and Lorenzo 2001; water: Bao et al. 2014). Furthermore, highly complex microbial processes at the root–soil interface influence root functionality, making the rhizosphere one of the least understood ecosystem spheres (Lazarovitch et al. 2018).

The traditional set of crop management measures (rotation, tillage, seeding time/density, and fertilization/irrigation) has only marginally allowed agronomy to shape site-optimized

The Root Systems in Sustainable Agricultural Intensification, First Edition.
Edited by Zed Rengel and Ivica Djalovic.

root systems in a targeted way (Rebetzke 2016). Therefore, expectations are high that plant breeding will facilitate improved agronomic control over root systems.

A number of points should be considered when defining strengths, weaknesses, opportunities, and threats (SWOT) of root improvement through breeding.

i) Roots determine the capacity of crops to explore and exploit soil resources (water and nutrients). Therefore, they are breeding targets of interest to enhance yield stability, resource use efficiency, and resistance against environmental stresses (e.g. Hammer et al. 2009; Siddique et al. 2015; Duncan et al. 2018). Root improvement can be a promising strategy (*strength*), particularly in the context of sustainable cropping system intensification (Collette et al. 2011), low-input agriculture (including organic farming), and climate change resilience.
ii) Genetic resources associated with root systems are highly diverse (Nakhforoosh et al. 2014), but that diversity is still underutilized in selection and breeding. Therefore, it can be expected there are still large margins to improve the efficiency of root systems when including it explicitly as a breeding target for novel cultivars (*opportunity*).
iii) The size of root systems in existing (modern) cultivars might be insufficient for an optimal utilization of available soil resources. This concern has been formulated for wheat in a study investigating cultivars that were developed in the context of high-input-driven Green Revolution and showed reduced rooting vigour compared to pre-Green Revolution cultivars and landraces (Waines and Ehdaie 2007). However, beyond the influence of population size and origin of the selected populations on a conclusive assessment, increased photosynthate investment into root systems did not necessarily translate into a yield advantage. For example, high tiller number in wheat wild relatives and landraces explains their extensive shoot-borne root system, whereas breeding for a reduced tiller number has optimized assimilate allocation to roots in high-yielding cultivars (Peng et al. 2011). Defining improved root systems is therefore more complex than modifying shoot-root dry matter partitioning and is largely dependent on site conditions in the target environment (Palta et al. 2011; Tardieu 2011; Bodner et al. 2015). This implies substantial uncertainty in defining appropriate target traits for root system improvement in a breeding programme (*weakness*).
iv) Root systems are complex and multivariate compound organs with high spatio-temporal variability and strong interaction with environment (de Dorlodot et al. 2007). Promising (composite) root traits (e.g. rooting depth) in a parental line might therefore easily fail to be introgressed into a progeny due to low heritability, polygenic control, and/or epistatic effects (*threat*).
v) Methodologies applied to study roots have to match the throughput requirements of breeding programmes. Typically, F2 crossing populations in a breeding nursery consist of several hundred individuals to be characterized (Witcombe and Virk 2001). This is in contrast with the throughput of the traditional, and most of the current, (field) root sampling, and measurement methods (*weakness*).

Research on roots was traditionally hampered by a "measurement bottleneck" restricting the quantification of the hidden network of root axes. The importance of roots for future crop production (Lynch 2007b; Den Herder et al. 2010) has motivated the development of new technologies (Smit et al. 2013; Downie et al. 2015; Kuijken et al. 2015; Atkinson et al. 2019) that overcome weaknesses and threats in root breeding and take better advantages of the potential strengths and opportunities outlined above. The recent boom of root phenotyping has provided new evidence for the high diversity among genotypes and its potential role in superior cultivar performance (e.g. Kamoshita et al. 2002; Ruta et al. 2010). However, except for some success stories regarding wheat, rice, and common bean (e.g. Richards 2008; Lynch 2011; Rose et al. 2012), an apparent gap exists between the number of basic scientific discoveries and cases where roots are included as targets in practical breeding. This indicates that increased data availability from phenotyping platforms alone is not sufficient to provide information relevant to breeders. We consider the following specific requirements have to be met in order to underpin usefulness of root phenotyping in breeding: (i) determining target root traits for specific environments, (ii) specifying the most appropriate method(s) to phenotype those traits and (iii) enhancing inference among platforms and towards field environment and yield relevance.

The objective of this review is to elaborate on the potential of root system modelling together with phenotyping and envirotyping to support the breeding process (Figure 15.1). The overall aim is thus to appraise the interfaces between phenotyping and modelling,

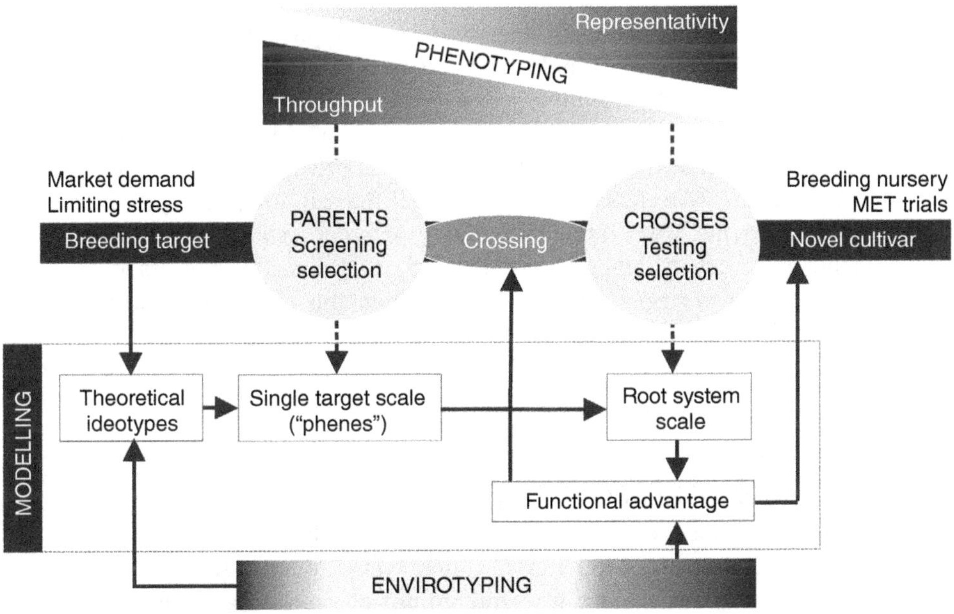

Figure 15.1 Suggested integration of root phenotyping and modelling in breeding for root traits. Theoretical (modelled) ideotypes provide quantitative concretization to the general breeding target (e.g. increased drought resistance), highlighting the contribution of single traits. With this background, an adequate high-throughput phenotyping platform is selected to screen breeding germplasm for targeted root traits in order to identify potential parents to be used in strategic crosses. Models upscale from phenotyped (single) screening traits to an estimate of their impacts at the root system architectural and functional scales. During testing of, and selection from, the crossing population, phenotyping advances to more representative conditions approximating field target environments and yield-relevant stages with outputs at higher scale (e.g. root distribution). Again, models facilitate *in silico* functional evaluation of the modified root systems to infer expected achievement of the breeding target prior to multi-environmental trials (MET) in breeding programmes. Envirotyping contributes to identification of limiting stresses in target environments and provides input (material properties and boundary conditions) for modelling root system functionality.

demonstrate benefits for breeding from their mutual integration, and address possible future developments beyond the current state of the art to facilitate the below-ground improvement of crops. Specifically, we first discuss approaches for defining distinctiveness between root systems (Section 15.2.1) and contributions from statistical models (Section 15.2.2). Introducing the trait-based structure of root architecture models, we outline how they provide a theoretical tool to guide phenotyping (Section 15.2.3). Delineating the reach of different root phenotyping platforms (Section 15.3.1), we provide examples how modelling can help overcome platform restrictions (Section 15.3.2). Dominant yield-limiting environmental stresses and the role of specific root traits in stress responses are discussed (Section 15.3.2.1), and examples are given of how models can serve as quantitative tools to design root system ideotypes for improved stress resistance (Section 15.4.2). Finally, we outline some aspects of envirotyping (Section 15.5) from the standpoint of both measurements for model inputs and modelling for target environment characterization.

15.2 Improving Root Systems – Which Traits to Target?

Crop improvement relies on the identification of desirable phenotypes within a breeding population. For root systems, this is not trivial as most of the root traits are not just genetically complex and controlled by multiple genes, but also highly influenced by environment (G × E interaction). A reductionist view of individual traits might be insufficient to capture the functional distinctiveness among rooting types in different genotypes. This points to a need for the comprehensive classification approaches that capture phenotypic differentiation among root systems. We will therefore first review root system classification and the traits used therein, and then highlight how modelling can support the process of identification of relevant traits to screen for by uncovering relations and dependencies between groups of traits.

15.2.1 Quantitative Parameters for Comparative Root System Classification

Phylogenetic relations used for plant systematics are hardly applicable to delineating distinctive groups of root systems. Fitter (1987) therefore advocated for a functional classification approach to capture different root system types. Functional classifications have been used widely in plant ecology to define groups of species based on a number of shared biological attributes that relate to function rather than phylogeny (Lavorel et al. 1997). An overview of common groups of parameters used for quantitative classification of root systems aimed at identifying distinctive functions is given in Figure 15.2 (see also Chapters 2, 4 and 13).

Historically, the first concept of systematization of root system was suggested by Goebel (1930) based on the anatomical origin of roots, thereby distinguishing between allorhizy in dicotyledons and homorhizy in monocotyledons (see also Chapter 13). The idea of root axes origin (and its relation to successional phenological events) was then further used to define a developmental approach to root system classification (Cannon 1949). Based on this, Zobel and Waisel (2010) suggested taxonomy for root system characterization

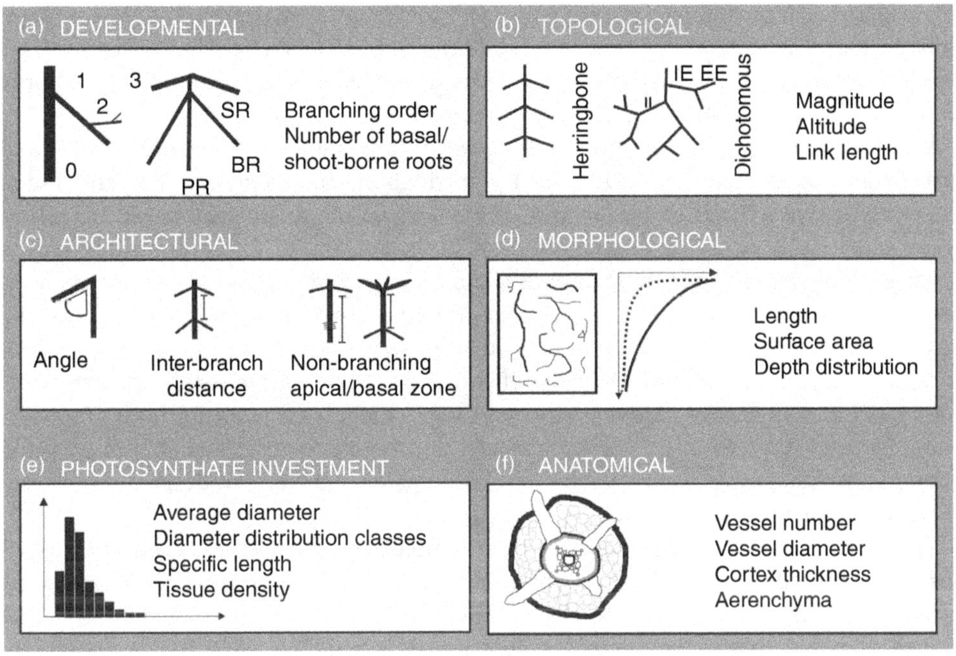

Figure 15.2 Approaches for identification and classification of distinctive groups of root systems with some key parameters. (a) developmental approach (PR primary root, BR basal root, SR shoot-borne root); (b) topological approach (II interior-interior link, IE interior-exterior link, EE exterior-exterior link; magnitude = number of exterior links; altitude = number of links in the longest path from an exterior link to the origin; cf. Fitter et al. 1991); (c) architectural approach; (d) morphological approach; (e) photosynthate investment approach; and (f) anatomical approach.

(Figure 15.2a), distinguishing between a primary root system (comprising the embryonic radicle, basal roots emerging from the hypocotyl/mesocotyl, and their respective laterals) and a secondary root system (mainly in monocotyledonous species) composed of shoot-borne axes emerging from stem nodes upon tillering (growth stage BBCH 20) and their laterals. A detailed description and discussion of the genetic background of, e.g. maize roots can be found in Hochholdinger et al. (2004).

Beyond the qualitative differentiation between root axes types, a functionally relevant quantitative parameter emerging from the developmental concept is the order of lateral branching. According to Kutschera et al. (2009), the number of lateral orders is dependent on both species and environmental stimuli, with annuals developing up to five orders of laterals and perennial woody species up to 16 orders. Characterization of root systems following a developmental concept requires observations over time to precisely track emergence of novel roots, allocate them to a specific axes type, and measure their morphological characteristics (e.g. diameter, length, and surface area). For soil-based 2D imaging systems, where root axes are only partially visible (e.g. in rhizobox images, around 30% of total root length; Nagel et al. 2012; Pfeifer et al. 2014; Bodner et al. 2018), as well as for field sampling, it is difficult, if not impossible, to distinguish the developmental origin of axes. In contrast, all axes are visible in high-throughput root phenotyping systems using agar plates

(Xu et al. 2013) or germination paper (Gioia et al. 2017); however, the short growth duration (mostly limited to the leaf development stage prior to BBCH 19) could be a constraint to a developmental classification, particularly for monocotyledonous species with shoot-borne axes emerging at later phenological stages (after BBCH 21).

A topological approach (Figure 15.2b) was initially suggested by Fitter (1987) and further developed by Berntson (1997), with herringbone and dichotomous branching as the extreme types. Fitter and Stickland (1991) derived the topological-based indices as quantitative surrogates for distinctive branching patterns to differentiate between a range of dicotyledonous species. With a focus on branching as a major grouping criterion, Fitter's classification has strong functional background oriented towards the resource uptake strategies (Taub and Goldberg 1996; Cruz et al. 2004). The segment parameters required for topological classification can be measured readily by using software such as WinRhizo (Régent Instrument Inc., Quebec, Canada) for intact root systems without the need to conserve their natural geometric location, e.g. using phenotyping images with fully visible root systems or scans of soil-derived and rinsed roots with intact connections of axes.

Several root phenotyping platforms, from simple seedling screens to advanced soil-based 3D methods, provide comprehensive sets of architectural parameters. Keeping the focus on the branching pattern, a range of image analysis tools are available that target biologically meaningful architectural descriptors (see Lobet 2017 for an overview; Figure 15.2c). For example, the program Root System Analyser, developed for neutron radiography images, provides a comprehensive structural description of root systems quantifying inter-branch distances, lengths of unbranched segments at the root apex and base, and the angles between branches of different orders (Leitner et al. 2014a; Lobet 2017).

The quantitative descriptors for destructively (field) sampled roots (root cores; e.g. Hodgkinson et al. 2017) and the minirhizotron data (Majdi 1996) with the architectural shape not preserved/visible rely on the morphological descriptors (Figure 15.2d). Their depth distribution (e.g. of length, surface area, and root numbers), described via empirical functions, has been used widely in models as well as for comparative purposes (Feddes and Raats 2004; Kashiwagi et al. 2006).

From basic morphological descriptors, the photosynthate investment strategies can be obtained and used for classification. For example, length distribution in various diameter classes indicates the proportion of fine vs. coarse axes. Adding root dry matter data, parameters such as specific root length (or surface) and tissue density can be derived (Ryser 2006; Ostonen et al. 2007; Figure 15.2e). Ma et al. (2018) recently demonstrated the relevance of these parameters for identification of plant adaptation strategies at the scale of vegetation zones. Several studies (Lynch 2007a; Rewald et al. 2016) pointed to the functional importance of the photosynthate investment costs when defining traits for the high-yield-compatible root architecture ideotypes.

The root anatomical traits (Figure 15.2f) have been studied relatively little regarding quantitative distinctions, in spite of their high functional importance for transport processes (e.g. Steudle 2000) and their sensitive response to adverse soil conditions (e.g. compaction and flooding; Justin and Armstrong 1987; Bengough et al. 2005). The classical root literature (Kutschera et al. 2009) provides extensive empirical evidence that root anatomy constitutes an important distinctive criterion for some species (Rewald et al. 2012). However, only recently, an approach has been developed that allows acquisition of anatomical root data at sufficient throughput for phenotyping, i.e. laser ablation tomography and specialized image analysis (Chimungu et al. 2015; Paez-Garcia et al. 2015).

15.2.2 Statistical Models to Identify Distinctiveness Among Root Systems

The composite-trait character of root systems suggests a need to use the data-driven statistical models to identify distinctive groups from a population. This is particularly relevant to exploiting the extended measurement capacities of 'big-data' phenotyping technologies (Chitwood and Topp 2015). The available phenotypic descriptors are mostly transversal to the groups of traits underlying the deductive (i.e. based on an *a priori* idea of traits to be used in grouping; Gitay and Noble 1997) classification schemes described above and therefore encourage data-driven approaches.

Statistical models for the identification of shared patterns among individuals (i.e. genotypes) in a multivariate dataset (i.e. a screening population) can be either unsupervised or supervised (e.g. Sathya and Abraham 2013). Unsupervised approaches such as principal component analysis and clustering methods explore the existence of distinctive features within a set of unlabelled input data without prior knowledge or hypotheses (Mitra et al. 2002). In contrast, supervised approaches, comprising numerous types of learning algorithms (e.g. support vector machines (SVMs), artificial neural networks, and decision trees), classify unknown data based on structures identified ('learned') from a labelled training dataset (Singh et al. 2016; Zhao et al. 2018).

Unsupervised statistical classification of root systems was used by Bodner et al. (2013). They combined principal component analysis and hierarchical clustering to explore two measured (cover crop species and wheat genetic resources) and a simulated dataset using morphological and topological descriptors. Root systems of the investigated samples clustered into four to six distinctive groups between two contrasting rooting types: (i) high density, topsoil-concentrated exploitative systems, and (ii) explorative systems with more even allocation of roots from the topsoil to deeper layers. Figure 15.3 exemplifies the steps of this unsupervised, data-driven classification for a simulated dataset (using RootBox; Leitner et al. 2010) of 288 root systems.

Supervised machine learning for root system classification has been used less frequently (Zhong et al. 2009; Iyer-Pascuzzi et al. 2010; Zhao et al. 2016). Unbiased and stable classification with high accuracy strongly depends on a precisely labelled and representative training dataset of sufficient size (Chen and Stow 2002) that allows the selected classifier to effectively learn the distinctive patterns in the data. Multi-trait pairwise comparisons via SVM classification (Zhong et al. 2009; Iyer-Pascuzzi et al. 2010; Zhao et al. 2016) and random forests (RF) combined with SVM classification (Zhao et al. 2016) were applied to identify the top-ranking root traits that most effectively differentiated between genotypes in an investigated population of root systems.

The high classification accurary achieved by machine learning methods underlines the effectiveness of supervised approaches to exploit multi-trait root datasets for identification of distinctive phenotypes (Pound et al. 2017). However, the results of Zhao et al. (2016) also demonstrated the dimensionality challenge for multi-trait classification approaches: classification

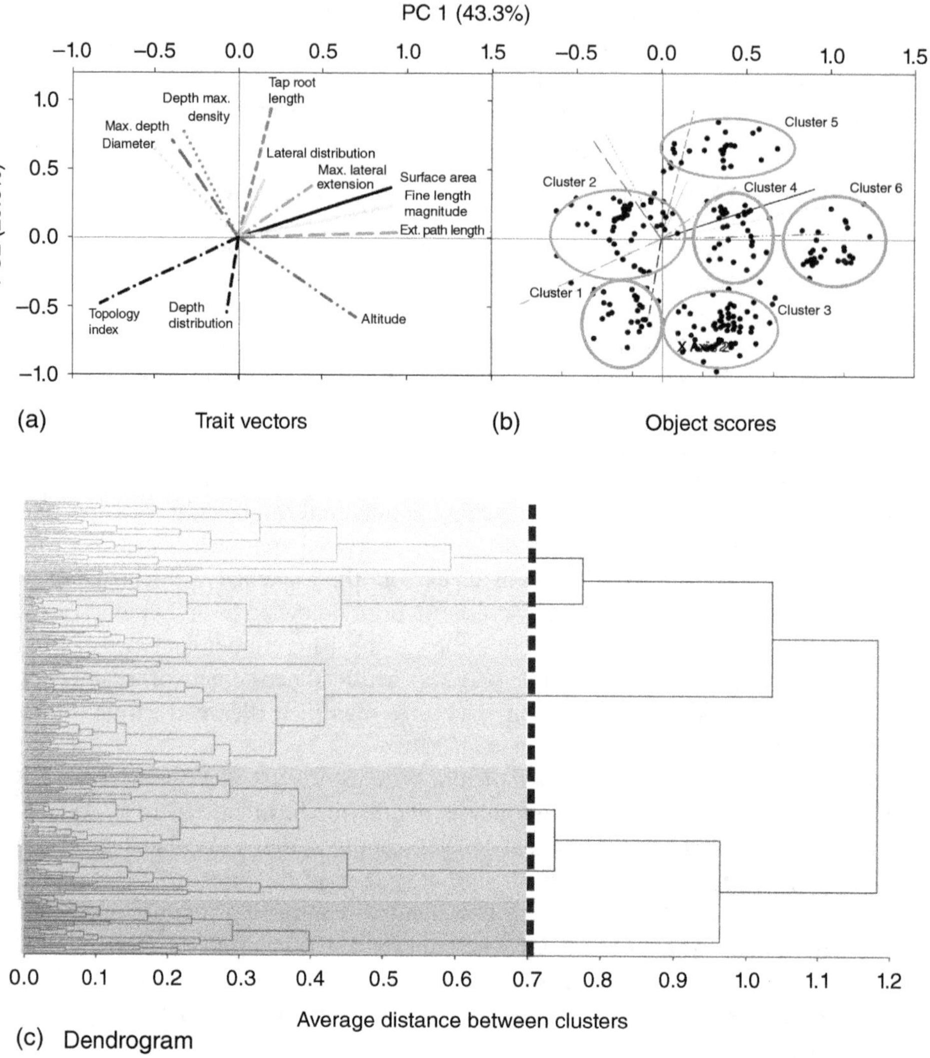

Figure 15.3 Unsupervised root system classification combining principal component analysis (PCA) and cluster analysis. (a) the parameter vectors; (b) the biplot with individuals ordered by their scores along the first two principal component axes; (c) dendrogram showing distance clustering according to the PC scores (clusters are identified by different grayscales; dashed line indicates the distance threshold for cluster number according to the cubic clustering criterion). Cluster groups one, two, and six differ mainly in rooting density (PC1 score), and cluster groups three to five differ in depth distribution (PC2 score). *Source:* modified from Bodner et al. 2013. An open access article under the terms of Creative Commons CC BY license.

accuracy of their pea root dataset was improved when using only the five top-ranking traits compared to all 36 traits they measured. This indicates that optimizing the performance of supervised learning algorithms requires variable selection to reduce data dimensionality (Guyon and Elisseeff 2003) and thereby avoid the overfitting and unstable solutions (Domingos 2012).

Further exploration of multi-trait root datasets according to the method of Zhao et al. (2016) could result in a database of important distinctive traits for different species/cultivars/landraces that facilitates targeted selection of phenotyping platforms. Furthermore, the use of mechanistic root architecture models, described below, might be tested to train classifiers with virtual data that contain information on trait relations, which are then applied to the measured testing datasets for classification.

15.2.3 Modelling as a Tool to Identify Target Traits

Root simulation models provide a versatile tool to support trait selection by decomposing the complexity of root system architecture and its related functioning (see also Chapter 14). Root models quantitatively describe the temporal sequence and the spatial location of different root types, ranging from growth and differentiation of the first-emerging primary axes to the complex composite structure of a mature root system. Translation of genetic programmes (and their interplay with environmental stimuli) of root system formation into mathematical formalism makes root modelling a tailored approach to study dependencies and relations among traits across different scales and identify key traits that shape the distinct root system architectures.

15.2.3.1 Components of Root Architecture Models

The structure of most mechanistic root architecture models follows a developmental concept: single axes types (embryonic, basal, and shoot-borne) emerge in a phenologically determined temporal sequence, elongates according to empirical growth functions (e.g. negative exponential) and regularly branches within a certain segment of a parental axis (Dunbabin et al. 2013). Differences among models originate from the underlying mathematics as well as their specific focus and level of detail. Starting with root growth and structure (architecture) to model 3D solute uptake from soil (Somma et al. 1997; Pagès et al. 2004), models have become increasingly more complex/realistic, for example, taking into account different types of tropisms to improve simulation of root plasticity in response to soil signals (Leitner et al. 2010). Further advances towards the comprehensive structural–functional root models (Ndour et al. 2017) have focused on a better representation of interfacial processes (e.g. mycorrhiza; Schnepf et al. 2016) as well as radial/axial transport into and inside the network of root axes (Couvreur et al. 2018).

From the point of view of descriptors to identify distinctive root phenotypes, four conceptual scales (i.e. functional, single axis, root system, and environmental integration) with their respective groups of model parameters representing measurable traits can be distinguished. The largest scale considers the root system and its interactions with the environment. The root system scale considers the topological structure of a root system, i.e. its composition of different types of root axes and their interconnectedness. Each root axis is thought to be built of many small root segments; thus, it is a discrete representation of the root axis with the segment length defining the spatial resolution along the root axis. Functions, such as root water uptake, are often described at the root segment level, i.e. each segment may have different values of parameters that affect its function. Figure 15.4 presents the integration of the different scales in the root architecture model CRootBox (Schnepf et al. 2018).

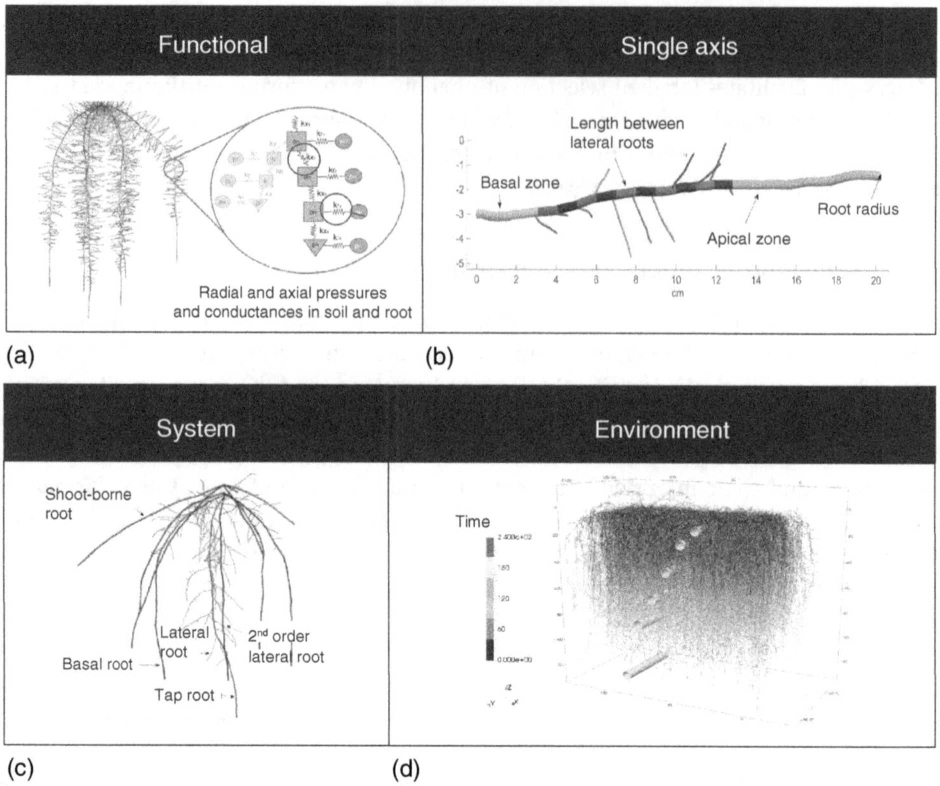

Figure 15.4 Four conceptual scales in CRootBox: (a) functionality of specific root segments; (b) single axes growth and morphology; (c) whole-root system composition and spatial arrangement; and (d) environmental integration and interactions (root growth over time [days] in a field soil with rhizotubes installed at different depth). *Source:* (a), (c), and (d): the authors (data from Schnepf et al. 2018; (b): from Zhao et al. 2017. An open access article under the terms of Creative Commons CC BY license.

15.2.3.1.1 Functionality Functional models are described at the segment scale. For simplicity, we exemplify this scale by the description of water movement within the root segments. Other important processes to be considered are phloem flux (Lacointe and Minchin 2008), nutrient transport, and/or root exudation (Hinsinger et al. 2009).

Water movement within the roots is described by the model of Doussan et al. (1998), whereby the net flux q of each segment is the sum of the radial flux q_r and axial flux q_z. The radial water flux q_r is given by

$$q_r = -2a\pi l k_r \left(p_s - p_r\right),$$

with unit [L^3 T^{-1}], where a is the segment radius [L], l is the segment length [L], k_r is the root radial conductance [L^2 T M^{-1}], p_s is the pressure of the soil surrounding the segment, and p_r is the xylem pressure within the root segment [M L^{-1} T^{-2}]. Axial water flow q_z is defined as

$$q_z = -k_z\left(\frac{\partial p_r}{\partial z} - pge_z \cdot \upsilon\right),$$

with the same unit [$L^3\,T^{-1}$], where k_z is the root axial conductance [$L^5\,T\,M^{-1}$], p_r the xylem pressure [$M\,L^{-1}\,T^{-2}$], ρ the density of water [$M\,L^{-3}$], g the gravitational acceleration [$L\,T^{-2}$], e_z the downward unit vector [1], and v the normed direction of the segment [1].

Fluxes in multiple connected segments are calculated using Kirchhoff's first law. The connected segments can be represented by nodes and edges of a mathematical graph, and according to Kirchhoff's first law, the sum of net fluxes for each node i must vanish:

$$\sum_{j \in N(i)} q_{ij} = 0,$$

where $N(i)$ are the neighbouring nodes that are connected to node i, and q_{ij} is the net flux of the segment connecting node i and node j.

For CRootBox, this model was implemented in Python and presented by Schnepf et al. (2018). For water movement, the most important plant parameters are root radial conductivity and root axial conductivity. Generally, these parameters can change with root age and can be altered by plant systemic responses (e.g. McGaughey et al. 2018).

15.2.3.1.2 Single-axis Growth Root architecture models often differ in the way they represent growth processes (see also Chapter 14). In CRootBox, a single root is composed of connected root segments. It is characterized by three zones, starting with the basal zone, followed by the branching zone (where lateral roots emerge) and the apical zone at the root tip. The length of the basal zone (l_b) and the length of the apical zone (l_a) can be determined easily from root images. The length of the branching zone is obtained from root images where the distance between laterals (l_n) and their number is determined. The maximal root axis length (l_{max}) requires images of fully developed root systems, which are often not available, and is thus frequently estimated using literature values (Landl et al. 2018).

Root growth type (*gf*) is either linear or negative exponential and is described by the initial elongation rate *r*. With negative exponential growth, the elongation rate gets smaller with root age, whereas it stays constant for linear growth. The suitable parameter values can be determined from single images if the age of roots is known or estimated from image sequences (e.g. Leitner et al. 2014a). These parameters are root radius, the angle between the root and its parent root (θ), and (for primary roots) the insertion angle, i.e. the angle between the root and the *Z*-axis. Radius and angles can be readily measured from root system images.

During growth, the root tip changes its growth direction due to a random optimization process. Therefore, *n* random axial changes are picked, whereby the axial change is sampled from a normal distribution $N(0,\sigma)$, and the best change according to an objective function is performed. In this way, various tropisms, such as gravi-, plagio-, hydro-, or chemotropism, are described. The strength of the tropism is controlled by *N,* and σ can be interpreted as the flexibility of the root to change direction.

In the branching zone, laterals may emerge not only in a strict topological order (in which a second order root follows a first-order root, etc.), but following the concept of

Table 15.1 Parameters of a single root in CRootBox.

Parameter name (units)	Description
type (1)	Index that determines the type of the root
l_a (cm)	Length of the apical zone
l_b (cm)	Length of the basal zone
l_{max} (cm)	Maximal length the root will reach
gf (1)	Type of the growth function (linear or negative exponential)
r (cm day^{-1})	Initial growth rate
a (cm)	Root radius
θ (1)	Insertion angle
Tropisms: type (1), n (1), and σ (cm^{-1})	Type of tropism, number of trials, stiffness factor
successors	A list of types of lateral roots
successorP	The probability of emergence of each type in the list *successor*

Pagès et al. (2004) in which a root may be followed by several possible root types (successor) at prescribed respective probabilities (*successorP*).

The set of parameters describing growth of a single root in CRootBox is summarized in Table 15.1. All parameters describe the root under optimal conditions and could be altered when coupled to an environmental interaction model. In CRootBox, all above values are given by mean and standard deviation. Two-dimensional images are a valid source for parameterization of 3D root architecture models (Landl et al. 2018). As high-throughput root phenotyping typically produces such 2D images, they represent an important source for root system architecture model parameterization.

15.2.3.1.3 Root System The two main root architectures distinguished by the model are the tap root and the fibrous root systems. In CRootBox, combinations of different root axes types compose these distinct root architectures. For each plant, the initial root type is the radicle that starts emerging from the seed at a given planting depth (*plantingdepth*). After some time, (*firstB*) basal roots may start to develop in a certain interval (*delayB*) until the maximum number of basal roots is reached (*maxB*). In monocotyledons, shoot-borne roots may emerge from root crowns, each having a fixed number of root axes that start to grow. New root crowns develop upwards along the shoot axis at prescribed inter-crown distance (*nz*) and time interval (*delayRC*). The root architecture parameters of CRootBox are summarized in Table 15.2.

15.2.3.1.4 Environmental Integration and Interactions For integration and interaction of the roots with the surrounding soil, root developmental and functional models are coupled to a soil model. For water movement in soil, the Richards equation is applied with the VanGenuchten–Mualem model used to describe soil hydraulic properties. Solutes are

Table 15.2 Root system architecture parameters of CRootBox.

Parameter name (units)	Description
plantingdepth (cm)	Depth of the seed placement
firstB (1)	Emergence time of the first basal root
delayB (days)	Time delay between successive basal root emergence
maxB (1)	Maximal number of basal roots
nC (1)	Number of roots per root crown
firstSB (days)	Emergence time of the first shoot-borne root
delaySB (days)	Time delay between successive shoot-borne root emergence
delayRC (days)	Time delay between the development of root crowns
nz (cm)	Distance between root crowns

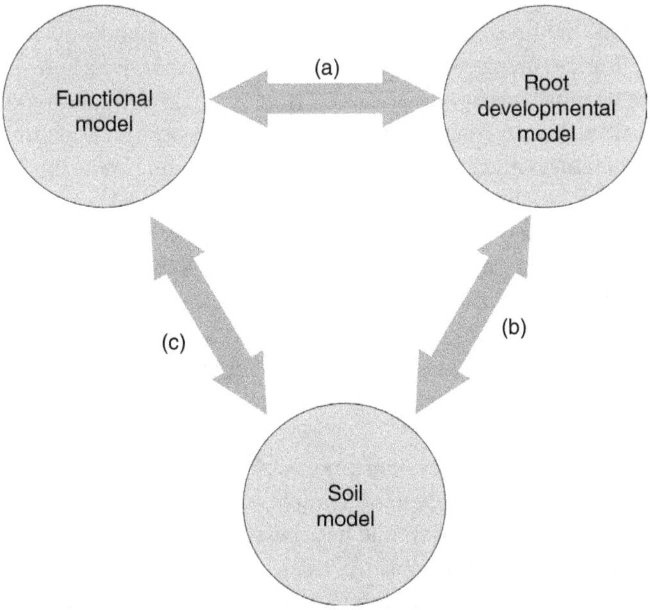

Figure 15.5 Interaction of the submodels in CRootBox: (a→) local root responses, (a←) topological updates, (b↗) root plasticity, (b↙) soil model parameters influenced by roots, and (c↔) coupled via a sink term.

modelled with an advection dispersion reaction equation. The functional model at the segment scale, the root growth developmental model, and the soil model mutually influence each other as visualized in Figure 15.5.

Root development influences the functional model by creating additional segments. The functional model on the other hand modifies root development responding to the segment state (e.g. increased radial flux or radial solute uptake). Root system development is strongly influenced by the soil state. In CRootBox, the following mechanisms of root plasticity are implemented: (i) hydro- or chemotropism is a directed growth in dependence on the soil model state; (ii) the root elongation is dependent of water saturation, soil density, and temperature; and (iii) branching patterns and insertion angles can be nutrient-dependent.

Depending on the specific soil model, soil parameters are changing in the presence of roots (e.g. soil density, soil hydraulic properties, or nutrient availability). Finally, the functional hydrological model is coupled to the soil domain via volumetric sink terms (Leitner et al. 2014b). In this way, the root system is described as highly branched 1-dimensional model that is coupled to a 3-dimensional model of the bulk soil.

15.2.3.2 Unifying Formats

Lobet et al. (2015) developed the Root System Markup Language (RSML) as a format to represent root systems, resulting from both model simulations and experiments. Specifically, RSML was designed to enable portability of root architecture data between different software tools in an easy and interoperable manner, allowing seamless collaborative work. Furthermore, it provides a standard format that facilitates links between experimental data (e.g. from phenotyping), image analysis pipelines, and mathematical models. RSML follows the XML standard and stores two- or three-dimensional image metadata, plant and root properties and geometries, continuous functions along individual root paths, and a suite of annotations at the image, plant, or root scale at one or several time points. Plant ontologies are used to describe botanical entities that are relevant at the scale of root system architecture.

An XML scheme describes the features and constraints of RSML and facilitates its implementation. Many (open source) packages have been developed already in several languages (R, Excel, Java, Python, and C#) to enable researchers to integrate RSML files into their research workflow. Up to now, RSML has been adopted in the root system architecture analysis and visualization pipelines (Shahzad et al. 2018; Ndour et al. 2017; Delory et al. 2016; Das et al. 2015; Cai et al. 2015), root architecture models (Schnepf et al. 2018), and applications of both (Wu et al. 2018; Lobet et al. 2017; Mairhofer et al. 2017).

15.2.3.3 Trait Dependencies and Relations

There are two main challenges for defining a root system target trait in the context of breeding. Firstly, the trait should be at a low level of genetic complexity in order to be better accessible for targeted crossing. Secondly, the trait should be accessible to measurement via (high-throughput) phenotyping to allow evaluation of the existing variation in a crossing population or in potential breeding germplasm. Root modelling allows exploration of trait relations at scale between simpler constitutive traits and more complex system traits. Schnepf et al. (2018), using a high number of simulations of the root architecture model CRootBox, computed the dependencies of characteristic root system measures (traits at the root system scale) on the individual root traits (model input parameters) of a maize root system.

Inter-branch spacing is a key parameter considered to be genetically fixed, with auxin controlling the formation of lateral root primordia along a primary axis (Guseman et al. 2015) (see also Chapter 2). The strong irregularity of lateral branches, often following a linearly decreasing pattern between basal and apical zones, is then a result of sensitive adaptation of lateral root emergence along the axis. Model analysis by Schnepf et al. (2018) revealed that inter-branch distance mainly controls root system parameters describing the total size of root system, such as root length or surface area, but also the volume of soil influenced by rhizosphere that increases non-linearly with decreasing inter-branch distance. In contrast, the shape of the root system (e.g. maximum rooting depth or lateral expansion) is not influenced by spacing of laterals. Spacing, via root density, is also related to parameters of functional relevance for uptake and transport processes, such as half-distance between roots and the root architecture parameters related to hydraulic properties.

Some studies revealed that there are major Quantitative Trait Loci (QTL) for the angle between root axes that could be exploited to achieve root systems of different shape and therefore differential access to distinct soil resources (e.g. Christopher et al. 2013). Modelling demonstrated that branching angle as a lower-scale model parameter controls the root system scale properties that describe the overall shape of the root system, such as maximum depth, lateral spread, or the convex hull volume as a descriptor used to quantify soil exploration. It also influences properties that are important for resource acquisition, such as the half-distance between roots or the rhizosphere volume. It does not, however, modify the parameters determining total size of the root system, such as root length or number of tips, the rhizosphere volume, or root hydraulic properties. The higher the insertion angle (i.e. more shallow roots), the more the mean depth of root water uptake is shifted upwards in the soil profile.

In a fibrous root system, such as that of maize, the number of primary roots can vary significantly. This strongly determines measures such as total root length. The half-distance between roots decreases with increasing number of primary roots, but the rhizosphere volume increases. The mean depth of standard root water uptake is not influenced by this parameter. The initial growth rate of primary roots significantly influences all the root system traits that were determined by Schnepf et al. (2018). Thus, it can be concluded that this parameter should be carefully considered in any phenotyping activity.

Furthermore, Schnepf et al. (2018) quantified correlations between different root architecture measurements and showed that some are variable across the parameter space. A root system that has a strong gravitropism, for example, generally shows a negative correlation between the convex hull volume and the maximum rooting depth, whereas for a root system without gravitropism, an increase in maximum rooting depth also results in an increase in the convex hull volume (positive correlation).

The statistical approach used by Schnepf et al. (2018) can be applied to the selected lower-scale root traits (model input parameters) to achieve a certain outcome (root system property). Compared with experiments, model simulation can usually be repeated much more often; in this work, the number of replications for each parameter set was 1000. Thus, experimental data could be used to inform and parameterize root architecture models, which can then in turn be used to create the realistic, data-driven scenarios for further investigations.

15.3 Modelling to Bridge Phenotyping Platforms

15.3.1 Phenotyping Platforms and Inference Limitations

Image-based root phenotyping has strongly improved measurement capacities compared to the traditional approaches (excavation, coring; see Böhm 1979) (see also Chapter 13). Major progress has been achieved in terms of throughput (e.g. Thomas et al. 2016 – more than 700 plants per experiment) and resolution (e.g. Rogers et al. 2016 – ~30 µm voxel size in 3D X-ray CT). The technology-driven dynamics of phenotyping impressively increased the number of available datasets (e.g. Fahlgren et al. 2015 reported 1.2 million root images of *Arabidopsis thaliana* and 58 000 for rice), shifting the focus of root research strongly towards greenhouse/growth chamber experiments. In the context of breeding, measurement capacity to handle large numbers of accessions is essential to discover superior genotypes and to balance phenomic with genomic information (Cabrera-Bosquet et al. 2012; Araus and Cairns 2014). However, limited *in situ* field root phenotyping capabilities impose inference restrictions that continuously challenge the applicability of phenotyping results to breeding (but see advances in the field of 'shovelomics'; Bucksch et al. 2014; Burridge et al. 2016; York et al. 2018).

To overcome these restrictions, root phenotyping, in analogy to the assessment of genotype x environment interaction in classical multi-environment breeding trials (MET), should strive to create multi-platform datasets allowing quantification of the role of platform specificity. This is particularly relevant considering that (i) some studies found contradictory outcomes between measurement approaches, e.g. depending on the medium used (Wojciechowski et al. 2009), (ii) relations between high-throughput root screening and field samples are often too weak for reliable extrapolation (Thomas et al. 2016), and (iii) inference from seedling screening towards fully developed root systems is questionable, particularly for monocotyledons with a dominant shoot-borne root system (Watt et al. 2013).

The type of inference restriction is largely set by the available phenotyping/measurement approach. An overview of platform restrictions due to (i) dimensionality, (ii) developmental stage, (iii) substrate for plant growth, and (iv) derived output parameter groups (see also Section 15.2.1) to classify root system differences in a population is given in Table 15.3.

The first key inference limitation of several phenotyping platforms is the phenological stage at which roots are measured. The majority of high-throughput systems are size-restricted (e.g. pouches 18 × 16.5 cm; Richard et al. 2015; agar plates 25 × 30 cm; Bengough et al. 2004) and therefore rely on seedlings/juvenile plants. The resulting inference limitation from such seedling measurements is particularly evident for monocotyledons with secondary root axes emerging from shoot nodes (Arihara and Crosbie 1982). These axes are still invisible in the seedling stage, but become dominant in the mature root system. Efforts are made to increase the size of high-throughput phenotyping systems to extend experimental duration and improve the breeding relevance of phenotyping datasets (e.g. rhizoslides 49 × 61 cm; Le Marié et al. 2014; rhizotubes 50 × 18 cm diameter; Jeudy et al. 2016).

Another restriction is the substrate used: artificial substrates such as hydroponics, agar, or germination paper (i.e. with physical properties largely different from soil) might limit inference to field conditions. For wheat, Watt et al. (2013) still successfully extrapolated root length from a germination paper seedling measurements to field observations of plants at the same development stage, but not to the later (e.g. flowering) stages. This suggested that phenotyping inference was more restricted by attainable phenological stage than substrate. However, root systems are probably less substrate-dependent in the seedling stage than later in their phenology because of seed reserves-driven root growth and thus reduced adaptive plasticity to different environments/substrates compared to the later stages. Therefore, substrate selection is likely more critical for phenotyping approaches targeting

Table 15.3 Inference restrictions associated with root phenotyping by different platforms/approaches.

Dimension	Stage	Substrate	Classification parameters	References
1D				
Coring	Maturity	Soil	Morphology (D), Investment (E)	Hodgkinson et al. (2017)
Columns	Flowering	Soil	Development (A), Morphology (D), Investment (E)	Zhao et al. (2016)
Minirhizotron	Maturity	Soil	Development (A, partially[a]), Architecture (C, partially), Morphology (D)	Ephrath et al. (1999)
2D				
Agar plates	Seedling	Transparent agar	Development (A, partially), Topology (B), Architecture (C), Morphology (D), Investment (E)	Bochicchio et al. (2015)
Pouches	Seedling	Germination paper	Development (A, partially), Topology (B), Architecture (C), Morphology (D), Investment (E)	Le Marié et al. (2014); Gioia et al. (2017)
Rhizoboxes	Flowering	Soil	Development (A, partially), Architecture (C, partially), Morphology (D, partially)	Nagel et al. (2012); Bodner et al. (2018)
3D				
Basket	Early vegetative	Soil	Architecture (C, partially)	Uga (2012)
Shovelomics	Maturity	Soil	Architecture (C, partially)	Trachsel et al. (2011)
CT[b]	Early vegetative	Soil	Development (A), Topology (B), Architecture (C), Morphology (D)	Metzner et al. (2015)
MRI[b]	Early vegetative	Soil	Development (A), Topology (B), Architecture (C), Morphology (D)	Schulz et al. (2013)

Systems differ by dimensionality (1D, 2D, and 3D), maximum experimental duration (final phenological stage tested), substrate used, and the type of classification parameters that can be measured (letters refer to classification approaches in Figure 15.2).

[a] 'Partially' means that not all parameters of a given classification type can be assessed because of non-visible root axes (minirhizotrons and rhizoboxes), early measurement stage (agar plates and pouches; not all axes types developed) or less detailed insight into the sample (basket and shovelomics).

[b] CT Computer tomography, MRI Magnetic Resonance Imaging.

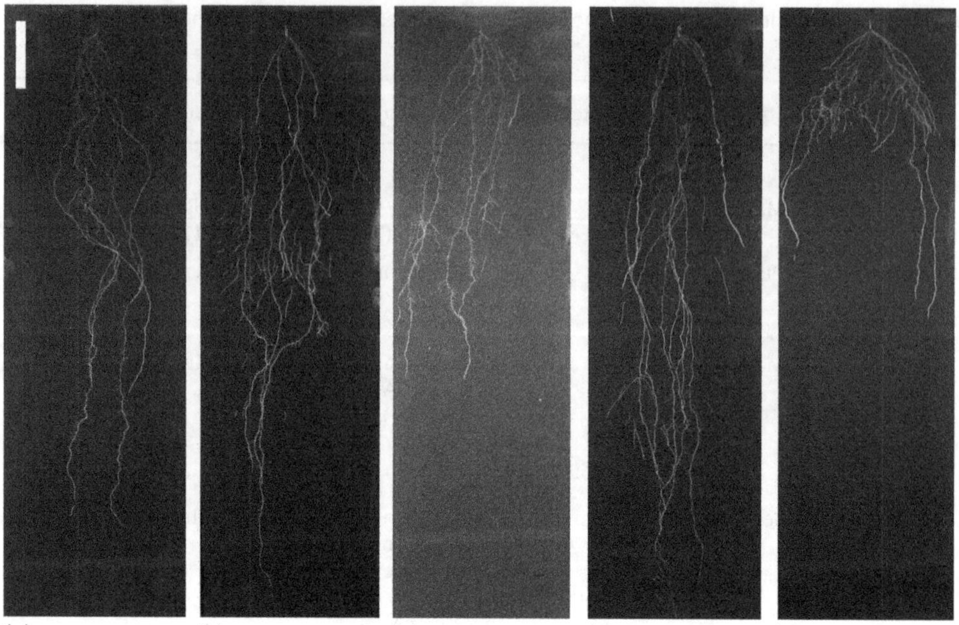

(a) (b) (c) (d) (e)

Figure 15.6 Durum wheat (*Triticum turgidum* subsp. *durum* cv. Floradur) grown in (a) dry topsoil (22% sand, 61% silt, and 17% clay; water content at matric suction −100 kPa $WC_{-100\,kPa} = 0.18\,cm^3\,cm^{-3}$); (b) moist topsoil ($WC_{-10\,kPa} = 0.27\,cm^3\,cm^{-3}$); (c) dry subsoil (26% sand, 70% silt, and 4% clay; $WC_{-100\,kPa} = 0.13\,cm^3\,cm^{-3}$); (d) moist subsoil ($WC_{-10\,kPa} = 0.33\,cm^3\,cm_{-}^{3}$); and (e) sand (100% $S_ø$ = 0.2–0.5 mm; $WC_{-10\,kPa} = 0.12\,cm^3\,cm^{-3}$). The RGB images were taken by a Canon EOS 60d using UV illumination for contrast enhancement between roots and background (conversion to grayscale was done in Adobe Photoshop CS5). See Table 15.3 for the corresponding morphological and branching parameters. White scale bar in (a) = 10 cm.

fully developed root systems. An example of substrate and moisture influences on root system formation of durum wheat (growth stage BBCH 44) using a rhizobox phenotyping platform is given in Figure 15.6.

Table 15.4 provides the corresponding morphological total root length and branching parameters (measured for the three main axes; Leitner et al. 2014a). As expected, low soil moisture resulted in a significant ($p = 0.029$) reduction of root growth in both topsoil and subsoil; the low water retention of sand did not allow a treatment at −100 kPa. Changes in total root length as well as different branching patterns (e.g. zero order: lateral length, and first order: inter-branch distance and number of branches) indicated significant effects of the substrate at similar soil moisture content (Figure 15.6). Root systems acclimated to the substrate, implying that straightforward extrapolation from phenotyping to field conditions depends on comparability of substrates. Interpretation of the substrate effect on root architecture is complex and probably a combined result of hydraulic (water retention and conductivity characteristics) and mechanical properties.

The third restriction is related to the attainable parameter set (which is partially linked to the different dimensionality of measurements). The capacity to identify distinctive genotypes might be method-dependent because of different degrees of phenotypic differentiation between traits and observation scales. For example, genotypes with similar total

Table 15.4 Root parameters of durum wheat (*Triticum turgidum* subsp. *durum* cv. Floradur; see Figure 15.6) in response to substrate and moisture, comprising total root length and root architecture of single axes (average of three longest axes).

	Topsoil		Subsoil			p-value
	Wet	Dry	Wet	Dry	Sand	Substrate
Total root length (cm)	3827	2615*a*	4821	1493*ab*	910*b*	0.016
Single axis						
Zero order length (cm)	24.9	26.4*a*	24.3	24.9*ab*	22.6*b*	0.002
Lateral length (cm)	174	208*a*	148	134*ab*	105*b*	0.007
Zero order / lateral length	0.14	0.13*a*	0.16	0.19*a*	0.22*b*	0.014
1st order axis						
Basal length (cm)	0.30	0.24	0.24	0.22	0.53	0.263
Apical length (cm)	4.28	2.81	3.58	3.59	3.67	0.653
Inter-branch distance (cm)	0.29	0.20*a*	0.28	0.21*a*	0.28*b*	0.005
Number of 1st order laterals	133	211*a*	139	185*a*	127*b*	<0.001
2nd order axis						
Basal length (cm)	0.41	0.37	0.38	0.33	0.28	0.106
Apical length(cm)	0.90	0.65*a*	0.75	0.56*a*	0.42*b*	0.001
Inter-branch distance (cm)	0.49	0.38	0.49	0.43	0.46	0.055
Number of 1st order laterals	13.7	14.5	2.3	6.0	4.7	0.712
3rd order axis						
Basal length (cm)	0.12	0.10*a*	0.12	0.11*ab*	0.15*b*	0.024
Apical length (cm)	0.21	0.20*a*	0.25	0.20*a*	0.27*b*	0.022
Inter-branch distance (cm)	0.30	0.23	0.28	0.41	0.39	0.189
Number of 1st order laterals	0.1	0.0	0.3	0.0	0.3	0.360

The p-values and comparison of means (Tukey test; same letter indicate no significant difference at $\alpha = 0.05$) refer to different substrates (topsoil, Figure 15.6a, b, vs. subsoil, Figure 15.6c, d, vs. sand, Figure 15.6e) in dry conditions (dry topsoil: 0.18 $cm^3\ cm^{-3}$, dry subsoil: 0.13 $cm^3\ cm^{-3}$, sand: 0.12 $cm^3\ cm^{-3}$).

root length might have significant differences in their depth distribution (Wasson et al. 2014) and the root types contributing to length. Also, heritability differs among root traits (and consequently the number of replicates required to detect distinctive genotypes is different; Adu et al. 2014) constituting an important criterion for root breeding to be taken into account when selecting a specific platform.

Currently, 3D phenotyping approaches such as X-ray computed tomography (CT) and magnetic resonance imaging (MRI) provide the most comprehensive root system description for soil-grown plants; prevailing restrictions of size (phenology) and data processing time (throughput) are expected to become lessened in the future (Wu et al. 2018; Xu et al. 2018) (see also Chapter 13). The 2D rhizobox systems allow phenotyping of fully developed root systems grown in soil with intermediate throughput. Restrictions are mainly due to only partial visibility of root axes, which is challenging for automated image analysis (Rellán-Álvarez et al. 2015) and restricts some types of classification (e.g. the segment-based topological approach).

Field methods are mostly restricted to the single time point 1D observations (except mini-rhizotrons) with a focus on morphology/depth distribution and photosynthate investment. Not surprisingly, the characterization of root systems at such a high scale of integration (e.g. total root length and root dry matter) with limited throughput often leads to low heritability estimates of root traits in field studies (Serraj et al. 2004; Richards 2008).

15.3.2 Modelling to Overcome Platform Restrictions

From the above overview, it can be concluded that each phenotyping approach provides a specific view of a root system, depending on dimensionality, phenological stage, substrate, and set of traits. In contrast to the rapidly developing phenotyping technologies and platforms, there is still a lack of platform inter-comparison. Only few studies used different approaches for the same set of genotypes, relating the results via linear correlation (e.g. Watt et al. 2013) or rank comparison (e.g. Hargreaves et al. 2009). Considering the diversity of phenotyping conditions and the plasticity of root systems, the use of simple linear models for platform inter-comparison, without explicitly taking the phenological stage and the phenotyping environment into account, is unlikely to result in a useful knowledge and may even complicate direct comparison further. We propose root models as mechanistic scaling tools that can be used readily for a more promising inter-platform comparison. Models might turn the problem of genotype x platform interaction into an advantage by revealing distinct root plasticity of genotypes and thereby providing more comprehensive root system characterization.

15.3.2.1 Bridging Phenology Restrictions

The first application of models to overcome phenotyping restrictions has been recently presented by Zhao et al. (2017) – improving inference from a seedling root platform to a later phenological stage. When directly comparing root parameters (length and inter-branch distance) of 16 pea genotypes measured in an agar-based platform for seedling measurements with a column-based platform for mature roots, only weak relationships between the two platforms were found ($R^2_{length} = 0.34$, $R^2_{branching} = 0.12$). Therefore, the root architecture model RootBox (Leitner et al. 2010) was parameterized using traits (elongation rates of primary and lateral axes and inter-branch distances) measured at the seedling platform 7–10 days after sowing. The model was then used to predict the root system at flowering (82 days after sowing) and compare the modelled traits to measurements (Figure 15.7a).

Thus, the empirical linear function to infer the column-grown root systems at 82 days after sowing was substituted by a model-based prediction – parameterized with the seedling root data – substantially increasing the fit between parameters obtained in the two phenotyping platforms (e.g. R^2 of 0.95 for root length). However, two additional empirical assumptions were included in the model: (i) the linear root elongation rate, measured in the early seedling stage, was smoothly reduced to zero between the onset and the end of flowering (elongation rate constraint), and (ii) the even distribution of lateral branches

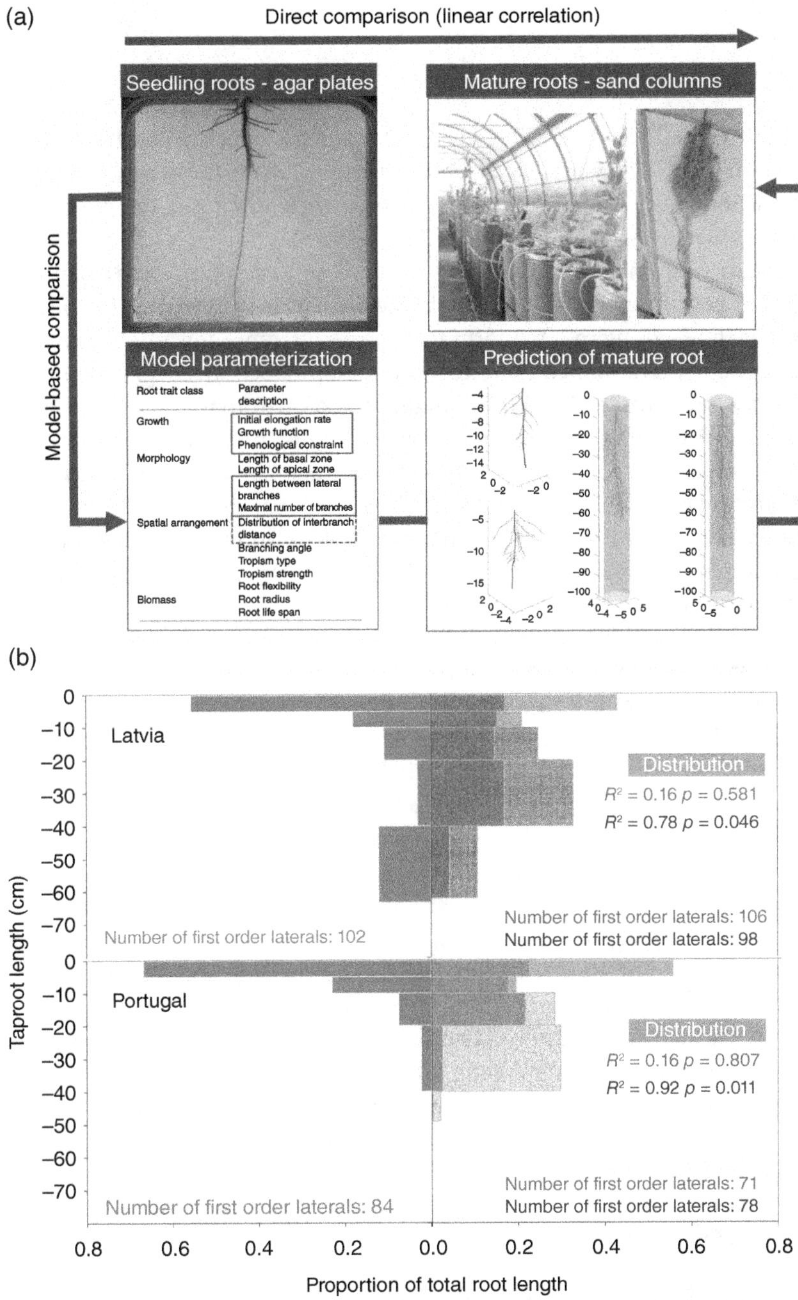

Figure 15.7 (a) Comparison of pea (*Pisum sativum*) root systems measured in two phenotyping platforms (agar plates vs. sand-filled columns) at different phenological stages (BBCH 12 vs. BBCH 69–71). Instead of direct comparison (top arrow), the root architecture model RootBox (Leitner et al. 2010) was parameterized with the seedling root data (left arrow; parameters set from measured traits in rectangular frame; distribution of inter-branch distance required model modification, dashed rectangular frame) to predict the parameters of mature roots in the columns (bottom right: example of two simulated root systems at 7 and 82 days after sowing). The seedling (agar plates) platform-based predictions of mature root systems were compared with the measured data at the same BBCH stage (right arrow). (b) Depth distribution of lateral root length along the taproot of two contrasting pea genotypes from Latvia (top) and Portugal (bottom). Left: measured root length distribution in a sand-filled column phenotyping system (BBCH 69-71). Right: predicted distribution based on a model parameterized with the seedling root data; light grey: with default (constant) inter-branch distance function; dark grey: with modified (a linear decrease from the basal to the apical end) inter-branch distance function. *Source:* adapted from Zhao et al. 2017. An open access article under the terms of Creative Commons CC BY license.

along parent axes in the model was modified to a linear decrease between basal and apical ends of the branching zone (inter-branch distance function). With these model adaptations, the depth distribution of lateral root lengths along the tap root and the number of first-order lateral axes were predicted accurately (Figure 15.7b).

To date, the publication by Zhao et al. (2017) has been the only example where a platform inter-comparison was performed via a root model. The focus was to overcome the phenological restrictions of the seedling root platform to improve inference towards breeding in relevant growth stages. It was assumed that the substrate influence was negligible due to ample supply of water and nutrients, and thus, predictions of mature root systems did not consider acclimations of the root architecture to water and/or nutrient distribution via tropisms. Providing the data on soil properties (including water content, matric potential, and nutrient dynamics) are available, more advanced predictions, including the substrate effects, would be feasible. Thereby, a model-based inter-platform comparison could not only bridge substrate restrictions/differences, but simultaneously reveal genotypic differences in substrate-induced root plasticity.

15.3.2.2 Bridging Parameter and Scale Restrictions

Phenotyping outputs range from the simple constitutive (e.g. number of seminal axes in cereals; Sanguineti et al. 2007) to the complex composite descriptors (e.g. root length density; Kashiwagi et al. 2006). For breeding, simple traits are advantageous because of higher heritability, but composite descriptors are more closely linked to whole-plant performance (e.g. stress resistance, yield, etc.). Accordingly, the platforms that minimize environmentally induced variation (e.g. homogeneous growth media and uniform seedlings) optimize detectability of genetically determined differences among genotypes during screening. In contrast, platforms using near-field conditions (soil, advanced phenological stages, etc.) are more reliable in (i) testing root system distinctiveness, uniformity, and stability (DUS) among genotypes and (ii) studying their potential yield impact. Ideally, a phenotyping platform would fulfil the both breeding requirements within a single experiment, i.e. efficient lower-scale (constitutive) trait quantification for selection with high genetic gain, as well as higher-scale (ambient-shaped) root system observations for validation of trait impact on the desired whole-plant breeding target. However, due to apparent contradictions in simultaneously optimizing phenotyping platforms for multiple scales, modelling as a tool to link root descriptors across the scales could (at least partially) help reduce the need for multi-platform assessments (Figure 15.8).

15.3.2.2.1 Forward Modelling from Simple Traits to Complex Descriptors Upscaling from detailed single root measurements towards the root system scale has been motivated originally by computational limitations to solving water/solute transport equations for a fully developed 3D root system. Transforming the 3D details of root architecture into a simpler sink-term function has allowed meaningful representation of root functionality in the Richards equation (water)/convection–dispersion solute-based simulations (Darrah et al. 2006; Roose and Schnepf 2008) at significantly reduced computational costs.

The study by Zhao et al. (2017) described above (Figure 15.7) represents forward modelling from detailed architecture to morphological averages in order to bridge differences in the detail/type of phenotyping outputs. Once the 3D root system is simulated from an

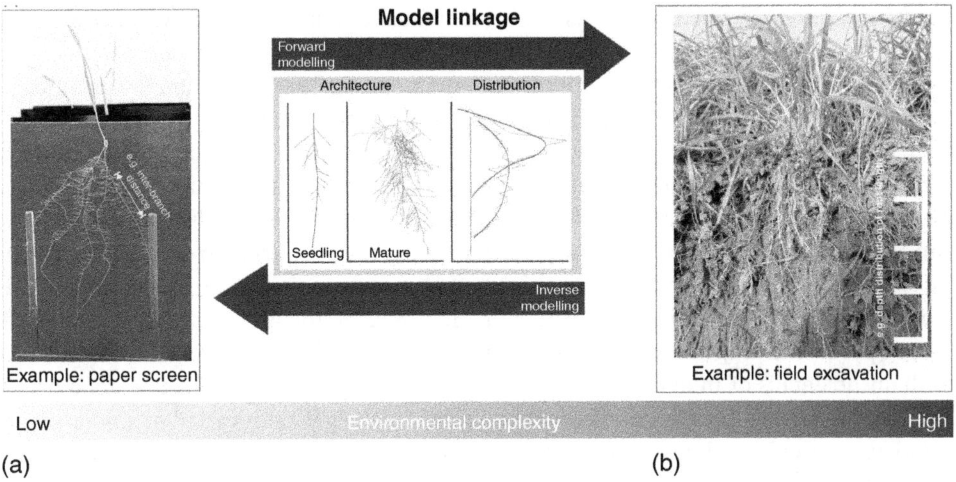

Figure 15.8 Models to bridge different scales of root system descriptions. (a) Simple traits with high heritability relevant to screening promising parental lines under controlled conditions. (b) Composite traits assessed in field conditions to evaluate distinctness, uniformity, and stability of pre-selected promising genotypes under complex environmental conditions. Forward modelling up-scales from the seedlings/mature root architectural details to the composite descriptors, whereas inverse modelling aims to find architectural constituents from the composite measures.

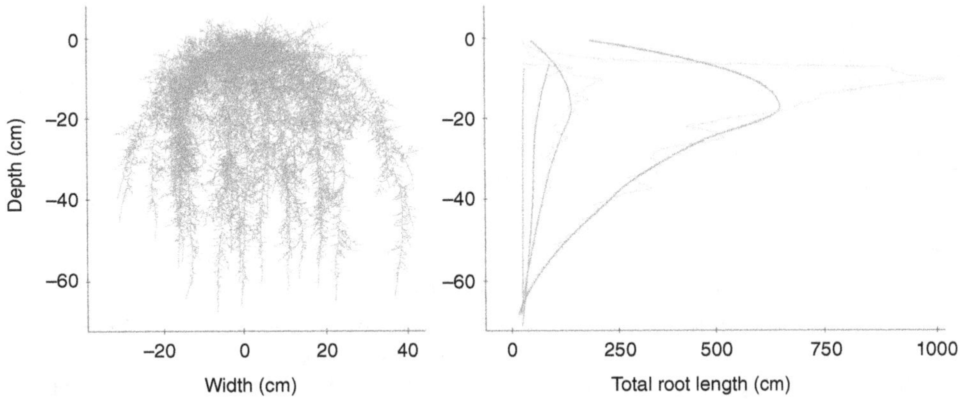

Figure 15.9 Upscaling of a 3D root system (left; lateral extension centred on zero) to a 1D depth distribution (in cm root length per plant) of different axes types (right; bold lines represent the smooth density profile; from left to right: primary root, seminal roots, first-order laterals, second order laterals). Example simulated via CRootBox online (https://plant-root-soil-interactions-modelling.github.io/CRootBox).

initial architectural parameter set, the higher-scale descriptors (e.g. field-coring to assess root distributions, minirhizotron intersection profiles, etc.) can be obtained easily by averaging over the relevant observation volume (Figure 15.9).

Linking platform-specific scales of root description, however, implies different challenges compared to the mathematical/computational questions in the context of water/solute transport simulations. Modelling capacity at scales from the simple to the complex traits depends on (i) reliable measurements of the (constitutive) initial set for growth and branching parameters, (ii) realistic representation of trait x environment interaction over time, and (iii) availability of quantitative information to parameterize the environmental modifiers. Incorporation of increasing mechanistic detail into a model (for requirement #2) simultaneously increases parameterization needs (requirements #1 and #3), which might become a bottleneck. With non-available/unreliable measurements, additional parameters could lead to over-parameterization, diminishing stability of model inference. In some cases, the empirical scaling factors (e.g. root vigour factor for log-normal variation of individual elongation rates in OpenSimRoot; Postma et al. 2017) or stochastic elements (e.g. standard deviation of growth and branching parameters in CRootBox; Schnepf et al. 2018) can be used. These approaches improve realistic representations of soil-grown root systems without mechanistic details of individual processes that are either insufficiently understood or too complex for detailed representation or parameterization (e.g. length variation among single lateral axes, inter-branch distance variation, etc.). In addition, the improved knowledge on trait interrelations (cf. Section 15.2.3.3) can contribute to target phenotyping towards sensitive traits that facilitate upscaling to the composite root system descriptors.

The progress in linking different scales of root system description between phenotyping platforms is mainly (i) a biological challenge of identifying root processes and key traits that bridge scales/environments and (ii) subsequently, development of the respective modules interlinked into existing simulation tools. For example, a challenge for realistic root system representation is the heterogeneous distribution of lateral roots along a parent axis. With the growing knowledge of auxin-induced regulation of lateral root initiation and emergence (e.g. Du and Scheres 2017; Laskowski and Kirsten 2017) (see also Chapter 2), an appropriate mathematical approximation/simplification could enhance upscaling accuracy. Integration of such single-process modules, such as auxin-induced lateral root patterning, into system-scale simulations, would facilitate understanding of their whole-plant functional impact and allow testing the effects of these single traits with comparatively well-defined molecular genetic basis (e.g. induction of the auxin influx and efflux carriers; Péret et al. 2013).

There are still numerous knowledge gaps hampering our understanding of the multiple processes, interactions, and environmental triggers regulating root architecture. Furthermore, novel processes require development of tailored phenotyping techniques building upon established physiological or molecular methods, particularly to meet throughput requirements of screening and selection.

15.3.2.2.2 Inverse Modelling from Average Descriptors to Single Traits Despite the advances in phenotyping, still the most common way for breeders to identify promising genotypes and distinctive traits is observations in a breeding nursery at yield-relevant stages/conditions. High priority is given to being able to approximate the phenotyping and field conditions, i.e. developing field-phenotyping techniques (Yang et al. 2020). Observations at this scale face the well-known problem of low heritability of complex traits (Tardieu and Tuberosa 2010), strong genotype x environment interaction and, in case of roots, unresolved throughput constraints.

Model inversion has been discussed as a tool for downscaling from composite descriptors to simple traits. Inverse modelling uses different optimization methods (Yeh 1986) to estimate effective (i.e. not directly measured) parameter values in an objective function that minimizes deviation between observed and simulated data. If high numbers of parameters have to be optimized simultaneously, however, the risk of non-unique (i.e. local minima) or unrealistic solutions increases – particularly with insufficient size of observed datasets as well as uncertainties in boundary conditions. For example, inverse modelling has been applied to parameterization of the root water uptake function (sink term) in water flow simulations (e.g. Schelle et al. 2012). Increasing efforts are made to infer root parameters from moisture depletion profiles measured by 3D electrical field methods (e.g. electrical resistivity tomography, electrical resistance tomography (ERT); Amato et al. 2009; Garré et al. 2011) or ground penetrating radar, GPR (Guo et al. 2013). However, root parameter optimization from soil moisture profiles still results in root descriptors at a high scale (mainly depth distribution). These can be related to differences in evapotranspiration among genotypes obtained from the soil moisture depletion profiles by either direct calculation of a simplified water balance or using a hydrological model.

It is of interest to obtain more detailed root parameters using inverse optimization. For example, Cai et al. (2017) combined soil moisture with rhizotube (minirhizotron) measurements and successfully estimated root stress response (Feddes model), water uptake compensation (Feddes-Jarvis model), and root conductance (Couvreur model). In contrast, Garré et al. (2011) showed that combining rhizotube data with soil moisture profiles was insufficient for a unique optimization result of architectural root traits (i.e. branching and elongation). The key problems in inversely estimating lower-scale architecture traits are (i) the increasing number of parameters that have to be optimized and (ii) the challenge of defining reasonable parameter constraints; to overcome these problems, Kalogiros et al. (2016) used a density-based model to estimate elongation rates and branching frequencies from higher-level root distribution measurements.

We expect that inverse parameter estimation from higher-scale root (and soil moisture) observations would progress further by (i) combining the soil moisture measurements of high spatial and temporal resolution with novel field-phenotyping datasets containing intermediate-scale architectural information (e.g. shape parameters from shovelomics), (ii) integrating different model types (e.g. density-based, 3D architecture, etc), and (iii) reducing/constraining optimization parameters by accounting for model parameter dependencies (cf. Section 15.2.3.3).

15.4 Root Systems Tailored to Specific Stress Environments

15.4.1 Limiting Stresses and Target Traits

Root breeding can be seen as a specific aspect of 'physiological breeding' (Reynolds and Langridge 2016), whereby secondary traits are targeted during selection based on the dissection of yield (as the principal complex trait of interest) into single (secondary) constituents. For example, in water-limited environments, yield is a function of water

Table 15.5 Global extent of abiotic stresses effecting arable land.

Stress	Extent (% global arable land)	References
Heat	12	Teixeira et al. (2013)
Cold/frost	12	Van Velthuizen (2007)
Water	26	Van Velthuizen (2007)
Phosphorus	30	MacDonald et al. (2011)
Acidity[a]	25	Lynch and Clair (2004)
Salinity	5.4 (50)[b]	Bai et al. (2008)
Compaction	4.8[c]	Oldeman et al. (2017)
Nitrogen	No estimates[d]	

[a] Refers to $pH_{water} < 5.5$.
[b] Number in parentheses refers to irrigated land (Lynch and Clair 2004).
[c] Authors' own calculation with a reference arable land area of 1420 Mha (FAOSTAT data for 2017).
[d] We did not find reliable estimates of global N-limitation in arable land – probably highest limitations are found in soils with low organic carbon content and low N mineralization.

uptake, water-use efficiency, and harvest index, with water uptake being essentially controlled by root system traits (e.g. Bänziger et al. 2000; Blum 2009). Also, at the level of root systems, trait-based selection requires (i) a precise definition of key constraints in a target environment, (ii) the specific root traits required to overcome the prevailing limitations, and (iii) an estimate of expected impact on whole-plant functioning (Tardieu 2011; cf. Figure 15.2). Table 15.5 provides a short overview of prevailing abiotic stresses currently limiting global crop production, with yield gap (deviation of the actual from potential yield) estimates of 43%, 60%, and 47% for wheat, maize, and rice, respectively (Neumann et al. 2010).

The spatio-temporal localization of resources is key to defining the promising root breeding traits (Bodner et al. 2015). For example, the advantage of deep rooting for water uptake can be expected to be higher in dry ecosystems with limited in-season rainfall and stored winter rainfall as the main source of plant water supply (terminal drought) compared to temperate summer-rainfall ecosystems with topsoil contributing water from incoming rainfall (Figure 15.10).

For non-adsorbing mobile nutrients, such as nitrate and sulfate, a leaching risk and thus relevance of deep root penetration for nutrient capture is dependent on the climate and edaphic site characteristics (Thorup-Kristensen 2001). However, it should be borne in mind that root uptake is not a passive response to site-specific spatio-temporal nutrient availability, but includes active modification of nutrient dynamics by roots e.g. via rhizodeposition-induced priming of microbial mineralization enhancing organically bound nutrient release (e.g. glucose-rich rhizodeposits increasing N mineralization by microorganisms; Cheng and Kuzyakov 2005).

Figure 15.11 highlights the root system properties associated with adaptation to three different types of stresses, i.e. drought (important for mobile and highly temporally-variable resources), phosphorus (representative of topsoil-localized immobile and relatively stable resources), and soil compaction.

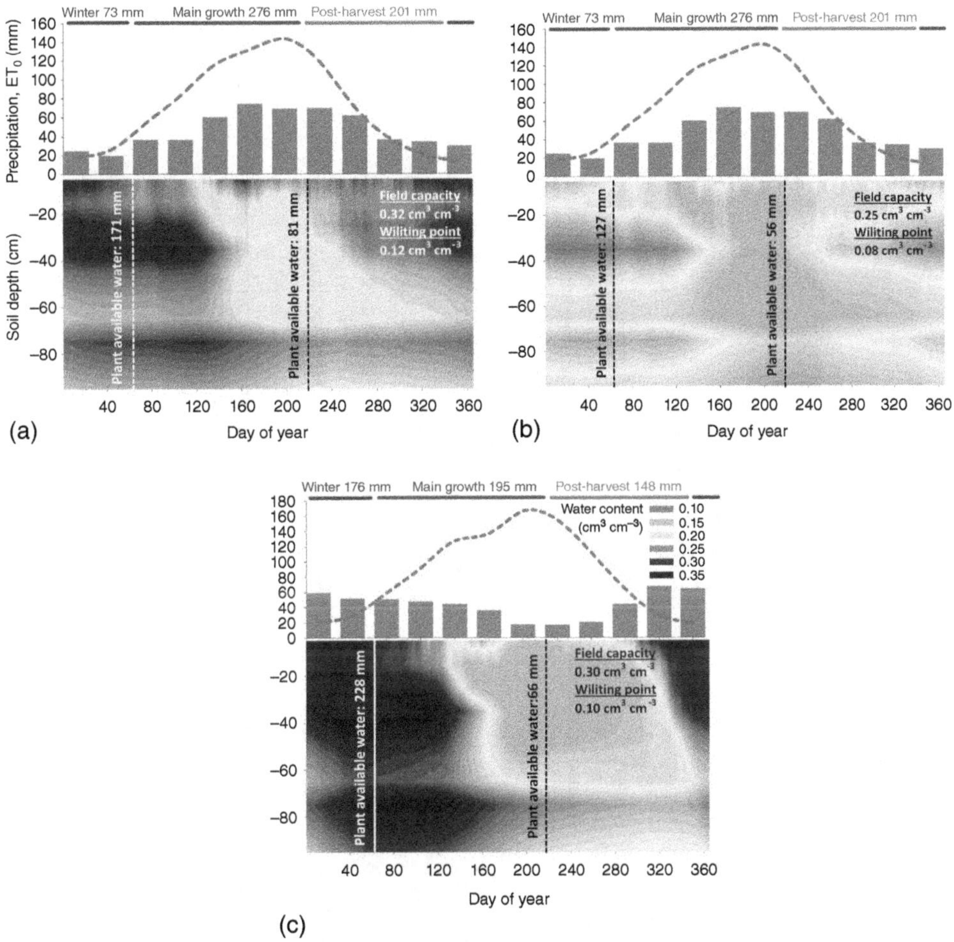

Figure 15.10 Target environment characteristics dictate an optimal rooting pattern. Example for soil water distribution at three sites with similar total rainfall. Long-term averages in (a) and (b) 550 mm and in (c) 519 mm. The sites differ in soil hydraulic properties, with higher water storage in (a) and (c) and low storage in (b), and rainfall (bars) and reference evapotranspiration (ET_0, Penman-Monteith; dashed lines) distribution, with temperate summer-rainfall climates in (a) and (b), and Mediterranean climate with winter rainfall in (c). Plant-available soil water in March (post-winter onset of growth) and July (post-harvest) are indicated by the two vertical dashed lines in each graph. Soil moisture distribution (contour plots at the bottom in each graph) is a 25-year average simulated with Hydrus 1D.

15.4.1.1 Root Traits Important for Acquisition of Mobile Resources

At the root system scale, the main factor associated with effective capture of convectively transported mobile resources is extensive soil profile exploration (Kage and Ehlers 1996). In contrast, rooting density is commonly considered non-limiting when exceeding about 0.5 cm cm^{-3} (Passioura 1983). Most root length/biomass is concentrated in the topsoil (e.g. 50% in the top 20 cm for temperate crops; Fan et al. 2016; 70% in the top 30 cm as a global estimate, Jackson et al. 1996). The maximum rooting depth in cropland is 2.1 ± 0.2 m (global average; Canadell et al. 1996).

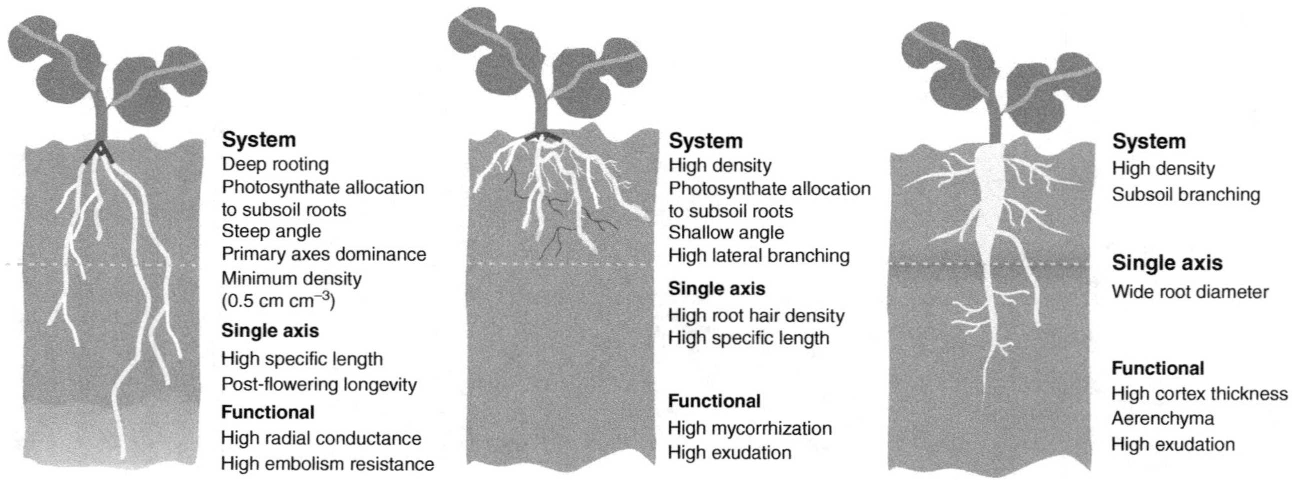

Figure 15.11 Potential target traits for root improvement at different scales of root system description in relation to the predominant limiting abiotic stress (left: low water availability in topsoil; middle: relatively good phosphorus availability in the fertilized topsoil; right: compaction).

Influential phenes underlying explorative root systems are steep root angles and high depth penetration driven by lower-order branches (primary, basal, and shoot-borne axes; York et al. 2013). Sanguineti et al. (2007) reported heritability of 88% for the angle between the primary and basal axes (i.e. spread of root angle) of a durum wheat genotype, which can be used in high-throughput seedling screening as a proxy for deep rooting.

In the later growth stages, a comprehensive shape parameter differentiating deep vs. shallow root systems is the convex hull (Mairhofer et al. 2012). Geometric shapes to describe 16 different rooting types (e.g. cylindrical, inverted conical, etc.) have already been used by Kutschera et al. (1995). These complex shapes are the product of directed growth of the primary and lateral axes driven by variable environmental stimuli (Malamy 2005). The root axes types differ in tropism: contrary to primary and basal roots, laterals initially grow agravitropically, later bending at a certain distance from their origin (gravitropic set-off point; Mullen and Hangarter 2003; Roychoudhry et al. 2013). Other signal-response pathways (e.g. hydrotropism and halotropism; Eapen et al. 2005; Galvan-Ampudia et al. 2013) further modify directionality of root growth beyond the gravity vector, resulting in the phenotypic diversity of root shapes.

Phytohormones mediate lateral branching (initiation, emergence, and elongation; Nibau et al. 2008). For example, lateral emergence is suppressed by water stress (Xiong et al. 2006; Robbins and Dinneny 2018), and osmotic adjustment enhances elongation of primary axes at low water potentials (but shoot growth is supressed; Saab et al. 1990). This results in a dominance of lower-order root axes in relation to laterals and, together with hydrotropism, in a shift from surface-localized to more evenly distributed, deep root systems. In contrast to water, a branching response to (equally mobile) nitrate differs between the systemic modifications (i.e. suppression of lateral root initiation at high shoot N concentration; Little et al. 2005) vs. localized modifications of root architecture (i.e. lateral branching in response to local soil patches at low shoot N concentration; Zhang and Forde 1998).

A common adaptation of plants to heterogeneity in resource availability is altered allometric relations (e.g. root-to-shoot dry matter ratio or root-to-leaf surface area ratio), with increased photosynthate allocation to roots in case of stress (e.g. in wheat under water-limited Mediterranean conditions; Siddique et al. 1990). Within the root system, increased allocation in response to abiotic stress tends to result in thiner axes, expressed as higher specific root length (Ryser 2006; Comas et al. 2013), increasing the physiologically active proportion of roots with higher root-soil contact and lower radial resistance.

There are only few studies on genotypic differences in the root system lifespan for annual crops – concerning either root longevity (Eissenstat et al. 2000) or post-flowering root growth (Ford et al. 2006; Manschadi et al. 2006). Due to the size and phenology restrictions, most root phenotyping platforms do not have the capacity to contribute novel datasets in this respect (even less so on over-wintering root growth for mid-latitude cropping systems; Thorup-Kristensen et al. 2009). Methods to estimate the effect of longevity/sustained growth on root carbon economics as well as effective resource uptake/remobilization during generative growth would be urgently needed (Wasson et al. 2012).

The functional scale of root traits is related to radial and axial transport of water and nutrients influenced by anatomy and physiology of carriers as well as rhizosphere modifications by roots and root-associated microbes that improve resource availability.

For drought stress, Vadez (2014) pointed to root hydraulics as a potential bottleneck that is still beyond current phenotyping capabilities. Some authors consider radial transport as the main uptake limitation (Steudle 2000), while others highlight the role of high axial resistances due to embolized root xylem vessels (e.g. as much as 50% of root xylem vessels interrupted due to embolism in maize; McCully 1999). Currently, there are still no phenotyping approaches to measure satisfactorily these functional-scale traits (with promising advances mainly from MRI and CT; Holbrook et al. 2001; Knipfer et al. 2016). For nitrate, enhanced uptake per unit root length suggests that functional traits are relevant drivers of improvements in modern wheat genotypes (Aziz et al. 2017), e.g. via activity of root nitrate transporters (Taulemesse et al. 2015).

15.4.1.2 Root Traits for Capturing Resources with Low Mobility

Root traits to optimize uptake of nutrients with low mobility in soil have been studied mainly for phosphorus (P) due to its important role in plant metabolism, insufficient availability in many soils, and lowest diffusion rate among major plant nutrients (Schachtman et al. 1998). Efficient P uptake is related to an exploitative rooting type with high topsoil density (Lynch and Brown 2001). The key architectural traits to optimize topsoil foraging are shallow angles, high numbers of basal and shoot-borne roots, as well as intense lateral branching (López-Bucio et al. 2002; Lynch 2011; Chen et al. 2018). Higher specific root length of low P plants results from altered proportions of various root types (increased number of laterals and basal roots with higher specific length; Lynch and Brown 2001), with Gahoonia and Nielsen (2004) finding no P effect on root diameter of single axes as a possible explanation for increased specific length.

Frequently reported single-axis adaptation to enhancing P uptake is root hairs: low P increases root hair density and/or length, thereby optimizing foraging for P resources (Bates and Lynch 1996; Ma et al. 2001; Niu et al. 2012). For soil-adsorbed nutrients, active mobilization by root exudates and/or microbially induced (e.g. mycorrhizas and/or P-solubilizing bacteria; Bolan 1991; Khan et al. 2007) chemical changes in the rhizosphere are of particular importance (Hinsinger 2001). Plants overcome prevalent soil-borne limitations at the cost of increased photosynthate investment into the root zone (Kuzyakov and Domanski 2000). However, establishment of adequate phenotyping techniques to screen for such mechanisms of enhanced root functionality is still at its infancy (Kuijken et al. 2015) and other approaches (e.g. metabolomics; Verma et al. 2018) are promising in this regard.

15.4.1.3 Root Traits for Dense Soils

Mechanical impedance of root growth is a widespread problem in soil types with dense horizons (IUSS Working Group WRB 2015) as well as management-induced compaction (Hamza and Anderson 2005). In addition to mechanical stress, plants growing in compacted soil suffer from anoxic conditions in case of insufficient drainage as well as water shortage due to limited root zone depth. A penetrometer resistance of 2 MPa is frequently used as a lower limit in the concept of least limiting water range (Silva et al. 1994). Penetration resistance increases with decreasing water content: thus, at low water content, mechanical root impedance can become the limiting stress for plant growth, even at intermediate bulk densities (in the range of 1.3–1.5 $g\,cm^{-3}$).

Clark and Barraclough (1999) reported values of maximum axial pressure exerted by seedling roots of various crop species between 0.24 (sunflower) and 0.58 (pea) MPa. An increase in root diameter in dense soil is a commonly observed response, enhancing root penetration capacity (Materechera et al. 1992). Genotypes with larger root diameter increase their bending stiffness and thus improve penetration of dense soil (e.g. rice; Clark et al. 2008). However, roots growing in dense soil tend to clump in cracks/biopores, leading to low density, heterogeneous distribution (White and Kirkegaard 2010) and thus reduced uptake of poorly mobile resources (e.g. up to −22% for P and K in maize; Dolan et al. 1992). An increase in lateral root density outside of dense soil layers has been documented as a compensatory architectural response, with significant differences among genotypes (e.g. maize and soybean; Bushamuka and Zobel 1998). This indicates a shift from exploration of a larger soil volume towards intensive exploitation of resources from a limited soil volume (Colombi et al. 2018).

At the functional scale, root anatomy (flattened shape, higher cortical proportion, and aerenchyma formation) is strongly modified by soil strength (Lipiec et al. 2012; Colombi and Walter 2016). Aerenchyma formation via programmed cell death has been documented for low oxygen and nutrient availability as well as mechanical stresses (e.g. in case of anoxia, facilitating maintenance of root metabolism; Drew et al. 2000). Ethylene has been suggested as a key regulator of both aerenchyma formation and root thichening (Clark et al. 2003). Cultivar difference in resistance to waterlogging suggests that aerenchyma might constitute a diverse trait of interest for improved adaptation that is also relevant to dense soils with periodic anoxic stress (e.g. in wheat; Hayashi et al. 2013). Another functional root response to high frictional resistance in dense soils is enhanced boarder cell sloughing and mucilage formation at the root cap (respectively, 58% and 42% reduction in penetration resistance; Iijima et al. 2004), and exudation (Oleghe et al. 2017). However, it remains unclear whether these rhizosphere stress responses can be considered relevant selection traits varying among cultivars, what would be their theoretical optimum in terms of rhizoeconomics (carbon costs; Lynch 2007a), and how they could be phenotyped.

15.4.2 Ideotype Modelling to Identify Functional Bottlenecks

Root architecture models have been developed with a main focus on testing the quantitative hypotheses regarding water and nutrient uptake under variable soil environmental conditions. Inserted in the breeding context, modelling insights into sensitive parameters to optimize root system functioning are an important decision support system to identify target traits for improving genotypes and eventually searching for relevant QTL (Yin et al. 2003; Hammer et al. 2006). Ideotyping promotes a model-assisted mechanistic/physiological approach to trait-based plant breeding (Martre et al. 2015). With the advance in root phenotyping and structural-functional root models, application has quickly advanced from canopy simulation models towards root ideotyping. The now widely used root ideotypes for mobile (mainly water and nitrate) and immobile resources (mainly phosphorus) (cf. Section 15.4.1, Figure 15.11) are essentially a result of quantitative integration of single trait measurements via 3D root architecture models and subsequent *in silico* simulation experiments.

Ideotype architecture modelling contributed to better understanding of the target environment dependence on rooting depth vs. density for water and nitrate uptake. Tron et al. (2015) simulated 48 root architectures in 16 drought environments differing in rainfall distribution, soil storage capacity, and initial soil moisture. Even though transpiration was optimized by deep root penetration in most environments, for coarse soils with low subsoil water storage and predominant in-season rainfall (e.g. continental summer-rainfall climates; cf. Figure 15.10) root architectures with high topsoil root density were superior in water extraction compared with deep-rooted types. Similarly, Dunbabin et al. (2003) found that, in low-storage soils, high topsoil root density was equally important as deep rooting for nitrate uptake because rapid depletion reduced nitrate leaching early in the season when root systems were still relatively small.

Root models with advanced radial and axial transport simulation provided quantitative estimates for relative contribution of structural vs. functional traits for drought resistance (Lobet et al. 2014). An early modelling study of Sperry et al. (1998) suggested that, except for coarse soils, xylem cavitation constituted the main limitation for transpiration rather than the extent of root surface in soil (relative to the transpiring leaf surface). A simulation study by Leitner et al. (2014b) indicated that distinction between water spenders and water savers among maize phenotypes was mainly the result of different hydraulic properties of the root system architecture. Combining modelling and neutron radiography, Ahmed et al. (2018) revealed that differences in hydraulic properties among root types (crown, seminal, and lateral roots) determined the spatio-temporal water uptake dynamics of maize root system. These findings demonstrate how root ideotypes selected from structural-functional root modelling can point to hydraulic target traits to overcome the functional bottleneck postulated by Vadez (2014).

Quantitative estimates of the optimum level of trait expression are the basis for selection. Based on phenotyping results of maize branching density (with the number of lateral branches ranging from 1 to 41 cm^{-1} of parent axis), Postma et al. (2014) determined optimum densities for nitrate (<7 cm^{-1}) and phosphorus (>9 cm^{-1}). Their modelling study highlighted a different impact of inter-root competition when targeting optimization of uptake of mobile vs. immobile resources with a differential magnitude of depletion zones.

In case of two traits influencing a targeted root system function, e.g. shallow angle and root hair density for P foraging, York et al. (2013) modelled responses that indicated optimum levels of the traits interacting to maximize uptake at the system level. Similarly, optimum trait expression for capturing two simultaneously limiting resources have been studied e.g. in the case of topsoil phosphorus and subsoil water, with the optimum root angle determined by the marginal rate of substitution of water for phosphorus availability over depth (Ho et al. 2004).

For breeding that focuses on maximizing yield, root ideotype modelling has to take carbon constraints into account. Although the extent of competition vs. coordination in photosynthate distribution between organs is controversial (Osaki et al. 1997; Reynolds et al. 2009), roots with a similar functional role regarding resource acquisition can significantly differ in carbon costs. Regarding P acquisition, for example, Lynch and Ho (2005) identified metabolically less expensive root types (adventitious roots), cortical aerenchyma, and root hairs as the carbon-efficient traits, whereas exudation and mycorrhizal symbionts implied higher carbon costs. Empirical evidence was reported on variability of root cortex anatomy in bean and maize (aerenchyma, cell size, cell file number, and senescence) and model-assisted assessment of their functional implication for crop performance (Lynch 2011; Schneider et al. 2017). In summary, these examples demonstrate that modelling facilitates quantitative inference on the optimum level of trait expression and provides an estimate of the effects of different root traits on the performance of specific genotypes.

15.5 Envirotyping

Modelling applications such as ideotyping critically depend on accurate representation of the environment. For theoretical hypotheses testing (e.g. to reveal sensitivity of single root traits or trait combinations), simplified assumptions on material properties (i.e. soil water retention, nutrient adsorption/desorption, etc.), boundary (i.e. meteorological settings) and initial conditions (i.e. resource distribution) are mostly sufficient. However, for *in silico* testing of genotypes with measured/phenotyped root traits differences, the accurate simulation of the target environment conditions becomes more critical. This points to the importance of characterizing the environment ('envirotyping'; Xu 2016) together with plant (root) phenotyping. Mutual benefits from linking environmental measurements and modelling are identified below.

The spatial and temporal resolution of environmental data should match the scales of root architecture simulation. For example, simulating water uptake at a single root scale (dimension: 10^{-3} m) with available soil data limited to a coarser texture (typically 1D 10^{-1} m depth increments for water content) largely hampers modelling possibilities e.g. in regard to hydrotropism. Environmental data are commonly assessed by point measurements (from weather stations, soil moisture sensors, etc.) at low spatial (for meteorological and soil moisture data) and/or temporal resolution (for nutrient availability). Besides scale aspects of root system simulations, the data resolution requirements are dictated mainly by the spatio-temporal variability of the environmental variable itself (Nielsen and Wendroth 2003). For example, soil moisture has a high temporal and vertical spatial resolution, whereas phosphorus is less variable in time and vertical gradients, but is distributed highly heterogeneously within a soil layer.

In addition to heterogeneity in availability of resources, basic soil properties such as soil aggregates and soil pores are often assumed to be time-invariant. However, depending on the process of interest, this implies several problems. For example, (i) it is well known that soil structure varies over time (at scales between hours and seasons) and interacts with roots, microbes, and agricultural management (Schwen et al. 2011b; Carminati and Vetterlein 2012; Bodner et al. 2014) thereby changing the relevant plant water dynamics. (ii) Mechanical effects on root growth direction (thigmotropism) cannot be accounted for in detail, but influence root architecture; for example, in dense subsoil, 85–100% of roots are clumped in the pre-existing pores and are thus highly heterogeneous in distribution (White and Kirkegaard 2010). Finally, (iii) pore scale properties (minimum aperture and tortuosity) are not represented, even though they could limit root/microbe access to nutrients (Borer et al. 2018), influence nutrient turnover (anaerobic microsites; Keiluweit et al. 2017), and water/nutrient transport to the root (hydraulic cut-off; Czyż and Dexter 2013).

Advances to resolve these challenges are achieved with novel sensors and more detailed soil models. Several spatially-resolved measurement approaches to map soil properties (e.g. moisture, density, and salinity) are currently available and are expected to be refined further regarding improved resolution in the future (Barnes et al. 2003; Adamchuk et al. 2004; Bogena et al. 2015; Klotzsche et al. 2018). Detailed description of the soil pore network has been achieved by micro-CT imaging (<3 μm; Tracy et al. 2015) and translated into models for water transport (Xiong et al. 2016) and root-soil interactions (Roose et al. 2016) at the pore scale. The extended Darcy-scale models accounting for time-variability in soil properties have been suggested (Schwen et al. 2011a; Ghanbarian et al. 2013; Alletto et al. 2015). Novel root modelling approaches to account for growth response to mechanical soil properties and preferential root growth in macropores have been presented recently (Landl et al. 2017) and are expected to become established models in the future.

Models can be used as a tool to simplify the complexity of environmental growth-influencing factors and to identify key limiting stressors at specific sites. The groups of target environments with distinctive stress characteristics (timing, intensity, and duration) can be identified, facilitating inference on relevant plant (root) traits for adaptation. This is particularly relevant for growth factors with high inter- and intra-annual variability (e.g. temperature and water), whereby the target environment definition for adapted crop genotypes requires probabilistic estimates based on long-term data time series (see the example in Figure 15.7). A qualitative approach, based on temperature (occurrence of heat/cold), rainfall (occurrence of droughts), soil (acidity), and management (irrigation/rainfed), has been used by CIMMYT to delineate 12 breeding mega-environments (Rajaram et al. 1993). These mega-environments were further refined via statistical models (Gauch and Zobel 1997; Yan et al. 2000) applied to yield data from multi-environment trials.

The predominant focus of model-based envirotyping has been on drought stress (but see e.g. Benbi and Richter 2002 for an overview of soil nitrogen models). Common assessments of water limitations are based on drought indices (Vicente-Serrano et al. 2012), calculated from simplified water balance models with rainfall and evapotranspiration as input data (some including also a soil storage term, e.g. PDSI; Alley 1984). More mechanistic hydrological models such as Hydrus require meteorological data and measured soil hydraulic properties e.g. simulating the frequency of water content or matric potential falling below a certain threshold (Novák 2009; Nakhforoosh et al. 2015). A different class of hydrological models using probabilistic formulations of water balance components has been proposed by Rodriguez-Iturbe et al. (2001) to characterize environmental constraints for plant growth in water-limited ecosystems. This type of model has also been used to obtain theoretical estimates of site-optimized rooting depth for different drought regimes (Laio et al. 2006). A mechanistic soil-plant-atmosphere model (APSIM) was used by Chenu et al. (2011) with a specific focus on the breeding target environment identification – delineating the low-stress, flowering-stress, and terminal-stress environments in the north-eastern Australian wheat production region.

The given examples highlight the importance of models as a tool for envirotyping by helping identify limiting stresses, providing representative boundary conditions for ideotyping, and specifying environmental variables that need to be measured.

15.6 Conclusions

Plant breeding is expected to optimize root systems for improved resource acquisition and use as well as resistance to abiotic stresses in crop production. Substantial advances in root measurements have been achieved by recent enhancements in image-based plant phenotyping platforms. Root phenotyping has led to the new knowledge on (i) diversity of root traits, (ii) developmental processes of root growth and branching, and (iii) the response pathways of single root traits to environmental signals and their role in the construction of complex root system architectures. However, a major breakthrough from phenotyping research towards breeding strategies for cultivars with tailored root systems for specific field target (stress) environments has not been achieved so far.

The main challenge for root breeding is to advance from phenotyping single distinctive traits and processes towards genetic crosses with superior performance under field conditions. The common scaling challenge of physiological breeding, i.e. bridging trait-based selection and yield improvement, is particularly complex for root systems that are multivariate and highly plastic organs. Furthermore, root phenotyping, which is still performed mostly under controlled conditions, includes the second scaling challenge to infer from the (artificial) phenotyping to the (field) target environment.

We have identified several aspects where root models can contribute to a solution. Root models have advanced from tools for basic root-soil interaction research towards breeding applications. In this context, the main achievement from root simulations so far has been the definition of root system ideotypes (mainly for water-, nitrate-, or phosphorus-limited environments) that demonstrate the main constituting traits to be targeted. Still, we conclude that there are several aspects inter-connecting root phenotyping and modelling where further advances could be expected.

i) *Identification of distinctive root traits from a diverse screening population.* Statistical learning methods are emerging as novel tools, particularly adapted to the multivariate datasets generated by root phenotyping, to detect distinctive phenotypes and identify relevant traits for root system classification. Mechanistic root models incorporating the state of the art knowledge of root biology can be used for training of root classifiers that are then applied to cluster distinctive genotypes from phenotyping datasets.
ii) *Integration of phytohormonal regulation into root architecture models.* There is ample literature on the environmental triggers of phytohormone synthesis, their transport to sites of action, and the induced developmental responses. This is particularly relevant for lateral rooting as well as directed growth (tropisms; still poorly understood for most types, except gravitropism). Simplifying the complex molecular processes into mathematical equations and integrating them into existing (modular) root architecture models would strongly enhance their capacity to scale between phenological stages and environments.
iii) *Downscaling the complex root system observations to the underlying growth and branching traits.* Application of inverse modelling is still in an initial stage. Better understanding of parameter-relations within models and parameter constraints, improved high-resolution soil data, and inter-comparison of different types of

models might lead to improved estimates of effective root architectural parameters serving as the target traits.

iv) *Novel soil measurement tools and modelling*. High-resolution soil measurements reveal spatio-temporal dynamics of soil stresses, substantially improving the accuracy of the representation of target environments in model simulations. A major challenge remaining is the integration of pore geometry and mechanical effects on root growth (mainly for dense soils) to advance the understanding of the consequences for root architecture and functioning.

Modern plant breeding is a strongly translational science based on the agronomic, ecological, physiological, genetic, and statistical knowledge, ranging from the lab to field trials. In the context of breeding for improved root systems, modelling can act as an important driver for integration between an increasing numbers of root phenotyping platforms and their various scales of observation, encouraging multi-platform trials and thereby accelerating the scientific knowledge and practical selection gains regarding the root zone.

Acknowledgments

Some results presented in this chapter were from a study partially funded by the European Union Seventh Framework Program for research, technological development and demonstration under grant agreement no 613781 (EUROLEGUME) as well as from the Austrian Science Fund under grant agreement P25190-B16.

References

Adamchuk, V.I., Hummel, J.W., Morgan, M.T. et al. (2004). On-the-go soil sensors for precision agriculture. *Computers and Electronics in Agriculture* 44 (1): 71–91.

Adu, M.O., Chatot, A., Wiesel, L. et al. (2014). A scanner system for high-resolution quantification of variation in root growth dynamics of *Brassica rapa* genotypes. *Journal of Experimental Botany* 65 (8): 2039–2048.

Ahmed, M.A., Zarebanadkouki, M., Meunier, F. et al. (2018). Root type matters: measurement of water uptake by seminal., crown., and lateral roots in maize. *Journal of Experimental Botany* 69 (5): 1199–1206.

Alletto, L., Pot, V., Giuliano, S. et al. (2015). Temporal variation in soil physical properties improves the water dynamics modeling in a conventionally-tilled soil. *Geoderma* 243: 18–28.

Alley, W.M. (1984). The Palmer drought severity index: limitations and assumptions. *Journal of Climate and Applied Meteorology* 23 (7): 1100–1109.

Amato, M., Bitella, G., Rossi, R. et al. (2009). Multi-electrode 3D resistivity imaging of alfalfa root zone. *European Journal of Agronomy* 31 (4): 213–222.

Araus, J.L. and Cairns, J.E. (2014). Field high-throughput phenotyping: the new crop breeding frontier. *Trends in Plant Science* 19 (1): 52–61.

Arihara, J. and Crosbie, T.M. (1982). Relationships among seedling and mature root system traits of maize. *Crop Science* 22 (6): 1197–1202.

Atkinson, J.A., Pound, M.P., Bennett, M.J. et al. (2019). Uncovering the hidden half of plants using new advances in root phenotyping. *Current Opinion in Biotechnology* 55: 1–8.

Aziz, M.M., Palta, J.A., Siddique, K.H.M. et al. (2017). Five decades of selection for yield reduced root length density and increased nitrogen uptake per unit root length in Australian wheat varieties. *Plant and Soil* 413 (1–2): 181–192.

Bai, Z.G., Dent, D.L., Olsson, L. et al. (2008). Proxy global assessment of land degradation. *Soil Use and Management* 24 (3): 223–234.

Bänziger, M., Edmeades, G.O., Beck, D. et al. (2000). Breeding for Drought and Nitrogen Stress Tolerance in Maize: from Theory to Practice. Mexico: CIMMYT.

Bao, Y., Aggarwal, P., Robbins, N.E. et al. (2014). Plant roots use a patterning mechanism to position lateral root branches toward available water. *Proceedings of the National Academy of Sciences of the USA* 111 (25): 9319.

Barnes, E.M., Sudduth, K.A., Hummel, J.W. et al. (2003). Remote-and ground-based sensor techniques to map soil properties. *Photogrammetric Engineering and Remote Sensing* 69 (6): 619–630.

Bates, T.R. and Lynch, J.P. (1996). Stimulation of root hair elongation in *Arabidopsis thaliana* by low phosphorus availability. *Plant, Cell & Environment* 19 (5): 529–538.

Benbi, D.K. and Richter, J. (2002). A critical review of some approaches to modelling nitrogen mineralization. *Biology and Fertility of Soils* 35 (3): 168–183.

Bengough, A.G., Gordon, D.C., Al-Menaie, H. et al. (2004). Gel observation chamber for rapid screening of root traits in cereal seedlings. *Plant and Soil* 262 (1): 63–70.

Bengough, A.G., Bransby, M.F., Hans, J. et al. (2005). Root responses to soil physical conditions; growth dynamics from field to cell. *Journal of Experimental Botany* 57 (2): 437–447.

Berntson, G.M. (1997). Topological scaling and plant root system architecture: developmental and functional hierarchies. *The New Phytologist* 135 (4): 621–634.

Blum, A. (2009). Effective use of water (EUW) and not water-use efficiency (WUE) is the target of crop yield improvement under drought stress. *Field Crops Research* 112 (2–3): 119–123.

Bochicchio, R., Sofo, A., Terzano, R. et al. (2015). Root architecture and morphometric analysis of *Arabidopsis thaliana* grown in Cd/Cu/Zn-gradient agar dishes: a new screening technique for studying plant response to metals. *Plant Physiology and Biochemistry* 91: 20–27.

Bodner, G., Leitner, D., Nakhforoosh, A. et al. (2013). A statistical approach to root system classification. *Frontiers in Plant Science* 4: 292.

Bodner, G., Leitner, D., and Kaul, H.P. (2014). Coarse and fine root plants affect pore size distributions differently. *Plant and Soil* 380 (1–2): 133–151.

Bodner, G., Nakhforoosh, A., and Kaul, H.-P. (2015). Management of crop water under drought: a review. *Agronomy for Sustainable Development* 35 (2): 401–442.

Bodner, G., Nakhforoosh, A., Arnold, T. et al. (2018). Hyperspectral imaging: a novel approach for plant root phenotyping. *Plant Methods* 14 (1): 84.

Bogena, H.R., Huisman, J.A., Güntner, A. et al. (2015). Emerging methods for noninvasive sensing of soil moisture dynamics from field to catchment scale: a review. *Wiley Interdisciplinary Reviews: Water* 2 (6): 635–647.

Böhm, W. (1979). Root parameters and their measurement. In: Methods of Studying Root Systems (ed. W. Böhm), 125–138. Berlin: Springer-Verlag.

Bolan, N.S. (1991). A critical review on the role of mycorrhizal fungi in the uptake of phosphorus by plants. *Plant and Soil* 134 (2): 189–207.

Borer, B., Tecon, R., and Or, D. (2018). Spatial organization of bacterial populations in response to oxygen and carbon counter-gradients in pore networks. *Nature Communications* 9 (1): 769.

Bucksch, A., Burridge, J., York, L.M. et al. (2014). Image-based high-throughput field phenotyping of crop roots. *Plant Physiology* 166: 470–486.

Burridge, J., Jochua, C.N., Bucksch, A. et al. (2016). Legume shovelomics: high-throughput phenotyping of common bean (*Phaseolus vulgaris* L.) and cowpea (*Vigna unguiculata* subsp., *unguiculata*) root architecture in the field. *Field Crops Research* 192: 21–32.

Bushamuka, V.N. and Zobel, R.W. (1998). Differential genotypic and root type penetration of compacted soil layers. *Crop Science* 38 (3): 776–781.

Cabrera-Bosquet, L., Crossa, J., von Zitzewitz, J. et al. (2012). High-throughput phenotyping and genomic selection: the frontiers of crop breeding converge. *Journal of Integrative Plant Biology* 54 (5): 312–320.

Cai, J.H., Zeng, Z.H., Connor, J.N. et al. (2015). RootGraph: a graphic optimization tool for automated image analysis of plant roots. *Journal of Experimental Botany* 66 (21): 6551–6562.

Cai, G., Vanderborght, J., Couvreur, V. et al. (2017). Parameterization of root water uptake models considering dynamic root distributions and water uptake compensation. *Vadose Zone Journal* 17: 1–21.

Canadell, J., Jackson, R.B., Ehleringer, J.B. et al. (1996). Maximum rooting depth of vegetation types at the global scale. *Oecologia* 108 (4): 583–595.

Cannon, W.A. (1949). A tentative classification of root systems. *Ecology* 30 (4): 542–548.

Carminati, A. and Vetterlein, D. (2012). Plasticity of rhizosphere hydraulic properties as a key for efficient utilization of scarce resources. *Annals of Botany* 112 (2): 277–290.

Chen, D. and Stow, D. (2002). The effect of training strategies on supervised classification at different spatial resolutions. *Photogrammetric Engineering and Remote Sensing* 68 (11): 1155–1162.

Chen, Y., Rengel, Z., Palta, J. et al. (2018). Efficient root systems for enhancing tolerance of crops to water and phosphorus limitation. *Indian Journal of Plant Physiology* 23: 689–696.

Cheng, W. and Kuzyakov, Y. (2005). Root effects on soil organic matter decomposition. In: Roots and Soil Management: Interactions between Roots and the Soil, Agronomy Series, vol. 48 (eds. R.W. Zobel and S.F. Wright), 119–143. Madison: Agronomy Society of America.

Chenu, K., Cooper, M., Hammer, G.L. et al. (2011). Environment characterization as an aid to wheat improvement: interpreting genotype-environment interactions by modelling water-deficit patterns in North-Eastern Australia. *Journal of Experimental Botany* 62 (6): 1743–1755.

Chimungu, J.G., Loades, K.W., and Lynch, J.P. (2015). Root anatomical phenes predict root penetration ability and biomechanical properties in maize (*Zea mays*). *Journal of Experimental Botany* 66 (11): 3151–3162.

Chitwood, D.H. and Topp, C.N. (2015). Revealing plant cryptotypes: defining meaningful phenotypes among infinite traits. *Current Opinion in Plant Biology* 24: 54–60.

Christopher, J., Christopher, M., Jennings, R. et al. (2013). QTL for root angle and number in a population developed from bread wheats (*Triticum aestivum*) with contrasting adaptation to water-limited environments. *Theoretical and Applied Genetics* 126 (6): 1563–1574.

Clark, L.J. and Barraclough, P.B. (1999). Do dicotyledons generate greater maximum axial root growth pressures than monocotyledons? *Journal of Experimental Botany* 50 (336): 1263–1266.

Clark, L.J., Whalley, W.R., and Barraclough, P.B. (2003). How do roots penetrate strong soil? In: Roots: The Dynamic Interface Between Plants and the Earth (ed. J.J. Abe), 93–104. Heidelberg: Springer.

Clark, L.J., Price, A.H., Steele, K.A. et al. (2008). Evidence from near-isogenic lines that root penetration increases with root diameter and bending stiffness in rice. *Functional Plant Biology* 35 (11): 1163–1171.

Collette, L., Hodgkin, T., Kassam, A.H., and Kenmore, P. (2011). Save and Grow: A Policymaker's Guide to Sustainable Intensification of Smallholder Crop Production. Rome: FAO.

Colombi, T. and Walter, A. (2016). Root responses of triticale and soybean to soil compaction in the field are reproducible under controlled conditions. *Functional Plant Biology* 43 (2): 114–128.

Colombi, T., Torres, L.C., Walter, A. et al. (2018). Feedbacks between soil penetration resistance., root architecture and water uptake limit water accessibility and crop growth – a vicious circle. *Science of the Total Environment* 626: 1026–1035.

Comas, L.H., Becker, S.R., Von Mark, V.C. et al. (2013). Root traits contributing to plant productivity under drought. *Frontiers in Plant Sciences* 4: 442.

Couvreur, V., Faget, M., Lobet, G. et al. (2018). Going with the flow: multiscale insights into the composite nature of water transport in roots. *Plant physiology* 178 (4): 1689–1703.

Cruz, C., Green, J.J., Watson, C.A. et al. (2004). Functional aspects of root architecture and mycorrhizal inoculation with respect to nutrient uptake capacity. *Mycorrhiza* 14 (3): 177–184.

Czyż, E.A. and Dexter, A.R. (2013). Plant wilting can be caused either by the plant or by the soil. *Soil Research* 50 (8): 708–713.

Darrah, P.R., Jones, D.L., Kirk, G.J.D. et al. (2006). Modelling the rhizosphere: a review of methods for 'upscaling' to the whole-plant scale. *European Journal of Soil Science* 57 (1): 13–25.

Das, A., Schneider, H., Burridge, J. et al. (2015). Digital imaging of root traits (DIRT): a high-throughput computing and collaboration platform for field-based root phenomics. *Plant Methods* 11: 51.

Delory, B.M., Baudson, C., Brostaux, Y. et al. (2016). ArchiDART: an R package for the automated computation of plant root architectural traits. *Plant and Soil* 398 (1–2): 351–365.

Den Herder, G., Van Isterdael, G., Beeckman, T. et al. (2010). The roots of a new green revolution. *Trends in Plant Science* 15 (11): 600–607.

Dolan, M.S., Dowdy, R.H., Voorhees, W.B. et al. (1992). Corn phosphorus and potassium uptake in response to soil compaction. *Agronomy Journal* 84 (4): 639–642.

Domingos, P. (2012). A few useful things to know about machine learning. *Communications of the ACM* 55 (10): 78–87.

de Dorlodot, S., Forster, B., Pagès, L. et al. (2007). Root system architecture: opportunities and constraints for genetic improvement of crops. *Trends in Plant Science* 12 (10): 474–481.

Doussan, C., Pagès, L., and Vercambre, G. (1998). Modelling of the hydraulic architecture of root systems: an integrated approach to water absorption – model description. *Annals of Botany* 81 (2): 213–223.

Downie, H.F., Adu, M.O., Schmidt, S. et al. (2015). Challenges and opportunities for quantifying roots and rhizosphere interactions through imaging and image analysis. *Plant, Cell & Environment* 38 (7): 1213–1232.

Drew, M.C., He, C.-J., and Morgan, P.W. (2000). Programmed cell death and aerenchyma formation in roots. *Trends in Plant Science* 5 (3): 123–127.

Du, Y. and Scheres, B. (2017). Lateral root formation and the multiple roles of auxin. *Journal of Experimental Botany* 69 (2): 155–167.

Dunbabin, V., Diggle, A., and Rengel, Z. (2003). Is there an optimal root architecture for nitrate capture in leaching environments? *Plant, Cell & Environment* 26 (6): 835–844.

Dunbabin, V.M., Postma, J.A., Schnepf, A. et al. (2013). Modelling root – soil interactions using three-dimensional models of root growth, architecture and function. *Plant and Soil* 372: 93–124.

Duncan, E.G., O'Sullivan, C.A., Roper, M.M. et al. (2018). Yield and nitrogen use efficiency of wheat increased with root length and biomass due to nitrogen, phosphorus, and potassium interactions. *Journal of Plant Nutrition and Soil Science* 181 (3): 364–373.

Eapen, D., Barroso, M.L., Ponce, G. et al. (2005). Hydrotropism: root growth responses to water. *Trends in Plant Science* 10 (1): 44–50.

Eissenstat, D.M., Wells, C.E., Yanai, R.D. et al. (2000). Building roots in a changing environment: implications for root longevity. *New Phytologist* 147 (1): 33–42.

Ephrath, J.E., Silberbush, M., and Berliner, P.R. (1999). Calibration of minirhizotron readings against root length density data obtained from soil cores. *Plant and Soil* 209: 201–208.

Fahlgren, N., Gehan, M.A., and Baxter, I. (2015). Lights, camera, action: high-throughput plant phenotyping is ready for a close-up. *Current Opinion in Plant Biology* 24: 93–99.

Fan, J., McConkey, B., Wang, H. et al. (2016). Root distribution by depth for temperate agricultural crops. *Field Crops Research* 189: 68–74.

Feddes, R.A. and Raats, P.A.C. (2004). Parameterizing the soil-water-plant root system. *Unsaturated-Zone Modeling: Progress, Challenges, Applications* 6: 95–141.

Fitter, A.H. (1987). An architectural approach to the comparative ecology of plant root systems. *New Phytologist* 106: 61–77.

Fitter, A.H. and Stickland, T.R. (1991). Architectural analysis of plant root systems. 2. Influence of nutrient supply on architecture in contrasting plant species. *New Phytologist* 118 (3): 383–389.

Fitter, A.H., Stickland, T.R., Harvey, M.L., and Wilson, G.W. (1991). Architectural analysis of plant root systems 1. Architectural correlates of exploitation efficiency. *New Phytologist* 118 (3): 375–382.

Ford, K.E., Gregory, P.J., Gooding, M.J. et al. (2006). Genotype and fungicide effects on late-season root growth of winter wheat. *Plant and Soil* 284 (1–2): 33–44.

Forde, B. and Lorenzo, H. (2001). The nutritional control of root development. *Plant and Soil* 232 (1–2): 51–68.

Gahoonia, T.S. and Nielsen, N.E. (2004). Root traits as tools for creating phosphorus efficient crop varieties. *Plant and Soil* 260 (1–2): 47–57.

Galvan-Ampudia, C.S., Julkowska, M.M., Darwish, E. et al. (2013). Halotropism is a response of plant roots to avoid a saline environment. *Current Biology* 23 (20): 2044–2050.

Garré, S., Javaux, M., Vanderborght, J. et al. (2011). Three-dimensional electrical resistivity tomography to monitor root zone water dynamics. *Vadose Zone Journal* 10 (1): 412–424.

Gauch, H. and Zobel, R.W. (1997). Identifying mega-environments and targeting genotypes. *Crop Science* 37 (2): 311–326.

Ghanbarian, B., Hunt, A.G., Ewing, R.P. et al. (2013). Tortuosity in porous media: a critical review. *Soil Science Society of America Journal* 77 (5): 1461–1477.

Gioia, T., Galinski, A., Lenz, H. et al. (2017). *GrowScreen-PaGe*, a non-invasive, high-throughput phenotyping system based on germination paper to quantify crop phenotypic diversity and plasticity of root traits under varying nutrient supply. *Functional Plant Biology* 44 (1): 76–93.

Gitay, H. and Noble, I.R. (1997). What are functional types and how should we seek them. In: Plant Functional Types: Their Relevance to Ecosystem Properties and Global Change (eds. T.M. Smith, H.H. Shugart and F.I. Woodward), 3–19. Cambridge: Cambridge University Press.

Goebel, K. (1930). Organographie der Pflanzen, 2e. Jena: G. Fischer.

Guo, L., Chen, J., Cui, X. et al. (2013). Application of ground penetrating radar for coarse root detection and quantification: a review. *Plant and Soil* 362 (1–2): 1–23.

Guseman, J.M., Hellmuth, A., Lanctot, A. et al. (2015). Auxin-induced degradation dynamics set the pace for lateral root development. *Development* 142 (5): 905–909.

Guyon, I. and Elisseeff, A. (2003). An introduction to variable and feature selection. *Journal of Machine Learning Research* 3: 1157–1182.

Hammer, G., Cooper, M., Tardieu, F. et al. (2006). Models for navigating biological complexity in breeding improved crop plants. *Trends in Plant Science* 11 (12): 587–593.

Hammer, G.L., Dong, Z., McLean, G. et al. (2009). Can changes in canopy and/or root system architecture explain historical maize yield trends in the US corn belt? *Crop Science* 49 (1): 299–312.

Hamza, M.A. and Anderson, W.K. (2005). Soil compaction in cropping systems: a review of the nature, causes and possible solutions. *Soil and Tillage Research* 82 (2): 121–145.

Hargreaves, C.E., Gregory, P.J., and Bengough, A.G. (2009). Measuring root traits in barley (*Hordeum vulgare* ssp *vulgare* and ssp *spontaneum*) seedlings using gel chambers, soil sacs and X-ray microtomography. *Plant and Soil* 316 (1–2): 285–297.

Hayashi, T., Yoshida, T., Fujii, K. et al. (2013). Maintained root length density contributes to the waterlogging tolerance in common wheat (*Triticum aestivum* L.). *Field Crops Research* 152: 27–35.

Hinsinger, P. (2001). Bioavailability of soil inorganic P in the rhizosphere as affected by root-induced chemical changes: a review. *Plant and Soil* 237 (2): 173–195.

Hinsinger, P., Bengough, A.G., Vetterlein, D. et al. (2009). Rhizosphere: biophysics, biogeochemistry and ecological relevance. *Plant and Soil* 321 (1–2): 117–152.

Ho, M.D., McCannon, B.C., and Lynch, J.P. (2004). Optimization modeling of plant root architecture for water and phosphorus acquisition. *Journal of Theoretical Biology* 226 (3): 331–340.

Hochholdinger, F., Woll, K., Sauer, M. et al. (2004). Genetic dissection of root formation in maize (*Zea mays*) reveals root-type specific developmental programmes. *Annals of Botany* 93 (4): 359–368.

Hodge, A. (2004). The plastic plant: root responses to heterogeneous supplies of nutrients. *New Phytologist* 162 (1): 9–24.

Hodgkinson, L., Dodd, I.C., Binley, A. et al. (2017). Root growth in field-grown winter wheat: some effects of soil conditions, season and genotype. *European Journal of Agronomy* 91: 74–83.

Holbrook, N.M., Ahrens, E.T., Burns, M.J. et al. (2001). *in vivo* observation of cavitation and embolism repair using magnetic resonance imaging. *Plant Physiology* 126 (1): 27–31.

Iijima, M., Higuchi, T., and Barlow, P.W. (2004). Contribution of root cap mucilage and presence of an intact root cap in maize (*Zea mays*) to the reduction of soil mechanical impedance. *Annals of Botany* 94 (3): 473–477.

IUSS Working Group WRB (2015). World reference base for soil resources 2014, update 2015. In: International soil classification system for naming soils and creating legends for soil maps. World Soil Resources Reports 106. Rome, Italy: FAO.

Iyer-Pascuzzi, A.S., Symonova, O., Mileyko, Y. et al. (2010). Imaging and analysis platform for automatic phenotyping and trait ranking of plant root systems. *Plant Physiology* 152 (3): 1148–1157.

Jackson, R.B., Canadell, J., Ehleringer, J.R. et al. (1996). A global analysis of root distributions for terrestrial biomes. *Oecologia* 108 (3): 389–411.

Jeudy, C., Adrian, M., Baussard, C. et al. (2016). RhizoTubes as a new tool for high throughput imaging of plant root development and architecture: test, comparison with pot grown plants and validation. *Plant Methods* 12 (1): 31.

Justin, S. and Armstrong, W. (1987). The anatomical characteristics of roots and plant response to soil flooding. *New Phytologist* 106 (3): 465–495.

Kage, H. and Ehlers, W. (1996). Does transport of water to roots limit water uptake of field crops? *Zeitschrift für Pflanzenernährung und Bodenkunde* 159 (6): 583–590.

Kalogiros, D.I., Adu, M.O., White, P.J. et al. (2016). Analysis of root growth from a phenotyping data set using a density-based model. *Journal of Experimental Botany* 67 (4): 1045–1058.

Kamoshita, A., Wade, L., Ali, M. et al. (2002). Mapping QTLs for root morphology of a rice population adapted to rainfed lowland conditions. *Theoretical and Applied Genetics* 104 (5): 880–893.

Kashiwagi, J., Krishnamurthy, L., Crouch, J.H. et al. (2006). Variability of root length density and its contributions to seed yield in chickpea (*Cicer arietinum* L.) under terminal drought stress. *Field Crops Research* 95 (2–3): 171–181.

Keiluweit, M., Wanzek, T., Kleber, M. et al. (2017). Anaerobic microsites have an unaccounted role in soil carbon stabilization. *Nature Communications* 8 (1): 1771.

Khan, M.S., Zaidi, A., and Wani, P.A. (2007). Role of phosphate-solubilizing microorganisms in sustainable agriculture – a review. *Agronomy for Sustainable Development* 27 (1): 29–43.

Klotzsche, A., Jonard, F., Looms, M.C. et al. (2018). Measuring soil water content with ground penetrating radar: a decade of progress. *Vadose Zone Journal* 17 (1): 1–9.

Knipfer, T., Cuneo, I., Brodersen, C. et al. (2016). In-situ visualization of the dynamics in xylem embolism formation and removal in the absence of root pressure: a study on excised grapevine stems. *Plant Physiology* 171: 1024–1036.

Kuijken, R.C.P., van Eeuwijk, F.A., Marcelis, L.F.M. et al. (2015). Root phenotyping: from component trait in the lab to breeding. *Journal of Experimental Botany* 66 (18): 5389–5401.

Kutschera, L., Sobotik, M., Lichtenegger, E. et al. (1995). Bewurzelung von Pflanzen in den verschiedenen Lebensräumen. Land Oberösterreich: OÖ. Landesmuseum.

Kutschera, L., Lichtenegger, E., and Sobotik, M. (2009). Wurzelatlas der Kulturpflanzen gemäßigter Gebiete mit Arten des Feldgemüsebaues. Frankfurt: DLG-Verlag.

Kuzyakov, Y. and Domanski, G. (2000). Carbon input by plants into the soil Review. *Journal of Plant Nutrition and Soil Science* 163 (4): 421–431.

Lacointe, A. and Minchin, P.E. (2008). Modelling phloem and xylem transport within a complex architecture. *Functional Plant Biology* 35 (10): 772–780.

Laio, F., D'Odorico, P., and Ridolfi, L. (2006). An analytical model to relate the vertical root distribution to climate and soil properties. *Geophysical Research Letters* 33: L18401.

Landl, M., Huber, K., Schnepf, A. et al. (2017). A new model for root growth in soil with macropores. *Plant and Soil* 415 (1–2): 99–116.

Landl, M., Schnepf, A., Vanderborght, J. et al. (2018). Measuring root system traits of wheat in 2D images to parameterize 3D root architecture models. *Plant and Soil* 425 (1–2): 457–477.

Laskowski, M. and Kirsten, H. (2017). Periodic lateral root priming: what makes it tick? *The Plant Cell* 29 (3): 432–444.

Lavorel, S., McIntyre, S., Landsberg, J. et al. (1997). Plant functional classifications: from general groups to specific groups based on response to disturbance. *Trends in Ecology and Evolution* 12 (12): 474–478.

Lazarovitch, N., Vanderborght, J., Jin, Y. et al. (2018). The root zone: soil physics and beyond. *Vadose Zone Journal* 17: 180002.

Le Marié, C., Kirchgessner, N., Marschall, D. et al. (2014). Rhizoslides: paper-based growth system for non-destructive, high throughput phenotyping of root development by means of image analysis. *Plant Methods* 10 (1): 1.

Leitner, D., Klepsch, S., Bodner, G. et al. (2010). A dynamic root system growth model based on L-systems. *Plant and Soil* 332 (1–2): 177–192.

Leitner, D., Felderer, B., Vontobel, P. et al. (2014a). Recovering root system traits using image analysis exemplified by two-dimensional neutron radiography images of lupine. *Plant Physiology* 164: 24–35.

Leitner, D., Meunier, F., Bodner, G. et al. (2014b). Impact of contrasted maize root traits at flowering on water stress tolerance – a simulation study. *Field Crops Research* 165: 125–137.

Lipiec, J., Horn, R., Pietrusiewicz, J. et al. (2012). Effects of soil compaction on root elongation and anatomy of different cereal plant species. *Soil and Tillage Research* 121: 74–81.

Little, D.Y., Rao, H., Oliva, S. et al. (2005). The putative high-affinity nitrate transporter NRT2.1 represses lateral root initiation in response to nutritional cues. *Proceedings of the National Academy of Sciences of the USA* 102 (38): 13693–13698.

Lobet, G. (2017). Image analysis in plant sciences: publish then perish. *Trends in Plant Science* 22 (7): 559–566.

Lobet, G., Couvreur, V., Meunier, F. et al. (2014). Plant water uptake in drying soils. *Plant Physiology* 164: 1619–1627.

Lobet, G., Pound, M.P., Diener, J. et al. (2015). Root system markup language: toward a unified root architecture description language. *Plant Physiology* 167 (3): 617–627.

Lobet, G., Koevoets, I.T., Noll, M. et al. (2017). Using a structural root system model to evaluate and improve the accuracy of root amage analysis pipelines. *Frontiers in Plant Science* 8: 447.

López-Bucio, J., Hernández-Abreu, E., Sánchez-Calderón, L. et al. (2002). Phosphate availability alters architecture and causes changes in hormone sensitivity in the Arabidopsis root system. *Plant Physiology* 129 (1): 244–256.

Lynch, J.P. (2007a). Rhizoeconomics: the roots of shoot growth limitations. *Hortscience* 42 (5): 1107–1109.

Lynch, J.P. (2007b). Roots of the second green revolution. *Australian Journal of Botany* 55 (5): 493–512.

Lynch, J.P. (2011). Root phenes for enhanced soil exploration and phosphorus acquisition: tools for future crops. *Plant Physiology* 156: 1041–1049.

Lynch, J.P. and Brown, K.M. (2001). Topsoil foraging – an architectural adaptation of plants to low phosphorus availability. *Plant and Soil* 237 (2): 225–237.

Lynch, J.P. and Clair, S.B.S. (2004). Mineral stress: the missing link in understanding how global climate change will affect plants in real world soils. *Field Crops Research* 90 (1): 101–115.

Lynch, J.P. and Ho, M.D. (2005). Rhizoeconomics: carbon costs of phosphorus acquisition. *Plant and Soil* 269 (1–2): 45–56.

Ma, Z., Bielenberg, D.G., Brown, K.M. et al. (2001). Regulation of root hair density by phosphorus availability in *Arabidopsis thaliana*. *Plant, Cell & Environment* 24 (4): 459–467.

Ma, Z., Guo, D., Xu, X. et al. (2018). Evolutionary history resolves global organization of root functional traits. *Nature* 555: 94–97.

MacDonald, G.K., Bennett, E.M., Potter, P.A. et al. (2011). Agronomic phosphorus imbalances across the world's croplands. *Proceedings of the National Academy of Sciences of the USA* 108 (7): 3086–3091.

Mairhofer, S., Zappala, S., Tracy, S.R. et al. (2012). RooTrak: automated recovery of three-dimensional plant root architecture in soil from x-ray microcomputed tomography images using visual tracking. *Plant Physiology* 158 (2): 561–569.

Mairhofer, S., Pridmore, T., Johnson, J. et al. (2017). X-ray computed tomography of crop plant root systems grown in soil. *Current Protocols in Plant Biology* 2 (4): 270–286.

Majdi, H. (1996). Root sampling methods-applications and limitations of the minirhizotron technique. *Plant and Soil* 185 (2): 255–258.

Malamy, J.E. (2005). Intrinsic and environmental response pathways that regulate root system architecture. *Plant, Cell & Environment* 28 (1): 67–77.

Manschadi, A.M., Christopher, J., and Hammer, G.L. (2006). The role of root architectural traits in adaptation of wheat to water-limited environments. *Functional Plant Biology* 33 (9): 823–837.

Martre, P., Quilot-Turion, B., Luquet, D. et al. (2015). Model-assisted phenotyping and ideotype design. In: Crop Physiology. Applications for Genetic Improvement and Agronomy, 2e (eds. V.O. Sadras and D.F. Calderini), 349–373. Amsterdam, The Netherlands: Elsevier.

Materechera, S.A., Alston, A.M., Kirby, J.M. et al. (1992). Influence of root diameter on the penetration of seminal roots into a compacted subsoil. *Plant and Soil* 144 (2): 297–303.

McCully, M.E. (1999). Root xylem embolisms and refilling. Relation to water potentials of soil, roots, and leaves, and osmotic potentials of root xylem sap. *Plant Physiology* 119 (3): 1001–1008.

McGaughey, S.A., Qiu, J., Tyerman, S.D., and Byrt, C.S. (2018). Regulating root aquaporin function in response to changes in salinity. *Annual Plant Reviews online*: 381–416.

Metzner, R., Eggert, A., van Dusschoten, D. et al. (2015). Direct comparison of MRI and X-ray CT technologies for 3D imaging of root systems in soil: potential and challenges for root trait quantification. *Plant Methods* 11 (1): 17.

Mitra, P., Murthy, C.A., and Pal, S.K. (2002). Unsupervised feature selection using feature similarity. *IEEE Transactions on Pattern Analysis and Machine Intelligence* 24 (3): 301–312.

Mullen, J.L. and Hangarter, R.P. (2003). Genetic analysis of the gravitropic set-point angle in lateral roots of Arabidopsis. *Advances in Space Research* 31 (10): 2229–2236.

Nagel, K.A., Putz, A., Gilmer, F. et al. (2012). GROWSCREEN-Rhizo is a novel phenotyping robot enabling simultaneous measurements of root and shoot growth for plants grown in soil-filled rhizotrons. *Functional Plant Biology* 39 (11): 891–904.

Nakhforoosh, A., Grausgruber, H., Kaul, H.-P. et al. (2014). Wheat root diversity and root functional characterization. *Plant and Soil* 380 (1–2): 211–229.

Nakhforoosh, A., Grausgruber, H., Kaul, H.-P. et al. (2015). Dissection of drought response of modern and underutilized wheat varieties according to Passioura's yield-water framework. *Frontiers in Plant Science* 6: 570.

Ndour, A., Vadez, V., Pradal, C., and Lucas, M. (2017). Virtual plants need water too: functional-structural root system models in the context of drought tolerance breeding. *Frontiers in Plant Science* 8: 1577.

Neumann, K., Verburg, P.H., Stehfest, E. et al. (2010). The yield gap of global grain production: a spatial analysis. *Agricultural Systems* 103 (5): 316–326.

Nibau, C., Gibbs, D.J., and Coates, J.C. (2008). Branching out in new directions: the control of root architecture by lateral root formation. *New Phytologist* 179 (3): 595–614.

Nielsen, D.R. and Wendroth, O. (2003). Spatial and Temporal Statistics: Sampling Field Soils and Their Vegetation. Catena Verlag.

Niu, Y.F., Chai, R.S., Jin, G.L. et al. (2012). Responses of root architecture development to low phosphorus availability: a review. *Annals of Botany* 112 (2): 391–408.

Novák, V. (2009). Physiological drought–how to quantify it? In: Bioclimatology and Natural Hazards (eds. K. Střelcová, C. Mátyás, A. Kleidon, et al.), 89–95. Heidelberg: Springer.

Oldeman, L.R., Hakkeling, R.T.A., and Sombroek, W.G. (2017). World Map of the Status of Human-Induced Soil Degradation: An Explanatory Note. Wageningen: International Soil Reference and Information Centre.

Oleghe, E., Naveed, M., Baggs, E.M. et al. (2017). Plant exudates improve the mechanical conditions for root penetration through compacted soils. *Plant and Soil* 421 (1–2): 19–30.

Osaki, M., Shinano, T., Matsumoto, M. et al. (1997). A root-shoot interaction hypothesis for high productivity of field crops. *Soil Science and Plant Nutrition* 43 (Suppl 1): 1079–1084.

Ostonen, I., Püttsepp, Ü., Biel, C. et al. (2007). Specific root length as an indicator of environmental change. *Plant Biosystems* 141 (3): 426–442.

Paez-Garcia, A., Motes, C., Scheible, W.-R. et al. (2015). Root traits and phenotyping strategies for plant improvement. *Plants* 4 (2): 334–355.

Pagès, L., Vercambre, G., Drouet, J.-L. et al. (2004). RootTyp: a generic model to depict and analyse the root system architecture. *Plant and Soil* 258: 103–119.

Palta, J.A., Chen, X., Milroy, S.P. et al. (2011). Large root systems: are they useful in adapting wheat to dry environments? *Functional Plant Biology* 38 (5): 347–354.

Passioura, J.B. (1983). Roots and drought resistance. *Agricultural Water Management* 7 (1–3): 265–280.

Peng, J.H., Sun, D., and Nevo, E. (2011). Domestication evolution, genetics and genomics in wheat. *Molecular Breeding* 28 (3): 281.

Péret, B., Middleton, A.M., French, A.P. et al. (2013). Sequential induction of auxin efflux and influx carriers regulates lateral root emergence. *Molecular Systems Biology* 9 (1): 699.

Pfeifer, J., Faget, M., Walter, A. et al. (2014). Spring barley shows dynamic compensatory root and shoot growth responses when exposed to localised soil compaction and fertilisation. *Functional Plant Biology* 41 (6): 581–597.

Postma, J.A., Dathe, A., and Lynch, J. (2014). The optimal lateral root branching density for maize depends on nitrogen and phosphorus availability. *Plant Physiology* 166: 590–602.

Postma, J.A., Kuppe, C., Owen, M.R. et al. (2017). OpenSimRoot: widening the scope and application of root architectural models. *New Phytologist* 215 (3): 1274–1286.

Pound, M.P., Atkinson, J.A., Townsend, A.J. et al. (2017). Deep machine learning provides state-of-the-art performance in image-based plant phenotyping. *GigaScience* 6 (10): 1–10.

Rajaram, S., Van Ginkel, M., Fischer, R.A. (1993). CIMMYT's wheat breeding mega-environments (ME). Proceedings of the 8th International Wheat Genetic Symposium, pp. 1101–1106.

Rebetzke, G.J. (2016). From inspiration to impact: delivering value from global root research. *Journal of Experimental Botany* 67 (12): 3601–3603.

Rellán-Álvarez, R., Lobet, G., Lindner, H. et al. (2015). GLO-Roots: an imaging platform enabling multidimensional characterization of soil-grown root systems. *Elife* 4: e07597.

Rewald, B., Meinen, C., Trockenbrodt, M. et al. (2012). Root taxa identification in plant mixtures–current techniques and future challenges. *Plant and Soil* 359 (1–2): 165–182.

Rewald, B., Kunze, M.E., and Godbold, D.L. (2016). NH 4: NO 3 nutrition influence on biomass productivity and root respiration of poplar and willow clones. *Gcb Bioenergy* 8 (1): 51–58.

Reynolds, M. and Langridge, P. (2016). Physiological breeding. *Current Opinion in Plant Biology* 31: 162–171.

Reynolds, M., Foulkes, M.J., Slafer, G.A. et al. (2009). Raising yield potential in wheat. *Journal of Experimental Botany* 60 (7): 1899–1918.

Richard, C.A.I., Hickey, L.T., Fletcher, S. et al. (2015). High-throughput phenotyping of seminal root traits in wheat. *Plant Methods* 11 (1): 13.

Richards, R.A. (2008). Genetic opportunities to improve cereal root systems for dryland agriculture. *Plant Production Science* 11 (1): 12–16.

Robbins, N.E. and Dinneny, J.R. (2018). Growth is required for perception of water availability to pattern root branches in plants. *Proceedings of the National Academy of Sciences of the USA* 115 (4): E822–E831.

Rodriguez-Iturbe, I., Porporato, A., Laio, F. et al. (2001). Plants in water-controlled ecosystems: active role in hydrologic processes and response to water stress: I. scope and general outline. *Advances in Water Resources* 24 (7): 695–705.

Rogers, E.D., Monaenkova, D., Mijar, M. et al. (2016). X-ray computed tomography reveals the response of root system architecture to soil texture. *Plant Physiology* 171: 2028–2040.

Roose, T. and Schnepf, A. (2008). Mathematical models of plant–soil interaction. *Philosophical Transactions of the Royal Society of London A: Mathematical, Physical and Engineering Sciences* 366 (1885): 4597–4611.

Roose, T., Keyes, S.D., Daly, K.R. et al. (2016). Challenges in imaging and predictive modeling of rhizosphere processes. *Plant and Soil* 407 (1–2): 9–38.

Rose, T.J., Impa, S.M., Rose, M.T. et al. (2012). Enhancing phosphorus and zinc acquisition efficiency in rice: a critical review of root traits and their potential utility in rice breeding. *Annals of Botany* 112 (2): 331–345.

Roychoudhry, S., Del Bianco, M., Kieffer, M. et al. (2013). Auxin controls gravitropic setpoint angle in higher plant lateral branches. *Current Biology* 23 (15): 1497–1504.

Ruta, N., Liedgens, M., Fracheboud, Y. et al. (2010). QTLs for the elongation of axile and lateral roots of maize in response to low water potential. *Theoretical and Applied Genetics* 120 (3): 621–631.

Ryser, P. (2006). The mysterious root length. *Plant and Soil* 286 (1–2): 1–6.

Saab, I.N., Sharp, R.E., Pritchard, J. et al. (1990). Increased endogenous abscisic acid maintains primary root growth and inhibits shoot growth of maize seedlings at low water potentials. *Plant Physiology* 93 (4): 1329–1336.

Sanguineti, M.C., Li, S., Maccaferri, M. et al. (2007). Genetic dissection of seminal root architecture in elite durum wheat germplasm. *Annals of Applied Biology* 151 (3): 291–305.

Sathya, R. and Abraham, A. (2013). Comparison of supervised and unsupervised learning algorithms for pattern classification. *International Journal of Advanced Research in Artificial Intelligence* 2 (2): 34–38.

Schachtman, D.P., Reid, R.J., and Ayling, S.M. (1998). Phosphorus uptake by plants: from soil to cell. *Plant Physiology* 116 (2): 447–453.

Schelle, H., Iden, S.C., Fank, J. et al. (2012). Inverse estimation of soil hydraulic and root distribution parameters from lysimeter data. *Vadose Zone Journal* 11 (4): vzj2011.0169.

Schenk, H.J. and Jackson, R.B. (2002). The global biogeography of roots. *Ecological Monographs* 72 (3): 311–328.

Schneider, H., Postma, J.A., Wojciechowski, T. et al. (2017). Root cortical senescence improves growth under suboptimal availability of N, P, and K. *Plant Physiology* 174: 2333–2347.

Schnepf, A., Leitner, D., Schweiger, P.F. et al. (2016). L-system model for the growth of arbuscular mycorrhizal fungi, both within and outside of their host roots. *Journal of The Royal Society Interface* 13 (117): 20160129.

Schnepf, A., Leitner, D., Landl, M. et al. (2018). CRootBox: a structural–functional modelling framework for root systems. *Annals of Botany* 121 (5): 1033–1053.

Schulz, H., Postma, J., Dusschoten, D. et al. (2013). Plant root system analysis from MRI images. In: Computer Vision, Imaging and Computer Graphics Theory and Application, vol. 359 (eds. G. Csurka, M. Kraus, R. Laramee, et al.), 411–425. Heidelberg: Springer.

Schwen, A., Bodner, G., and Loiskandl, W. (2011a). Time-variable soil hydraulic properties in near-surface soil water simulations for different tillage methods. *Agricultural Water Management* 99 (1): 42–50.

Schwen, A., Bodner, G., Scholl, P. et al. (2011b). Temporal dynamics of soil hydraulic properties and the water-conducting porosity under different tillage. *Soil and Tillage Research* 113 (2): 89–98.

Serraj, R., Krishnamurthy, L., Kashiwagi, J. et al. (2004). Variation in root traits of chickpea (*Cicer arietinum* L.) grown under terminal drought. *Field Crops Research* 88 (2–3): 115–127.

Shahzad, Z., Kellermeier, F., Armstrong, E.M. et al. (2018). EZ-root-VIS: a software pipeline for the rapid analysis and visual reconstruction of root system architecture. *Plant physiology* 177 (4): 1368–1381.

Siddique, K.H.M., Belford, R.K., and Tennant, D. (1990). Root: shoot ratios of old and modern., tall and semi-dwarf wheats in a Mediterranean environment. *Plant and Soil* 121 (1): 89–98.

Siddique, K.H.M., Chen, Y.L., and Rengel, Z. (2015). Efficient root system for abiotic stress tolerance in crops. *Procedia Environmental Sciences* 29: 295.

Silva, A.P., Kay, B.D., and Perfect, E. (1994). Characterization of the least limiting water range of soils. *Soil Science Society of America Journal* 58 (6): 1775–1781.

Singh, A., Ganapathysubramanian, B., Singh, A.K. et al. (2016). Machine learning for high-throughput stress phenotyping in plants. *Trends in Plant Science* 21 (2): 110–124.

Smit, A.L., Bengough, A.G., Engels, C. et al. (2013). Root Methods: A Handbook. Berlin: Springer.

Somma, F., Clausnitzer, V., and Hopmans, J.W. (1997). An algorithm for three-dimensional simultaneous modeling of root growth, transient water flow, and transport and uptake. *Plant Soil* 202: 281–293.

Sperry, J.S., Adler, F.R., Campbell, G.S. et al. (1998). Limitation of plant water use by rhizosphere and xylem conductance: results from a model. *Plant, Cell & Environment* 21 (4): 347–359.

Steudle, E. (2000). Water uptake by plant roots: an integration of views. *Plant and Soil* 226 (1): 45–56.

Sultan, S.E. (2000). Phenotypic plasticity for plant development, function and life history. *Trends in Plant Science* 5 (12): 537–542.

Tardieu, F. (2011). Any trait or trait-related allele can confer drought tolerance: just design the right drought scenario. *Journal of Experimental Botany* 63 (1): 25–31.

Tardieu, F. and Tuberosa, R. (2010). Dissection and modelling of abiotic stress tolerance in plants. *Current Opinion in Plant Biology* 13 (2): 206–212.

Taub, D.R. and Goldberg, D. (1996). Root system topology of plants from habitats differing in soil resource availability. *Functional Ecology* 10 (2): 258–264.

Taulemesse, F., Le Gouis, J., Gouache, D. et al. (2015). Post-flowering nitrate uptake in wheat is controlled by N status at flowering, with a putative major role of root nitrate transporter NRT2.1. *PLoS ONE* 10 (3): e0120291.

Teixeira, E.I., Fischer, G., van Velthuizen, H. et al. (2013). Global hot-spots of heat stress on agricultural crops due to climate change. *Agricultural and Forest Meteorology* 170: 206–215.

Thomas, C.L., Graham, N.S., Hayden, R. et al. (2016). High-throughput phenotyping (HTP) identifies seedling root traits linked to variation in seed yield and nutrient capture in field-grown oilseed rape (*Brassica napus* L.). *Annals of Botany* 118 (4): 655–665.

Thorup-Kristensen, K. (2001). Are differences in root growth of nitrogen catch crops important for their ability to reduce soil nitrate-N content, and how can this be measured? *Plant and Soil* 230 (2): 185–195.

Thorup-Kristensen, K., Cortasa, M.S., and Loges, R. (2009). Winter wheat roots grow twice as deep as spring wheat roots, is this important for N uptake and N leaching losses? *Plant and Soil* 322 (1–2): 101–114.

Trachsel, S., Kaeppler, S.M., Brown, K.M. et al. (2011). Shovelomics: high throughput phenotyping of maize (*Zea mays* L.) root architecture in the field. *Plant and Soil* 341 (1–2): 75–87.

Tracy, S.R., Daly, K.R., Sturrock, C.J. et al. (2015). Three-dimensional quantification of soil hydraulic properties using X-ray computed tomography and image-based modeling. *Water Resources Research* 51 (2): 1006–1022.

Tron, S., Bodner, G., Laio, F. et al. (2015). Can diversity in root architecture explain plant water use efficiency? A modeling study. *Ecological Modelling* 312: 200–210.

Uga, Y. (2012). Quantitative measurement of root growth angle by using the basket method. In: Methodologies for Root Drought Studies in Rice, 22–26. The Philippines: *International Rice Research Institute.*

Vadez, V. (2014). Root hydraulics: the forgotten side of roots in drought adaptation. *Field Crops Research* 165: 15–24.

Van Velthuizen, H. (2007). Mapping Biophysical Factors that Influence Agricultural Production and Rural Vulnerability. Rome: FAO.

Verma, A., Kumar, S., Kumar, G. et al. (2018). Rhizosphere metabolite profiling: an opportunity to understand plant-microbe interactions for crop improvement. In: Crop Improvement Through Microbial Biotechnology (eds. R. Prasad, S.S. Gill and N. Tuteja), 343–361. Amsterdam, The Netherlands: Elsevier.

Vicente-Serrano, S.M., Beguería, S., Lorenzo-Lacruz, J. et al. (2012). Performance of drought indices for ecological, agricultural, and hydrological applications. *Earth Interactions* 16 (10): 1–27.

Waines, J.G. and Ehdaie, B. (2007). Domestication and crop physiology: roots of green-revolution wheat. *Annals of Botany* 100 (5): 991–998.

Wasson, A.P., Richards, R.A., Chatrath, R. et al. (2012). Traits and selection strategies to improve root systems and water uptake in water-limited wheat crops. *Journal of Experimental Botany* 63 (9): 3485–3498.

Wasson, A.P., Rebetzke, G.J., Kirkegaard, J.A. et al. (2014). Soil coring at multiple field environments can directly quantify variation in deep root traits to select wheat genotypes for breeding. *Journal of Experimental Botany* 65 (21): 6231–6249.

Watt, M., Moosavi, S., Cunningham, S.C. et al. (2013). A rapid, controlled-environment seedling root screen for wheat correlates well with rooting depths at vegetative, but not reproductive, stages at two field sites. *Annals of Botany* 112 (2): 447–455.

Weaver, J.E. and Bruner, W.E. (1926). Root Development of Field Crops. New York: McGraw-Hill Book Company, Inc.

White, R.G. and Kirkegaard, J.A. (2010). The distribution and abundance of wheat roots in a dense., structured subsoil – implications for water uptake. *Plant, Cell & Environment* 33 (2): 133–148.

Witcombe, J.R. and Virk, D.S. (2001). Number of crosses and population size for participatory and classical plant breeding. *Euphytica* 122 (3): 451–462.

Wojciechowski, T., Gooding, M.J., Ramsay, L. et al. (2009). The effects of dwarfing genes on seedling root growth of wheat. *Journal of Experimental Botany* 60 (9): 2565–2573.

Wu, D., Guo, Z., Ye, J. et al. (2018). *Combining high-throughput micro-CT-RGB phenotyping and genome-wide association study to dissect the genetic architecture of tiller growth in rice. bioRxiv* 70 (2): 545–561.

Xiong, L., Wang, R.-G., Mao, G. et al. (2006). Identification of drought tolerance determinants by genetic analysis of root response to drought stress and abscisic acid. *Plant Physiology* 142 (3): 1065–1074.

Xiong, Q., Baychev, T.G., and Jivkov, A.P. (2016). Review of pore network modelling of porous media: experimental characterisations, network constructions and applications to reactive transport. *Journal of Contaminant Hydrology* 192: 101–117.

Xu, Y. (2016). Envirotyping for deciphering environmental impacts on crop plants. *Theoretical and Applied Genetics* 129 (4): 653–673.

Xu, W., Ding, G., Yokawa, K. et al. (2013). An improved agar-plate method for studying root growth and response of *Arabidopsis thaliana*. *Scientific Reports* 3: 1273.

Xu, Z., Valdes, C., and Clarke, J. (2018). Existing and potential statistical and computational approaches for the analysis of 3D CT images of plant roots. *Agronomy* 8 (5): 71.

Yan, W., Hunt, L.A., Sheng, Q. et al. (2000). Cultivar evaluation and mega-environment investigation based on the GGE biplot. *Crop Science* 40 (3): 597–605.

Yang, W., Feng, H., Zhang, X. et al. (2020). Crop phenomics and high-throughput phenotyping: past decades, current challenges, and future perspectives. *Molecular Plant* 13 (2): 187–214.

Yeh, W.W.G. (1986). Review of parameter identification procedures in groundwater hydrology: the inverse problem. *Water Resources Research* 22 (2): 95–108.

Yin, X., Stam, P., Kropff, M.J. et al. (2003). Crop modeling., QTL mapping., and their complementary role in plant breeding. *Agronomy Journal* 95 (1): 90–98.

York, L.M., Nord, E., and Lynch, J. (2013). Integration of root phenes for soil resource acquisition. *Frontiers in Plant Science* 4: 355.

York, L.M., Slack, S., Bennett, M.J. et al. (2018). Wheat shovelomics I: a field phenotyping approach for characterising the structure and function of root systems in tillering species. https://doi.org/10.1101/280875bioRxiv280875.

Zhang, H. and Forde, B.G. (1998). An Arabidopsis MADS box gene that controls nutrient-induced changes in root architecture. *Science* 279 (5349): 407–409.

Zhao, J., Bodner, G., and Rewald, B. (2016). Phenotyping: using machine learning for improved pairwise genotype classification based on root traits. *Frontiers in Plant Science* 7: 1864.

Zhao, J., Bodner, G., Rewald, B. et al. (2017). Root architecture simulation improves the inference from seedling root phenotyping towards mature root systems. *Journal of Experimental Botany* 68 (5): 965–982.

Zhao, J., Sykacek, P., Bodner, G. et al. (2018). Root traits of European *Vicia faba* cultivars – using machine learning to explore adaptations to agroclimatic conditions. *Plant, Cell & Environment* 41 (9): 1984–1996.

Zhong, D., Novais, J., Grift, T.E. et al. (2009). Maize root complexity analysis using a support vector machine method. *Computers and Electronics in Agriculture* 69 (1): 46–50.

Zobel, R.W. and Waisel, Y. (2010). A plant root system architectural taxonomy: a framework for root nomenclature. *Plant Biosystems* 144 (2): 507–512.

Index

The Root Systems in Sustainable Agricultural Intensification, First Edition.
Edited by Zed Rengel and Ivica Djalovic.

www.ingramcontent.com/pod-product-compliance
Lightning Source LLC
LaVergne TN
LVHW062108200926
848484LV00007B/1620